Analytic Methods for Coagulation-Fragmentation Models, Volume II

Monographs and Research Notes in Mathematics

Series Editors:
John A. Burns, Thomas J. Tucker, Miklos Bona, Michael Ruzhansky

Monomial Algebras, Second Edition
Rafael Villarreal

Matrix Inequalities and Their Extensions to Lie Groups
Tin-Yau Tam, Xuhua Liu

Elastic Waves
High Frequency Theory
Vassily Babich, Aleksei Kiselev

Difference Equations
Theory, Applications and Advanced Topics, Third Edition
Ronald E. Mickens

Sturm-Liouville Problems
Theory and Numerical Implementation
Ronald. B. Guenther, John. W. Lee

Analysis on Function Spaces of Musielak-Orlicz Type
Jan Lang, Osvaldo Mendez

Analytic Methods for Coagulation-Fragmentation Models, Volume I
Jacek Banasiak, Wilson Lamb, Philippe Laurencot

Analytic Methods for Coagulation-Fragmentation Models, Volume II
Jacek Banasiak, Wilson Lamb, Philippe Laurencot

Analytic Methods for Coagulation-Fragmentation Models, Volume II

Jacek Banasiak
Wilson Lamb
Philippe Laurençot

CRC Press
Taylor & Francis Group
Boca Raton London New York

CRC Press is an imprint of the
Taylor & Francis Group, an **informa** business

CRC Press
Taylor & Francis Group
6000 Broken Sound Parkway NW, Suite 300
Boca Raton, FL 33487-2742

International Standard Book Number-13:   978-0-367-23548-2 (Hardback)

Library of Congress Cataloging-in-Publication Data

Names: Banasiak, J., author. | Lamb, Wilson, author. | Laurencot, Philippe, author.
Title: Analytic methods for coagulation-fragmentation models / Jacek Banasiak, Wilson Lamb, Philippe Laurencot.
Other titles: Analytic methods for coagulation fragmentation models
Description: Boca Raton, Florida : CRC Press, [2019]- | Includes bibliographical references and index.
Identifiers: LCCN 2019004825| ISBN 9781498772655 (hardback : alk. paper : v. 1) | ISBN 9780367235482 (hardback : v. 2) | ISBN 9780429280320 (v. 2) | ISBN 9781315154428 (ebook : v. 1)
Subjects: LCSH: Coagulation. | Aggregation (Chemistry) | Semigroups. | Fragmentation reactions.
Classification: LCC QD547 .B36 2019 | DDC 541/.3415015118--dc23
LC record available at https://lccn.loc.gov/2019004825

Visit the Taylor & Francis Web site at
http://www.taylorandfrancis.com

and the CRC Press Web site at
http://www.crcpress.com

Printed and bound in Great Britain by
TJ International Ltd, Padstow, Cornwall

To the memory of Marian von Smoluchowski

Contents

Preface

In this two-volume monograph devoted to coagulation-fragmentation (C-F) models, the first part of Volume I presented details of the physical and biological background of C-F equations, together with a historical account of the mathematical contributions found in the physical and mathematical literature up to the 1980s. Following this, the first volume focussed largely on fragmentation, and it was pointed out that, due to the linearity of the fragmentation operator, it is natural to investigate the fragmentation equation using the well-developed theory of semigroups of linear operators. Relevant aspects of the latter were provided, culminating with the application of semigroup theory to establish, under suitable conditions, the well-posedness of the fragmentation and growth-fragmentation equations in appropriately chosen weighted L_1-spaces, and also the analyticity, honesty or dishonesty of the associated fragmentation semigroup.

Now, in Volume II, coagulation comes very much to the fore as we consider equations in which both fragmentation and coagulation feature. In the first half, we address the question of the well-posedness of C-F equations. In cases where fragmentation is the dominant mechanism, semigroup theory can once again be applied, and we show that the results obtained on the fragmentation semigroup in Volume I can be used to establish the existence and uniqueness of classical solutions to the C-F equation. When fragmentation has a weaker influence, or is even absent, a different strategy, relying rather on approximation and compactness arguments, is required. Consequently, we go on to describe this approach, pioneered in [15, 189, 252], and give a detailed account of how it enables additional results on the existence of solutions to be established. In contrast to the semigroup approach, which yields, simultaneously, existence and uniqueness of classical solutions, the compactness technique leads only to the existence of weak solutions and so uniqueness and regularity issues have to be investigated subsequently and independently. Our investigations into well-posedness also lead naturally to the question of mass conservation, and we show that this is answered positively when expected. We next turn to the loss of matter in finite time, due either to a runaway growth driven by coagulation (gelation) or to an explosive disintegration triggered by fragmentation (shattering). The occurrence of these two phenomena is investigated herein with the help of integral inequalities [112]. The results thus obtained supplement the explicit formulas derived in the physical literature, reviewed in Chapter 2 and discussed in Chapter 5 in connection with the dishonesty of the fragmentation semigroup.

Besides being of mathematical interest, a feature of C-F models that is highly relevant for applications is their predictive power, particularly in enabling the eventual outcome of the process being modelled to be predicted from the long-term behaviour of solutions. While explicit solutions, when available, readily provide such information in a very precise way, the complexity of the C-F model means that only a few exact solutions are known and, as discussed in Section 2.3.2 of Volume I, these are primarily for the fragmentation equation with homogeneous coefficients – a few exceptions being the coagulation equations with constant, additive and multiplicative coagulation kernels and the C-F equation with constant coefficients. Numerical simulations and formal asymptotic techniques can also be employed to shed some light on the predictive behaviour of C-F models, but it is clearly desirable that these can be backed up by rigorous analysis, and this is discussed in the later parts of the volume. Not surprisingly, it is the self-similar dynamical behaviour of the continuous fragmentation equation with homogeneous coefficients which is best understood, as is the long-term behaviour of the discrete growth-fragmentation equations (see Chapter 5 in Volume I). For the coagulation equation (without fragmentation) with a homogeneous coagulation kernel, self-similarity is also expected on physical grounds, either as time increases to infinity in the mass-conserving regime, or as time approaches the gelation time when gelation occurs in finite time. In the mass-conserving regime, this conjecture is supported by the existence of mass-conserving self-similar solutions, which has been established recently for several coagulation kernels. Still, the stability of these solutions and their role in the dynamics of the coagulation equation are far from being understood, as recent results raise some doubts about the universality of scaling behaviour for coagulation equations. Indeed, according to formal asymptotic analysis and the complete classification of the long-term behaviour of the coagulation equation with the so-called diagonal kernel, it is possible that oscillations occur. Finally, when both coagulation and fragmentation are turned on, a balance between these two counteracting mechanisms may lead to steady-state behaviour. For this to be possible, it is important to know that time-independent solutions of the C-F equation exist. For instance, when the coagulation and fragmentation coefficients satisfy the so-called detailed balance condition, the existence of stationary solutions is automatically guaranteed, as well as the availability of a Lyapunov functional, with the latter paving the way for the study of the stability of the former, as shown in this volume. The detailed balance condition is, however, not a genuine property of coagulation and fragmentation coefficients and, in the general case, the best result that has yet been achieved is the existence of stationary solutions – a topic which is also dealt with in this volume.

About the Authors

Jacek Banasiak is a Professor of Mathematics at the University of Pretoria, South Africa, where he holds a DST/NRF Research Chair in Mathematical Models and Methods in Biosciences and Bioengineering, and at Łódź University of Technology, Poland. His main research areas are functional analytic methods in kinetic theory and mathematical biology, singular perturbations, general applied analysis and partial differential equations and evolution problems. He is the author/co-author of 5 monographs and over 120 papers in these fields. He is also Editor-in-Chief of Afrika Matematika (Springer). In 2012 he received the South African Mathematical Society Award for Research Distinction and in 2013 he was awarded the Cross of Merit (Silver) of the Republic of Poland.

Wilson Lamb is an Honorary Research Fellow at the University of Strathclyde, Scotland, having retired recently from his position there as a Senior Lecturer in Mathematics. He is also a Member of the Associate Faculty of the African Institute for Mathematical Sciences. His main research interests lie in applicable functional analysis, evolution equations and the mathematical analysis of coagulation and fragmentation processes. He has published over 45 refereed research publications, and has given lecture courses at all levels to undergraduate and postgraduate students. These include courses on differential equations, functional analysis, dynamical systems and mathematical biology. He was nominated for the University of Strathclyde Students' Association Teaching Excellence Awards in 2012, 2013 and 2014; in 2013, he was shortlisted for the category of "Best in Science Faculty".

Philippe Laurençot is Directeur de Recherche (senior researcher) at the Centre National de la Recherche Scientifique (CNRS) and is affiliated with the Institut de Mathématiques de Toulouse, France. His main research interests include the mathematical analysis of evolution partial differential equations, dynamical system approach to evolution partial differential equations, coagulation equations and mathematical models in biology. He is the author of over 170 scientific publications and has given invited talks all over the world.

Symbol Description

ACP	Abstract Cauchy Problem
C-F	Coagulation-Fragmentation
$(G_A(t))_{t\geq 0}$	C_0-semigroup generated by the operator A
X_w	weighted space $L_1((0,\infty), w(x)\mathrm{d}x)$
$X_{0,m}$	weighted space $L_1((0,\infty), (1+x^m)\mathrm{d}x)$
$\|\cdot\|_{[0,m]}$	norm in $X_{0,m}$
X_m	weighted space $L_1((0,\infty), x^m\mathrm{d}x)$
$\|\cdot\|_{[m]}$	norm in X_m
$X_{m,+}$	positive cone of X_m
$M_m(f)$	moment of order m of $f \in X_m$
$X_{m,w}$	space X_m endowed with its weak topology
$X_{,w}$	Banach space X endowed with its weak topology
$\mathcal{L}(X,Y)$	space of bounded linear operators from X to Y
$\mathcal{L}(X)$	space of bounded linear operators on X
$\overline{A}$	closure of the operator A
$\rho(A)$	the resolvent set of operator A
$\sigma(A)$	the spectrum of operator A
$\sigma_p(A)$	the point spectrum of operator A
$\sigma_c(A)$	the continuous spectrum of operator A
$\sigma_r(A)$	the residual spectrum of operator A
$s(A)$	the spectral bound of operator A
$\sigma_{per,s(A)}$	the peripheral spectrum of operator A
$Span X$	the set of all linear combinations of elements of the set X
$AC(I)$	the set of absolutely continuous functions on the interval I
m	the symbol denoting either m, or the pair $0, m$
∂_x	the derivative, partial or ordinary, with respect to x
∂	the ordinary derivative if there is no need to specify the variable of differentiation
$B_X(x,r)$	the open ball with centre x and radius r in X
$B_X(r)$	the open ball with centre 0 and radius r in X
F_m	the fragmentation operator in X_m
A_m	the loss operator in the fragmentation process in X_m
B_m	the gain operator in the fragmentation process in X_m
$\mathbb{R}_+$	the open positive half-line $(0,\infty)$
$\hookrightarrow$	continuously embedded in
$\rightharpoonup$	weak convergence
$\overset{\star}{\rightharpoonup}$	$\star$-weak convergence
$\propto$	proportional
$\int_{0^+}$	integral in a right neighbourhood of 0
$\mathbf{1}_E$	indicator function of the set E

r_+	positive part $\max\{0, r\}$ of $r \in \mathbb{R}$
k_+	additive kernel $k_+(x, y) = x + y$
$k_\times$	multiplicative kernel $k_\times(x, y) = xy$
$\mathbb{N}$	set of positive integers
$\mathbb{N}_0$	set of nonnegative integers
$x \wedge y$	infimum of x and y
$x \vee y$	supremum of x and y
$\binom{m}{l}$	binomial coefficient

Chapter 6

Introduction to Volume II

6.1 Introduction

In this volume, the focus is on the well-posedness and qualitative properties of the continuous-size C-F equation

$$\partial_t f(t,x) = \mathcal{F} f(t,x) + \mathcal{C} f(t,x) \,, \qquad (t,x) \in (0,\infty)^2 \,, \tag{6.1.1a}$$

$$f(0,x) = f^{in}(x) \,, \qquad x \in (0,\infty) \,, \tag{6.1.1b}$$

where

$$\mathcal{F} f(t,x) = -a(x)f(t,x) + \int_x^\infty a(y)b(x,y)f(t,y)\,\mathrm{d}y \,, \tag{6.1.2}$$

and

$$\mathcal{C} f(t,x) = \frac{1}{2}\int_0^x k(x-y,y)f(t,x-y)f(t,y)\,\mathrm{d}y - f(t,x)\int_0^\infty k(x,y)f(t,y)\,\mathrm{d}y \tag{6.1.3}$$

model fragmentation and coagulation, respectively. We recall that the basic assumption underlying this model is that only a single size variable, such as cluster mass or volume, is required to differentiate between clusters, with $f(t,x)$, regarded as a density function, denoting the density of clusters of size $x > 0$ at time $t \geq 0$. The interpretation of the various model coefficients is then as follows. The coagulation kernel, $k(x,y)$, gives the rate at which clusters of size x coalesce with clusters of size y and $a(x)$ represents the overall rate of fragmentation of x-sized clusters. The coefficient $b(x,y)$ is often called the daughter distribution function, or the fragmentation kernel, and, in broad terms, gives the number of size x clusters produced by the fragmentation of a size y cluster. The daughter distribution function b is assumed to be nonnegative, with $b(x,y) = 0$ for $x > y$, and to satisfy the mass-conservation condition

$$\int_0^y x b(x,y)\,\mathrm{d}x = y, \quad \text{for each } y > 0. \tag{6.1.4}$$

This condition is simply the mathematical formulation of the physical requirement that the total mass of all daughter clusters formed in each fragmentation event must be the same as the mass of the fragmenting parent cluster. When (6.1.4) holds, and the coagulation kernel satisfies the natural constraint $k(x,y) = k(y,x)$ for $(x,y) \in (0,\infty)^2$, then, as shown in Section 2.3.1, a formal calculation leads to

$$\frac{d}{dt}\int_0^\infty x f(t,x)\,\mathrm{d}x = 0, \tag{6.1.5}$$

when f is a solution of (6.1.1). The integral on the left-hand side of (6.1.5) represents the total mass in the system at time t and is usually referred to as the first moment of the solution f. Thus, on setting

$$M_1(f(t)) := \int_0^\infty x f(t, x) \, \mathrm{d}x, \qquad t \in [0, \infty), \tag{6.1.6}$$

it follows that solutions of (6.1.1) are expected to satisfy the mass-conservation property

$$M_1(f(t)) = M_1(f^{in}), \qquad t \in [0, \infty) . \tag{6.1.7}$$

Clearly, as in Section 2.3.1, other moments associated with solutions can be defined by

$$M_m(f(t)) := \int_0^\infty x^m f(t, x) \, \mathrm{d}x, \qquad t \in [0, \infty), \ m \in \mathbb{R},$$

and these also feature in investigations into C-F equations. For example, the zeroth moment, $M_0(f(t))$, is of interest as it gives the total number of clusters at time t. Note that, due to (6.1.7), the weighted L_1-space

$$X_1 := \left\{ f : (0, \infty) \to \mathbb{R} \ \text{ such that } \ \|f\|_{[1]} := \int_0^\infty x |f(x)| \, \mathrm{d}x \ < \infty \right\}$$

is the most natural functional setting to study the well-posedness of (6.1.1) but other weighted L_1-spaces such as

$$X_{0,1} := \left\{ f : (0, \infty) \to \mathbb{R} \ \text{ such that } \ \|f\|_{[0,1]} := \int_0^\infty (1+x) |f(x)| \, \mathrm{d}x \ < \infty \right\}$$

and

$$X_m := \left\{ f : (0, \infty) \to \mathbb{R} \ \text{ such that } \ \|f\|_{[m]} := \int_0^\infty x^m |f(x)| \, \mathrm{d}x \ < \infty \right\} ,$$

$$X_{0,m} := \left\{ f : (0, \infty) \to \mathbb{R} \ \text{ such that } \ \|f\|_{[0,m]} := \int_0^\infty (1+x^m) |f(x)| \, \mathrm{d}x \ < \infty \right\} ,$$

where $m \in \mathbb{R}$, play an important role as well.

An in-depth study of the linear fragmentation equation

$$\partial_t f(t, x) = \mathscr{F} f(t, x) , \qquad (t, x) \in (0, \infty)^2, \tag{6.1.8}$$

$$f(0, x) = f^{in}(x) , \qquad x \in (0, \infty), \tag{6.1.9}$$

can be found in Volume I, where it is shown that, in appropriately chosen state spaces and under suitable assumptions on the fragmentation coefficients, there is an operator realisation, F of $\mathscr{F}$, with domain $D(F)$, which generates a positive, and possibly analytic, semigroup $(G(t))_{t \geq 0}$. The solution to (6.1.8)–(6.1.9) is then given by $f(t, x) = [G(t)f^{in}](x)$ for $t \in [0, \infty)$ and $x \in (0, \infty)$. Once the existence of the fragmentation semigroup $(G(t))_{t \geq 0}$ has been established, a natural approach for tackling the initial-value problem for the combined C-F equation is to rewrite (6.1.1) as

$$f(t, x) = [G(t) f^{in}](x) + \int_0^t [G(t-s) \mathcal{C} f(s, \cdot)](x) \mathrm{d}s , \qquad (t, x) \in (0, \infty)^2 . \tag{6.1.10}$$

As (6.1.10) takes the form of a fixed-point equation, there are a number of associated fixed-point methods available for proving the existence of a solution. This semigroup-based approach is detailed in Section 8.1, where it will be seen that the established properties of the fragmentation semigroup $(G(t))_{t \geq 0}$ are essential for proving the solvability of (6.1.10). For instance, if $(G(t))_{t \geq 0}$ is analytic, then (6.1.10) can be solved even for unbounded coagulation rates, provided they are controlled by the fragmentation rates. This requirement, that fragmentation must dominate coagulation, is a weakness of approaching the full equation (6.1.1) from the linear side and thus, although this method yields classical, unique and mass-conserving solutions, if the fragmentation is weak or even absent, other strategies have to be employed. One of these strategies, that relies heavily on approximation and compactness arguments, is described in detail in Section 8.2.

In essence, the compactness method can be viewed as a two-step process. First, a sequence of truncated, and more tractable, C-F problems is investigated and this leads to a corresponding sequence of solutions that can be regarded as approximate solutions of the full, untruncated C-F equation (6.1.1). Truncated versions of (6.1.1) may be obtained in a routine manner by simply restricting the coagulation and fragmentation rate coefficients to compact supports, provided that the resulting truncated rates are bounded. For example, when k is locally bounded in $[0, \infty)^2$ and a is locally bounded in $[0, \infty)$, the truncated coagulation kernel, k_j, and fragmentation rate, a_j, can be given by

$$k_j(x, y) = k(x, y) \mathbf{1}_{[0,j]}(x + y) , \quad a_j(x) = a(x) \mathbf{1}_{[0,j]}(x) , \qquad (x, y) \in (0, \infty)^2 , \quad (6.1.11)$$

for each $j \in \mathbb{N}$. Other possible choices include $k_j(x, y) = \min\{k(x, y), j\}$ and $a_j(x) = \min\{a(x), j\}$, $(x, y) \in (0, \infty)^2$, which is useful for coefficients featuring a singularity for small sizes. The next step is to show that this sequence of approximate solutions converges to a function that satisfies (6.1.1) in an appropriately defined manner. Applications of this approach to C-F problems originate from work carried out on the discrete analogue of (6.1.1), in which the integrals are replaced by series. In connection with this, we should cite the contributions made by McLeod [189], White [273], Spouge [250], Ball, Carr and Penrose [15], and Ball and Carr [14]. For the continuous coagulation equation, pioneering work can be traced back to Galkin [123]. However, the first person to apply this method to the continuous C-F equation was Stewart [252], who established the existence of solutions to (6.1.1) for suitably restricted, but possibly unbounded, coagulation and fragmentation rate coefficients. Subsequent developments of this approach are gathered in Section 8.2, together with uniqueness results which have to be established independently as already mentioned. Roughly speaking, the existence results that are proved yield weak solutions which are either mass-conserving, in that they satisfy (6.1.7), or have a total mass that is known only to be nonincreasing. In the latter case, the analysis used does not establish that loss of mass actually occurs, and so a separate, detailed investigation of this phenomenon, referred to as gelation or shattering, is carried out in Chapter 9.

The fact that mass conservation may not hold for some choices of coagulation and fragmentation coefficients has been known for some time due to the availability of closed-form solutions exhibiting mass-loss behaviour for particular examples of the coagulation equation

$$\partial_t f(t, x) = \mathcal{C} f(t, x) , \qquad (t, x) \in (0, \infty)^2, \qquad (6.1.12)$$

$$f(0, x) = f^{in}(x) , \qquad x \in (0, \infty), \qquad (6.1.13)$$

[183, 263] or fragmentation equation (6.1.8)–(6.1.9) [116, 188]. In other words, there are coagulation kernels k for which solutions to (6.1.12)–(6.1.13) are such that

$$T_{gel} := \inf \left\{ t > 0 \ : \ M_1(f(t)) < M_1(f^{in}) \right\} < \infty ,$$

and this phenomenon is referred to as gelation. In the same vein, there are fragmentation rates a for which solutions to the fragmentation equation (6.1.8)–(6.1.9) satisfy

$$T_{sh} := \inf \left\{ t > 0 \ : \ M_1(f(t)) < M_1(f^{in}) \right\} < \infty ,$$

a phenomenon referred to as shattering. The occurrence of gelation for the C-F equation (6.1.1) is shown in Chapter 9, the proof being based on integral inequalities introduced in [112, 183]. Also included is the possibility of instantaneous gelation, that is, $T_{gel} = 0$. As for the shattering phenomenon, it is closely connected to the dishonesty of the semigroup $(G(t))_{t \geq 0}$ associated to the fragmentation equation (6.1.8)–(6.1.9), see Chapter 5, but an alternative proof can be designed along the lines of [112] and is also provided in Chapter 9.

The aim of Chapter 10 is to provide a deeper insight into the dynamics of C-F models. We begin with a study of self-similar behaviour of solutions to the fragmentation equation (6.1.8)–(6.1.9) for the specific case when the fragmentation coefficients, a and b, are homogeneous; that is, there exist positive constants a_0 and γ, and a nonnegative function, h, such that

$$a(x) = a_0 x^\gamma , \qquad b(x, y) = \frac{1}{y} h\left(\frac{x}{y}\right) , \qquad 0 < x < y , \qquad (6.1.14)$$

and

$$\int_0^1 z h(z) \, \mathrm{d}z = 1 , \qquad (6.1.15)$$

the last property following from (6.1.4). The positivity of γ is required here to prevent the occurrence of the shattering phenomenon and thanks to this assumption, the fragmentation equation possesses a unique scale invariance which complies with mass conservation. The existence, uniqueness, and stability of mass-conserving self-similar solutions are studied in Chapter 10, building upon the contributions of [43, 113, 116, 188, 202].

Similarly, the coagulation equation (6.1.12)–(6.1.13) possesses some scale invariance when the coagulation kernel k satisfies the homogeneity property

$$k(\xi x, \xi y) = \xi^\lambda k(x, y) , \qquad (\xi, x, y) \in (0, \infty)^3 , \qquad (6.1.16)$$

for some $\lambda \in \mathbb{R}$, including one that leaves the total mass invariant. As gelation is expected to take place for coagulation kernels with homogeneity $\lambda > 1$, searching for mass-conserving self-similar solutions is only relevant for $\lambda \in (-\infty, 1]$ and we may thus study the existence of such solutions to the coagulation equation for this range of the parameter λ. Besides the well-known explicit solutions available for the constant coagulation kernel $k \equiv 2$ and the additive kernel $k_+(x, y) := x + y$, for which the stability issue is thoroughly investigated in [198], several existence results are now available and are described in Chapter 10. The presentation is based on [45, 113, 118, 219] and includes the few uniqueness results available so far [163, 214, 220]. Moreover, we also highlight the very special case of the diagonal coagulation kernel for which the dynamics is completely understood [172], before concluding our discussion on coagulation dynamics by looking into the particular case of coagulation kernels with homogeneity $\lambda = 1$, and also the existence of self-similar solutions with infinite mass, two topics which have received several interesting contributions in recent years [46, 141, 213, 217, 218].

The remainder of Chapter 10 deals with the full C-F equation (6.1.1). When both coagulation and fragmentation are turned on, the latter decreases the mean size while the former increases it, and we may reach a final state where a balance arises from these two processes. The analysis begins with the so-called coagulation equation with binary fragmentation, in which the breakup of a particle can yield only two daughter clusters, and consequently the daughter distribution function b must satisfy also

$$\int_0^y b(x, y) \mathrm{d}x = 2 , \qquad y \in (0, \infty) . \qquad (6.1.17)$$

As shown in Section 2.2.2, the C-F equation (6.1.1a) can then be written in the equivalent form

$$\partial_t f(t,x) = \mathcal{F}_b f(t,x) + \mathcal{C} f(t,x) \,, \tag{6.1.18}$$

where $\mathcal{C}$ is still defined by (6.1.3), but fragmentation is now represented by

$$\mathcal{F}_b f(t,x) = -\frac{1}{2} \int_0^x F(y, x-y) f(t,x) \, \mathrm{d}y + \int_x^\infty F(x, y-x) f(t,y) \, \mathrm{d}y \,, \tag{6.1.19}$$

with

$$F(x,y) = F(y,x) = a(x+y)b(x, x+y) \,, \qquad (x,y) \in (0,\infty)^2 \,.$$

The function F is referred to as the binary fragmentation kernel, representing the splitting of a cluster of size $x+y$ into clusters of sizes x and y. A particularly instructive, though not genuine, case where stationary solutions exist is when the coagulation kernel, k, and the binary fragmentation kernel, F, that features in equation (6.1.18), satisfy the detailed balance condition

$$k(x,y)Q(x)Q(y) = F(x,y)Q(x+y) \,, \qquad (x,y) \in (0,\infty)^2 \,, \tag{6.1.20}$$

for some function $Q : (0,\infty) \to [0,\infty)$. A typical example is the case of constant coagulation and fragmentation coefficients $k \equiv 2$, $F \equiv 2$, with the function Q being given by $Q \equiv 1$ [2]. The specific structure (6.1.20) of the coefficients guarantees simultaneously the existence of stationary solutions and that of a Lyapunov functional and we provide a detailed analysis of the long-term dynamics for this case. In the general case, the existence of stationary solutions is already a challenging issue and we report the available results. As for the construction of self-similar solutions to the fragmentation equation or the coagulation equation, the proof relies on a dynamical systems approach, in which either the Schauder or Tychonov fixed-point theorem is used. Unfortunately, this approach only provides the existence of a stationary solution and gives no information about its uniqueness and possible stability, two issues which are still open as far as we know.

We conclude Chapter 10 by returning to the situation where both coagulation and fragmentation coefficients are homogeneous and satisfy (6.1.16) and (6.1.14), respectively. Assuming further that $\gamma = \lambda - 1 \in (0,1]$, the C-F equation (6.1.1) possesses a scale invariance which complies with mass conservation and paves the way towards the existence of mass-conserving self-similar solutions, a question which is discussed at the end of the chapter.

6.2   Chapter Summaries

VOLUME I

In Chapter 2, we begin by highlighting the ubiquity of coagulation and fragmentation processes in nature and technology. A number of areas in which coagulation and fragmentation play an important role are described. This is then followed by a detailed presentation of relevant earlier developments in the field, produced primarily in mathematical, physical and engineering studies. The discussion that is given ranges from a derivation of the original Smoluchowski equation to an account of some key results pertaining to moments and exact solutions, which have been obtained for specific coagulation and fragmentation kernels. The phenomena of gelation and shattering are also discussed in Chapter 2, and a review is given of some of the older investigations into the existence, uniqueness and asymptotic behaviour of solutions, together with a prerequisite treatment of steady-state and self-similar solutions.

The purpose of Chapter 3 is to present essential functional analytic terminology that will be needed throughout the book. In the first part we introduce function spaces. In particular, we discuss spaces of function-valued functions and their relation to spaces of scalar-valued multivariate functions, that later plays an important role in the interpretation of solutions of abstract Cauchy problems (ACPs). We also introduce the Banach spaces that are used in the analysis. The second part of Chapter 3 is devoted to a survey of the abstract concept of order in Banach spaces and its properties that are relevant to applications in the C-F theory, where the requirement that the solutions are nonnegative is crucial for both the mathematical analysis of the problem and the physical interpretation of the solutions. We cover concepts such as the relation between order and norm, positive operators, ideals and irreducibility, that are used later in the analysis of the long-term behaviour of solutions to fragmentation equations. The material in Chapter 3 is mostly based on [26, 201, 281].

In Chapter 4 we continue our survey of functional analytic methods used in the theory of C-F equations but now focus on the theory of strongly continuous semigroups of operators. As described above, this theory is a perfect tool for the analysis of the linear part of (6.1.1); that is, the fragmentation equation, but can also provide valuable insights into several classes of nonlinear models investigated in Chapter 8. We discuss the formulation of fragmentation equations that enables them to be cast in the form of an ACP in a suitable Banach space, so that they can then be analysed by semigroup theory. Further we provide a survey of classical results pertaining to the generation of semigroups and their properties, mostly based on [106, 227], but we also touch upon relevant aspects of analytic semigroups, and interpolation spaces related to them, that play an essential role in the solvability of semilinear equations, see [187, 242], of which the C-F equation is an example. Positivity of semigroups is discussed at length, following [9, 40, 79, 107], as this has a bearing on the long-term asymptotics of solutions to several classes of fragmentation equations. Since one of the basic methods for proving that a given operator generates a semigroup is by using perturbation theory, we provide an exhaustive survey of perturbation techniques that are suitable for fragmentation models, see [8, 26, 30, 255, 268, 269]. The last part of Chapter 4 is devoted to the characterisation of the domain of the generator that determines whether the generated semigroup is honest or not; that is, whether, in our context, the solutions have the crucial physical property of mass conservation, or whether there is a loss of mass that is unaccounted for. We also show that providing the characterisation of the generator settles the question of the uniqueness of the solutions to the Cauchy problem at hand. This part is mostly based on [26] with new developments due to [206, 259, 275, 276].

Chapter 5 contains the applications of the theory presented in Chapters 3 and 4 to linear fragmentation type models that include pure fragmentation and fragmentation with growth and decay. The chapter begins with classical results, see the monograph [26] and the articles [17, 18, 20] preceding it, in the standard state space $X_1 = L_1((0, \infty), x\mathrm{d}x)$, that, since the first moment $M_1(f(t))$ gives the total mass, describes all configurations with finite mass. We provide a comprehensive description of the models that conserve mass and that are uniquely solvable, as well as of those that are dishonest, or admit multiple solutions. Further, following [34, 38], we analyse the evolution of the number of particles; that is, of the zeroth-order moment, $M_0(f(t))$, of the solution, and describe other invariant subspaces of the fragmentation dynamics. A substantial part of the chapter is devoted to the analysis of the fragmentation equation in the spaces with finite moments of orders zero and some $m > 1$. The reason for this is that under relatively mild assumptions on the fragmentation kernel, the fragmentation semigroup in such spaces is analytic and this allows for better generation results to be proved and, in particular, provides an explicit characterisation of the domain of the generator. This is particularly fruitful in the context of discrete fragmentation models, where similar techniques also make it possible to prove the compactness of the fragmentation semigroup, [24, 29, 35]. Analyticity and compactness, together with the theory of positive semigroups developed in Chapter 4, enable us to provide a comprehensive analysis of the long-term behaviour of the solutions to the pure fragmentation models, as well as for the fragmentation models with growth and decay that are particularly relevant in biological sciences, [22, 23, 27, 29, 36]. The remaining part of the chapter, based on [11, 31, 33, 36], is concerned with the analysis of various aspects of the continuous fragmentation equation with growth or decay, such as their well-posedness and honesty.

VOLUME II

Chapter 6 is essentially an introduction to Volume II, and, as such, it contains a brief review of the material presented in Volume I together with a summary of the main topics covered in Volume II. This is followed by Chapter 7, which can be regarded as a top-up to the Mathematical Toolbox (Chapter 3) which features in Volume I. It is in Chapter 7 that we give a collection of mathematical prerequisites for Volume II that were not needed in Volume I. In particular, some key results on weak convergence and compact embeddings are provided, as well as a number of very useful inequalities. Some aspects of the theory of dynamical systems are also given.

In Chapter 8, nonlinearity, which played a minimal role in Volume I, enters the game, as we begin our study of the full C-F equation (6.1.1). The chapter primarily is devoted to the question of well-posedness of the models and is divided into two parts, describing the different approaches to the problem that were alluded to above. In the first part, we build upon the semigroup theory developed in Chapter 5 by treating the nonlinear coagulation part of the C-F equation as a perturbation of the linear part and using the linear semigroup theory to re-write the original problem as the integral equation (6.1.10). Then we establish the existence of the solution as the fixed-point of this equation. Using this approach we prove the existence of globally defined classical solutions to the growth–coagulation–fragmentation equation with bounded coagulation kernel, see [25, 32, 33, 34]. If the fragmentation semigroup is analytic, then a stronger result is possible. By using the interpolation spaces related to the analytic fragmentation semigroup, described in Chapter 4, we establish the existence of global solutions to the C-F equation with an unbounded coagulation kernel with arbitrary growth rate provided it is bounded by the fragmentation part. The fact that the solutions are global follows from careful moment estimates that are also used later, in a similar way, to prove the existence of a weak solution, see [35, 37], and [24, 28] for the case of discrete models. The advantage of this approach is that the abstract theory simultaneously delivers classical, unique and mass-conserving solutions but, for it to be applicable, the whole process must be dominated by the fragmentation part.

In the second part of Chapter 8, an alternative method, which relies on compactness arguments, is presented for analysing (6.1.1), without the drawback of requiring a dominant linear term to be present in the equation. Pioneering work in this direction includes [14, 183, 190, 250, 273] for the discrete C-F system of equations, and [124, 252] for the continuous C-F equation. As discussed above, the basic principle underlying this approach is not to solve directly the evolution equation under consideration but, instead, a suitably designed approximation which depends on a large parameter j, and for which straightforward arguments lead to the existence of a unique solution, f_j, which, of course, depends on the parameter j. The ultimate goal being to pass to the limit as $j \to \infty$, the j-dependent approximating equation has to be constructed in such a way that, at least formally, it gives back the original equation in the limit as $j \to \infty$. Typically, for C-F equations, truncating the coagulation kernel and the overall fragmentation rate in an appropriate way, such as (6.1.11), provides adequate approximating equations. The next step is to derive estimates on the family $(f_j)_{j \geq 1}$ which do not depend on j and are sufficient to guarantee not only that $(f_j)_{j \geq 1}$ has cluster points as $j \to \infty$, but also that any of these cluster points is a weak solution to the original equation. In the landmark work [252], it was shown that, due to the nonlocal structure of the coagulation operator, it is sufficient to obtain weak compactness of $(f_j)_{j \geq 1}$ in a well-chosen weighted L_1-space. The latter actually varies according to the magnitude of both the coagulation kernel and the overall fragmentation rate, and the integrability properties of the daughter distribution function. We shall investigate various situations, including coagulation kernels with linear or quadratic growth, strong fragmentation, singular daughter distribution functions, and coagulation kernels featuring singularities for small sizes. We also provide conditions on the coagulation and fragmentation coefficients that guarantee the existence of at least one mass-conserving solution. Unlike the semigroup approach, which provides simultaneously the existence and uniqueness of a solution to the C-F equation, the compactness method only yields the existence of a weak solution to the C-F equation, so that the uniqueness issue has to be studied independently. We conclude the chapter by giving an in-depth discussion on results that have been established on the uniqueness of weak solutions to (6.1.1).

Further information on the issue of mass conservation is provided in Chapter 9. In contrast to Chapter 8, where the aim was to identify coagulation and fragmentation coefficients for which mass-conserving solutions do exist, here we focus on the failure of mass conservation and use the techniques developed in [112, 183] to establish the occurrence of the gelation and shattering phenomena, which were mentioned in Section 6.1. Though of a different nature, both reflect a loss of matter in the system due to the creation of a new phase, involving clusters, or particles, of infinite size (or gel or giant particles) for the former, and particles of size zero (or dust) for the latter. As mentioned earlier, it was conjectured and discussed at length by physicists in the 1980s that coagulation kernels existed for which solutions to the coagulation equation have a finite gelation time. Nevertheless, it took some time before a complete mathematical proof of the occurrence of gelation was provided, with the arguments used relying either on a stochastic approach [144], or on a deterministic approach [112]; we shall focus on the latter. In the absence of fragmentation, the possibility of having a vanishing gelation time ($T_{gel} = 0$) is not excluded and results in that direction are obtained in [66, 260]. The finiteness of the shattering time for solutions to the fragmentation equation, when the overall fragmentation rate is unbounded for small sizes, is shown in [116] by a stochastic approach, and is also readily seen in [188] due to the computation of explicit solutions. It is strongly tied to the dishonesty of the semigroup associated with the fragmentation equation as discussed thoroughly in Chapter 5, but it turns out that the argument developed in [112] to prove the onset of gelation is also suitable to study the occurrence of shattering, and this is described at the end of the chapter.

The aim of Chapter 10 is to provide a deeper insight into the dynamics of C-F models. It is obvious that, for continuous size models where there is no minimum particle size, fragmentation alone decreases the mean size of the particles to zero, at least when the overall fragmentation rate a is positive, while coagulation alone increases it to infinity, at least when the coagulation kernel k is positive. Only a combination of coagulation and fragmentation can, in principle, lead to the stabilisation of the mean size at a finite value in the long term. We begin by concentrating on processes that involve only fragmentation or coagulation, dealing first with fragmentation. Roughly speaking, if we assume that the overall fragmentation rate a and the daughter distribution function b are positive, then it is expected that the size distribution function $f(t)$ converges to zero as $t \to \infty$. As this property is shared by a large variety of fragmentation coefficients, it does not reveal much about the size and time scales of the fragmentation dynamics under study. Thanks to the homogeneity property (6.1.14), and assuming that $\gamma > 0$, the fragmentation equation possesses a unique scale invariance which complies with mass conservation. It is then tempting to look for particular solutions to the fragmentation equation that are invariant with respect to this specific change of scale. Such solutions are usually referred to as mass-conserving, self-similar solutions to the fragmentation equation and results on the existence of such solutions date back to [116], with explicit formulas being available for some specific choices of the profile h that defines the daughter distribution function b in (6.1.14), see [116, 146, 188, 230, 258, 284, 285]. In Chapter 10, we provide an alternative construction of self-similar solutions to the fragmentation equation, following the approach of [113, 202], but pushing it a little further in order to construct self-similar solutions under the optimal condition on h found in [43, 116] by a stochastic approach. We supplement this existence result with conditions that guarantee the uniqueness of the self-similar solution (for a given mass) and show that the latter is an attractor for the dynamics of the fragmentation equation. In other words, all solutions to the fragmentation equation emanating from an initial condition having total mass $\varrho > 0$ converge in a suitable sense, as $t \to \infty$, to the unique self-similar solution to the fragmentation equation with the same total mass ϱ. We mainly follow the approach in [113, 202] to investigate the above-mentioned uniqueness and stability issues.

Not surprisingly, it turns out that the coagulation equation also possesses some scale invariance when the coagulation kernel k satisfies the homogeneity property (6.1.16), including one that leaves the total mass invariant. As gelation is expected to take place for coagulation kernels with homogeneity $\lambda > 1$, searching for mass-conserving self-similar solutions is only relevant for $\lambda \in (-\infty, 1]$ and we may thus study the existence of such solutions to the coagulation equation for this range of the parameter λ. Though the existence and properties of mass-conserving self-similar solutions to the coagulation equation with homogeneous coagulation kernels satisfying (6.1.16) for some $\lambda \in (-\infty, 1]$ have been extensively discussed in the physical literature in the 1980s, see [180, 263] and the references therein, the only available results up to 2005 dealt with the constant coagulation kernel $k \equiv 2$ and the additive kernel $k_+(x,y) := x + y$, for which explicit formulas for mass-conserving, self-similar solutions are available, being derived with the help of the Laplace transform. The same approach can be used to study the role of these special solutions in the dynamics of the coagulation equation, and these particular cases, which have been thoroughly studied in [198], will be examined. Far less is known about general homogeneous coagulation kernels and, to the best of our knowledge, up to now only the existence issue for several classes of coagulation kernels has received a satisfactory answer [45, 113, 118, 219]. We shall provide the corresponding results, along with the few uniqueness results available so far. Moreover, we also highlight the very special case of the diagonal coagulation kernel for which the dynamics is completely understood, before concluding our discussion on coagulation dynamics by looking into the particular case of coagulation kernels with homogeneity $\lambda = 1$, and also the existence of self-similar solutions with infinite mass, two topics which have received several interesting contributions in recent years [46, 141, 213, 217, 218].

The remainder of Chapter 10 deals with the full C-F equation (6.1.1). When both coagulation and fragmentation are turned on, the latter decreases the mean size of clusters while the former increases it, and we may reach a final state where a balance arises from these two processes. A particularly instructive, though not genuine, case where such a situation occurs is when the coagulation kernel, k, and the binary fragmentation kernel, F, that features in equation (6.1.19), satisfy the detailed balance condition (6.1.20). The specific structure of the coefficients guarantees simultaneously the existence of stationary solutions and that of a Lyapunov functional and we provide a detailed analysis of the long-term dynamics in that case. As mentioned in Section 6.1, the detailed balance condition (6.1.20) is satisfied by the constant coagulation and fragmentation coefficients $k \equiv 2$, $F \equiv 2$, with the function Q being given by $Q \equiv 1$, and this special case will be examined. We shall also report on the currently available results on the more challenging issue of the existence of stationary solutions for more general kernels. As in the construction of self-similar solutions to the fragmentation equation or the coagulation equation, the proof relies on a dynamical systems approach, in which either the Schauder, or the Tychonov fixed-point theorem is used. Unfortunately, this approach only provides the existence of a stationary solution and gives no information about its uniqueness and possible stability, two issues which are still open as far as we know.

We conclude the chapter by returning to the situation where both coagulation and fragmentation coefficients are homogeneous and satisfy (6.1.16) and (6.1.14), respectively. Assuming further that $\gamma = \lambda - 1 \in (0, 1]$, the C-F equation possesses a scale invariance which complies with mass conservation and paves the way towards the existence of mass-conserving self-similar solutions, a question which is discussed at the end of the chapter.

In the closing Chapter 11, we provide a short survey of results that have been established on C-F equations with spatial diffusion, and also on the Becker–Döring equations. Both topics are related to the main theme of the book, and there have been many important investigations into each, yielding a wealth of results that really deserve a more comprehensive treatment. However, in an attempt to keep this book to a reasonable length, we have refrained from including such a detailed account, and instead we simply review the outcome of the many contributions that have been made in recent years.

Chapter 7

Mathematical Toolbox II

7.1 Weak Topology and Related Results

We begin by recalling the notation introduced in Chapter 3 for spaces of integrable functions. Given a σ-finite measure space $(\Omega, \mathcal{B}, \mu)$, we denote the space of measurable functions on Ω (with respect to the σ-algebra $\mathcal{B}$) by $L_0(\Omega, \mathcal{B})$. For $p \in [1, \infty]$, $L_p(\Omega, \mathcal{B}, \mu)$ denotes the classical Lebesgue space with norm $\| \cdot \|_p$ defined by

$$\|f\|_p := \left(\int_\Omega |f(x)|^p \, \mathrm{d}\mu(x) \right)^{1/p} , \qquad f \in L_p(\Omega, \mathcal{B}, \mu) ,$$

for $p \in [1, \infty)$, and

$$\|f\|_\infty := \inf \{ M > 0 \; : \; |f| \le M \;\; \mu\text{-a.e. in } \Omega \} , \qquad f \in L_\infty(\Omega, \mathcal{B}, \mu) .$$

If there is no possibility of misunderstanding, we shall simply write $L_p(\Omega)$, or $L_p(\Omega, \mu)$. Moreover, when $\Omega = (a, b) \subset \mathbb{R}$, we shall use the notation $L_p(a, b)$, and shall abbreviate $\mathrm{d}\mu(x)$ by $\mathrm{d}x$ when μ is the Lebesgue measure.

Besides the so-called strong topology induced by the norm of $L_p(\Omega)$, $p \in [1, \infty]$, another classical topology with which $L_p(\Omega)$ may be endowed is the weak topology for $p \in [1, \infty)$ and the weak-$\star$ topology for $p = \infty$. For a sequence of functions, weak and weak-$\star$ convergence are defined as follows.

Definition 7.1.1.

(a) Let $p \in [1, \infty)$. A sequence $(f_n)_{n \ge 1}$ in $L_p(\Omega)$ converges weakly to f (written as $f_n \rightharpoonup f$) in $L_p(\Omega)$ if

$$\lim_{n \to \infty} \int_\Omega f_n(x) \vartheta(x) \, \mathrm{d}\mu(x) = \int_\Omega f(x) \vartheta(x) \, \mathrm{d}\mu(x)$$

for all $\vartheta \in L_{p'}(\Omega)$ where $p' := \infty$ when $p = 1$ and $p' := p/(p-1)$ when $p \in (1, \infty)$.

(b) A sequence $(f_n)_{n \ge 1}$ in $L_\infty(\Omega)$ converges $\star$-weakly to f (written as $f_n \overset{\star}{\rightharpoonup} f$) in $L_\infty(\Omega)$ if

$$\lim_{n \to \infty} \int_\Omega f_n(x) \vartheta(x) \, \mathrm{d}\mu(x) = \int_\Omega f(x) \vartheta(x) \, \mathrm{d}\mu(x)$$

for all $\vartheta \in L_1(\Omega)$.

Given $p \in (1, \infty)$ and a sequence of functions $(f_n)_{n \ge 1}$ in $L_p(\Omega)$, it is well known that Kakutani's theorem [51, Theorem 3.17] and the reflexivity of $L_p(\Omega)$ [51, Theorem 4.10] ensure that any bounded sequence in $L_p(\Omega)$ has a subsequence that converges weakly in $L_p(\Omega)$. A result in the same spirit is available in $L_\infty(\Omega)$: since $L_\infty(\Omega)$ is the dual of the separable space $L_1(\Omega)$, it follows from the Banach–Alaoglu theorem that a bounded sequence of functions in $L_\infty(\Omega)$ has a weakly-$\star$ convergent subsequence [51, Theorem 3.16].

Unfortunately, such a nice feature is not shared by the space $L_1(\Omega)$ and a bounded sequence in $L_1(\Omega)$ need not have a subsequence which converges weakly in $L_1(\Omega)$. Obstructions are in particular due to

(a) *concentration*: for instance, the sequence $(f_n)_{n \geq 1}$, defined in $\Omega = \mathbb{R}$ by $f_n(x) = n e^{-n^2 x^2}$, is clearly bounded in $L_1(\mathbb{R})$ but converges to the Dirac mass centred at $x = 0$ in the sense of distributions;

(b) *vanishing*: for instance, the sequence $(g_n)_{n \geq 1}$, defined in $\Omega = \mathbb{R}$ by $g_n(x) = e^{-|x-n|}$, is clearly bounded in $L_1(\mathbb{R})$ with the norm of each term equal to 2, but it converges to zero in the sense of distributions.

Preventing the occurrence of these phenomena leads to a result on sequential weak compactness of subsets of $L_1(\Omega)$, known as the Dunford–Pettis theorem, which we state below. First we introduce the concept of the modulus of uniform integrability of a bounded subset $\mathcal{E}$ of $L_1(\Omega)$ which provides a control on how elements of $\mathcal{E}$ concentrate on sets of small measure.

Definition 7.1.2. *Let $\mathcal{E}$ be a bounded set of $L_1(\Omega)$. For $\varepsilon > 0$ we set*

$$\eta\{\mathcal{E}, \varepsilon\} := \sup \left\{ \int_E |f| \, \mathrm{d}\mu \ : \ f \in \mathcal{E}, \ E \in \mathcal{B}, \ \mu(E) \leq \varepsilon \right\}. \tag{7.1.1}$$

The modulus of uniform integrability $\eta\{\mathcal{E}\}$ of $\mathcal{E}$ is then

$$\eta\{\mathcal{E}\} := \lim_{\varepsilon \to 0} \eta\{\mathcal{E}, \varepsilon\} = \inf_{\varepsilon > 0} \eta\{\mathcal{E}, \varepsilon\}. \tag{7.1.2}$$

The notation

$$\eta\{\mathcal{E}, \varepsilon; L_1(\Omega, \mathcal{B}, \mu)\} \quad and \quad \eta\{\mathcal{E}; L_1(\Omega, \mathcal{B}, \mu)\}$$

will be used whenever the L_1-space involved is not clear from the context.

Owing to the assumed boundedness of $\mathcal{E}$ in $L_1(\Omega)$, the quantity $\eta\{\mathcal{E}, \varepsilon\}$ is finite for all $\varepsilon > 0$, so that the modulus of uniform integrability $\eta\{\mathcal{E}\}$ is well defined. With this definition, we can state the Dunford–Pettis theorem which provides a necessary and sufficient condition for the sequential weak compactness in $L_1(\Omega)$ of a bounded subset of $L_1(\Omega)$.

Theorem 7.1.3 (Dunford–Pettis). *Let $\mathcal{E}$ be a bounded set of $L_1(\Omega)$. The following two statements are equivalent:*

(a) $\mathcal{E}$ is relatively sequentially weakly compact in $L_1(\Omega)$;

(b) $\mathcal{E}$ satisfies the following two properties:

$$\eta\{\mathcal{E}\} = 0 \tag{7.1.3}$$

and, for every $\varepsilon > 0$, there is $\Omega_\varepsilon \in \mathcal{B}$ such that $\mu(\Omega_\varepsilon) < \infty$ and

$$\sup_{f \in \mathcal{E}} \int_{\Omega \setminus \Omega_\varepsilon} |f| \, \mathrm{d}\mu \leq \varepsilon. \tag{7.1.4}$$

It is easily seen that the above counterexamples of bounded sequences in $L_1(\mathbb{R})$ having no weakly convergent subsequence in $L_1(\mathbb{R})$ fail to satisfy (7.1.3) and (7.1.4), respectively. In fact, condition (7.1.3) guarantees that concentration on sets of small measure cannot occur, while condition (7.1.4) prevents the escape to infinity. The latter is obviously always satisfied when $\mu(\Omega)$ is finite, as we can simply take $\Omega = \Omega_\varepsilon$ for each $\varepsilon > 0$.

Checking condition (7.1.3) can be troublesome when dealing with equations involving derivatives. Fortunately, alternative formulations are available which may be more tractable depending on the problem under study. In particular, we recall the definition of a uniformly integrable subset of $L_1(\Omega)$.

Definition 7.1.4 (Uniform integrability). *A subset $\mathcal{E}$ of $L_1(\Omega)$ is uniformly integrable in $L_1(\Omega)$ if $\mathcal{E}$ is a bounded subset of $L_1(\Omega)$ such that*

$$\lim_{M \to \infty} \sup_{f \in \mathcal{E}} \int_{\{|f| \geq M\}} |f| \, d\mu = 0 \ . \tag{7.1.5}$$

The next result provides a connection between (7.1.3) and uniform integrability.

Lemma 7.1.5. *Let $\mathcal{E}$ be a bounded subset of $L_1(\Omega)$. Then*

$$\eta\{\mathcal{E}\} = \lim_{M \to \infty} \sup_{f \in \mathcal{E}} \int_{\{|f| \geq M\}} |f| \, d\mu \ .$$

Proof. On the one hand, for $\varepsilon > 0$, $M > 0$, $E \in \mathcal{B}$, and $f \in \mathcal{E}$, there holds

$$\int_E |f| \, d\mu \leq M\mu(E) + \int_{\{|f| \geq M\}} |f| \, d\mu \ ,$$

from which we deduce that

$$\eta\{\mathcal{E}, \varepsilon\} \leq M\varepsilon + \sup_{f \in \mathcal{E}} \int_{\{|f| \geq M\}} |f| \, d\mu \ .$$

Letting first $\varepsilon \to 0$, and then $M \to \infty$, gives

$$\eta\{\mathcal{E}\} \leq \lim_{M \to \infty} \sup_{f \in \mathcal{E}} \int_{\{|f| \geq M\}} |f| \, d\mu \ .$$

On the other hand, for $M > 0$ and $f \in \mathcal{E}$, the classical inequality

$$\mu(\{x \in \Omega \ : \ |f(x)| \geq M\}) \leq \frac{\|f\|_1}{M}$$

entails that

$$\int_{\{|f| \geq M\}} |f| \, d\mu \leq \eta\left\{\mathcal{E}, \frac{1}{M} \sup_{g \in \mathcal{E}} \|g\|_1\right\} \ ,$$

hence

$$\lim_{M \to \infty} \sup_{f \in \mathcal{E}} \int_{\{|f| \geq M\}} |f| \, d\mu \leq \eta\{\mathcal{E}\} \ ,$$

and the proof is complete. $\qquad\qquad\square$

In other words controlling the behaviour of elements of $\mathcal{E}$ on sets of small measure is equivalent to controlling the contribution of their large values to the L_1-norm. An alternative version of uniform integrability is due to de la Vallée-Poussin [93] and we summarise the three equivalent formulations of uniform integrability in the next result.

Theorem 7.1.6. *Let $\mathcal{E} \subset L_1(\Omega)$ be bounded. The following statements are equivalent.*

(a) $\eta\{\mathcal{E}\} = 0$.

(b) $\mathcal{E}$ is uniformly integrable in $L_1(\Omega)$.

(c) *There is a convex function* $\Phi \in C^\infty([0,\infty))$ *such that*

$$\sup_{f \in \mathcal{E}} \int_\Omega \Phi(|f|) \, \mathrm{d}\mu < \infty \,, \qquad (7.1.6)$$

and Φ *has the following properties:* $\Phi(0) = \Phi'(0) = 0$, Φ' *is a concave function which is positive in* $(0, \infty)$, *and*

$$\lim_{r \to \infty} \Phi'(r) = \lim_{r \to \infty} \frac{\Phi(r)}{r} = \infty \,.$$

Moreover, Φ *can be constructed so as to satisfy the additional growth property*

$$\lim_{r \to \infty} \frac{\Phi'(r)}{r^{p-1}} = 0 \quad \text{for all} \quad p \in (1, 2] \,. \qquad (7.1.7)$$

The equivalence between the assertions (b) and (c) in Theorem 7.1.6 is established in [93] when $\mathcal{E}$ is a sequence of functions, and in [96, 234] when $\mu(\Omega) < \infty$ but without the concavity of the function Φ'. We also refer to [150, 205] when $\mathcal{E}$ is reduced to a single function $f \in L_1(\Omega)$. The fact that Φ can be chosen with a concave derivative is shown in [177] and turns out to be helpful as we shall see later. In fact, as noticed in [161], there is some flexibility in the construction of Φ performed in [177] which allows the function Φ to be endowed with additional properties, such as (7.1.7) which limits its growth at infinity. A version of Theorem 7.1.6 is proved in [161, Theorem 8] but the function Φ constructed therein need not satisfy (7.1.7). We revisit here the proof from [161, Theorem 8] and modify it so that Φ also satisfies (7.1.7). Let us begin with the following auxiliary result.

Lemma 7.1.7. *Let* $\Phi \in C^1([0, \infty))$ *be a nonnegative and convex function with* $\Phi(0) = \Phi'(0) = 0$ *and consider a nondecreasing sequence of integers* $(i_k)_{k \geq 0}$ *such that* $i_0 = 1$, $i_1 \geq 2$ *and* $i_k \to \infty$ *as* $k \to \infty$. *Given* $f \in L_1(\Omega)$ *and* $k \geq 1$, *we have the following inequality:*

$$\int_{\{|f|<i_k\}} \Phi(|f|) \, \mathrm{d}\mu \leq \Phi'(1) \int_\Omega |f| \, \mathrm{d}\mu + \sum_{m=0}^{k-1} (\Phi'(i_{m+1}) - \Phi'(i_m)) \int_{\{|f| \geq i_m\}} |f| \, \mathrm{d}\mu \,.$$

Proof. Since $\Phi(0) = 0$, the convexity of Φ entails that $\Phi(r) \leq r\Phi'(r)$ for $r \geq 0$. Combining this property with the monotonicity of Φ' gives

$$\int_{\{|f|<i_k\}} \Phi(|f|) \, \mathrm{d}\mu \leq \int_{\{|f|<i_k\}} \Phi'(|f|) \, |f| \, \mathrm{d}\mu$$

$$= \int_{\{0 \leq |f|<1\}} \Phi'(|f|) \, |f| \, \mathrm{d}\mu + \sum_{m=0}^{k-1} \int_{\{i_m \leq |f|<i_{m+1}\}} \Phi'(|f|) \, |f| \, \mathrm{d}\mu$$

$$\leq \Phi'(1) \int_{\{0 \leq |f|<1\}} |f| \, \mathrm{d}\mu + \sum_{m=0}^{k-1} \Phi'(i_{m+1}) \int_{\{i_m \leq |f|<i_{m+1}\}} |f| \, \mathrm{d}\mu$$

$$= \Phi'(1) \int_{\{0 \leq |f|<1\}} |f| \, \mathrm{d}\mu + \sum_{m=0}^{k-1} \Phi'(i_{m+1}) \int_{\{|f| \geq i_m\}} |f| \, \mathrm{d}\mu$$

$$- \sum_{m=1}^{k} \Phi'(i_m) \int_{\{|f| \geq i_m\}} |f| \, \mathrm{d}\mu$$

$$\leq \Phi'(1) \int_\Omega |f| \, \mathrm{d}\mu + \sum_{m=0}^{k-1} (\Phi'(i_{m+1}) - \Phi'(i_m)) \int_{\{|f| \geq i_m\}} |f| \, \mathrm{d}\mu \,,$$

as claimed. $\qquad\qquad\qquad\qquad\qquad\qquad\qquad\qquad\qquad\qquad\qquad\qquad\qquad\qquad\square$

Proof of Theorem 7.1.6. The equivalence between the assertions (a) and (b) is a straightforward consequence of Lemma 7.1.5. The fact that the assertion (c) implies the assertion (b) follows readily from the unboundedness of $r \mapsto \Phi(r)/r$ as $r \to \infty$, since

$$\lim_{M \to \infty} \sup_{f \in \mathcal{E}} \int_{\{|f| \geq M\}} |f| \, d\mu \leq \left(\sup_{f \in \mathcal{E}} \int_{\Omega} \Phi(|f|) \, d\mu \right) \lim_{M \to \infty} \sup_{r \geq M} \left\{ \frac{r}{\Phi(r)} \right\} = 0 \ .$$

It remains to prove the converse statement, that is, to show that the uniform integrability of the set $\mathcal{E}$ allows us to construct a convex function having the claimed properties and such that (7.1.6) holds true. To this end, we note that, for a convex function Φ and $f \in L_1(\Omega)$, Lemma 7.1.7 provides a connection between the L_1-norm of $\Phi(|f|)$ and the integrals of $|f|$ on sets where $|f|$ is large, and thereby, some guidelines on how to construct the desired function Φ. More specifically, it follows from the uniform integrability of the set $\mathcal{E}$ that we can construct, by induction, a sequence $(j_m)_{m \geq 0}$ of positive integers such that

$$j_0 = 1 \ , \ j_{m+1} \geq \max\{2j_m, e^{m+1}\} \ , \qquad m \geq 0 \ , \tag{7.1.8}$$

and

$$\sup_{f \in \mathcal{E}} \int_{\{|f| \geq j_m\}} |f| \, d\mu \leq \frac{1}{m^2} \ , \qquad m \geq 1 \ . \tag{7.1.9}$$

We next define a piecewise quadratic C^1-smooth function Φ_0 by

$$\Phi_0'(r) := \begin{cases} \dfrac{r}{j_1 - j_0} \ , & r \in [0, j_1) \ , \\[3mm] \dfrac{r - j_m}{j_{m+1} - j_m} + m + \dfrac{1}{j_1 - j_0} \ , & r \in [j_m, j_{m+1}) \ , \ m \geq 1 \ , \end{cases}$$

and

$$\Phi_0(r) := \int_0^r \Phi_0'(s) \, ds \ , \qquad r \geq 0 \ .$$

Let us first check that Φ_0 possesses the requested properties. Clearly, $\Phi_0(0) = \Phi_0'(0) = 0$, $\Phi_0' > 0$ in $(0, \infty)$, and

$$\lim_{r \to j_m -} \Phi_0'(r) = m + \frac{1}{j_1 - j_0} = \Phi_0'(j_m) \ , \qquad m \geq 1 \ , \tag{7.1.10}$$

so that $\Phi' \in C([0, \infty))$ and $\Phi \in C^1([0, \infty))$. In addition, Φ_0 is twice differentiable on $[0, j_1)$ and (j_m, j_{m+1}), $m \geq 1$, with

$$\Phi_0''(r) := \begin{cases} \dfrac{1}{j_1 - j_0} \ , & r \in [0, j_1) \ , \\[3mm] \dfrac{1}{j_{m+1} - j_m} \ , & r \in (j_m, j_{m+1}) \ , \ m \geq 1 \ . \end{cases}$$

Since $j_{m+1} - j_m \geq j_m \geq j_m - j_{m-1} > 0$ for $m \geq 1$ by (7.1.8), we realise that Φ_0'' is nonnegative and nonincreasing; that is, Φ_0 is convex, and Φ_0' is concave and nondecreasing. Finally, on the one hand, it follows from (7.1.10) and the monotonicity of Φ_0' that

$$\lim_{r \to \infty} \Phi_0'(r) = \infty \ ,$$

which implies that $\Phi_0(r)/r \to \infty$ as $r \to \infty$ by L'Hospital's rule. On the other hand, given $p \in (1,2]$, we infer from (7.1.8) that, for $m \geq 1$ and $r \in [j_m, j_{m+1})$,

$$
\begin{aligned}
\frac{\Phi_0'(r)}{r^{p-1}} &= \frac{r - j_m}{r^{p-1}(j_{m+1} - j_m)} + \frac{m}{r^{p-1}} + \frac{1}{r^{p-1}(j_1 - j_0)} \\
&\leq \frac{(r - j_m)^{2-p}}{j_{m+1} - j_m}\left(\frac{r - j_m}{r}\right)^{p-1} + \frac{m+1}{r^{p-1}} \\
&\leq \left(\frac{r - j_m}{r}\right)^{p-1}\frac{(j_{m+1} - j_m)^{2-p}}{j_{m+1} - j_m} + \frac{m+1}{j_m^{p-1}} \\
&\leq \frac{1}{(j_{m+1} - j_m)^{p-1}} + \frac{m+1}{j_m^{p-1}} \\
&\leq \frac{m+2}{j_m^{p-1}} \leq \frac{1 + \ln(j_m)}{j_m^{p-1}} \ .
\end{aligned}
$$

Since the right-hand side of the previous inequality converges to zero as $m \to \infty$, we have thus established that

$$
\lim_{r \to \infty} \frac{\Phi_0'(r)}{r^{p-1}} = 0 \ .
$$

Finally, thanks to (7.1.9), (7.1.10) and Lemma 7.1.7, we obtain for $f \in \mathcal{E}$ and $m \geq 1$,

$$
\begin{aligned}
\int_{\{|f| < j_m\}} \Phi_0(|f|) \, \mathrm{d}\mu &\leq (1 + \Phi_0'(1)) \int_{\Omega} |f| \, \mathrm{d}\mu + \sum_{i=1}^{m-1} \frac{1}{i^2} \\
&\leq (1 + \Phi_0'(1)) \sup_{g \in \mathcal{E}} \int_{\Omega} |g| \, \mathrm{d}\mu + \frac{\pi^2}{6} \ .
\end{aligned}
$$

Since $\mathcal{E}$ is bounded in $L_1(\Omega)$, the right-hand side of the previous inequality is bounded. We then let $m \to \infty$ and use Fatou's lemma to conclude that

$$
\int_{\Omega} \Phi_0(|f|) \, \mathrm{d}\mu \leq (1 + \Phi_0'(1)) \sup_{g \in \mathcal{E}} \int_{\Omega} |g| \, \mathrm{d}\mu + \frac{\pi^2}{6} \ ,
$$

hence the boundedness of $\{\Phi_0(|f|) : f \in \mathcal{E}\}$ in $L_1(\Omega)$.

We have thus constructed a function Φ_0 that has all the properties listed in Theorem 7.1.6 except for the C^∞-smoothness. The latter can be achieved by a suitable mollification of Φ_0 and we refer to the proof of [161, Theorem 8] for details. $\qquad\square$

For future use we introduce the following notation.

Definition 7.1.8. *We define* $\mathcal{C}_{VP}$ *as the set of convex functions* $\Phi \in C^\infty([0, \infty))$ *satisfying the following properties:*

$$
\Phi(0) = \Phi'(0) = 0 \ , \quad \Phi'(r) > 0 \ \text{ for } \ r > 0 \ , \tag{7.1.11a}
$$

$$
\Phi' \ \text{is a concave function}, \tag{7.1.11b}
$$

$$
\lim_{r \to \infty} \frac{\Phi'(r)}{r^{p-1}} = 0 \quad \text{for all} \quad p \in (1,2] \ . \tag{7.1.11c}
$$

We also define $\mathcal{C}_{VP,\infty}$ *as the subset of functions* $\Phi \in \mathcal{C}_{VP}$ *satisfying additionally*

$$
\lim_{r \to \infty} \Phi'(r) = \lim_{r \to \infty} \frac{\Phi(r)}{r} = \infty \ . \tag{7.1.12}
$$

For instance, $\Phi(r) = r \ln(1 + r)$, $r \in [0, \infty)$, belongs to $\mathcal{C}_{VP,\infty}$. Several useful properties of functions in $\mathcal{C}_{VP}$ are presented in the next result.

Proposition 7.1.9. *Let* $\Phi \in \mathcal{C}_{VP}$. *Then:*

(a) *for* $r \geq 0$, $\Phi(r) \leq r\Phi'(r) \leq 2\Phi(r)$;

(b) *for* $r \geq 0$ *and* $s \geq 0$, $s\Phi'(r) \leq \Phi(r) + \Phi(s)$;

(c) *the function* $\Phi_1 : r \mapsto \Phi(r)/r$ *(with* $\Phi_1(0) = 0$*) is concave and nondecreasing on* $[0, \infty)$;

(d) *for* $r \geq 0$ *and* $\lambda \geq 0$, $\Phi(\lambda r) \leq \max\{1, \lambda^2\}\Phi(r)$;

(e) *for* $r \geq 0$ *and* $s \geq 0$,

$$(r + s)[\Phi(r + s) - \Phi(r) - \Phi(s)] \leq 2[s\Phi(r) + r\Phi(s)] \ ;$$

(f) *for* $r \geq 0$,

$$r\Phi'(r) - \Phi(r) \geq \frac{[\Phi'(r)]^2}{2\Phi''(0)} \ ;$$

(g) *for all* $p \in (1, 2]$, *the function* $r \mapsto \Phi(r)r^{-p}$ *belongs to* $L_\infty(0, \infty)$.

Proof. First, we note that, from the convexity of Φ and the concavity of Φ', we have

$$\Phi(s) - \Phi(r) \geq (s - r)\Phi'(r) \ , \qquad (r, s) \in [0, \infty)^2 \ , \tag{7.1.13}$$

$$\Phi'(s) - \Phi'(r) \leq (s - r)\Phi''(r) \ , \qquad (r, s) \in [0, \infty)^2 \ . \tag{7.1.14}$$

(a) Since $\Phi(0) = 0$, taking $s = 0$ in (7.1.13) readily gives the first inequality. Thanks to (7.1.14) with $s = 0$, the function Ψ defined, for $r \geq 0$, by $\Psi(r) := 2\Phi(r) - r\Phi'(r)$ satisfies

$$\Psi'(r) = \Phi'(r) - r\Phi''(r) \geq 0$$

with $\Psi(0) = 0$. Therefore Ψ is nonnegative in $[0, \infty)$, and the second inequality follows.

(b) For $r \geq 0$ and $s \geq 0$, we infer from (a) and (7.1.13) that

$$s\Phi'(r) \leq r\Phi'(r) - \Phi(r) + \Phi(s) \leq \Phi(r) + \Phi(s) \ .$$

(c) On differentiating Φ_1 once, and using (a), we obtain

$$\Phi_1'(r) = \frac{r\Phi'(r) - \Phi(r)}{r^2} \geq 0 \ , \qquad r > 0 \ ,$$

and thus the claimed monotonicity of Φ_1. We next differentiate Φ_1 twice and deduce from the concavity of Φ' that

$$\Phi_1''(r) = \frac{r^2\Phi''(r) - 2r\Phi'(r) + 2\Phi(r)}{r^3} = \frac{1}{r^3} \int_0^r s^2\Phi'''(s) \, \mathrm{d}s \leq 0 \ ,$$

and the concavity of Φ_1 follows.

(d) We first claim that Φ' is sublinear, that is,

$$\Phi'(\lambda s) \leq \max\{1, \lambda\}\Phi'(s) \ , \qquad (\lambda, s) \in [0, \infty)^2 \ . \tag{7.1.15}$$

Indeed, for $\lambda \in [0, 1]$, the claim (7.1.15) is a straightforward consequence of the monotonicity of Φ'. For $\lambda > 1$, it follows from the concavity of Φ', and the property $\Phi'(0) = 0$, that $\Phi'(s) \geq \Phi'(\lambda s)/\lambda$. Now, given $r \geq 0$, we deduce the estimate (d) by integrating each side of (7.1.15) with respect to s over $[0, r]$ and using the property $\Phi(0) = 0$.

(e) Let $r \geq 0$ and $s \geq 0$. Since $\Phi(0) = 0$, there holds

$$(r + s)\Phi(r + s) - r\Phi(r) - s\Phi(s)$$
$$= \int_0^r \int_0^s [(\rho + \sigma)\Phi''(\rho + \sigma) + 2\Phi'(\rho + \sigma)] \, d\sigma \, d\rho \, . \qquad (7.1.16)$$

It follows from the concavity (7.1.14) of Φ' (with $r = \rho + \sigma$ and $s = 0$) and (7.1.16) that

$$(r + s)\Phi(r + s) - r\Phi(r) - s\Phi(s) \leq 3 \int_0^r \int_0^s \Phi'(\rho + \sigma) \, d\sigma \, d\rho \, .$$

Next, for $\rho \in [0, r]$ and $\sigma \in [0, s]$, we infer from (7.1.15) that

$$\Phi'(\rho + \sigma) \leq \frac{\rho + \sigma}{\rho}\Phi'(\rho) \quad \text{and} \quad \Phi'(\rho + \sigma) \leq \frac{\rho + \sigma}{\sigma}\Phi'(\sigma) \, ,$$

so that $\Phi'(\rho + \sigma) \leq \Phi'(\rho) + \Phi'(\sigma)$. Consequently,

$$(r + s)\Phi(r + s) - r\Phi(r) - s\Phi(s) \leq 3 \int_0^r \int_0^s [\Phi'(\rho) + \Phi'(\sigma)] \, d\sigma \, d\rho$$
$$\leq 3 \left[s\Phi(r) + r\Phi(s) \right] \, ,$$

which completes the proof of (e).

(f) Thanks to the concavity of Φ' we have

$$\Phi'(r) = \int_0^r \Phi''(s) \, ds \leq \Phi''(0)r \, , \qquad r \geq 0 \, .$$

Let $r \geq 0$. Combining the above estimate with the monotonicity of Φ' gives

$$r\Phi'(r) - \Phi(r) = \int_0^r [\Phi'(r) - \Phi'(s)] \, ds \geq \int_0^{\Phi'(r)/\Phi''(0)} [\Phi'(r) - \Phi'(s)] \, ds$$
$$\geq \int_0^{\Phi'(r)/\Phi''(0)} [\Phi'(r) - \Phi''(0)s] \, ds = \frac{[\Phi'(r)]^2}{2\Phi''(0)} \, ,$$

as claimed.

(g) Let $p \in (1, 2]$. On the one hand, by (7.1.11c), there is $R > 0$ such that $\Phi'(r) \leq pr^{p-1}$ for $r \geq R$, hence $\Phi(r) \leq \Phi(R) + r^p - R^p$ for $r \geq R$. On the other hand, it follows from the concavity of Φ that, for $r \in [0, R]$,

$$\Phi(r) = \int_0^r (r - s)\Phi''(s) \, ds \leq \frac{\Phi''(0)}{2}r^2 \leq \frac{\Phi''(0)R^{2-p}}{2}r^p \, ,$$

which completes the proof.   □

We end this section with a discussion on some connections between weak convergence in L_1 and almost everywhere convergence, and first recall Vitali's theorem, see [149, Proposition 10.10]. This theorem provides a useful connection (in fact, the equivalence) between weak convergence in L_1, combined with μ-almost everywhere convergence, and norm convergence in L_1.

Theorem 7.1.10 (Vitali). *Let $(f_n)_{n\geq 1}$ be a sequence in $L_1(\Omega)$ and assume that there is a function $f \in L_1(\Omega)$ such that $(f_n)_{n\geq 1}$ converges μ-a.e. in Ω to f. Then the following statements are equivalent:*

(a) $(f_n)_{n\geq 1}$ converges (strongly) to f in $L_1(\Omega)$;

(b) the set $\mathscr{E} := \{f_n : n \geq 1\}$ is bounded in $L_1(\Omega)$ and satisfies (7.1.3) and (7.1.4).

Another outcome of a suitable joint use of weak convergence in L_1 and μ-almost everywhere convergence is a result (Proposition 7.1.12) that can be used to establish the weak limit of the product of a bounded sequence having a μ-almost everywhere limit with a sequence converging weakly in L_1. It is a consequence of the Dunford–Pettis theorem (Theorem 7.1.3) and Egorov's theorem, see [149, Proposition 10.11], which we recall now.

Theorem 7.1.11 (Egorov). *Assume that $\mu(\Omega) < \infty$ and let $(g_n)_{n\geq 1}$ be a sequence of measurable functions in Ω such that $(g_n)_{n\geq 1}$ converges μ-a.e. in Ω to a measurable function g. Then, for any $\delta > 0$, there is a measurable subset $E_\delta \in \mathcal{B}$ such that*

$$\mu(E_\delta) \leq \delta \quad and \quad \lim_{n\to\infty} \sup_{x\in\Omega\setminus E_\delta} |g_n(x) - g(x)| = 0 .$$

Proposition 7.1.12. *Let $(f_n)_{n\geq 1}$ be a sequence in $L_1(\Omega)$ which converges weakly in $L_1(\Omega)$ to $f \in L_1(\Omega)$, and let $(g_n)_{n\geq 1}$ be a bounded sequence in $L_\infty(\Omega)$ which converges μ-a.e. to a function $g \in L_\infty(\Omega)$. Then the sequence $(f_n g_n)_{n\geq 1}$ converges weakly in $L_1(\Omega)$ to fg and*

$$\lim_{n\to\infty} \int_\Omega |f_n||g_n - g| \, \mathrm{d}\mu = 0 .$$

Proposition 7.1.12 appears implicitly in [100, 252] and may be found in [117, Proposition 2.61] and [165, Lemma A.2]. We reproduce below the proof from [165, Lemma A.2].

Proof of Proposition 7.1.12. Set $\mathscr{E} := \{f_n : n \geq 1\}$ and

$$M := \sup_{n\geq 1} \|f_n\|_1 + \sup_{n\geq 1} \|g_n\|_\infty \in (0,\infty) . \tag{7.1.17}$$

Let $\varepsilon \in (0,1)$. Owing to the weak convergence of $(f_n)_{n\geq 1}$ in $L_1(\Omega)$, we infer from Theorem 7.1.3 that there are $\delta_\varepsilon > 0$ and $\Omega_\varepsilon \in \mathcal{B}$ such that $\mu(\Omega_\varepsilon) < \infty$,

$$\sup_{n\geq 1} \int_{\Omega\setminus\Omega_\varepsilon} |f_n| \, \mathrm{d}\mu \leq \frac{\varepsilon}{4M} \quad and \quad \eta\{\mathscr{E}, \delta_\varepsilon\} \leq \frac{\varepsilon}{4M} , \tag{7.1.18}$$

where $\eta\{\mathscr{E}, \delta_\varepsilon\}$ being defined in (7.1.1). Since $\mu(\Omega_\varepsilon) < \infty$ and $(g_n)_{n\geq 1}$ also converges μ-a.e. in Ω_ε, Egorov's theorem (Theorem 7.1.11) ensures that there is $E_\varepsilon \in \mathcal{B}$ such that $E_\varepsilon \subset \Omega_\varepsilon$,

$$\mu(E_\varepsilon) \leq \delta_\varepsilon \quad and \quad \lim_{n\to\infty} \sup_{x\in\Omega_\varepsilon\setminus E_\varepsilon} |g_n(x) - g(x)| = 0 . \tag{7.1.19}$$

Consequently, from (7.1.17), (7.1.18) and (7.1.19),

$$\|f_n(g_n - g)\|_1 = \int_{\Omega\setminus\Omega_\varepsilon} |f_n||g_n - g| \, \mathrm{d}\mu + \int_{\Omega_\varepsilon\setminus E_\varepsilon} |f_n||g_n - g| \, \mathrm{d}\mu + \int_{E_\varepsilon} |f_n||g_n - g| \, \mathrm{d}\mu$$

$$\leq 2M \int_{\Omega\setminus\Omega_\varepsilon} |f_n| \, \mathrm{d}\mu + \|f_n\|_1 \sup_{x\in\Omega_\varepsilon\setminus E_\varepsilon} |g_n(x) - g(x)| + 2M \, \eta\{\mathscr{E}, \delta_\varepsilon\}$$

$$\leq \varepsilon + M \sup_{x\in\Omega_\varepsilon\setminus E_\varepsilon} |g_n(x) - g(x)| .$$

Thanks to (7.1.19), we may pass to the limit as $n \to \infty$ in the above inequality and obtain

$$\limsup_{n \to \infty} \|f_n(g_n - g)\|_1 \leq \varepsilon \; .$$

Hence $\|f_n(g_n - g)\|_1 \longrightarrow 0$ as $n \to \infty$ since $\varepsilon \in (0,1)$ is arbitrary. Now, given $\vartheta \in L_\infty(\Omega)$,

$$\left| \int_\Omega (f_n g_n - fg)\vartheta \; \mathrm{d}\mu \right| \leq \|\vartheta\|_\infty \|f_n(g_n - g)\|_1 + \left| \int_\Omega (f_n - f)g\vartheta \; \mathrm{d}\mu \right| \; .$$

We have just shown that the first term in the right-hand side of the above inequality converges to zero as $n \to \infty$ while the weak convergence in $L_1(\Omega)$ of $(f_n)_{n \geq 1}$ to f guarantees the convergence to zero of the second term since $g\vartheta \in L_\infty(\Omega)$. □

Remark 7.1.13. *Proposition 7.1.12 may be viewed as an extension of the continuity of the pointwise product $(f,g) \mapsto fg$ from $L_p(\Omega) \times L_{p/(p-1),w}(\Omega)$ in $L_{1,w}(\Omega)$ for $p \in (1,\infty)$. Here and below, $X_{,w}$ denotes the Banach space X endowed with its weak topology.*

As we shall see later on, Proposition 7.1.12 is at the heart of the proof that quadratic nonlinear nonlocal operators, such as the coagulation operator, are continuous with respect to the weak topology of $L_1 \times L_1$. In fact, an illustrative application of Proposition 7.1.12 is the sequential continuity of the convolution operator $(f,g) \mapsto f*g$ from $L_{1,w}(\mathbb{R}^d) \times L_{1,w}(\mathbb{R}^d)$ into itself. This is established in the following corollary.

Corollary 7.1.14. *Let $d \in \mathbb{N}$, and consider two sequences $(f_n)_{n \geq 1}$ and $(g_n)_{n \geq 1}$ in $L_1(\mathbb{R}^d)$ which converge weakly in $L_1(\mathbb{R}^d)$ to f and g, respectively, as $n \to \infty$. Then $(f_n * g_n)_{n \geq 1}$ converges weakly in $L_1(\mathbb{R}^d)$ to $f * g$ as $n \to \infty$.*

Proof. Let $\vartheta \in L_\infty(\mathbb{R}^d)$ and define

$$F_n(y) := \int_{\mathbb{R}^d} f_n(x)\vartheta(x + y) \; \mathrm{d}x \; , \quad y \in \mathbb{R}^d \; , \; n \geq 1 \; .$$

On the one hand, it follows from Fubini's theorem that

$$\int_{\mathbb{R}^d} (f_n * g_n)(x)\vartheta(x) \; \mathrm{d}x = \int_{\mathbb{R}^d} F_n(y)g_n(y) \; \mathrm{d}y \; , \qquad n \geq 1 \; . \tag{7.1.20}$$

On the other hand,

$$|F_n(y)| \leq \|\vartheta\|_\infty \|f_n\|_1 \leq \|\vartheta\|_\infty \sup_{m \geq 1} \|f_m\|_1 \quad \text{for a.e.} \; y \in \mathbb{R}^d \; ,$$

so that $(F_n)_{n \geq 1}$ is bounded in $L_\infty(\mathbb{R}^d)$. Furthermore, the function $x \mapsto \vartheta(x + y)$ belongs to $L_\infty(\mathbb{R}^d)$ for a.e. $y \in \mathbb{R}^d$, and the weak convergence of $(f_n)_{n \geq 1}$ in $L_1(\mathbb{R}^d)$ implies that

$$\lim_{n \to \infty} F_n(y) = F(y) := \int_{\mathbb{R}^d} f(x)\vartheta(x + y) \; \mathrm{d}x \quad \text{for a.e.} \; y \in \mathbb{R}^d \; .$$

We are thus in a position to apply Proposition 7.1.12 to $(F_n g_n)_{n \geq 1}$ and conclude that $(F_n g_n)_{n \geq 1}$ converges weakly in $L_1(\mathbb{R}^d)$ towards Fg. We then let $n \to \infty$ in (7.1.20) to obtain

$$\lim_{n \to \infty} \int_{\mathbb{R}^d} (f_n * g_n)(x)\vartheta(x) \; \mathrm{d}x = \int_{\mathbb{R}^d} F(y)g(y) \; \mathrm{d}y = \int_{\mathbb{R}^d} (f * g)(x)\vartheta(x) \; \mathrm{d}x \; ,$$

the second identity being a consequence of Fubini's theorem. □

We end this section with a variant of the Arzelà–Ascoli theorem for time-dependent functions with values in a Banach space X, but which are only continuous with respect to the weak topology of X [270, Appendix A]. Given two real numbers $a < b$, the set $C([a,b], X_{,w})$ of weakly continuous functions from $[a,b]$ to $X_{,w}$ is the set of functions $f : [a,b] \to X$ such that $t \mapsto \langle \Lambda, f(t) \rangle$ belongs to $C([a,b])$ for all $\Lambda \in X^*$, where $\langle \cdot, \cdot \rangle$ denotes the duality product between X and its topological dual space X^*. As in the classical Arzelà–Ascoli theorem, an appropriate notion of time equicontinuity is needed which we define now.

Definition 7.1.15 (Weak equicontinuity). *Let $\mathcal{E}$ be a subset of $C([a,b], X_{,w})$ and $t_0 \in [a,b]$. The set $\mathcal{E}$ is weakly equicontinuous at t_0 if the set $\{\langle \Lambda, f \rangle : f \in \mathcal{E}\}$ is equicontinuous at t_0 for each $\Lambda \in X^*$. Specifically, for each $\Lambda \in X^*$ and $\varepsilon > 0$, there is $\delta = \delta(t_0, \Lambda, \varepsilon) > 0$ such that*
$$|\langle \Lambda, f(t) \rangle - \langle \Lambda, f(t_0) \rangle| \leq \varepsilon$$
for all $t \in [t_0 - \delta, t_0 + \delta] \cap [a,b]$ and $f \in \mathcal{E}$.

With this definition, a characterisation of relatively sequentially weakly compact subsets of $C([a,b], X_{,w})$ is available [270, Theorem A.3.1].

Theorem 7.1.16 (Arzelà–Ascoli). *Let X be a sequentially weakly complete Banach space and $\mathcal{E}$ a subset of $C([a,b], X_{,w})$. The set $\mathcal{E}$ is relatively sequentially weakly compact in $C([a,b], X_{,w})$ if and only if*

(a) *$\mathcal{E}$ is weakly equicontinuous at each $t_0 \in [a,b]$,*

(b) *there is a dense subset $\mathcal{D}$ of $[a,b]$ such that $\{f(t) : f \in \mathcal{E}\}$ is relatively sequentially weakly compact in X for all $t \in \mathcal{D}$.*

For $p \in (1, \infty)$, the Lebesgue space $L_p(\Omega, \mathcal{B}, \mu)$ is sequentially weakly complete as reflexive Banach spaces are sequentially weakly complete Banach spaces. For the applications we have in mind it is worth mentioning that $L_1(\Omega, \mathcal{B}, \mu)$ is also a sequentially weakly complete Banach space [274, Corollary III.C.14].

7.2 Continuous and Compact Embeddings

In accordance with the notation introduced in Section 3.1.4 and recalled in Section 6.1, we set $X_m := L_1((0, \infty), x^m \mathrm{d}x)$ and $X_{0,m} := L_1((0, \infty), (1 + x^m) \mathrm{d}x) = X_0 \cap X_m$, for $m \in \mathbb{R}$. The natural norms on X_m and $X_{0,m}$ are denoted by $\| \cdot \|_{[m]}$ and $\| \cdot \|_{[0,m]}$, respectively. We shall also use the notation

$$W_1^1((0, \infty), x^m \mathrm{d}x) := \{f \in X_m : f' \in X_m\} \, .$$

Proposition 7.2.1. *Let $(m_1, m_2) \in [0, \infty)^2$ be such that $m_1 < m_2$. Then $X_{0,m_2} \hookrightarrow X_{0,m_1}$.*

Proof. Let $0 \leq m_1 < m_2$ and $f \in X_{0,m_2}$. Then

$$\|f\|_{[0,m_1]} = \int_0^\infty |f(x)| \mathrm{d}x + \int_0^1 |f(x)| x^{m_1} \mathrm{d}x + \int_1^\infty |f(x)| x^{m_1} \mathrm{d}x$$
$$\leq \int_0^\infty |f(x)| \mathrm{d}x + \int_0^1 |f(x)| \mathrm{d}x + \int_1^\infty |f(x)| x^{m_2} \mathrm{d}x \leq 2 \int_0^\infty |f(x)| (1 + x^{m_2}) \mathrm{d}x = 2\|f\|_{[0,m_2]},$$
$$\tag{7.2.1}$$

and the stated result follows immediately. $\qquad \square$

Proposition 7.2.2. *Let* $(m_1, m_2, m_3) \in \mathbb{R}^3$ *be such that* $m_1 < m_2 < m_3$, *and let* $\mathcal{E}$ *be a bounded subset of* $X_{m_1} \cap X_{m_3} \cap W_1^1((0, \infty), x^{m_2}\mathrm{d}x)$. *Then* $\mathcal{E}$ *is relatively compact in* X_m *for all* $m \in (m_1, m_3)$.

Proof. Let $M > 0$ be such that

$$\int_0^\infty (x^{m_1} + x^{m_3})|f(x)|\mathrm{d}x + \int_0^\infty x^{m_2}\left(|f(x)| + |f'(x)|\right) \, \mathrm{d}x \le M \qquad (7.2.2)$$

for all $f \in \mathcal{E}$. We first observe that, for $f \in \mathcal{E}$ and $x > 0$,

$$\int_x^\infty y^{m_2} f'(y) \, \mathrm{d}y = -x^{m_2} f(x) - m_2 \int_x^\infty y^{m_2-1} f(y) \, \mathrm{d}y \, .$$

Consequently,

$$x^{m_2}|f(x)| \le \int_0^\infty y^{m_2}|f'(y)| \, \mathrm{d}y + \frac{|m_2|}{x} \int_x^\infty y^{m_2}|f(y)| \, \mathrm{d}y \, ,$$

and thus, thanks to (7.2.2),

$$|f(x)| \le \frac{M}{x^{m_2}}\left(1 + \frac{|m_2|}{x}\right) \, , \qquad x \in (0, \infty) \, . \qquad (7.2.3)$$

Now, let $m \in (m_1, m_3)$, $f \in \mathcal{E}$, and consider a measurable subset E of $(0, \infty)$. Then, for $R \ge 1$,

$$\int_E x^m|f(x)| \, \mathrm{d}x \le \int_0^{1/R} x^m|f(x)| \, \mathrm{d}x + \int_{E\cap(1/R,R)} x^m|f(x)| \, \mathrm{d}x + \int_R^\infty x^m|f(x)| \, \mathrm{d}x$$

$$\le R^{m_1-m} \int_0^\infty x^{m_1}|f(x)| \, \mathrm{d}x + R^{|m-m_2|} \int_{E\cap(1/R,R)} x^{m_2}|f(x)| \, \mathrm{d}x$$

$$+ R^{m-m_3} \int_R^\infty x^{m_3}|f(x)| \, \mathrm{d}x \, .$$

It then follows from (7.2.2) and (7.2.3) that

$$\int_E x^m|f(x)| \, \mathrm{d}x \le M\left(R^{m_1-m} + R^{m-m_3}\right) + MR^{|m-m_2|}(1 + |m_2|R)|E| \, ,$$

from which we deduce that

$$\eta\{\mathcal{E}, \delta; X_m\} \le M\left(R^{m_1-m} + R^{m-m_3}\right) + MR^{|m-m_2|}(1 + |m_2|R)\delta \, ,$$

recalling that the modulus of uniform integrability is defined in (7.1.1). Letting first $\delta \to 0$, and then $R \to \infty$, ensures that $\eta\{\mathcal{E}; X_m\} = 0$. In addition, it follows readily from (7.2.2) that, for $R \ge 1$ and $f \in \mathcal{E}$,

$$\int_R^\infty x^m|f(x)| \, \mathrm{d}x \le R^{m-m_3} \int_R^\infty x^{m_3}|f(x)| \, \mathrm{d}x \le MR^{m-m_3} \, ,$$

so that

$$\lim_{R\to\infty} \sup_{f\in\mathcal{E}} \int_R^\infty x^m|f(x)| \, \mathrm{d}x = 0 \, .$$

Therefore, according to the Dunford–Pettis theorem (Theorem 7.1.3), we have established that

$$\mathcal{E} \text{ is relatively sequentially weakly compact in } X_m \, . \qquad (7.2.4)$$

Consider now a sequence $(f_n)_{n \geq 1}$ in E. Since $W_1^1(1/R, R)$ is compactly embedded in $L_1(1/R, R)$ for all $R \geq 1$ [51, Theorem 8.8], we use a diagonal process to construct a function $f \in L_{1,loc}(0, \infty)$ and a subsequence of $(f_n)_{n \geq 1}$ (not relabelled) such that $(f_n)_{n \geq 1}$ converges to f a.e. in $(0, \infty)$. This property, along with (7.2.4), allows us to deduce from Vitali's theorem (Theorem 7.1.10) that $(f_n)_{n \geq 1}$ converges strongly to f in X_m. □

7.3 Dynamical Systems

In this section, we recall a handful of results on dynamical systems which are to be used in Chapter 10 to study qualitative properties of C-F equations.

7.3.1 Basic Concepts

Definition 7.3.1 (Dynamical system). *Let X be a topological space and J an interval of $\mathbb{R}$ containing 0. A continuous map $\Psi : J \times X \to X$ is a (continuous) dynamical system if*

(a) $\Psi(0, x) = x$ for all $x \in X$,

(b) $\Psi(s, \Psi(t, x)) = \Psi(s + t, x)$ for all $x \in X$, $s \in J$, and $t \in J$ such that $s + t \in J$.

Given a dynamical system $\Psi : J \times X \to X$ on the topological space X and $x \in X$, the trajectory $\Psi(\cdot, x)$ emanating from x is the curve $\{\Psi(t, x) \; : \; t \in J\}$. If $J = [0, \infty)$ and $x \in X$ is such that $\Psi(\cdot, x) = \{x\}$, then x is called a stationary solution.

An important notion, being at the heart of the study of the long-term behaviour of dynamical systems, is that of an invariant set.

Definition 7.3.2 (Invariant set). *Let $\Psi : J \times X \to X$ be a dynamical system on the topological space X. A subset Z of X is said to be positively invariant under Ψ if $\Psi(t, x) \in Z$ for all $t \in J \cap [0, \infty)$ and $x \in Z$.*

A particular invariant set is the so-called ω-limit set of a trajectory $\Psi(\cdot, x)$, starting at $x \in X$, which is the set of all cluster points of the function $t \mapsto \Psi(t, x)$ as $t \to \infty$.

Definition 7.3.3 (ω-limit set). *Let $\Psi : [0, \infty) \times X \to X$ be a dynamical system on the topological space X. Given $x \in X$, the ω-limit set $\omega(x)$ of x is the set of cluster points of the trajectory $\Psi(\cdot, x)$ as $t \to \infty$; that is, $z \in \omega(x)$ if and only if there is a positive sequence, $(t_n)_{n \geq 1}$, such that $t_n \to \infty$ and $\Psi(t_n, x) \longrightarrow z$ in X as $n \to \infty$.*

For $x \in X$, the ω-limit set $\omega(x)$ is a closed subset of X which is positively invariant under Ψ. It is also non-empty whenever the trajectory $\Psi(\cdot, x)$ is relatively compact in X.

A useful concept in the analysis of ω-limit sets is that of Lyapunov functionals.

Definition 7.3.4 (Lyapunov functional). *Let $\Psi : [0, \infty) \times X \to X$ be a dynamical system on the topological space X.*

(a) A function $L \in C(X; \mathbb{R})$ is a Lyapunov functional for Ψ if the inequality $L(\Psi(t_2, x)) \leq L(\Psi(t_1, x))$ holds true for all $t_2 \geq t_1 \geq 0$ and $x \in X$.

(b) A function $L \in C(X; \mathbb{R})$ is a strict Lyapunov functional if it is a Lyapunov functional for which any trajectory satisfying $L(\Psi(t_2, x)) = L(\Psi(t_1, x))$ for all $t_2 > t_1 \geq 0$ is stationary; that is, $\Psi(t, x) = x$ for $t \geq 0$.

The existence of a Lyapunov functional, equipped with suitable properties, provides information on the structure of ω-limit sets.

Theorem 7.3.5 (Invariance Principle). *Let $\Psi : [0, \infty) \times X \to X$ be a dynamical system on the topological space X. Assume that there is a nonnegative Lyapunov functional L for Ψ. Then, for any $x \in X$, there is $\ell_x \geq 0$ such that $L(z) = \ell_x$ for all $z \in \omega(x)$.*

Introducing the set $\mathcal{S}$ of stationary solutions

$$\mathcal{S} := \{x \in X \ : \ \Psi(t, x) = x \text{ for all } \ t \geq 0\} \,,$$

if we further assume that L is a strict Lyapunov functional for Ψ, then $\omega(x) \subset \mathcal{S}$ for all $x \in X$.

7.3.2   Stationary Solutions and Fixed-Point Theorems

As we shall see below, an important issue in the study of C-F equations is the existence of either stationary or self-similar solutions. Investigating this question on the associated stationary problems proves to be rather intractable, but, fortunately, progress can be made by utilising dynamical systems theory. The purpose of this section is thus to describe the dynamical systems approach that will be used in Chapter 10.

Theorem 7.3.6. *Let X be a Banach space, Y a subset of X, and $\Psi : [0, \infty) \times Y \to Y$ a dynamical system with a positively invariant set $Z \subset Y$ which is a non-empty compact and convex subset of X. Then there is $x_0 \in Z$ such that $\Psi(t, x_0) = x_0$ for all $t \geq 0$.*

When X is a finite-dimensional vector space, Theorem 7.3.6 may be found in [4, Theorem 22.13] while the general case is given in [125, Proof of Theorem 5.2]. The proof comprises two steps: first, it is shown, with the help of Schauder's fixed-point theorem [238, Theorem 5.28], that, given an arbitrary $T > 0$, there is at least one time-periodic solution with period T and that the set of such solutions is a compact subset of X. The second step exploits the fact that stationary solutions belong to the intersection of these sets, which is non-empty as they form a decreasing sequence of compact sets in X.

We end this section with a version of Theorem 7.3.6 that holds when the invariant set Z is only weakly compact in the Banach space X [113, Theorem 1.2].

Theorem 7.3.7. *Let X be a Banach space, Y a subset of X, and $\Psi : [0, \infty) \times Y \to Y$ a dynamical system with a positively invariant set $Z \subset Y$ which is a non-empty convex subset of X. Assume further that:*

(a) if $(x_n)_{n \geq 1}$ is a sequence in Y which converges weakly in X to $x \in Y$ then $(\Psi(t, x_n))_{n \geq 1}$ converges weakly in X to $\Psi(t, x)$ for all $t \geq 0$,

(b) Z is sequentially weakly compact in X.

Then there is $x_0 \in Z$ such that $\Psi(t, x_0) = x_0$ for all $t \geq 0$.

The proof of Theorem 7.3.7 follows the same lines as that of Theorem 7.3.6, except that it is Tychonov's fixed-point theorem [238, Theorem 5.28] which is used instead of Schauder's fixed-point theorem.

7.4   Algebraic Inequalities

In this section, we gather together several algebraic inequalities that we shall use in Chapter 8 when deriving moment estimates. Some of these inequalities will also prove to be helpful for the analysis carried out in Chapter 10. We begin with the classical result that compares the powers of a sum with the corresponding sum of powers.

Lemma 7.4.1. *For $m \in (0,1)$ there holds*

$$2^{m-1}(x^m + y^m) \le (x+y)^m \le x^m + y^m , \qquad (x,y) \in [0,\infty)^2 . \tag{7.4.1}$$

For $m \ge 1$, there holds

$$x^m + y^m \le (x+y)^m \le 2^{m-1}(x^m + y^m) , \qquad (x,y) \in [0,\infty)^2 . \tag{7.4.2}$$

Proof. For $m \in (0,1)$, Lemma 7.4.1 follows from a study of the sign of the first derivatives of the functions $z \mapsto 1 + z^m - (1+z)^m$ and $z \mapsto (1+z)^m - 2^{m-1}(1+z^m)$ on $[0,1]$. The proof of Lemma 7.4.1 for $m \ge 1$ relies on a similar argument. □

We next report a refinement of the second inequality in (7.4.2) that is derived in [78, p. 216], but with a different constant.

Lemma 7.4.2. *Let $m > 1$. Then*

$$(x+y)^m - x^m - y^m \le C_m(xy^{m-1} + x^{m-1}y) , \qquad (x,y) \in (0,\infty)^2 ,$$

with $C_m := 2^{m-1} - 1$ for $m \in (1,2] \cup [3,\infty)$ and $C_m := m$ for $m \in (2,3)$.

Proof. For further use, we note that

$$m2^{m-3} > 2^{m-1} - 1 \quad \text{if and only if} \quad m \in (1,2) \cup (3,\infty) , \tag{7.4.3}$$

and

$$m - 1 \ge 2^{m-2} , \qquad m \in [2,3] . \tag{7.4.4}$$

Lemma 7.4.2 is obviously true for $m = 2$ with $C_2 = 1$. We next handle separately the cases $m \in (1,2)$, $m \in (2,3]$, and $m > 3$.

Case 1: $m \in (1,2)$. We define the function

$$g_1(z) := 1 + z^m + (2^{m-1} - 1)(z + z^{m-1}) - (1+z)^m , \qquad z \in [0,1] .$$

Then, for $z \in (0,1]$,

$$g_1'(z) = mz^{m-1} - m(1+z)^{m-1} + (2^{m-1} - 1)(1 + (m-1)z^{m-2}) ,$$

$$g_1''(z) = (m-1)z^{m-2}\left(m + \frac{(2^{m-1} - 1)(m-2)}{z} - m\left(1 + \frac{1}{z}\right)^{m-2}\right)$$

$$= (m-1)z^{m-2}G_1(1/z) ,$$

with $G_1(Z) := m + (2^{m-1} - 1)(m-2)Z - m(1+Z)^{m-2}$, $Z \in [1,\infty)$. In particular,

$$g_1(0) = g_1(1) = 0 , \qquad \lim_{z \to 0} g_1'(z) = \infty , \qquad g_1'(1) = 0 . \tag{7.4.5}$$

Now, for $Z \ge 1$,

$$G_1'(Z) = -(2-m)(2^{m-1} - 1) + m(2-m)(1+Z)^{m-3} .$$

Then G_1' is decreasing on $[1,\infty)$ and, since $G_1'(1) = (2-m)(m2^{m-3} - 2^{m-1} + 1) > 0$ by (7.4.3) and $G_1'(Z) \to -(2-m)(2^{m-1} - 1) < 0$ as $Z \to \infty$, there is a unique $Z_1 \in (1,\infty)$

such that $G_1' > 0$ on $(1, Z_1)$ and $G_1' < 0$ on (Z_1, ∞). As $G_1(1) = 2(1 - 2^{m-1} + m2^{m-3}) > 0$ by (7.4.3), and $G_1(Z) \to -\infty$ as $Z \to \infty$, we conclude that there is a unique $Z_0 \in (Z_1, \infty)$ such that $G_1 > 0$ on $(1, Z_0)$ and $G_1 < 0$ on (Z_0, ∞). Recalling the connection between g_1'' and G_1 and setting $z_0 := 1/Z_0$, we have shown that $g_1'' < 0$ on $(0, z_0)$ and $g_1'' > 0$ on $(z_0, 1)$. Due to (7.4.5), these properties imply that there is a unique $z_1 \in (0, z_0)$ such that $g_1' > 0$ on $(0, z_1)$ and $g_1' < 0$ on $(z_1, 1)$. Therefore, for $z \in [0, 1]$, $g_1(z) \geq \min\{g_1(0), g_1(1)\} = 0$ and we have shown that

$$(2^{m-1} - 1)(z + z^{m-1}) \geq (1 + z)^m - 1 - z^m , \qquad z \in [0, 1] .$$

To complete the proof of Lemma 7.4.2 in this case, consider $(x, y) \in (0, \infty)^2$. Then $\max\{x, y\}^m g_1(\min\{x, y\}/\max\{x, y\}) \geq 0$, from which the stated inequality follows.

Case 2: $m \in (2, 3]$. Define

$$g_2(z) := 1 + z^m + m(z + z^{m-1}) - (1 + z)^m , \qquad z \in [0, 1] .$$

Then, for $z \in [0, 1]$,

$$g_2'(z) = m \left[1 + z^{m-1} + (m - 1)z^{m-2} - (1 + z)^{m-1} \right] ,$$
$$g_2''(z) = m(m - 1) \left[z^{m-2} + (m - 2)z^{m-3} - (1 + z)^{m-2} \right]$$
$$= m(m - 1)z^{m-2} G_2(1/z) ,$$

where $G_2(Z) := 1 + (m - 2)Z - (1 + Z)^{m-2}$, $Z \in [1, \infty)$. Observing that

$$G_2'(Z) = (m - 2)[1 - (1 + Z)^{m-3}] > 0 , \qquad Z \geq 1 ,$$

and $G_2(1) = m - 1 - 2^{m-2} \geq 0$ by (7.4.4), we conclude that $G_2 > 0$ on $(1, \infty)$ and thus $g_2'' > 0$ on $(0, 1)$. As $g_2(0) = g_2'(0) = 0$, we finally obtain

$$m(z + z^{m-1}) \geq (1 + z)^m - 1 - z^m , \qquad z \in [0, 1] ,$$

and complete the proof as in the previous case.

Case 3: $m \in (3, \infty)$. We define g_1 and G_1 as in Case 1 and note that

$$g_1(0) = g_1(1) = 0 , \qquad g_1'(0) = 2^{m-1} - m - 1 , \qquad g_1'(1) = 0 . \qquad (7.4.6)$$

Also, G_1' is decreasing on $(1, \infty)$ with $G_1'(1) > 0$ by (7.4.3) and $G_1'(Z) \to -\infty$ as $Z \to \infty$. There is thus a unique $Z_1 \in (1, \infty)$ such that $G_1' > 0$ on $(1, Z_1)$ and $G_1' < 0$ on (Z_1, ∞). Again, $G_1(1) = 2(1 - 2^{m-1} + m2^{m-3}) > 0$ by (7.4.3) while $G_1(Z) \to -\infty$ as $Z \to \infty$ and we infer from these properties that there is a unique $Z_0 \in (Z_1, \infty)$ such that $G_1 > 0$ on $(1, Z_0)$ and $G_1 < 0$ on (Z_0, ∞). Consequently, $g_1'' < 0$ on $(0, z_0)$ and $g_1'' > 0$ on $(z_0, 1)$ with $z_0 := 1/Z_0$ and we argue as in Case 1 to complete the proof of Lemma 7.4.2. $\square$

Remark 7.4.3. *The constant C_m computed in Lemma 7.4.2 is optimal. Indeed, taking $x = y$ gives equality when $m \in (1, 2] \cup (3, \infty)$. For $m \in (2, 3]$, there holds*

$$\lim_{y \to 0} \frac{(x + y)^m - x^m - y^m}{y} = mx^{m-1} = \lim_{y \to 0} C_m \frac{xy^m + x^m y}{y} .$$

In the same vein, the following inequality is proved in [65, Lemma 2.3].

Lemma 7.4.4. *Let $m > 1$. Then*

$$(x + y) [(x + y)^m - x^m - y^m] \leq c_m (x^m y + x y^m) , \qquad (x, y) \in (0, \infty)^2 ,$$

where $c_m := m$ if $m \in (1, 2]$ and $c_m := 2^m - 2$ if $m > 2$.

Proof. For $(x, y) \in (0, \infty)^2$, we infer from Lemma 7.4.2 that

$$(x + y) [(x + y)^m - x^m - y^m] = (x + y)^{m+1} - x^{m+1} - y^{m+1} - x^m y - x y^m$$
$$\leq (C_{m+1} - 1)(x^m y + x y^m) ,$$

as claimed. □

Lemma 7.4.4 is actually valid for any nonnegative convex function $\Phi \in C^2([0, \infty))$ satisfying $\Phi(0) = \Phi'(0) = 0$ and having a convex derivative enjoying the Δ_2-condition (also referred to as the doubling condition) $\Phi'(2r) \leq C\Phi'(r)$ [160]. The constant of course depends on Φ.

We next state another refinement of the second inequality in (7.4.2).

Lemma 7.4.5. *Let $m \in (1, 2]$. Then*

$$(x + y)^m - x^m - y^m \leq (2^m - 2)(xy)^{m/2} , \qquad (x, y) \in (0, \infty)^2 .$$

Proof. Lemma 7.4.5 being again obvious for $m = 2$, we assume that $m \in (1, 2)$ and define the function

$$g(z) := (2^m - 2)z^{m/2} + 1 + z^m - (1 + z)^m , \qquad z \in [0, 1] .$$

Observe that $g(0) = g(1) = 0$. Next, for $z \in (0, 1]$,

$$g'(z) = m(2^{m-1} - 1)z^{(m-2)/2} + m z^{m-1} - m(1 + z)^{m-1} = m z^{m-1} G(1/z) ,$$

with $G(Z) := (2^{m-1} - 1)Z^{m/2} + 1 - (1 + Z)^{m-1}$, $Z \in [1, \infty)$. Now, for $Z \in [1, \infty)$,

$$G'(Z) = \frac{m(2^{m-1} - 1)}{2} Z^{(m-2)/2} - (m - 1)(1 + Z)^{m-2}$$
$$= (m - 1)Z^{(m-2)/2} G_1 \left(\frac{1}{\sqrt{Z}} + \sqrt{Z} \right) ,$$

with $G_1(Y) := (m(2^{m-1} - 1)/2(m - 1)) - Y^{m-2}$, $Y \in [2, \infty)$. Since G_1 is an increasing function satisfying

$$G_1(2) = \frac{2^{m-1} - m}{2(m - 1)} < 0 \quad \text{and} \quad \lim_{Y \to \infty} G_1(Y) = \frac{m(2^{m-1} - 1)}{2(m - 1)} > 0 ,$$

there exists $Y_1 \in (2, \infty)$ such that $G_1 < 0$ on $[2, Y_1)$ and $G_1 > 0$ on (Y_1, ∞). This implies the existence of $Z_1 > 1$ such that $G' < 0$ on $[1, Z_1)$ and $G' > 0$ on (Z_1, ∞). Noting that $G(1) = 0$ and $G(Z) \to \infty$ as $Z \to \infty$, we conclude that there is $Z_0 > Z_1$ such that $G < 0$ on $(1, Z_0)$ and $G > 0$ on (Z_0, ∞). Equivalently, $g' > 0$ on $(0, 1/Z_0)$ and $g' < 0$ on $(1/Z_0, 1]$, a property which readily implies that $g(z) \geq \min\{g(0), g(1)\} = 0$.

To complete the proof, consider $(x, y) \in (0, \infty)^2$ and assume without loss of generality that $x \leq y$. Then we have just proved that $g(x/y) \geq 0$, from which the stated inequality follows. □

Letting $m \to 1$ in Lemma 7.4.5 provides the following inequality.

Corollary 7.4.6. *There holds*

$$(x+y)\ln(x+y) - x\ln(x) - y\ln(y) \leq 2\ln 2\sqrt{xy}, \qquad (x,y) \in (0,\infty)^2.$$

Proof. Consider $(x,y) \in (0,\infty)^2$ and $m \in (1,2)$. According to Lemma 7.4.5,

$$\frac{(x+y)^m - x^m - y^m}{m-1} \leq 2\frac{2^{m-1}-1}{m-1}(xy)^{m/2}. \qquad (7.4.7)$$

On the one hand,

$$\lim_{m\to 1} 2\frac{2^{m-1}-1}{m-1}(xy)^{m/2} = 2\ln(2)\sqrt{xy}.$$

On the other hand,

$$\frac{(x+y)^m - x^m - y^m}{m-1} = x\frac{(x+y)^{m-1} - x^{m-1}}{m-1} + y\frac{(x+y)^{m-1} - y^{m-1}}{m-1},$$

hence

$$\lim_{m\to 1} \frac{(x+y)^m - x^m - y^m}{m-1} = x\ln(x+y) - x\ln(x) + y\ln(x+y) - y\ln(y).$$

Letting $m \to 1$ in (7.4.7) gives the stated inequality. □

We finally recall the classical Young inequality.

Proposition 7.4.7. *Consider $p \in (1,\infty)$. Then*

$$xy \leq \frac{1}{p}x^p + \frac{p-1}{p}y^{p/(p-1)}, \qquad (x,y) \in [0,\infty)^2.$$

7.5 The Gronwall–Henry Inequality

In this section we shall give an elementary proof of a version of Gronwall's inequality that is due to Henry; see [140, p. 188] as well as [5, Theorem 3.3.1] and [73, Lemma 8.1.1] for alternative proofs and [233, Section 1.4] for several variants. This inequality is crucial for certain estimates that are used later in the proof of Theorem 8.1.2 and is also referred to as the singular Gronwall lemma in the literature.

Lemma 7.5.1. *Let $u \in L_{\infty,loc}((0,T]) \cap L_1(0,T)$, $0 < T < \infty$, be a nonnegative function satisfying*

$$u(t) \leq \frac{c}{t^\gamma} + c\int_0^t u(\tau)(t-\tau)^{-\alpha}d\tau, \qquad t \in (0,T], \qquad (7.5.1)$$

where $\gamma < 1$, $0 < \alpha < 1$ and $c > 0$. Then there is a constant $C = cC(\gamma,\alpha,T)$ such that

$$u(t) \leq \frac{C}{t^\gamma}, \qquad t \in (0,T]. \qquad (7.5.2)$$

Proof. First we observe that, for any $\beta < 1, \delta < 1$ and $a < b < \infty$, we have

$$\int_a^b (b-t)^{-\beta}(t-a)^{-\delta}dt = (b-a)^{-\beta-\delta+1}\int_0^1 (1-v)^{-\beta}v^{-\delta}dv$$
$$= B(1-\beta, 1-\delta)(b-a)^{-\beta-\delta+1}, \qquad (7.5.3)$$

where B denotes the Beta function. Given that u satisfies (7.5.1), it follows from (7.5.3) that

$$\int_0^t u(\tau)(t-\tau)^{-\alpha}d\tau \leq c\int_0^t \tau^{-\gamma}(t-\tau)^{-\alpha}d\tau + c\int_0^t (t-\tau)^{-\alpha}\left(\int_0^\tau u(s)(\tau-s)^{-\alpha}ds\right)d\tau$$
$$= c(\theta_\gamma * \theta_\alpha)(t) + c_{1,\alpha}\int_0^t u(s)(t-s)^{1-2\alpha}ds, \qquad (7.5.4)$$

where $*$ denotes convolution, $\theta_\kappa(t) = t^{-\kappa}$ and $c_{1,\alpha} = cB(1-\alpha, 1-\alpha)$. Inserting (7.5.4) into (7.5.1), we obtain

$$u(t) \leq c\theta_\gamma(t) + c^2(\theta_\gamma * \theta_\alpha)(t) + c_{2,\alpha}\int_0^t u(\tau)(t-\tau)^{1-2\alpha}d\tau, \qquad t \in (0,T], \qquad (7.5.5)$$

with $c_{2,\alpha} = c\,c_{1,\alpha}$. Note that the convolution $\theta_\gamma * \theta_\kappa$ exists for any choice of $\gamma < 1$ and $\kappa < 1$, since

$$(\theta_\gamma * \theta_\kappa)(t) = B(1-\gamma, 1-\kappa)\,t^{1-\gamma-\kappa} = B(1-\gamma, 1-\kappa)\,\theta_{\gamma+\kappa-1}(t). \qquad (7.5.6)$$

Furthermore,

$$(\theta_\gamma * \theta_\kappa)(t) \leq \frac{C(\gamma, \kappa, T)}{t^\gamma}, \qquad t \in (0,T], \qquad (7.5.7)$$

where $C(\gamma, \kappa, T)$ is a positive constant.

If $1 - 2\alpha \geq 0$, then we can infer from (7.5.5) and (7.5.7) that

$$u(t) \leq \frac{c + C(\gamma, \alpha, T)}{t^\gamma} + c_{2,\alpha}t^{1-2\alpha}\int_0^t u(\tau)d\tau,$$

and then apply the standard arguments used to establish Gronwall-type inequalities (see below) to obtain the desired result. Otherwise, by induction, for $r \geq 2$,

$$u(t) \leq c\theta_\gamma(t) + \sum_{j=1}^{r-1} c_{j,\alpha}(\theta_\gamma * \theta_{j\alpha-j+1})(t) + \bar{c}_{r,\alpha}(u * \theta_{r\alpha-r+1})(t), \qquad t \in (0,T]. \qquad (7.5.8)$$

Indeed, by (7.5.1) and (7.5.3),

$$(u * \theta_{r\alpha-r+1})(t) \leq \int_0^t (t-\tau)^{-r\alpha+r-1}\left[c\tau^{-\gamma} + c\int_0^\tau (\tau-s)^{-\alpha}u(s)\,ds\right]d\tau$$
$$= c(\theta_\gamma * \theta_{r\alpha-r+1})(t) + c\int_0^t u(s)\int_s^t (t-\tau)^{-r\alpha+r-1}(\tau-s)^{-\alpha}d\tau ds$$
$$= c(\theta_\gamma * \theta_{r\alpha-r+1})(t) + cB(r(1-\alpha), 1-\alpha)(u * \theta_{(r+1)\alpha-r})(t).$$

Inserting this estimate in (7.5.8) gives

$$u(t) \leq c\theta_\gamma(t) + \sum_{j=1}^{r-1} c_{j,\alpha}(\theta_\gamma * \theta_{j\alpha-j+1})(t) + c\bar{c}_{r,\alpha}(\theta_\gamma * \theta_{r\alpha-r+1})(t)$$

$$+ c\bar{c}_{r,\alpha} B(r(1-\alpha), 1-\alpha)(u * \theta_{(r+1)\alpha-(r+1)+1})(t),$$

hence (7.5.8) for $r+1$ with $c_{r,\alpha} := c\bar{c}_{r,\alpha}$ and $\bar{c}_{r+1,\alpha} := c\bar{c}_{r,\alpha} B(r(1-\alpha), 1-\alpha)$.

We now use (7.5.8) with the smallest integer $r \geq 2$ such that $r\alpha - r + 1 < 0$ and thus obtain

$$u(t) \leq c\theta_\gamma(t) + \sum_{j=1}^{r-1} c_{j,\alpha}(\theta_\gamma * \theta_{j\alpha-j+1})(t) + \bar{c}_{r,\alpha} \int_0^t u(\tau)(t-\tau)^{-r\alpha+r-1}d\tau$$

$$\leq \Theta(t) + C_0(\alpha, T) \int_0^t u(\tau)d\tau, \tag{7.5.9}$$

with $\Theta := c\theta_\gamma + \sum_{j=1}^{r-1} c_{j,\alpha}(\theta_\gamma * \theta_{j\alpha-j+1})$ and $C_0(\alpha, T) := \bar{c}_{r,\alpha} T^{r-1-r\alpha}$. It follows from (7.5.9) that

$$\frac{\mathrm{d}}{\mathrm{d}t} \left(e^{-C_0(\alpha,T)t} \int_0^t u(\tau)d\tau \right) \leq e^{-C_0(\alpha,T)t} \Theta(t), \qquad t \in (0, T),$$

and, after integration,

$$\int_0^t u(\tau)d\tau \leq e^{C_0(\alpha,T)t} \int_0^t \Theta(\tau)d\tau.$$

Consequently, by (7.5.9),

$$u(t) \leq \Theta(t) + C_0(\alpha, T)e^{C_0(\alpha,T)t} \int_0^t \Theta(\tau)d\tau, \qquad t \in (0, T). \tag{7.5.10}$$

Since $\alpha < 1$, we have $j\alpha - j + 1 < 1$ for any $j \geq 1$ and therefore, from (7.5.7), there exists a constant $C_1(\gamma, \alpha, T) > 0$ such that

$$\Theta(t) = c\theta_\gamma(t) + \sum_{j=1}^{r-1} c_{j,\alpha}(\theta_\gamma * \theta_{j\alpha-j+1})(t) \leq \frac{C_1(\gamma, \alpha, T)}{t^\gamma}.$$

Hence (7.5.10) can be written as

$$u(t) \leq \frac{C_1(\gamma, \alpha, T)}{t^\gamma} + C_2(\gamma, \alpha, T) \leq \frac{C(\gamma, \alpha, T)}{t^\gamma},$$

and this gives (7.5.2). $\qquad\qquad\qquad\qquad\qquad\qquad\qquad\qquad\qquad\qquad\qquad\qquad\qquad\qquad\qquad \square$

Chapter 8

Solvability of Coagulation-Fragmentation Equations

Our aim in this chapter is to present a number of results that have been established, over the past twenty years or so, on the existence and uniqueness of solutions to the continuous C-F equation

$$\partial_t f(t, x) = \mathcal{C}f(t, x) + \mathcal{F}f(t, x) \,, \qquad (t, x) \in (0, \infty)^2, \qquad (8.0.1a)$$

with initial condition

$$f(0, x) = f^{in}(x) \,, \qquad x \in (0, \infty) \,. \qquad (8.0.1b)$$

As explained in Section 2.2.2 of Volume I, the coagulation operator $\mathcal{C}$ and the fragmentation operator $\mathcal{F}$ are given by

$$\mathcal{C}f(x) = \frac{1}{2} \int_0^x k(x - y, y) f(x - y) f(y) \, \mathrm{d}y - \int_0^\infty k(x, y) f(y) f(x) \, \mathrm{d}y \qquad (8.0.2)$$

and

$$\mathcal{F}f(x) = -a(x)f(x) + \int_x^\infty a(y) b(x, y) f(y) \, \mathrm{d}y \qquad (8.0.3)$$

for $x \in (0, \infty)$. In (8.0.2) and (8.0.3) the coagulation kernel k is a nonnegative and measurable symmetric function defined on $(0, \infty)^2$, the overall fragmentation rate a is a nonnegative measurable function on $(0, \infty)$, and the daughter distribution function b is a nonnegative and measurable function possessing the following properties: for a.e. $y > 0$,

$$\int_0^y x b(x, y) \, \mathrm{d}x = y \quad \text{and} \quad b(x, y) = 0 \ \text{ for a.e. } \ x > y. \qquad (8.0.4)$$

We recall that the first condition in (8.0.4) ensures that there is no loss of matter during fragmentation events.

Section 8.1.1 will actually provide results on the more general initial-value problem

$$\partial_t f(t, x) = \pm \partial_x [r(x) f(t, x)] - \mu(x) f(t, x) + \mathcal{C}f(t, x) + \mathcal{F}f(t, x) \,, \qquad (t, x) \in (0, \infty)^2, \quad (8.0.5)$$

in which the standard C-F equation is augmented by the inclusion of transport and death terms. Here, the function μ represents the death/annihilation rate of particles, while the transport term $\pm \partial_x (rf)(x)$ accounts for the decay $(+)$ or growth $(-)$, at rate $r(x) \geq 0$, of particles of size x.

Two strategies will be used in this chapter to investigate the solvability of (8.0.1). The first relies heavily on the theory of substochastic semigroups of operators that is discussed in Chapters 4 and 5 of Volume I in connection with fragmentation problems. The second exploits the results on weak compactness presented in Sections 7.1 and 7.2 of this volume. Detailed accounts of these strategies will be given in the following two sections. In preparation for these, we pause briefly to describe, in simple terms, the underlying ideas behind each approach.

The Semigroup Method. Existence and uniqueness of solutions to (8.0.1) are obtained by first using semigroup-based techniques to deal with the linear fragmentation part of the equation, and then treating coagulation as a nonlinear perturbation. This strategy for dealing with C-F problems can be traced back to the seminal paper [2]. There are several advantages to be gained by adopting this approach. Firstly, existence and uniqueness of classical differentiable solutions are established simultaneously. Secondly, the theory of substochastic semigroups enables a wide range of unbounded fragmentation kernels to be catered for, and, thirdly, some C-F problems involving transport and annihilation terms can also be dealt with, see Section 8.1. Unfortunately, these advantages come with a cost in that the coagulation process has to be subordinate to fragmentation in a manner that will be explained later. In particular, this means that fragmentation must always feature in the problem.

The Compactness Method. As pointed out above, a disadvantage of the semigroup approach is that it cannot handle the pure coagulation equation, and proves similarly to be unsuitable for C-F processes in which the fragmentation is nonlinear, see Section 2.4.5 of Volume I. This is not the case with the second method which utilises compactness results to tackle the question of existence of solutions. The strategy now begins with the construction of a one-parameter family of approximations to the C-F equation. These approximating equations are devised in such a way that the well-posedness of each is easy to prove, with, in addition, the corresponding solutions possessing several regularity properties (mostly with regard to moment estimates and L_p-regularity). The next step is to establish that a solution of the original problem can be obtained from these approximate solutions by passing to the limit with respect to the approximation parameter. To this end, estimates on the approximating solutions have to be derived so that certain compactness properties are guaranteed; these properties are required to justify the limiting procedure. Clearly, to be handled properly, the nonlinear coagulation term requires stronger estimates than the linear fragmentation term. Therefore, in contrast to the semigroup approach, it is fragmentation, and not coagulation, that plays the role of a perturbation. As we shall see later, the derivation of appropriate estimates depends strongly on the assumptions on the coagulation and fragmentation coefficients, and several different cases will be investigated in this chapter, with a particular emphasis being placed on the conservation of matter whenever possible. Finally, it turns out that the appropriate compactness to work with is the sequential weak compactness in L_1, as realised in the landmark work [252]. That only weak convergence is needed stems from the fact that the nonlinear terms in (8.0.1) are nonlocal, a feature which is typical in kinetic equations such as the homogeneous Boltzmann equation [75, 76, 267]. It should be noted that the compactness method that has just been described only establishes the existence of at least one weak solution to the C-F equation, in a sense that will be made precise in Section 8.2. In contrast to the semigroup approach, the uniqueness issue has to be investigated independently, see Section 8.2.5.

For convenience, we recall (from Section 6.1, see also Section 3.1.4 of Volume I) that the function spaces that feature prominently in the analysis of C-F equations are $X_m = L_1((0, \infty), x^m \mathrm{d}x)$ and $X_{0,m} = X_0 \cap X_m$ for $m \in \mathbb{R}$. All are Banach spaces with respective norms $\| \cdot \|_{[m]}$ and $\| \cdot \|_{[0,m]}$ defined by

$$\|f\|_{[m]} = \int_0^\infty x^m |f(x)| \, \mathrm{d}x \ \text{ and } \ \|f\|_{[0,m]} = \int_0^\infty (1 + x^m)|f(x)| \, \mathrm{d}x.$$

We denote the moment of order m of $f \in X_m$ by $M_m(f)$, where

$$M_m(f) = \int_0^\infty x^m f(x) \, \mathrm{d}x \ , \tag{8.0.6}$$

and note that $M_m(f) = \|f\|_{[m]}$ when f lies in the positive cone $X_{m,+}$ of X_m.

8.1 Coagulation-Fragmentation Equations via Semigroups

8.1.1 Global Classical Solutions of Transport-Coagulation-Fragmentation Equations with Bounded Coagulation Kernels

We begin with the full transport-coagulation-fragmentation equation (8.0.5) and recall the definition of the expressions $\mathcal{T}^{\pm}$ and $\mathcal{B}$ introduced in Volume I via equations (5.2.2) and (5.2.3), respectively:

$$[\mathcal{T}^{\pm}f](x) = \pm\partial_x[r(x)f(x)] - (a(x) + \mu(x))f(x) , \qquad x \in (0,\infty) , \tag{8.1.1}$$

$$[\mathcal{B}f](x) = \int_x^{\infty} a(y)b(x,y)f(y) \, \mathrm{d}y , \qquad x \in (0,\infty) . \tag{8.1.2}$$

We also define the bilinear expression

$$
\begin{aligned}
{[\tilde{C}(f,g)](x)} &= [\tilde{C}_1(f,g)](x) - [\tilde{C}_2(f,g)](x) \\
&= \frac{1}{2}\int_0^x k(x-y,y)f(x-y)g(y)\mathrm{d}y - f(x)\int_0^{\infty} k(x,y)g(y)\mathrm{d}y,
\end{aligned}
\tag{8.1.3}
$$

for $x \in (0,\infty)$, and the corresponding quadratic form

$$Cf = \tilde{C}(f,f). \tag{8.1.4}$$

In addition to (8.0.4), we assume that there is $j > 0$, $l \ge 0$, $a_0 > 0$, and $b_0 > 0$ such that

$$
\begin{aligned}
0 &\le a(x) \le a_0(1 + x^j) , \\
n_0(x) &= \int_0^x b(y,x) \, \mathrm{d}y \le b_0(1 + x^l) , \qquad x \in (0,\infty) ,
\end{aligned}
\tag{8.1.5}
$$

and that

$$0 \le \mu \in L_{\infty,loc}([0,\infty)) , \tag{8.1.6}$$

see (5.1.150) and (5.2.4). Concerning the transport term, we require the function r to satisfy (5.2.5), that is,

$$r \in AC((0,\infty)) \quad \text{and} \quad r(x) > 0 , \quad x \in (0,\infty) , \tag{8.1.7}$$

as well as (5.2.35) in the growth case, that is,

$$\tilde{r} = \sup_{x\in(0,\infty)} \frac{r(x)}{x} < \infty . \tag{8.1.8}$$

Finally, besides being symmetric, the coagulation kernel k is assumed to satisfy

$$0 \le k \in L_{\infty}((0,\infty)^2). \tag{8.1.9}$$

The initial-value problem (8.0.5) is posed in the Banach space $X_{0,m}$, where m is assumed to satisfy

$$m > 1 \quad \text{when } j + l \le 1, \quad \text{and} \quad m \ge j + l \quad \text{when } j + l > 1. \tag{8.1.10}$$

This assumption means that there is an operator realisation, $K_{0,m}^{\pm}$, of the transport-fragmentation part of the problem that generates a positive quasi-contractive honest semi-group $(G_{K_{0,m}^{\pm}}(t))_{t\ge 0}$ on $X_{0,m}$, see Propositions 5.2.19, 5.2.20 and Theorems 5.2.31, 5.2.34 in

Volume I. If we now use the coagulation expression (8.1.4) to define an associated operator, $C_{0,m}$, on $X_{0,m}$ by

$$C_{0,m}f := \mathcal{C}f, \quad f \in X_{0,m}, \tag{8.1.11}$$

then it is natural to interpret (8.0.5) as the semilinear abstract Cauchy problem

$$\frac{d}{dt}f(t) = K_{0,m}^{\pm}f(t) + C_{0,m}f(t), \quad t > 0, \quad f(0) = f^{in}. \tag{8.1.12}$$

Note that here, and throughout Section 8.1, we shall use upper case C, accompanied by appropriate subscripts, to denote operators, with lower case c reserved for constants.

We are now in a position to establish existence and uniqueness results for solutions of (8.1.12). For convenience, we recall the notions of classical, strict and mild solutions that are given in Section 4.8 of Volume I. A (globally-defined) classical solution of the abstract formulation (8.1.12) of the C-F problem is a function $f \in C([0,\infty), X_{0,m}) \cap C^1((0,\infty), X_{0,m})$ satisfying (8.1.12). A classical solution f with the additional property that $f \in C^1([0,\infty), X_{0,m})$ is referred to as a strict solution, and a mild solution of (8.1.12) is a function $f \in C([0,\infty), X_{0,m})$ that satisfies the associated, but weaker, integral equation formulation

$$f(t) = G_{K_{0,m}^{\pm}}(t)f^{in} + \int_0^t G_{K_{0,m}^{\pm}}(t-s)C_{0,m}f(s)\,\mathrm{d}s.$$

Theorem 8.1.1. *Let the coagulation and fragmentation coefficients k, a and b satisfy (8.1.9), (8.0.4) and (8.1.5), while the death rate μ and transport velocity r satisfy (8.1.6), (8.1.7) and (8.1.8). Moreover, let m satisfy (8.1.10). Then, for any $0 \le f^{in} \in D(K_{0,m}^{\pm})$, there exists a unique, globally-defined, nonnegative strict solution, and for any $0 \le f^{in} \in X_{0,m}$ there exists a unique, globally-defined, nonnegative mild solution, $f : [0,\infty) \to X_{0,m}$ of (8.1.12).*

Proof. The proof follows in two steps. First we establish the existence of nonnegative solutions on maximal intervals of their existence. In the second step, we show that the solutions can be extended to $[0,\infty)$.

For future use, we consider the integral

$$\int_0^\infty \theta(x)\,[\tilde{C}(f,g)](x)\,\mathrm{d}x\,,$$

where θ, f, g are appropriately restricted real-valued functions. Then routine calculations, as in (2.3.6), show that

$$\int_0^\infty \theta(x)\,[\tilde{C}(f,g)](x)\,\mathrm{d}x = \frac{1}{2}\int_0^\infty \int_0^\infty \theta(x+y)k(x,y)f(x)g(y)\,\mathrm{d}x\mathrm{d}y$$
$$- \int_0^\infty \int_0^\infty \theta(x)k(x,y)f(x)g(y)\,\mathrm{d}x\mathrm{d}y\,. \tag{8.1.13}$$

Similarly, when θ is nonnegative,

$$\int_0^\infty \theta(x)\,|[\tilde{C}(f,g)](x)|\,\mathrm{d}x \le \frac{1}{2}\int_0^\infty \int_0^\infty \theta(x+y)k(x,y)|f(x)|\,|g(y)|\,\mathrm{d}x\mathrm{d}y$$
$$+ \int_0^\infty \int_0^\infty \theta(x)k(x,y)|f(x)|\,|g(y)|\,\mathrm{d}x\mathrm{d}y\,. \tag{8.1.14}$$

In particular, on choosing $\theta(x) = 1 + x^m$, and using the elementary inequality

$$(x+y)^m \le 2^m(x^m + y^m), \qquad m \ge 0, \tag{8.1.15}$$

we obtain

$$\|\tilde{C}(f,g)\|_{[0,m]} = \int_0^\infty (1+x^m) \left|[\tilde{C}(f,g)](x)\right| \, dx$$

$$\leq \frac{1}{2}\|k\|_\infty \left(\|f\|_{[0]}\|g\|_{[0]} + 2^m\|f\|_{[m]}\|g\|_{[0]} + 2^m\|f\|_{[0]}\|g\|_{[m]} + 2\|f\|_{[0,m]}\|g\|_{[0]} \right)$$

$$\leq K\|f\|_{[0,m]}\|g\|_{[0,m]},$$

for all $f, g \in X_{0,m}$, where K is a positive constant. Hence, $C_{0,m}$ is a bounded quadratic form on $X_{0,m}$.

Also, the bilinearity of $\tilde{C}$ leads to

$$C_{0,m}(f+g) = \tilde{C}(f,f) + \tilde{C}(f,g) + \tilde{C}(g,f) + \tilde{C}(g,g),$$

for all $f, g \in X_{0,m}$, from which it is straightforward to deduce that $C_{0,m}$ is Fréchet differentiable at each $f \in X_{0,m}$, with Fréchet derivative given by

$$[\partial C_{0,m} f]g := \tilde{C}(f,g) + \tilde{C}(g,f), \quad g \in X_{0,m}.$$

Consequently, for any $f, v \in X_{0,m}$, and for any $g \in B_{X_{0,m}}(f,\rho), \rho > 0$, we have

$$\|[\partial C_{0,m} g]v\|_{[0,m]} \leq 2K\|g\|_{[0,m]}\|v\|_{[0,m]} \leq L(f,\rho)\|v\|_{[0,m]},$$

where

$$L(f,\rho) = 2K(\|f\|_{[0,m]} + \rho). \tag{8.1.16}$$

Moreover, for any $f, g, v \in X_{0,m}$,

$$\|[\partial C_{0,m} f]v - [\partial C_{0,m} g]v\|_{[0,m]} = \|\tilde{C}(f-g,v) + \tilde{C}(v,f-g)\|_{[0,m]}$$
$$\leq 2K\|v\|_{[0,m]}\|f-g\|_{[0,m]} \to 0 \text{ as } \|f-g\|_{[0,m]} \to 0.$$

Hence, the Fréchet derivative is continuous with respect to f. These results establish, in particular, that $C_{0,m}$ is locally Lipschitz continuous. Precisely, for any given $f \in X_{0,m}, \rho > 0$ and $u, v \in B_{X_{0,m}}(f,\rho)$, we have

$$\|C_{0,m}u - C_{0,m}v\|_{[0,m]} = \|\tilde{C}(u-v,u) + \tilde{C}(v,u-v)\|_{[0,m]} \leq L(f,\rho)\|u-v\|_{[0,m]}. \tag{8.1.17}$$

Therefore, by Theorem 4.8.1 and the comments below it, for any $f^{in} \in X_{0,m}$ there exists a maximal $\tau(f^{in})$ and a mild solution $f \in C([0, \tau(f^{in})), X_{0,m})$ of (8.1.12). If, in addition, $f^{in} \in D(K_{0,m}^\pm)$, then $f \in C([0, \tau(f^{in})), D(K_{0,m}^\pm)) \cap C^1([0, \tau(f^{in})), X_{0,m})$ is a strict solution to (8.1.12). Furthermore, by construction of the solution, see [227, Theorem 6.1.4], for any $t_0 \in (0, \tau(f^{in}))$ there is $\rho > 0$ such that $f(t) \in B_{X_{0,m}}(f^{in}, \rho)$ for any $t \in [0, t_0]$.

Next we show that there is $t^* > 0$ such that both mild and strict solutions are in $X_{0,m,+}$ for all $t \in [0, t^*)$ if $f^{in} \in X_{0,m,+}$ (respectively $f^{in} \in D(K_{0,m}^\pm)_+$). We use the idea introduced at the end of Section 4.8 and focus on mild solutions. To this end, we note first that the solution f of (8.1.12) is also the unique solution to

$$f(t) = e^{-\alpha t} G_{K_{0,m}^\pm}(t) f^{in} + \int_0^t e^{-\alpha(t-s)} G_{K_{0,m}^\pm}(t-s) C_{0,m,\alpha} f(s) \, ds, \ 0 \leq t < \tau(f^{in}), \tag{8.1.18}$$

where $C_{0,m,\alpha} := C_{0,m} + \alpha I$ and $\alpha \in \mathbb{R}$.

By definition,

$$C_{0,m,\alpha} f = \alpha f + \tilde{C}_1(f,f) - \tilde{C}_2(f,f).$$

Clearly $\tilde{C}_1(f,f) \in X_{0,m,+}$ for any $f \in X_{0,m,+}$. Also, for $f \in B_{X_{0,m}}(f^{in}, \rho) \cap X_{0,m,+}$,

$$f(x) \int_0^\infty k(x,y) f(y) \, dy \leq \|k\|_\infty \|f\|_{[0]} \, f(x) \leq \|k\|_\infty \left(\|f^{in}\|_{[0,m]} + \rho \right) f(x) \,.$$

Hence, if

$$\alpha \geq \|k\|_\infty \left(\|f^{in}\|_{[0,m]} + \rho \right), \tag{8.1.19}$$

then

$$\alpha f(x) - \tilde{C}_2(f,f)(x) \geq \alpha f(x) - \|k\|_\infty \left(\|f^{in}\|_{[0,m]} + \rho \right) f(x) \geq 0 \,.$$

This suggests that the fixed-point problem (8.1.18) should be considered in

$$\Sigma(t_1) := \{ f \in Y(t_1) : f(t) \in \overline{B}(f^{in}, \rho_1) \cap X_{0,m,+} \text{ for all } t \in [0, t_1] \},$$

where $Y(t_1) := C([0,t_1], X_{0,m})$, for some $t_1 \in (0, t_0]$ and $\rho_1 \in (0, \rho)$. Let

$$Qf(t) := e^{-\alpha t} G_{K_{0,m}^\pm}(t) f^{in} + \int_0^t e^{-\alpha(t-s)} G_{K_{0,m}^\pm}(t-s) C_{0,m,\alpha} f(s) \, ds, \ 0 \leq t \leq t_1,$$

with α satisfying (8.1.19). Then $Q(\Sigma(t_1)) \subset C([0,t_1], X_{0,m})$ and $Qf(t) \in X_{0,m,+}$ for all $f \in \Sigma(t_1)$ and $t \in [0, t_1]$. Standard calculations show that Q is a Lipschitz continuous operator; precisely, for all $f, g \in \Sigma(t_1)$,

$$\|Qf - Qg\|_{Y(t_1)} \leq \zeta_1(t_1) \|f - g\|_{Y(t_1)}, \tag{8.1.20}$$

where

$$\zeta_1(t_1) := t_1 \left(L(f^{in}, \rho_1) + \alpha \right) \max\{1, e^{(\omega_{0,m}^\pm - \alpha)t_1}\},$$

with $L(f^{in}, \rho_1)$ defined via (8.1.16), $\omega_{0,m}^\pm$ are the growth bounds associated with the semigroups $(G_{K_{0,m}^\pm}(t))_{t \geq 0}$ (see (5.2.66) and (5.2.69) in Volume I), and

$$\|f\|_{Y(t_1)} := \sup_{t \in [0,t_1]} \|f(t)\|_{[0,m]}.$$

Similarly, from

$$\|Qf(t) - f^{in}\|_{[0,m]} \leq \|e^{-\alpha t} G_{K_{0,m}^\pm}(t) f^{in} - f^{in}\|_{[0,m]} + \int_0^t e^{(\omega_m^\pm - \alpha)(t-s)} \|[C_{0,m,\alpha} f](s)\|_{[0,m]} \, ds \,,$$

using

$$\|[C_{0,m,\alpha} f](s)\|_{[0,m]} \leq \|[C_{0,m,\alpha} f](s) - C_{0,m,\alpha} f^{in}\|_{[0,m]} + \|C_{0,m,\alpha} f^{in}\|_{[0,m]},$$

we obtain by (8.1.20)

$$\|Qf(t) - f^{in}\|_{[0,m]} \leq \rho_1 \zeta(t_1), \quad t \in [0, t_1],$$

where

$$\zeta(t_1) := \frac{1}{\rho_1} \|e^{-\alpha t} G_{K_{0,m}^\pm}(t) f^{in} - f^{in}\|_{Y(t_1)} + \zeta_1(t_1) + \frac{t_1 \max\{1, e^{(\omega_m^\pm - \alpha)t_1}\}}{\rho_1} \|C_{0,m,\alpha} f^{in}\|_{[0,m]}.$$

Since $\zeta(t_1) \to 0^+$ as $t_1 \to 0^+$, we can choose t^* so that $0 < \zeta(t^*) < 1$, in which case $Q(\Sigma(t^*)) \subset \Sigma(t^*)$. Hence there exists a unique solution $\bar{f} \in \Sigma(t^*)$ of $\bar{f} = Q\bar{f}$. Clearly, $\bar{f}$ is also a solution to $\bar{f} = Q\bar{f}$ in $C([0,t^*], X_{0,m})$ and thus it must coincide with the solution f of (8.1.18). Hence, $f \in C([0,t^*], X_{0,m,+})$. Then, by a standard extension argument, we see that $f(t)$ can be extended to the maximal interval of existence $[0, \tau^+(f^{in}))$ as a positive solution.

Next we prove that any strict solution $t \mapsto f(t)$, with $0 \leq f^{in} \in D(K_{0,m}^{\pm})$, is global in time. First we observe that, due to the construction of the solution, (8.1.12) can be integrated termwise with respect to the measure $(1 + x^m)dx$ and thus with respect to any measure $x^p dx$ with $0 \leq p \leq m$. Thus

$$\frac{d}{dt} M_p(f(t)) = \int_0^\infty [K_{0,m}^{\pm} f](t) x^p dx + \int_0^\infty [C_{0,m} f](t) x^p dx, \quad t \in [0, t^*],$$

where $M_p(f(t))$ denotes the p-th moment of the solution f at time t, see (8.0.6). To shorten notation, we also introduce

$$M_{0,p}(f(t)) = M_0(f(t)) + M_p(f(t)). \tag{8.1.21}$$

We now establish estimates on the specific moments $M_0(f(t))$, $M_1(f(t))$ and $M_{0,m}(f(t))$ which will, in turn, yield the desired global existence of the strict solution f. To this end, we note that (8.1.13), and the symmetry of the coagulation kernel k, give

$$\int_0^\infty \theta(x) \, [\mathcal{C} f](x) \, dx = \frac{1}{2} \int_0^\infty \int_0^\infty \chi_\theta(x,y) k(x,y) f(x) f(y) \, dx \, dy, \tag{8.1.22}$$

where

$$\chi_\theta(x,y) = \theta(x+y) - \theta(x) - \theta(y),$$

as defined in Section 2.3.1.

Equation (8.1.22) is the key to determining bounds on the contributions to the moments from only coagulation. As $X_{0,m}$ is a space of type L, we are able to use the results given in Section 3.1.2 to conclude that the vector function $f : [0, t^*] \mapsto X_{0,m}$ can be identified with a scalar function, which we shall denote by $\tilde{f}$, defined on $(0, t^*) \times (0, \infty)$. Clearly, we have

$$\int_0^\infty [C_{0,m} f(t)](x) \, dx = -\frac{1}{2} \int_0^\infty k(x,y) \tilde{f}(t,x) \tilde{f}(t,y) \, dx dy \leq 0$$

thanks to the nonnegativity of $\tilde{f}$, and

$$\int_0^\infty [C_{0,m} f(t)](x) \, x dx = 0.$$

Moreover, from Lemmas 7.4.1 and 7.4.2, we know that

$$0 \leq (x+y)^m - x^m - y^m \leq c_m \left(xy^{m-1} + x^{m-1} y \right), \qquad (x,y) \in (0,\infty)^2, \tag{8.1.23}$$

and this, coupled with the inequality,

$$x^{m-1} \leq 1 + x^m, \quad x \in (0, \infty), \tag{8.1.24}$$

which holds for any $m \geq 1$, leads to

$$\int_0^\infty [C_{0,m} f(t)](x) \, (1 + x^m) dx \leq c_m \|k\|_\infty M_1(f(t)) \, M_{0,m}(f(t)).$$

For the contributions made to the moments by the linear terms in the C-F equation, we need only refer to results that have already been established in Section 5.2.3. Of particular relevance are Propositions 5.2.15, 5.2.16, 5.2.19 and 5.2.20. On using these propositions, together with the fact that $K_{0,m}^-$ and $K_{0,m}^+$ are, respectively, the parts of K_m^- and K_m^+ in $X_{0,m}$ for each $m \geq 1$, we obtain the following estimates for the nonnegative solution $f : [0, t^*] \to X_{0,m}$:

$$
\int_0^\infty [K_{0,m}^- f(t)](x)\, x^m \mathrm{d}x \;\leq\; -\int_0^\infty \mu(x)\tilde{f}(t,x)\, x^m \mathrm{d}x
$$
$$
+m\int_0^\infty r(x)\tilde{f}(t,x)x^{m-1}\mathrm{d}x - \int_0^\infty N_m(x)a(x)\tilde{f}(t,x)\mathrm{d}x;
$$
$$
\int_0^\infty [K_{0,m}^+ f(t)](x)\, x^m \mathrm{d}x \;\leq\; -\int_0^\infty \mu(x)\tilde{f}(t,x)\, x^m \mathrm{d}x
$$
$$
-m\int_0^\infty r(x)\tilde{f}(t,x)x^{m-1}\mathrm{d}x - \int_0^\infty N_m(x)a(x)\tilde{f}(t,x)\mathrm{d}x;
$$

and

$$
\int_0^\infty [K_{0,m}^- f(t)](x)\, (1+x^m)\mathrm{d}x \;\leq\; -\int_0^\infty \mu(x)\tilde{f}(t,x)\, (1+x^m)\mathrm{d}x
$$
$$
+m\int_0^\infty r(x)\tilde{f}(t,x)x^{m-1}\mathrm{d}x
$$
$$
-\int_0^\infty (N_0(x)+N_m(x))a(x)\tilde{f}(t,x)\mathrm{d}x;
$$
$$
\int_0^\infty [K_{0,m}^+ f(t)](x)\, (1+x^m)\mathrm{d}x \;\leq\; -\int_0^\infty \mu(x)\tilde{f}(t,x)\, (1+x^m)\mathrm{d}x
$$
$$
-m\int_0^\infty r(x)\tilde{f}(t,x)x^{m-1}\mathrm{d}x
$$
$$
-\int_0^\infty (N_0(x)+N_m(x))a(x)\tilde{f}(t,x)\mathrm{d}x.
$$

Moreover, from (5.2.67) and (5.2.70), when $f(t)$ is in $D(K_{0,m}^\pm)$, $t \geq 0$, we have

$$
\mu f(t) \in X_{0,m},\; N_0\, af(t) \in X_0,\; N_m\, af(t) \in X_0,\; rf(t) \in X_{m-1},\; t \geq 0.
$$

Consequently, it follows that

$$
\int_0^\infty [K_{0,m}^\pm f(t)](x)\, \mathrm{d}x \leq -\int_0^\infty N_0(x)a(x)\tilde{f}(t,x)\mathrm{d}x \leq \int_0^\infty n_0(x)a(x)\tilde{f}(t,x)\mathrm{d}x.
$$

Combining the fragmentation and coagulation estimates, dropping all negative terms, and using the inequalities

$$
(1+x^l)(1+x^j) \leq 2(1+x^{l+j}) \leq 4(1+x^m),
$$

leads, finally, to:

(a) **the decay case** $(K_{0,m}^+)$:

$$
\frac{d}{dt}M_0(f(t)) \;\leq\; \int_0^\infty n_0(x)a(x)\tilde{f}(t,x)\mathrm{d}x \leq C_0 M_{0,m}(f(t)),
$$
$$
\frac{d}{dt}M_1(f(t)) \;\leq\; 0,
$$
$$
\frac{d}{dt}M_{0,m}(f(t)) \;\leq\; (c_0 + c_m\|k\|_\infty M_1(f(t)))M_{0,m}(f(t));
$$

(b) **the growth case** $(K_{0,m}^-)$:

$$\frac{d}{dt}M_0(f(t)) \leq \int_0^\infty n_0(x)a(x)\tilde{f}(t,x)\mathrm{d}x \leq C_0 M_{0,m}(f(t)),$$

$$\frac{d}{dt}M_1(f(t)) \leq \int_0^\infty r(x)\tilde{f}(t,x)\mathrm{d}x \leq \tilde{r}M_1(f(t)),$$

$$\frac{d}{dt}M_{0,m}(f(t)) \leq (c_0 + c_m\|k\|_\infty M_1(f(t)))M_{0,m}(f(t)) + m\int_0^\infty r(x)\tilde{f}(t,x)x^{m-1}\mathrm{d}x$$

$$\leq (c_0 + c_m\|k\|_\infty M_1(f(t)) + m\tilde{r})M_{0,m}(f(t)),$$

where $c_0 = 4a_0 b_0$ and $\tilde{r}$ is given by (8.1.8).

We observe that in both cases the differential inequality for $M_1(f(t))$ is decoupled from the others and can be solved independently giving $M_1(f(t)) \leq M_1(f(0))$ in the first case and $M_1(f(t)) \leq M_1(f(0))e^{\tilde{r}t}$ in the second. Thus, in both cases the inequality for $M_{0,m}(f(t))$ becomes linear with globally-defined coefficients. Hence $M_{0,m}(f(t))$, and consequently $M_0(f(t))$, exist for all t which proves the thesis for strict solutions.

To extend the global solvability result to mild solutions, we observe that any strict solution is also a mild solution. Moreover, $D(K_{0,m}^\pm)$ is dense in $X_{0,m}$. The stated result then follows from the continuous dependence of mild solutions on the initial condition. Indeed, first we note that, by the above estimates on the moments, for any $\bar{t} > 0$ and $g^{in} \in D(K_{0,m}^\pm)_+$ we have

$$\frac{d}{dt}M_{0,m}(g(t,g^{in})) \leq c_1 M_{0,m}(g(t,g^{in})) + c_2, \qquad t \in [0,\bar{t}],$$

for some $c_1, c_2 > 0$ (that can depend on $\bar{t}$), where as in Definition 7.3.1, $(t,g^{in}) \mapsto g(t,g^{in})$ is the solution originating from g^{in}. Thus, on $[0,\bar{t}]$,

$$\|g(t,g^{in})\|_{[0,m]} = M_{0,m}(g(t,g^{in})) \leq e^{c_1 t}\|g^{in}\|_{[0,m]} + \frac{c_2}{c_1}(e^{c_1 t} - 1)$$

$$\leq c_3\|g^{in}\|_{[0,m]} + c_4,$$

for some constants c_3, c_4. It follows that, for any $\rho > 0$, there is a constant c_5 such that $\|g(t,g^{in})\|_{[0,m]} \leq c_5$ on $[0,\bar{t}]$ for all $g^{in} \in B_{X_{0,m}}(f^{in},\rho) \cap D(K_{0,m}^\pm)$. Now, if a mild solution $t \mapsto f(t,f^{in})$ with $f^{in} \in X_{0,m,+}$ was not defined on $[0,\infty)$, then there would be $\bar{t} < \infty$ such that $\limsup_{t\to\bar{t}}\|f(t,f^{in})\|_{[0,m]} = \infty$, see Theorem 4.8.1. Thus, for any $L > 0$ there is $t' < \bar{t}$ with $\|f(t',f^{in})\|_{[0,m]} > L$. Let $L = 2c_5$ and let

$$\|f(t',f^{in})\|_{[0,m]} > 2c_5. \tag{8.1.25}$$

On the other hand, by the continuous dependence, there exists $\epsilon \in (0,\rho)$ such that if $\|f^{in} - g^{in}\|_{[0,m]} < \epsilon$, then

$$\|f(t,f^{in}) - g(t,g^{in})\|_{[0,m]} \leq \frac{c_5}{2}, \qquad t \in [0,t'].$$

For $g^{in} \in B_{X_{0,m}}(f^{in},\epsilon) \cap D(K_{0,m}^\pm)$ the above yields

$$\|f(t',f^{in})\|_{[0,m]} \leq \|g(t',g^{in})\|_{[0,m]} + \frac{c_5}{2} \leq \frac{3c_5}{2},$$

contradicting (8.1.25). $\qquad\qquad\qquad\qquad\qquad\qquad\qquad\qquad\qquad\qquad\qquad\square$

8.1.2   Global Classical Solutions of Coagulation-Fragmentation Equations with Unbounded Coagulation Kernels

In this section we return to the classical C-F equation (8.0.1):

$$\partial_t f(t,x) = \mathcal{C}f(t,x) + \mathcal{F}f(t,x) , \qquad (t,x) \in (0,\infty)^2 , \qquad (8.1.26)$$

the coagulation and fragmentation operators $\mathcal{C}$ and $\mathcal{F}$ being defined by (8.0.2) and (8.0.3), respectively. We retain the assumptions (8.0.4) and (8.1.5) on the fragmentation coefficients a and b. In particular, this means that there is $j > 0$, $l \geq 0$, $a_0 > 0$, and $b_0 > 0$ such that

$$a(x) \leq a_0(1+x^j), \qquad n_0(x) = \int_0^x b(y,x) \, dy \leq b_0(1+x^l), \qquad x \in (0,\infty) . \qquad (8.1.27)$$

Furthermore, we assume that there are constants $c_m > 0$ and $m > 1$, the latter also satisfying $m \geq j+l$, such that

$$\liminf_{x\to\infty} \frac{N_m(x)}{x^m} = c_m , \quad \text{where} \quad N_m(x) = x^m - \int_0^x y^m b(y,x) \, dy . \qquad (8.1.28)$$

Thanks to (8.1.28), we can apply Theorem 5.1.48(b) to deduce that the fragmentation operator $(F_{0,m}, D(A_{0,m})) = (A_{0,m} + B_{0,m}, D(A_{0,m}))$ is the generator of an analytic semigroup, denoted by $(G_{F_{0,m}}(t))_{t\geq 0}$ on $X_{0,m}$. Here, in accordance with Section 5.1.7, $A_{0,m}$ and $B_{0,m}$ are defined in $X_{0,m}$ by

$$[A_{0,m}f](x) := -a(x)f(x), \quad D(A_{0,m}) := \{f \in X_{0,m} : A_{0,m}f \in X_{0,m}\}, \quad (8.1.29)$$

$$[B_{0,m}f](x) := \int_x^\infty a(y)b(x,y)f(y)\,dy, \quad D(B_{0,m}) := D(A_{0,m}). \qquad (8.1.30)$$

Also, as we did in Section 5.1.7, it is convenient to use w_m for the weight function in $X_{0,m}$, and therefore in the analysis that follows

$$w_m(x) := 1 + x^m, \quad x \in (0,\infty).$$

The coagulation kernel k is assumed to be a measurable symmetric function such that, for some $K > 0$ and $0 < \alpha < 1$,

$$0 \leq k(x,y) \leq K(1+a(x))^\alpha(1+a(y))^\alpha, \quad (x,y) \in (0,\infty)^2. \qquad (8.1.31)$$

This assumption will be shown below to be sufficient for the local-in-time solvability of (8.1.26). However, to prove that solutions are global in time we need to strengthen (8.1.31) to

$$0 \leq k(x,y) \leq K\big((1+a(x))^\alpha + (1+a(y))^\alpha\big), \quad (x,y) \in (0,\infty)^2, \qquad (8.1.32)$$

again for $K > 0$ and $0 < \alpha < 1$.

To highlight the applicability of these assumptions, we note that they enable global existence of classical solutions to be established for all the coagulation kernels mentioned in [127], such as the shear kernel, [3, 246],

$$k(x,y) = k_0(x^{1/3} + y^{1/3})^{7/3}, \quad (x,y) \in (0,\infty)^2,$$

or the modified Smoluchowski kernel, [147],

$$k(x,y) = k_0 \frac{(x^{1/3} + y^{1/3})^2}{x^{1/3}y^{1/3} + c}, \quad (x,y) \in (0,\infty)^2, \ c > 0,$$

provided that $a(x) \geq a_1 x^\delta$, a_1 a constant, with $\delta > 7/9$ in the first case and $\delta > 2/3$ in the second case. Note that in the third example in [127],

$$k(x,y) = 8 y_c k_0 \frac{(x^{1/3} + y^{1/3})^q}{8 y_c + (x^{1/3} + y^{1/3})^3}, \quad (x,y) \in (0,\infty)^2, \ 0 \leq q < 3,$$

the kernel is bounded at infinity and thus is covered by the classical solvability results discussed in the previous section.

It is worth emphasising that the arguments used in this section to yield local classical solvability are valid provided only that the rate of growth of k is controlled by a small power of the fragmentation rate a. This shows that fast fragmentation of large clusters plays a stabilizing role in the C-F process, which agrees with physical intuition and is also consistent with a number of results on weak solvability that have been established, such as those presented in [109] and also later in this chapter.

Thus, using the linear operators $A_{0,m}$ and $B_{0,m}$ defined by (8.1.29) and (8.1.30), and the nonlinear operator $C_{0,m}$, defined via (8.1.11) but now only for f in the maximal domain

$$D(C_{0,m}) := \{ f \in X_{0,m} : C_{0,m} f \in X_{0,m} \},$$

the initial-value problem (8.0.1) can be written as the following abstract semilinear Cauchy problem in $X_{0,m}$:

$$\frac{df}{dt} = A_{0,m} f + B_{0,m} f + C_{0,m} f, \qquad f(0) = f^{in}. \tag{8.1.33}$$

We recall that the assumptions imposed on the fragmentation coefficients and m guarantee that $(F_{0,m}, D(A_{0,m})) = (A_{0,m} + B_{0,m}, D(A_{0,m}))$ is the generator of an analytic semigroup, denoted by $(G_{F_{0,m}}(t))_{t \geq 0}$, on $X_{0,m}$. However, in general, $0 \notin \rho(F_{0,m})$ and therefore to enable us to define appropriate intermediate spaces, of the type discussed in Section 4.4.2, we consider

$$F_{0,m,\omega} := F_{0,m} - \omega I = A_{0,m} - \omega I + B_{0,m} = A_{0,m,\omega} + B_{0,m},$$

where

$$A_{0,m,\omega} := A_{0,m} - \omega I,$$

and we assume that $\omega > 4 a_0 b_0$ (see Theorem 5.1.48). Clearly, $(F_{0,m,\omega}, D(A_{0,m}))$ is also the generator of an analytic semigroup, namely $(G_{F_{0,m,\omega}}(t))_{t \geq 0} = (e^{-\omega t} G_{F_{0,m}}(t))_{t \geq 0}$, but now we have the desired property that $0 \in \rho(F_{0,m,\omega})$. Thus, as discussed in Section 4.4.2 and Example 4.4.3, for each $\alpha \in [0,1]$, we have

$$D_{F_{0,m,\omega}}(\alpha, 1) = D_{A_{0,m,\omega}}(\alpha, 1) = X_{0,m}^\alpha, \qquad \alpha \in (0,1),$$

where

$$X_{0,m}^\alpha := \left\{ f \in X_{0,m} : \int_0^\infty |f(x)| (\omega + a(x))^\alpha w_m(x) \, dx < \infty \right\} \tag{8.1.34}$$

and equality of the spaces is interpreted in terms of equivalent norms. The natural norm on $X_{0,m}^\alpha$ will be represented by $\|\cdot\|_{[0,m]}^{(\alpha)}$, and we note that $X_{0,m}^0 = X_{0,m}$, and $X_{0,m}^1 = D(A_{0,m,\omega})$. Referring once again to Section 4.4.2, we see that we can assume that for each fixed $\alpha \in (0,1)$ there is a constant $c_1 \geq 1$ such that

$$c_1^{-1} \|f\|_{[0,m]}^{(\alpha)} \leq \|f\|_{D_{F_{0,m,\omega}}(\alpha,1)} \leq c_1 \|f\|_{[0,m]}^{(\alpha)}, \qquad f \in D_{F_{0,m,\omega}}(\alpha, 1). \tag{8.1.35}$$

Theorem 8.1.2. *Assume that a and b satisfy (8.1.27) and (8.1.28), where the constant m in (8.1.28) is such that $m > 1$ and $m \geq j + l$.*

1. If k satisfies (8.1.31), then, for each $f^{in} \in X^{\alpha}_{0,m,+}$, there is $\tau(f^{in}) > 0$ such that the initial-value problem (8.1.33) has a unique nonnegative classical solution

$$f \in C\left([0, \tau(f^{in})), X^{\alpha}_{0,m}\right) \cap C^1\left((0, \tau(f^{in})), X^{\alpha}_{0,m}\right) \cap C\left((0, \tau(f^{in})), D(A_{0,m})\right).$$

   Furthermore, there is a measurable representation of f that is absolutely continuous in $t \in (0, \tau(f^{in}))$ for any $x \in (0, \infty)$ and which satisfies (8.1.26) almost everywhere on $(0, \tau(f^{in})) \times (0, \infty)$.

2. If k satisfies (8.1.32), then, for each $f^{in} \in X^{\alpha}_{0,m,+}$, the corresponding local nonnegative classical solution is global in time.

Proof. To shorten notation, we now fix $\omega > \max\{4a_0b_0, 1\}$ and define the function a_ω on $(0, \infty)$ by

$$a_\omega(x) := \omega + a(x).$$

Then it is straightforward to show that

$$\frac{a^{\alpha}_{\omega}(x)}{\omega^{\alpha}} \leq (1 + a(x))^{\alpha} \leq a^{\alpha}_{\omega}(x). \tag{8.1.36}$$

As in the proof of Theorem 8.1.1, we face the problem that the operator $C_{0,m}$ is not positive. Again we use the idea introduced at the end of Section 4.8 and consider (8.1.26) written in the form

$$
\begin{aligned}
\partial_t f(t,x) \;=\; & -(a_\omega(x) + \gamma a^{\alpha}_{\omega}(x))f(t,x) + \int_x^{\infty} a(y)b(x,y)f(t,y)\mathrm{d}y \\
& + (\gamma a^{\alpha}_{\omega}(x) + \omega)f(t,x) - f(t,x)\int_0^{\infty} k(x,y)f(t,y)\mathrm{d}y \\
& + \frac{1}{2}\int_0^x k(x-y,y)f(t,x-y)f(t,y)\mathrm{d}y,
\end{aligned}
\tag{8.1.37}
$$

where γ is a constant that will be determined below, and α is the index appearing in (8.1.31).
   If we define an operator $A^{\alpha}_{0,m,\omega}$ by

$$A^{\alpha}_{0,m,\omega}f := -a^{\alpha}_{\omega}f, \qquad D(A^{\alpha}_{0,m,\omega}) := \{f \in X_{0,m} : A^{\alpha}_{0,m,\omega}f \in X_{0,m}\},$$

then we clearly have $D(A^{\alpha}_{0,m,\omega}) = X^{\alpha}_{0,m}$. Furthermore, $A^{\alpha}_{0,m,\omega} \in \mathscr{L}\left(X^{\alpha}_{0,m}, X_{0,m}\right)$. Therefore, from [187, Proposition 2.4.1], the operator

$$(F_\gamma, D(F_\gamma)) := (F_{0,m,\omega} + \gamma A^{\alpha}_{0,m,\omega}, D(A_{0,m}))$$

generates an analytic semigroup, say $(G_{F_\gamma}(t))_{t \geq 0}$, on $X_{0,m}$. Hence the spaces $D_{F_\gamma}(\alpha, 1)$ are also equivalent to $X^{\alpha}_{0,m}$. Since $(G_{F_{0,m,\omega}}(t))_{t \geq 0}$ and $(G_{\gamma A^{\alpha}_{0,m,\omega}}(t))_{t \geq 0}$ are positive and contractive, we can use the Trotter product formula [106, Corollary III.5.8], to deduce that $(G_{F_\gamma}(t))_{t \geq 0}$ is also a semigroup of positive contractions on $X_{0,m}$. Consequently, when $f \in X^{\alpha}_{0,m}$, it follows from (4.4.12), and (8.1.35) for F_γ, that

$$
\begin{aligned}
\|G_{F_\gamma}(t)f\|^{(\alpha)}_{[0,m]} \;\leq\; & c_{1,\gamma}\left(\|G_{F_\gamma}(t)f\|_{[0,m]} + \int_0^1 s^{-\alpha}\|F_\gamma G_{F_\gamma}(s)G_{F_\gamma}(t)f\|_{[0,m]}\mathrm{d}s\right) \\
\leq\; & c_{1,\gamma}\left(\|f\|_{[0,m]} + \int_0^1 s^{-\alpha}\|F_\gamma G_{F_\gamma}(s)f\|_{[0,m]}\mathrm{d}s\right) \\
=\; & c_{1,\gamma}\|f\|_{D_{F_\gamma}(\alpha,1)} \leq c^2_{1,\gamma}\|f\|^{(\alpha)}_{[0,m]},
\end{aligned}
\tag{8.1.38}
$$

for some constant $c_{1,\gamma} \geq 1$. Next consider the set

$$\mathcal{U} := \{f \in X_{0,m,+}^\alpha : \|f\|_{[0,m]}^{(\alpha)} \leq 1 + b\}, \tag{8.1.39}$$

for some arbitrarily fixed $b > 0$. The assumption (8.1.31) can be used to show that, when $f \in \mathcal{U}$, we have

$$
\begin{aligned}
\int_0^\infty k(x,y)f(y)\mathrm{d}y &\leq K(1+a(x))^\alpha \int_0^\infty (1+a(y))^\alpha f(y)\mathrm{d}y \\
&\leq K(1+a(x))^\alpha \|f\|_{[0,m]}^{(\alpha)} \leq Ka_\omega^\alpha(x)(1+b),
\end{aligned}
$$

for any $x > 0$. Thus, on setting

$$\gamma = K(1+b), \tag{8.1.40}$$

we have, for all $f \in \mathcal{U}$,

$$
\begin{aligned}
[C_\gamma f](x) :=\ & -f(x)\int_0^\infty k(x,y)f(y)\mathrm{d}y \\
& +(\gamma a_\omega^\alpha(x)+\omega)f(x) + \frac{1}{2}\int_0^x k(x-y,y)f(x-y)f(y)\mathrm{d}y \tag{8.1.41} \\
\geq\ & \omega f(x) + \frac{1}{2}\int_0^x k(x-y,y)f(x-y)f(y)\mathrm{d}y \geq 0.
\end{aligned}
$$

Note also that, for $f, g \in X_{0,m}^\alpha$, we have

$$\|\gamma A_{0,m,\omega}^\alpha f + \omega f\|_{[0,m]} \leq (\omega+\gamma)\|f\|_{[0,m]}^{(\alpha)} = (\omega + K(1+b))\,\|f\|_{[0,m]}^{(\alpha)}, \tag{8.1.42}$$

$$\int_0^\infty |f(x)| \left(\int_0^\infty k(x,y)|g(y)|\mathrm{d}y\right) w_m(x)\mathrm{d}x \leq K\|f\|_{[0,m]}^{(\alpha)}\|g\|_{[0,m]}^{(\alpha)} \tag{8.1.43}$$

and, in a similar way,

$$
\begin{aligned}
\frac{1}{2}\int_0^\infty &\left(\int_0^x k(x-y,y)|f(y)||g(x-y)|\mathrm{d}y\right) w_m(x)\mathrm{d}x \\
&= \frac{1}{2}\int_0^\infty \int_0^\infty k(x,y)|f(y)||g(x)|w_m(x+y)\mathrm{d}x\mathrm{d}y \leq 2^m K\|f\|_{[0,m]}^{(\alpha)}\|g\|_{[0,m]}^{(\alpha)}, \tag{8.1.44}
\end{aligned}
$$

where we have used (8.1.15).

Also, recalling from (8.1.11) that $C_{0,m}f = \tilde{\mathcal{C}}(f,f)$, where $\tilde{\mathcal{C}}$ is the bilinear expression given by (8.1.3), it follows from (8.1.43) and (8.1.44) that, for any $f \in X_{0,m}^\alpha$,

$$\|C_{0,m}f\|_{[0,m]} \leq K(1+2^m)\|f\|_{[0,m]}^{(\alpha)}\|f\|_{[0,m]}^{(\alpha)},$$

and, combining this with (8.1.42), establishes that C_γ is a continuous function from $X_{0,m}^\alpha$ into $X_{0,m}$. Moreover, for each $f \in \mathcal{U}$ we have

$$
\begin{aligned}
\|C_\gamma f\|_{[0,m]} &\leq (\omega + K(1+b))(1+b) + K(1+b)^2 + 2^m K(1+b)^2 \\
&\leq \omega(1+b)\left(1 + (2^m+2)K(1+b)\right) \tag{8.1.45} \\
&\leq \omega(1+b)\max\{1, K(1+b)\}(2^m+3) =: K(\mathcal{U}).
\end{aligned}
$$

Next we establish the Lipschitz continuity of C_γ. First we observe, from (8.1.42), that the linear component of C_γ, namely $-\gamma A_{0,m,\omega}^\alpha + \omega I$, satisfies

$$\|(-\gamma A_{0,m,\omega}^\alpha + \omega I)f - (-\gamma A_{0,m,\omega}^\alpha + \omega I)g\|_{[0,m]} \leq (\omega+\gamma)\|f-g\|_{[0,m]}^{(\alpha)}, \quad f,g \in X_{0,m}^\alpha. \tag{8.1.46}$$

Moreover, it follows from (8.1.43) and (8.1.44) that, for all $f, g \in \mathcal{U}$,

$$
\begin{aligned}
\|C_{0,m}f - C_{0,m}g\|_{[0,m]} &= \|\tilde{C}(f-g,f) + \tilde{C}(g, f-g)\|_{[0,m]} \\
&\leq K\left(\|f\|_{[0,m]}^{(\alpha)} + \|g\|_{[0,m]}^{(\alpha)}\right)(1 + 2^m)\|f - g\|_{[0,m]}^{(\alpha)}.
\end{aligned}
$$

Consequently,

$$
\|C_\gamma f - C_\gamma g\|_{[0,m]} \leq L(\mathcal{U})\|f - g\|_{[0,m]}^{(\alpha)}, \quad \text{for all } f, g \in \mathcal{U}, \tag{8.1.47}
$$

where

$$
L(\mathcal{U}) = \omega + \gamma + K\left(\|f\|_{[0,m]}^{(\alpha)} + \|g\|_{[0,m]}^{(\alpha)}\right)(1 + 2^m).
$$

In principle, the local existence and uniqueness of a nonnegative, classical solution for each $f^{in} \in X_{0,m,+}^\alpha$ now follow from Theorem 4.8.3 combined with (4.8.11). However, to make this a little more apparent, we shall provide a few additional technical details based on arguments given in [227, 242].

Let $f^{in} \in X_{0,m,+}^\alpha$ be such that

$$
\|f^{in}\|_{[0,m]}^{(\alpha)} \leq c_{1,\gamma}^{-2} b, \tag{8.1.48}
$$

where $c_{1,\gamma}$ and b are the constants in (8.1.38) and (8.1.39) respectively, and consider the equation

$$
\mathcal{T}f(t) = G_{F_\gamma}(t)f^{in} + \int_0^t G_{F_\gamma}(t-s)C_\gamma f(s)\mathrm{d}s \tag{8.1.49}
$$

posed in the space $Y = C([0,\tau], \mathcal{U})$, with $\mathcal{U}$ defined by (8.1.39). To enable us to apply the contraction mapping method to (8.1.49) we have to determine the value of τ so that $\mathcal{T}$ is a contraction on Y, when Y is equipped with the metric induced by the norm from $C([0,\tau], X_{0,m}^\alpha)$. Thanks to the choice of $\gamma = K(1 + b)$, we note that the operator C_γ maps $X_{0,m,+}^\alpha$ into $X_{0,m,+}$, and therefore $\mathcal{T}f(t) \in X_{0,m,+}$ provided that $f(s) \in X_{0,m,+}^\alpha$ for all $s \in [0,t]$.

Next, we apply a combination of estimates that can be found in [227] and [242]. As these are fairly standard we shall omit some steps in the following calculations. However, it is important to point out that some care must be taken as the spaces and semigroups that are used in our analysis are different, although equivalent, to those that feature in the corresponding arguments presented in [227, 242]. First, from the proof of [187, Proposition 2.2.9], together with [227, Equation (2.6.5)], we note that

$$
\|G_{F_\gamma}(t)\|_{\mathcal{L}(X_{0,m}, D_{F_\gamma}(\alpha,1))} \leq M_{\alpha,\gamma}t^{-\alpha}e^{-\delta t},
$$

for some $\delta > 0$, where we have used the fact that, up to equivalent norms, $X_{0,m} = D_{F_\gamma}(0,1)$. Therefore

$$
\|G_{F_\gamma}(t)\|_{\mathcal{L}(X_{0,m}, X_{0,m}^{(\alpha)})} \leq c_{1,\gamma}\|G_{F_\gamma}(t)\|_{\mathcal{L}(X_{0,m}, D_{F_\gamma}(\alpha,1))} \leq c_{1,\gamma}M_{\alpha,\gamma}t^{-\alpha}e^{-\delta t}. \tag{8.1.50}
$$

Following [242, Lemma 47.1], we now use (8.1.38) to establish that

$$
\begin{aligned}
\|\mathcal{T}f(t)\|_{[0,m]}^{(\alpha)} &\leq \|G_{F_\gamma}(t)f^{in}\|_{[0,m]}^{(\alpha)} + \int_0^t \|G_{F_\gamma}(t-s)C_\gamma f(s)\|_{[0,m]}^{(\alpha)}\mathrm{d}s \\
&\leq c_{1,\gamma}^2\|f^{in}\|_{[0,m]}^{(\alpha)} + c_{1,\gamma}M_{\alpha,\gamma}\int_0^t s^{-\alpha}e^{-\delta s}\|C_\gamma f(s)\|_{[0,m]}\,\mathrm{d}s \quad (8.1.51) \\
&\leq b + c_{1,\gamma}M_{\alpha,\gamma}K(\mathcal{U})(1-\alpha)^{-1}t^{1-\alpha},
\end{aligned}
$$

with $K(\mathcal{U})$ given by (8.1.45), provided that $f(s) \in \mathcal{U}$ for $0 \le s \le t$. It follows that $\mathcal{T}$ maps Y into Y whenever

$$\tau^{1-\alpha} < \frac{1-\alpha}{c_{1,\gamma} M_{\alpha,\gamma} K(\mathcal{U})}. \tag{8.1.52}$$

To prove that $t \mapsto \mathcal{T}f(t)$ is a Hölder continuous function we proceed initially in a similar manner to [227, Theorem 2.6.13]. However, in contrast to [227], we need to use interpolation norms instead of norms involving fractional powers of the generators, and this creates some complications. First we note that, by Proposition 4.4.2, the part $\hat{F}_\gamma$ of F_γ in $X_{0,m}^\alpha$, with domain $D(\hat{F}_\gamma) = X_{0,m}^{\alpha+1}$, generates an analytic semigroup that is the restriction of $(G_{F_\gamma}(t))_{t\ge 0}$. Thus we can construct the family of intermediate spaces $D_{\hat{F}_\gamma}(\beta, 1), 0 < \beta < 1$, in the same way as the family $D_{F_\gamma}(\alpha, 1)$. Then, on using Proposition 4.2.4(d) from Volume I, we obtain, for each $f \in X_{0,m}^{\alpha+1}$ and $0 < h < 1$,

$$\begin{aligned}\|(G_{F_\gamma}(h) - I)f\|_{[0,m]}^{(\alpha)} = \|(G_{\hat{F}_\gamma}(h) - I)f\|_{[0,m]}^{(\alpha)} &= \left\| \int_0^h \hat{F}_\gamma G_{\hat{F}_\gamma}(s)f \, ds \right\|_{[0,m]}^{(\alpha)} \tag{8.1.53}\\[6pt] &\le h^\beta \int_0^1 s^{-\beta} \|\hat{F}_\gamma G_{\hat{F}_\gamma}(s)f\|_{[0,m]}^{(\alpha)} ds \le h^\beta \|f\|_{D_{\hat{F}_\gamma}(\beta,1)}.\end{aligned}$$

Using Example 4.4.3, we see that $D_{\hat{F}_\gamma}(\beta, 1)$ is equivalent to $X_{0,m}^{\alpha+\beta}$ and hence to $D_{F_\gamma}(\alpha+\beta, 1)$ provided that $\alpha + \beta < 1$.

Thus, for $0 < h < 1$, $0 < \beta < 1 - \alpha$, $s < t$ and $f \in X_{0,m}^{\alpha+1}$, we can apply (8.1.53) and (8.1.50) to obtain

$$\begin{aligned}\|(G_{F_\gamma}(t + h - s) - G_{F_\gamma}(t - s))f\|_{[0,m]}^{(\alpha)} &= \|(G_{F_\gamma}(h) - I)G_{F_\gamma}(t-s)f\|_{[0,m]}^{(\alpha)} \\ &\le c_{1,\gamma} h^\beta \|G_{F_\gamma}(t - s)f\|_{D_{F_\gamma}(\alpha+\beta,1)} \\ &\le c h^\beta (t - s)^{-(\alpha+\beta)} \|f\|_{[0,m]},\end{aligned}$$

for some constant c. Hence

$$\|G_{F_\gamma}(t + h - s) - G_{F_\gamma}(t - s)\|_{\mathcal{L}(X_{0,m}, X_{0,m}^\alpha)} \le c h^\beta (t - s)^{-(\alpha+\beta)}. \tag{8.1.54}$$

Similarly, for $s < t - h$,

$$\|G_{F_\gamma}(t - h - s) - G_{F_\gamma}(t - s)\|_{\mathcal{L}(X_{0,m}, X_{0,m}^\alpha)} \le c h^\beta (t - h - s)^{-(\alpha+\beta)}, \tag{8.1.55}$$

for some constant c.

The estimates obtained above now lead readily to the desired Hölder continuity of the function $t \mapsto \mathcal{T}f(t)$. To show this, we use a convenient, and natural, splitting of the terms involved in the norm $\|\mathcal{T}f(t + h) - \mathcal{T}f(t)\|_{[0,m]}^{(\alpha)}$. Specifically, for $t > 0$, $0 < h < 1$ and $0 < \beta < 1 - \alpha$, we have

$$\|\mathcal{T}f(t + h) - \mathcal{T}f(t)\|_{[0,m]}^{(\alpha)} \le I_1 + I_2 + I_3$$

where, by (8.1.54) with $s = 0$,

$$I_1 = \|G_{F_\gamma}(t + h)f^{in} - G_{F_\gamma}(t)f^{in}\|_{[0,m]}^{(\alpha)} \le c h^\beta t^{-(\alpha+\beta)} \|f^{in}\|_{[0,m]}.$$

Further, again by (8.1.54),

$$I_2 \le \int_0^t \|(G_{F_\gamma}(t + h - s) - (G_{F_\gamma}(t - s))C_\gamma f(s)\|_{[0,m]}^{(\alpha)} ds \le K(\mathcal{U}) c h^\beta \frac{t^{1-(\alpha+\beta)}}{1 - (\alpha + \beta)}$$

and, by (8.1.38),

$$I_3 \;\leq\; \int_t^{t+h} \|G_{F_\gamma}(t+h-s)C_\gamma f\|(s)\|_{[0,m]}^{(\alpha)}\mathrm{d}s \;\leq\; \frac{c_{1,\gamma}K(\mathcal{U})M_{\alpha,\gamma}h^{1-\alpha}}{1-\alpha} \;\leq\; c_{\alpha,\gamma,u}h^\beta,$$

where $K(\mathcal{U})$ and $M_{\alpha,\gamma}$ are the constants that appear in (8.1.45) and (8.1.50), respectively. The calculations for $t > 0$ and $t - h > 0$ are similar.

To establish that $\mathscr{T}$ is a contraction on $Y = C([0,\tau],\mathcal{U})$ when τ is sufficiently small, we use (8.1.47) to obtain

$$
\begin{aligned}
\|\mathscr{T}f(t) - \mathscr{T}g(t)\|_{[0,m]}^{(\alpha)} \;&\leq\; \int_0^t \|G_{F_\gamma}(t-s)(C_\gamma f(s) - C_\gamma g(s))\|_{[0,m]}^{(\alpha)}\mathrm{d}s \\
&\leq\; c_{1,\gamma}M_{\alpha,\gamma}L(\mathcal{U}) \sup_{0\leq s\leq\tau} \|f(s) - g(s)\|_{[0,m]}^{(\alpha)} \int_0^t (t-s)^{-\alpha}\mathrm{d}s \\
&\leq\; \frac{c_{1,\gamma}M_{\alpha,\gamma}L(\mathcal{U})t^{1-\alpha}}{1-\alpha} \sup_{0\leq s\leq\tau} \|f(s) - g(s)\|_{[0,m]}^{(\alpha)}.
\end{aligned}
$$

Hence, by taking τ such that

$$\tau^{1-\alpha} < \min\left\{\frac{1-\alpha}{c_{1,\gamma}M_{\alpha,\gamma}L(\mathcal{U})}, \frac{1-\alpha}{c_{1,\gamma}M_\alpha K(\mathcal{U})}\right\}, \tag{8.1.56}$$

we see that $\mathscr{T}$ is a contractive mapping on Y.

We can now proceed as in [242, Lemmas 47.1 and 47.2], or [227, Theorem 6.3.1], to ascertain that, for any $f^{in} \in X_{0,m,+}^\alpha$, there is a unique mild solution f to (8.1.33) in $X_{0,m,+}^\alpha$ which, by the estimates obtained for I_1, I_2, I_3, is locally Hölder continuous in time on $(0,\tau)$. Hence, as in [227, Corollary 6.3.2], f is a classical solution; that is, $f \in C^1((0,\tau(f^{in})), X_{0,m}^\alpha) \cap C((0,\tau(f^{in})), D(A_{0,m}))$ for some $\tau(f^{in}) > 0$.

To complete the proof of Part 1 of Theorem 8.1.2, we note that $X_{0,m}$ is a space of type L, as discussed in Section 3.1.2 of Volume I, and thus any $f \in C^1((0,\tau(f^{in})), X_{0,m})$ has a measurable representation $(0,\tau(f^{in})) \times (0,\infty) \ni (t,x) \mapsto \tilde{f}(t,x)$ which, moreover, is absolutely continuous in t for any x and is such that the partial derivative of $\tilde{f}(t,x)$ exists almost everywhere on $(0,\tau(f^{in})) \times (0,\infty)$ and equals $\frac{d}{dt}f$ in $X_{0,m}$ for almost any $t \in (0,\tau(f^{in}))$. Since $D(A_{0,m}) \subset D(A_{0,m}^\alpha) \subset X_{0,m}$, we see that each term in (8.1.26) is well defined and thus the solution f has a representation which satisfies (8.1.26) almost everywhere on $(0,\tau(f^{in})) \times (0,\infty)$.

Let us turn to the proof of Part 2 of Theorem 8.1.2 and assume that the hypotheses of the theorem hold in a specific space X_{0,m_0}^α, where $m_0 > 1$ and $m_0 \geq j + l$. Then, from Part 1, the initial-value problem (8.1.33) has a unique nonnegative classical solution

$$f \in C\left([0,\tau(f^{in})), X_{0,m_0}^\alpha\right) \cap C^1\left((0,\tau(f^{in})), X_{0,m_0}^\alpha\right) \cap C\left((0,\tau(f^{in})), D(A_{0,m_0})\right),$$

for each $f^{in} \in X_{0,m_0,+}^\alpha$. By standard arguments [227, Theorem 6.1.4], we can extend this local solution to its maximal forward interval of existence $[0, \tau_{\max}(f^{in}))$. By Theorem 4.8.3, if $\tau_{\max}(f^{in}) < \infty$, then $t \mapsto \|f(t)\|_{[0,m_0]}^{(\alpha)}$ is unbounded as $t \to \tau_{\max}(f^{in})$. Thus, to prove that f is globally defined, we need to show that $t \mapsto \|f(t)\|_{[0,m_0]}^{(\alpha)}$ is *a priori* bounded on bounded time intervals. To assist us with this, we recall from Proposition 7.2.1 that, if $0 \leq m_1 \leq m_2$, then $X_{0,m_2} \hookrightarrow X_{0,m_1}$, and so, if we are able to prove that $\|f(t)\|_{[0,m_2]} < \infty$ on finite time intervals for some m_2, then the same is true for $\|f(t)\|_{[0,m_1]}$ for all $m_1 \in [0,m_2]$. Another important observation is that Theorem 5.1.47 c), combined with Theorem 5.1.48, establishes that when Part 1 holds in X_{0,m_0}^α for a specific m_0, then it is also valid in the scale of spaces

$X_{0,m}^\alpha$ with $m \geq m_0$. It follows from this that we can always choose an $m \geq \max\{2, m_0\}$ such that Part 1 of Theorem 8.1.2 holds.

Now, by inequalities (8.1.23), (7.2.1) and assumption (8.1.32), we can deduce that, for each $i \geq 2$, $1 < r \leq i$ and $f \in X_{0,i,+}$,

$$\int_0^\infty x^r [C_{0,i}f](x)\mathrm{d}x = \frac{1}{2}\int_0^\infty \int_0^\infty ((x+y)^r - x^r - y^r)k(x,y)f(x)f(y)\mathrm{d}x\mathrm{d}y \qquad (8.1.57)$$
$$\leq K_r(\|f\|_{[r-1]}^{(\alpha)}\|f\|_{[1]} + \|f\|_{[r-1]}\|f\|_{[1]}^{(\alpha)}),$$

where K_r is a positive constant, and, for each r,

$$\|f\|_{[r]}^{(\alpha)} = \int_0^\infty (\omega + a(x))^\alpha x^r |f(x)|\,\mathrm{d}x.$$

For the case $r = 1$ we have, $\int_0^\infty x[C_{0,i}f](x)\mathrm{d}x = 0$.

Turning now to the linear terms in the C-F equation, we recall from Theorem 5.1.47 c) that, if $N_{m_0}(x)/x^{m_0} \geq c'_{m_0}$ holds for some $m_0 > 1$ and c'_{m_0}, then there is $c'_i > 0$ such that $N_i(x)/x^i \geq c'_i > 0$ for all $i > 1$. Hence, for $f \in D(A_{0,i})_+$, (5.1.157) and (5.1.158) lead to

$$\int_0^\infty ([A_{0,i}f](x) + [B_{0,i}f](x))w_i(x)\mathrm{d}x = -\int_0^\infty (N_0(x) + N_i(x))a(x)f(x)\mathrm{d}x$$
$$= \int_0^\infty a(x)(n_0(x) - 1)f(x)\mathrm{d}x - \int_0^\infty (a(x) + \omega)f(x)x^i \frac{N_i(x)}{x^i}\mathrm{d}x + \omega \int_0^\infty f(x)N_i(x)\mathrm{d}x$$
$$\leq 4a_0b_0\|f\|_{[0,i]} - c'_i\|f\|_{[i]}^{(1)} + \omega\|f\|_{[i]} \leq 2\omega\|f\|_{[0,i]} - c'_i\|f\|_{[i]}^{(1)},$$

since $\omega > 4a_0b_0$. For the first moment of $(A_{0,i} + B_{0,i})f$, the situation is similar to that for the coagulation terms, and we have

$$\int_0^\infty x([A_{0,i}f](x) + [B_{0,i}f](x))\mathrm{d}x = 0.$$

Consequently, for a fixed $m \geq \max\{2, m_0\}$, let us take $f^{in} \in X_{0,m+j\alpha,+}^\alpha$. Then the function $t \mapsto \|f(t)\|_{[0,m+j\alpha]}^{(\alpha)}$ is differentiable by Part 1 and, by $X_{0,m+j\alpha} \hookrightarrow X_{0,m+j\alpha}^\alpha$, this yields the differentiability of $t \mapsto \|f(t)\|_{[0,m+j\alpha]}^{(\alpha)}$, and hence the differentiability of $t \mapsto \|f(t)\|_{[0,i]}^{(\alpha)}$ for $i \in [2, m+j\alpha]$. Then the associated local solution $t \mapsto f(t)$ in $X_{0,m+j\alpha,+}^\alpha$ satisfies

$$\frac{d}{dt}\|f(t)\|_{[0,i]} \leq 2\omega\|f(t)\|_{[0,i]} - c'_i\|f(t)\|_{[i]}^{(1)} + K_i(\|f(t)\|_{[i-1]}^{(\alpha)}\|f(t)\|_{[1]} + \|f(t)\|_{[i-1]}\|f(t)\|_{[1]}^{(\alpha)}).$$
$$(8.1.58)$$

To simplify (8.1.58), we use the following inequalities. For $f \in X_{0,m+j\alpha,+}$, $2 \leq i \leq m+j\alpha$ and $1 \leq r \leq i-1$, we apply Hölder's inequality, with $p = 1/\alpha$ and $q = 1/(1-\alpha)$, to obtain

$$\|f\|_{[r]}^{(\alpha)} = \int_0^\infty x^r a_\omega^\alpha(x)f(x)\mathrm{d}x = \int_0^1 x^r a_\omega^\alpha(x)f(x)\mathrm{d}x + \int_1^\infty x^r a_\omega^\alpha(x)f(x)\mathrm{d}x$$
$$\leq c_a \int_0^1 xf(x)\mathrm{d}x + \int_1^\infty x^{(i-1)/q}f^{1/q}(x)x^{(qr-i+1)/q}a_\omega^\alpha(x)f^{1/p}(x)\mathrm{d}x$$
$$\leq c_a\|f\|_{[1]} + \left(\int_0^\infty x^{i-1}f(x)\mathrm{d}x\right)^{1-\alpha}\left(\int_1^\infty x^{(r-(i-1)(1-\alpha))/\alpha}a_\omega(x)f(x)\mathrm{d}x\right)^\alpha$$
$$\leq c_a\|f\|_{[1]} + \|f\|_{[i-1]}^{1-\alpha}\left(\|f\|_{[i]}^{(1)}\right)^\alpha.$$
$$(8.1.59)$$

Note that the above derivation of (8.1.59) uses the fact that $(r - (i-1)(1-\alpha))/\alpha \leq i - 1 < i$ for $\alpha \in (0, 1)$ and $r \leq i - 1$, and hence

$$x^{(r-(i-1)(1-\alpha))/\alpha} \leq x^i, \qquad x \in [1, \infty).$$

Young's inequality, with $p = 1/\alpha$ and $q = 1/(1-\alpha)$, then leads to

$$\|f\|_{[i-1]}^{(\alpha)} \|f\|_{[1]} \leq c_a \|f\|_{[1]}^2 + \|f\|_{[1]} \|f\|_{[i-1]}^{1-\alpha} \left(\|f\|_{[i]}^{(1)} \right)^\alpha$$
$$\leq c_a \|f\|_{[1]}^2 + \|f\|_{[1]} \left((1-\alpha)\epsilon^{1/(\alpha-1)} \|f\|_{[i-1]} + \alpha\epsilon^{1/\alpha} \|f\|_{[i]}^{(1)} \right) \qquad (8.1.60)$$

and

$$\|f\|_{[i-1]} \|f\|_{[1]}^{(\alpha)} \leq c_a \|f\|_{[1]} \|f\|_{[i-1]} + \|f\|_{[1]}^{2-\alpha} \left(\|f\|_{[i]}^{(1)} \right)^\alpha$$
$$\leq c_a \|f\|_{[1]} \|f\|_{[i-1]} + \left((1-\alpha)\epsilon^{1/(\alpha-1)} \|f\|_{[i-1]}^{(2-\alpha)/(1-\alpha)} + \alpha\epsilon^{1/\alpha} \|f\|_{[i]}^{(1)} \right). \tag{8.1.61}$$

We now apply these inequalities to the solution $t \mapsto f(t) \in X_{0,m+j\alpha,+}$. Since $\|f(t)\|_{[1]} = \|f^{in}\|_{[1]}$ is constant on $[0, \tau_{\max}(f^{in}))$, by choosing ϵ so that $\alpha\epsilon^{1/\alpha} K_i(\|f\|_{[1]} + 1) \leq c_i'$, we see that there are positive constants $D_{0,i}, D_{1,i}, D_{2,i}, D_{3,i}$ such that (8.1.58) can be written as

$$\frac{d}{dt}\|f(t)\|_{[0,i]} \leq D_{0,i} + D_{1,i}\|f(t)\|_{[0,i]} + D_{2,i}\|f(t)\|_{[i-1]} + D_{3,i}\|f(t)\|_{[i-1]}^{(2-\alpha)/(1-\alpha)}. \tag{8.1.62}$$

In particular, for $i = 2$ we obtain

$$\frac{d}{dt}\|f(t)\|_{[0,2]} \leq D_{0,2} + D_{1,2}\|f(t)\|_{[0,2]} + D_{2,2}\|f^{in}\|_{[1]} + D_{3,2}\|f^{in}\|_{[1]}^{(2-\alpha)/(1-\alpha)}, \tag{8.1.63}$$

and thus $t \mapsto \|f(t)\|_{[0,2]}$ is bounded on $[0, \tau_{\max}(f^{in}))$. By (7.2.1), $t \mapsto \|f(t)\|_{[0,r]}$ is also bounded on $[0, \tau_{\max}(f^{in}))$ for all $r \in [0, 2]$. Since the boundedness of $t \mapsto \|f(t)\|_{[0,i]}$ implies the same property for $t \mapsto \|f(t)\|_{[i]}$ for any $i \geq 0$, we can use (8.1.62) to proceed inductively to establish the boundedness of $t \mapsto \|f(t)\|_{[0,i]}$ for all i in the interval $[2, m+j\alpha]$ provided that $f^{in} \in X_{0,m+j\alpha,+}^\alpha$. Thus, again by (7.2.1) and the continuous embedding $X_{0,m+j\alpha} \hookrightarrow X_{0,m}^\alpha$, it follows that $t \mapsto \|f(t)\|_{0,m}^{(\alpha)}$ does not blow up as $t \to \tau_{\max}(f^{in})$. Hence, in particular, $t \mapsto \|f(t)\|_{0,m_0}^{(\alpha)}$ is finite as $t \to \tau_{\max}(f^{in})$. Thus $t \mapsto f(t)$ is a global classical solution in any X_{0,m_0}^α for which the assumptions of the theorem hold.

To prove the global existence of solutions emanating from any initial condition $f^{in} \in X_{0,m_0,+}^\alpha$ we observe that since $X_{0,m+j\alpha}^\alpha$ is dense in X_{0,m_0}^α, a finite-time blow-up of such a solution would contradict the theorem on the continuous dependence of solutions on the initial data (which, in this case, follows from the Gronwall–Henry inequality, Lemma 7.5.1, see also [242, Theorem 47.5]), along the lines of the proof of Theorem 8.1.1. □

Remark 8.1.3. *A similar approach is used in Section 8.2.2.4 to prove the global existence of weak solutions under a different set of assumptions.*

8.2 Coagulation-Fragmentation Equations via Weak Compactness

The main aim of this section is to prove the existence of weak solutions to the C-F equation (8.0.1) and the starting point of the analysis is the well-posedness of the C-F equation with truncated kernels which is established in Section 8.2.1.1. We also collect in Section 8.2.1 various results which are needed later on, including a weak stability principle (Section 8.2.1.2), bounds on the coagulation operator according to the behaviour of the coagulation kernel k (Sections 8.2.1.3, 8.2.1.4, and 8.2.1.5), and positivity of weak solutions (Section 8.2.1.6). Existence results are then provided in the next three sections, with a focus on mass-conserving solutions in Section 8.2.2 and on singular coagulation coefficients in Section 8.2.4. Though different classes of coagulation and fragmentation coefficients are handled, we point out that the existence proofs based on the compactness approach share a common structure. More precisely, the starting point is the derivation of estimates for large and small sizes, thereby guaranteeing that no singularity forms as $x \to 0$ and that the escape of matter as $x \to \infty$ (if any) is somewhat under control. Depending on the assumptions on the coagulation and fragmentation coefficients, one has to begin with either the behaviour as $x \to 0$ or that as $x \to \infty$. The second step prevents concentration of matter somewhere in $(0, \infty)$ and is achieved either by estimating directly the behaviour on sets with small measures or by L_p-estimates. Estimates derived in the previous step usually play an important role here and are used to control what is going on for large and small sizes. Owing to the Dunford–Pettis theorem (Theorem 7.1.3), these two steps ensure weak compactness in L_1 with respect to the size variable x. The final step is devoted to the time compactness which is usually obtained thanks to a bound on the time derivative of solutions in a weighted L_1-space. Collecting all these properties allows one to apply the variant of the Arzelà–Ascoli theorem stated in Theorem 7.1.16 and thus to establish the sought-for compactness.

Once the existence of a weak solution is established, several issues are discussed, including that of uniqueness (Section 8.2.5), the infringement of mass conservation being studied in Chapter 9. Concerning the latter, let us recall that loss of matter after a finite time occurs for some classes of coagulation and fragmentation coefficients, resulting either from a runaway growth due to a coagulation kernel k growing sufficiently fast for large sizes or from the formation of dust (particles with zero size) due to an overall fragmentation rate a being unbounded for small sizes. The outcome of the existence proof, however, only provides weak solutions for which the total mass is a nonincreasing function of time and more refined arguments are needed to show that loss of matter does take place.

Coming back to the existence issue, we first define the notions of weak solution to the C-F equation (8.0.1) to be used in the sequel and begin with the classical notion of weak solutions to (8.0.1) which requires an additional assumption on the daughter distribution function b, namely, that the total number of particles resulting from every single fragmentation event is finite:

$$n_0(y) := \int_0^y b(x, y) \, dx < \infty \quad \text{for a.e.} \quad y \in (0, \infty) \ . \tag{8.2.1}$$

We also recall, from Section 2.3.1, the following notation which will be used throughout this section: given a measurable function ϑ in $(0, \infty)$ we set

$$\chi_\vartheta(x, y) := \vartheta(x + y) - \vartheta(x) - \vartheta(y) \ , \qquad (x, y) \in (0, \infty)^2 \ , \tag{8.2.2}$$

and

$$N_\vartheta(y) := \vartheta(y) - \int_0^y \vartheta(x)b(x, y) \, dx \ , \qquad y \in (0, \infty) \ , \tag{8.2.3}$$

whenever it makes sense. When $\vartheta(x) = \vartheta_m(x) := x^m$, $x \in (0, \infty)$, for some $m \in \mathbb{R}$, we set $\chi_m := \chi_{\vartheta_m}$ and $N_m := N_{\vartheta_m}$ to simplify notation. We also recall that we denote a Banach space X equipped with its weak topology by $X_{,w}$.

Definition 8.2.1 (Weak Solution). *Assume that b satisfies (8.0.4) and (8.2.1) and consider*

$f^{in} \in X_{0,1,+}$ and $T \in (0, \infty]$. A weak solution to the C-F equation (8.0.1) on $[0, T)$ is a nonnegative function

$$f \in C([0, t], X_{0,w}) \cap L_\infty((0, t), X_1) , \tag{8.2.4a}$$

such that $f(0) = f^{in}$ a.e. in $(0, \infty)$,

$$(s, x, y) \longmapsto k(x, y)f(s, x)f(s, y) \in L_1((0, t) \times (0, \infty)^2) , \tag{8.2.4b}$$

$$(s, y) \longmapsto a(y)(1 + n_0(y))f(s, y) \in L_1((0, t) \times (0, \infty)) , \tag{8.2.4c}$$

and

$$\int_0^\infty \vartheta(x)(f(t, x) - f^{in}(x)) \, \mathrm{d}x = \frac{1}{2} \int_0^t \int_0^\infty \int_0^\infty k(x, y)\chi_\vartheta(x, y)f(s, x)f(s, y) \, \mathrm{d}y\mathrm{d}x\mathrm{d}s$$

$$- \int_0^t \int_0^\infty a(y)N_\vartheta(y)f(s, y) \, \mathrm{d}y\mathrm{d}s \tag{8.2.4d}$$

for all $\vartheta \in L_\infty(0, \infty)$ and $t \in (0, T)$.

A *mass-conserving weak solution* to the C-F equation (8.0.1) on $[0, T)$ is a weak solution on $[0, T)$ that, in addition, satisfies

$$f \in C([0, T), X_{1,w}) \quad \text{and} \quad M_1(f(t)) = M_1(f^{in}) , \quad t \in [0, T) . \tag{8.2.5}$$

An almost immediate consequence of Definition 8.2.1 is that weak solutions possess better continuity properties with respect to time.

Proposition 8.2.2. *Assume that b satisfies (8.0.4) and (8.2.1) and consider a weak solution f to the C-F equation (8.0.1) on $[0, T)$, $T \in (0, \infty]$, associated with an initial condition $f^{in} \in X_{0,1,+}$. Then $f \in C([0, T), X_0)$.*

Proof. Let $t \in (0, T)$ and $s \in (0, t)$. Owing to Fubini–Tonelli's theorem,

$$\int_0^\infty \int_0^x k(x - y, y)f(s, x - y)f(s, y) \, \mathrm{d}y\mathrm{d}x$$

$$= \int_0^\infty \int_y^\infty k(x - y, y)f(s, x - y)f(s, y) \, \mathrm{d}x\mathrm{d}y$$

$$= \int_0^\infty \int_0^\infty k(x, y)f(s, x)f(s, y) \, \mathrm{d}x\mathrm{d}y ,$$

and

$$\int_0^\infty \int_x^\infty a(y)b(x, y)f(s, y) \, \mathrm{d}y\mathrm{d}x = \int_0^\infty a(y)n_0(y)f(s, y) \, \mathrm{d}y .$$

We infer from the previous two identities and the integrability properties (8.2.4b) and (8.2.4c) of weak solutions that both $\mathcal{C}f$ and $\mathcal{F}f$ belong to $L_1((0, t) \times (0, \infty))$ for all $t \in (0, T)$. Since $\partial_s f(s) = \mathcal{C}f(s) + \mathcal{F}f(s)$ for all $s \in (0, t)$ by (8.2.4d), we realise that $\partial_s f$ belongs to $L_1((0, t) \times (0, \infty))$ for all $t \in (0, T)$ and thus that $f \in C([0, T), X_0)$. $\square$

The additional integrability condition (8.2.1) is clearly needed for the second term in the right-hand side of the weak formulation (8.2.4d) of (8.0.1) to make sense, as it ensures that N_ϑ defined in (8.2.3) is well-defined for all $\vartheta \in L_\infty(0, \infty)$. However assumption (8.2.1) is obviously not satisfied by the daughter distribution function

$$b_\nu(x, y) := (\nu + 2)x^\nu y^{-\nu-1} , \qquad 0 < x < y , \tag{8.2.6}$$

when $\nu \in (-2, -1]$ [188]. An extended notion of weak solution is needed in that case and a weaker integrability condition is required on the daughter distribution function b. More precisely, we shall assume either that b only satisfies (8.0.4) or that, besides satisfying (8.0.4), there is $m \in (0, 1)$ such that

$$n_m(y) := \int_0^y x^m b(x, y) \, \mathrm{d}x < \infty \quad \text{for a.e.} \quad y \in (0, \infty) \ . \tag{8.2.7}$$

Clearly, b_ν satisfies (8.2.7) for any $m \in (-\nu - 1, 1)$ and $-\nu - 1 \geq 0$ for $\nu \in (-2, -1]$.

Definition 8.2.3 (Extended Weak Solution). *Let $m \in (0, 1]$ and assume that, either $m \in (0, 1)$ and b satisfies (8.0.4) and (8.2.7), or $m = 1$ and b satisfies (8.0.4). Consider $f^{in} \in X_{m,+} \cap X_1$ and $T \in (0, \infty]$. An extended weak solution to the C-F equation (8.0.1) on $[0, T)$ is a nonnegative function*

$$f \in C([0, t], X_{m,w}) \cap L_\infty((0, t), X_1) \ , \tag{8.2.8a}$$

such that $f(0) = f^{in}$ a.e. in $(0, \infty)$,

$$(s, x, y) \longmapsto \min\{x, y\}^m k(x, y) f(s, x) f(s, y) \in L_1((0, t) \times (0, \infty)^2) \ , \tag{8.2.8b}$$
$$(s, y) \longmapsto a(y)[y^m + n_m(y)] f(s, y) \in L_1((0, t) \times (0, \infty)) \ , \tag{8.2.8c}$$

for all $t \in (0, T)$, and the weak formulation (8.2.4d) is satisfied for all functions $\vartheta \in \Theta^m$, where

$$\Theta^m := \left\{ \vartheta \in C^{0,m}([0, \infty)) \cap L_\infty(0, \infty) \ : \ \vartheta(0) = 0 \right\}$$

for $m \in (0, 1)$ and

$$\Theta^1 := \left\{ \vartheta \in C^{0,1}([0, \infty)) \ : \ \vartheta(0) = 0 \right\} \ .$$

A mass-conserving extended weak solution to the C-F equation (8.0.1) on $[0, T)$ is an extended weak solution on $[0, T)$ that, in addition, satisfies (8.2.5).

The conditions (8.2.8b) and (8.2.8c) guarantee that (8.2.4d) is indeed meaningful for all functions $\vartheta \in \Theta^m$ thanks to the following lemma.

Lemma 8.2.4. *Let $m \in (0, 1]$ and assume that, either $m \in (0, 1)$ and b satisfies (8.2.7), or $m = 1$ and b satisfies (8.0.4). Consider a function $\vartheta \in C^{0,m}([0, \infty))$ satisfying $\vartheta(0) = 0$. Then*

$$|\chi_\vartheta(x, y)| \leq 2\|\vartheta\|_{C^{0,m}} \min\{x, y\}^m \ , \qquad (x, y) \in (0, \infty) \ , \tag{8.2.9}$$
$$|N_\vartheta(y)| \leq \|\vartheta\|_{C^{0,m}} [y^m + n_m(y)] \ , \qquad y \in (0, \infty) \ . \tag{8.2.10}$$

Proof. Consider $(x, y) \in (0, \infty)^2$. Owing to the symmetry of χ_ϑ, we may assume without loss of generality that $x \leq y$. Since $\vartheta \in C^{0,m}([0, \infty))$ with $\vartheta(0) = 0$, we observe that

$$|\chi_\vartheta(x, y)| \leq |\vartheta(x + y) - \vartheta(y)| + |\vartheta(x) - \vartheta(0)| \leq 2\|\vartheta\|_{C^{0,m}} x^m = 2\|\vartheta\|_{C^{0,m}} \min\{x, y\}^m$$

and

$$|N_\vartheta(y)| \leq |\vartheta(y) - \vartheta(0)| + \int_0^y |\vartheta(x) - \vartheta(0)| b(x, y) \, \mathrm{d}x \leq \|\vartheta\|_{C^{0,m}} [y^m + n_m(y)] \ ,$$

and the proof is complete. □

Similarly to weak solutions, extended weak solutions may have continuity properties with respect to the strong topology of X_m, but this however requires a slightly stronger integrability of the coagulation term.

Proposition 8.2.5. *Let $m \in (0,1]$ and assume that either $m \in (0,1)$ and b satisfies (8.0.4) and (8.2.7), or $m = 1$ and b satisfies (8.0.4). Consider an extended weak solution f to the C-F equation (8.0.1) on $[0,T)$, $T \in (0,\infty]$, associated with an initial condition $f^{in} \in X_{m,+} \cap X_1$. If f satisfies in addition*

$$(s,x,y) \longmapsto \max\{x,y\}^m k(x,y)f(s,x)f(s,y) \in L_1((0,T) \times (0,\infty)^2) , \qquad (8.2.11)$$

then $f \in C([0,T), X_m)$.

Proof. Let $t \in (0,T)$ and $s \in (0,t)$. It follows from Fubini–Tonelli's theorem that

$$\int_0^\infty x^m \int_0^x k(x-y,y)f(s,x-y)f(s,y) \, dydx$$
$$= \int_0^\infty \int_0^\infty (x+y)^m k(x,y)f(s,x)f(s,y) \, dydx$$
$$\leq 2^m \int_0^\infty \int_0^\infty \max\{x,y\}^m k(x,y)f(s,x)f(s,y) \, dydx ,$$

and

$$\int_0^\infty x^m \int_x^\infty a(y)b(x,y)f(s,y) \, dydx = \int_0^\infty a(y)n_m(y)f(s,y) \, dy .$$

It readily follows from (8.2.8c) and (8.2.11) that $\mathcal{C}f$ and $\mathcal{F}f$ belong to $L_1((0,t), X_m)$ for all $t \in (0,T)$ and (8.2.4d) implies that $\partial_t f$ lies in the same space. Proposition 8.2.5 is a straightforward consequence of this last property. $\qquad \square$

Since the class of test functions used in the weak formulation for weak solutions and extended weak solutions are of a different nature, we gather their definitions in the next statement.

Definition 8.2.6. *Let $m \in [0,1]$. We define the set of admissible test functions Θ^m by $\Theta^0 := L_\infty(0,\infty)$,*

$$\Theta^1 := \left\{ \vartheta \in C^{0,1}([0,\infty)) \; : \; \vartheta(0) = 0 \right\} ,$$

and

$$\Theta^m = \left\{ \vartheta \in C^{0,m}([0,\infty)) \cap L_\infty(0,\infty) \; : \; \vartheta(0) = 0 \right\} \quad for \quad m \in (0,1) .$$

Throughout the remainder of this chapter, as well as in Chapters 9 and 10, we shall generically refer to weak solutions instead of extended weak solutions, except when it is not clear from the context.

8.2.1   Truncated Kernels and Basic Estimates

8.2.1.1   Truncated Kernels

As the first step towards a proof of existence of weak solutions to the C-F equation (8.0.1) with general coefficients, we investigate its well-posedness when the coagulation kernel k and the overall fragmentation rate a are compactly supported. More precisely, we assume that there is $m_0 \in [0,1]$, $L > 0$, and $r > 0$ such that

$$k(x,y) \leq L(\min\{x,y\})^{m_0} , \qquad a(x) \leq L , \qquad n_{m_0}(x) \leq L x^{m_0} , \qquad (8.2.12a)$$

for $(x,y) \in (0,r)^2$, the function n_{m_0} being defined in (8.2.7), and

$$k(x,y) = 0 \ \ \text{if} \ \ x+y > r \ \ \text{and} \ \ a(x) = 0 \ \ \text{if} \ \ x > r . \qquad (8.2.12b)$$

Proposition 8.2.7. *Assume that k, a and b satisfy (8.0.4) and (8.2.12) for some $m_0 \in [0,1]$ and $r > 0$. Given a nonnegative function $f^{in} \in L_1((0,r), x^{m_0} dx)$ there is a unique weak solution*

$$f \in C^1([0,\infty), L_1((0,r), x^{m_0} dx))$$

to (8.0.1) on $[0,\infty)$. Setting $f(t,x) = 0$ for $(t,x) \in [0,\infty) \times (r,\infty)$, it satisfies

$$\frac{d}{dt} \int_0^\infty \vartheta(x) f(t,x) \, dx = \frac{1}{2} \int_0^\infty \int_0^\infty k(x,y) \chi_\vartheta(x,y) f(t,x) f(t,y) \, dy dx$$
$$- \int_0^\infty a(y) N_\vartheta(y) f(t,y) \, dy \tag{8.2.13}$$

for all $t > 0$ and $\vartheta \in \Theta^{m_0}$, as well as

$$\int_0^\infty x f(t,x) \, dx = \int_0^\infty x f^{in}(x) \, dx , \qquad t \geq 0 . \tag{8.2.14}$$

The proof of Proposition 8.2.7 relies on Banach's fixed-point theorem and by now is classical when $m_0 = 0$ [252, 272]. We show below how to adapt it to the general framework of Proposition 8.2.7.

Proof. Set $\mathcal{Y} := L_1((0,r), x^{m_0} dx)$ and denote its norm by $\|\cdot\|_{\mathcal{Y}}$. Let us first check that the coagulation and fragmentation operators are locally Lipschitz continuous maps from $\mathcal{Y}$ to $\mathcal{Y}$. To this end, consider $f_i \in \mathcal{Y}$, $i = 1, 2$, and compute

$$\|\mathscr{F} f_1 - \mathscr{F} f_2\|_{\mathcal{Y}} \leq \int_0^r x^{m_0} a(x) |(f_1 - f_2)(x)| \, dx$$
$$+ \int_0^r x^{m_0} \int_x^r a(y) b(x,y) |(f_1 - f_2)(y)| \, dy dx .$$

Owing to (8.2.12a) and Fubini–Tonelli's theorem,

$$\|\mathscr{F} f_1 - \mathscr{F} f_2\|_{\mathcal{Y}} \leq L \|f_1 - f_2\|_{\mathcal{Y}} + L \int_0^r n_{m_0}(y) |(f_1 - f_2)(y)| \, dy \tag{8.2.15}$$
$$\leq L(1 + L) \|f_1 - f_2\|_{\mathcal{Y}} .$$

Similarly, we infer from (8.2.12a) and Fubini–Tonelli's theorem that

$$\|\mathcal{C} f_1 - \mathcal{C} f_2\|_{\mathcal{Y}} \leq \frac{1}{2} \int_0^r \int_y^r x^{m_0} k(x-y,y) |f_1(x-y)| |(f_1 - f_2)(y)| \, dx dy$$
$$+ \frac{1}{2} \int_0^r \int_y^r x^{m_0} k(x-y,y) |f_2(y)| |(f_1 - f_2)(x-y)| \, dx dy$$
$$+ \int_0^r \int_0^r x^{m_0} k(x,y) [|f_1(x)| |(f_1 - f_2)(y)| + |f_2(y)| |(f_1 - f_2)(x)|] \, dy dx$$
$$\leq \frac{1}{2} \int_0^r \int_0^{r-y} (x+y)^{m_0} k(x,y) |f_1(x)| |(f_1 - f_2)(y)| \, dx dy$$
$$+ \frac{1}{2} \int_0^r \int_0^{r-y} (x+y)^{m_0} k(x,y) |f_2(y)| |(f_1 - f_2)(x)| \, dx dy$$
$$+ L \int_0^r \int_0^r x^{m_0} y^{m_0} [|f_1(x)| |(f_1 - f_2)(y)| + |f_2(y)| |(f_1 - f_2)(x)|] \, dy dx .$$

We again use (8.2.12a), along with the elementary inequality

$$(x + y)^{m_0} \le x^{m_0} + y^{m_0} \le 2(\max\{x, y\})^{m_0} , \qquad (x, y) \in (0, \infty)^2 ,$$

resulting from the subadditivity of $x \mapsto x^{m_0}$, to obtain

$$\|\mathcal{C}f_1 - \mathcal{C}f_2\|_Y \le L \int_0^r \int_0^{r-y} (\max\{x, y\})^{m_0} (\min\{x, y\})^{m_0} |f_1(x)| |(f_1 - f_2)(y)| \, \mathrm{d}x\mathrm{d}y$$

$$+ L \int_0^r \int_0^{r-y} (\max\{x, y\})^{m_0} (\min\{x, y\})^{m_0} |f_2(y)| |(f_1 - f_2)(x)| \, \mathrm{d}x\mathrm{d}y \qquad (8.2.16)$$

$$+ L \left[\|f_1\|_Y \|f_1 - f_2\|_Y + \|f_2\|_Y \|f_1 - f_2\|_Y \right] \le 2L \left(\|f_1\|_Y + \|f_2\|_Y \right) \|f_1 - f_2\|_Y .$$

Consider now $f^{in} \in Y$ such that $f^{in} \ge 0$ a.e. in $(0, r)$. Introducing

$$\bar{\mathcal{F}} f(x) := -a(x)f(x) + \left(\int_x^r a(y)b(x, y)f(y) \, \mathrm{d}y \right)_+$$

and

$$\bar{\mathcal{C}} f(x) := \frac{1}{2} \left(\int_0^x k(x - y, y)f(x - y)f(y) \, \mathrm{d}y \right)_+ - f(x) \int_0^r k(x, y)f(y) \, \mathrm{d}y$$

for $f \in Y$, the modified fragmentation and coagulation operators $\bar{\mathcal{F}}$ and $\bar{\mathcal{C}}$ are also locally Lipschitz continuous maps from Y to Y and a classical application of Banach's fixed-point theorem guarantees that there is $T_* \in (0, \infty]$, and a function $f \in C^1([0, T_*), Y)$ such that f solves

$$\partial_t f = \bar{\mathcal{C}} f + \bar{\mathcal{F}} f \quad \text{in } (0, T_*) \times (0, r) , \qquad (8.2.17)$$

with initial condition $f(0) = f^{in}$. Moreover, one has the following alternative: either $T_* = \infty$ or $T_* < \infty$ and $\|f(t)\|_Y \to \infty$ as $t \to T_*$, see [4, Theorem (7.6) & Remarks (7.10) (b)] for instance.

Now, on the one hand, it follows from (8.2.12a) and (8.2.17) that, for $t \in (0, T_*)$,

$$\frac{d}{dt} \int_0^r x^{m_0} (-f(t, x))_+ \, \mathrm{d}x = - \int_0^r x^{m_0} \mathbf{1}_{(0, \infty)} (-f(t, x)) \partial_t f(t, x) \, \mathrm{d}x$$

$$\le \int_0^r x^{m_0} \mathbf{1}_{(0, \infty)} (-f(t, x)) a(x) f(t, x) \, \mathrm{d}x$$

$$+ \int_0^r x^{m_0} \mathbf{1}_{(0, \infty)} (-f(t, x)) f(t, x) \int_0^r k(x, y) f(t, y) \, \mathrm{d}y\mathrm{d}x$$

$$\le \int_0^r x^{m_0} (-f(t, x))_+ \int_0^r k(x, y) |f(t, y)| \, \mathrm{d}y\mathrm{d}x$$

$$\le L \|f(t)\|_Y \|(-f(t))_+\|_Y .$$

Consequently, for $t \in (0, T_*)$,

$$\|(-f(t))_+\|_Y \le \|(-f^{in})_+\|_Y \exp \left(L \int_0^t \|f(s)\|_Y \, \mathrm{d}s \right) = 0 ,$$

from which we deduce that $f(t) \ge 0$ a.e. in $(0, r)$ for all $t \in [0, T_*)$. This in turn implies that $\bar{\mathcal{C}} f = \mathcal{C} f$ and $\bar{\mathcal{F}} f = \mathcal{F} f$ in $(0, T_*) \times (0, r)$, so that f solves (8.0.1) on $[0, T_*)$.

On the other hand, we infer from (8.0.1), (8.2.12a), the subadditivity of $x \mapsto x^{m_0}$, and the just established nonnegativity of f that, for $t \in (0, T_*)$,

$$\frac{d}{dt}\|f(t)\|_Y = \int_0^r \int_0^r k(x,y)\chi_{m_0}(x,y)f(t,x)f(t,y)\,\mathrm{d}y\mathrm{d}x - \int_0^r a(y)N_{m_0}(y)f(t,y)\,\mathrm{d}y$$

$$\leq \int_0^r a(y)n_{m_0}(y)f(t,y)\,\mathrm{d}y \leq L^2\|f(t)\|_Y\ ,$$

which excludes finite-time blow-up of $\|f(t)\|_Y$. Consequently, $T_* = \infty$. It is then straightforward to derive the weak formulation (8.2.13) and the mass conservation (8.2.14), the latter following from (8.2.12b) and the former with the choice $\vartheta(x) = x\mathbf{1}_{(0,r)}(x)$ for $x \in (0,\infty)$. □

We next show that L_p-integrability of the initial condition propagates through time.

Proposition 8.2.8. *Assume that k, a and b satisfy (8.0.4) and (8.2.12) for some $m_0 \in [0,1]$ and $r > 0$. Assume further that*

$$a(x) \leq Lx^{m_0}\ , \qquad x \in (0,r)\ . \tag{8.2.18}$$

Next, let $f^{in} \in L_1((0,r),x^{m_0}\mathrm{d}x)_+$ and denote the corresponding solution to (8.0.1) on $[0,\infty)$ by f. If there is a nonnegative and convex function $\Psi \in C^1([0,\infty))$, and a constant $C_\Psi > 0$ satisfying $\Psi(0) = 0$,

$$y \longmapsto \int_0^y x^{m_0}\Psi(b(x,y))\,\mathrm{d}x \in L_\infty(0,r)\ , \tag{8.2.19}$$

and

$$z\Psi'(z) \leq (1+C_\Psi)\Psi(z)\ , \qquad z \in [0,\infty)\ , \tag{8.2.20}$$

and such that $\Psi(f^{in}) \in L_1((0,r),x^{m_0}\mathrm{d}x)$, then $\Psi(f(t))$ belongs to $L_1((0,r),x^{m_0}\mathrm{d}x)$ for all $t \geq 0$. More precisely, $\Psi(f)$ belongs to $L_\infty((0,T),L_1((0,r),x^{m_0}\mathrm{d}x))$ for all $T > 0$.

According to Proposition 7.1.9 (a), any function $\Psi \in \mathcal{C}_{VP}$ satisfies (8.2.20) with $C_\Psi = 1$. The condition (8.2.20) is also satisfied by $\Psi(z) = z^p$, $z \geq 0$, for any $p > 1$ but not by $z \mapsto e^z$.

Proof. For $R \geq 1$, we define the convex function Ψ_R by $\Psi_R(0) = 0$ and $\Psi_R'(z) = \min\{\Psi'(z),\Psi'(R)\}$ for $z \in [0,\infty)$ and observe that Ψ_R belongs to $C^1([0,\infty))$ with $\Psi_R' \in L_\infty(0,\infty)$ and $\Psi_R \leq \Psi$. Consequently, $\Psi_R(f)$ belongs to $C^1([0,\infty),L_1((0,r),x^{m_0}\mathrm{d}x))$ and it follows from (8.0.1a), (8.2.18), and the nonnegativity of Ψ_R' that

$$\frac{d}{dt}\int_0^r x^{m_0}\Psi_R(f(t,x))\,\mathrm{d}x = \int_0^r x^{m_0}\Psi_R'(f(t,x))\partial_t f(t,x)\,\mathrm{d}x$$

$$\leq \frac{1}{2}\int_0^r \int_0^x x^{m_0}k(x-y,y)f(t,x-y)f(t,y)\Psi_R'(f(t,x))\,\mathrm{d}y\mathrm{d}x$$

$$+ \int_0^r \int_x^r x^{m_0}a(y)b(x,y)f(t,y)\Psi_R'(f(t,x))\,\mathrm{d}y\mathrm{d}x$$

$$\leq \frac{1}{2}\int_0^r \int_y^r x^{m_0}k(x-y,y)f(t,x-y)\Psi_R'(f(t,x))f(t,y)\,\mathrm{d}x\mathrm{d}y$$

$$+ L\int_0^r y^{m_0}f(t,y)\int_0^y x^{m_0}b(x,y)\Psi_R'(f(t,x))\,\mathrm{d}x\mathrm{d}y\ .$$

Owing to (8.2.20) we infer from the convexity and nonnegativity of Ψ_R that

$$f(t,x-y)\Psi_R'(f(t,x)) \leq \Psi_R(f(t,x-y)) + f(t,x)\Psi_R'(f(t,x)) - \Psi_R(f(t,x))$$
$$\leq \Psi_R(f(t,x-y)) + C_\Psi\Psi_R(f(t,x))\ ,$$

so that, thanks to (8.2.12a),

$$\int_0^r \int_y^r x^{m_0} k(x-y,y) f(t,x-y) \Psi_R'(f(t,x)) f(t,y) \, \mathrm{d}x \mathrm{d}y$$

$$\leq \int_0^r \int_y^r x^{m_0} k(x-y,y) \left[\Psi_R(f(t,x-y)) + C_\Psi \Psi_R(f(t,x)) \right] f(t,y) \, \mathrm{d}x \mathrm{d}y$$

$$\leq \int_0^r \int_0^{r-y} (x+y)^{m_0} k(x,y) \Psi_R(f(t,x)) f(t,y) \, \mathrm{d}x \mathrm{d}y$$

$$+ C_\Psi \int_0^r \int_y^r x^{m_0} k(x-y,y) \Psi_R(f(t,x)) f(t,y) \, \mathrm{d}x \mathrm{d}y$$

$$\leq (2 + C_\Psi) L \left(\int_0^r y^{m_0} f(t,y) \, \mathrm{d}y \right) \int_0^r x^{m_0} \Psi_R(f(t,x)) \, \mathrm{d}x \ .$$

Similarly,

$$b(x,y) \Psi_R'(f(t,x)) \leq \Psi(b(x,y)) + f(t,x) \Psi_R'(f(t,x)) - \Psi_R(f(t,x))$$

$$\leq \Psi(b(x,y)) + C_\Psi \Psi_R(f(t,x)) \ ,$$

and thus, by (8.2.19),

$$L \int_0^r y^{m_0} f(t,y) \int_0^y x^{m_0} b(x,y) \Psi_R'(f(t,x)) \, \mathrm{d}x \mathrm{d}y$$

$$\leq L \int_0^r y^{m_0} f(t,y) \int_0^y x^{m_0} \left[\Psi_R(b(x,y)) + C_\Psi \Psi_R(f(t,x)) \right] \, \mathrm{d}x \mathrm{d}y$$

$$\leq L \sup_{y>0} \left\{ \int_0^y x^{m_0} \Psi(b(x,y)) \, \mathrm{d}x \right\} \int_0^r y^{m_0} f(t,y) \, \mathrm{d}y$$

$$+ L C_\Psi \left(\int_0^r y^{m_0} f(t,y) \, \mathrm{d}y \right) \int_0^r x^{m_0} \Psi_R(f(t,x)) \, \mathrm{d}x \ .$$

Gathering the above estimates we conclude that

$$\frac{d}{dt} \int_0^r x^{m_0} \Psi_R(f(t,x)) \, \mathrm{d}x \leq C \left(\int_0^r y^{m_0} f(t,y) \, \mathrm{d}y \right) \left(1 + \int_0^r x^{m_0} \Psi_R(f(t,x)) \, \mathrm{d}x \right) \ ,$$

with

$$C := L \left(2 + C_\Psi + \sup_{y>0} \left\{ \int_0^y x^{m_0} \Psi(b(x,y)) \, \mathrm{d}x \right\} \right) \ .$$

Since $f \in C([0,\infty), L_1((0,r), x^{m_0} \mathrm{d}x))$, Gronwall's lemma implies that, for each $T > 0$, there is a positive constant $C(T)$, which does not depend on R, such that

$$\int_0^r x^{m_0} \Psi_R(f(t,x)) \, \mathrm{d}x \leq C(T) \left(1 + \int_0^r x^{m_0} \Psi_R(f^{in}(x)) \, \mathrm{d}x \right) \ , \qquad t \in [0,T] \ .$$

Since $\Psi_R \leq \Psi$ and $\Psi(f^{in}) \in L_1((0,r), x^{m_0} \mathrm{d}x)$, we further obtain

$$\int_0^r x^{m_0} \Psi_R(f(t,x)) \, \mathrm{d}x \leq C(T) \left(1 + \int_0^r x^{m_0} \Psi(f^{in}(x)) \, \mathrm{d}x \right) \ , \qquad t \in [0,T] \ .$$

Finally, $(\Psi_R)_{R \geq 1}$ converges pointwise to Ψ, so that $\Psi_R(f(t,x)) \to \Psi(f(t,x))$ a.e. in $(0,\infty)$

for all $t \in [0, T]$. We then use Fatou's lemma to let $R \to \infty$ in the previous inequality and deduce that

$$\int_0^r x^{m_0} \Psi(f(t,x)) \, \mathrm{d}x \leq \liminf_{R \to \infty} \int_0^r x^{m_0} \Psi_R(f(t,x)) \, \mathrm{d}x$$

$$\leq C(T) \left(1 + \int_0^r x^{m_0} \Psi(f^{in}(x)) \, \mathrm{d}x \right) , \qquad t \in [0, T] ,$$

thereby completing the proof of Proposition 8.2.8. □

Remark 8.2.9. *Although we barely use the outcome of Proposition 8.2.8 in the sequel, its inclusion is merited as the forthcoming derivation of L_p-estimates, or similar, follows the lines of its proof.*

We supplement Proposition 8.2.8 with a result on the propagation of positive upper and lower bounds.

Proposition 8.2.10. *Assume that k, a and b satisfy (8.0.4) and (8.2.12) for some $r > 0$ and $m_0 \in [0,1]$. Let $f^{in} \in L_1((0,r), x^{m_0}\mathrm{d}x)_+$ and denote the corresponding solution to the C-F equation (8.0.1) on $[0, \infty)$ by f.*

(a) If $f^{in} \in L_\infty(0,r)$ and there is $\bar{b} > 0$ such that

$$\int_x^r a(y)b(x,y) \, \mathrm{d}y \leq \bar{b} , \qquad x \in (0,r) , \tag{8.2.21}$$

then there is a positive function $W \in C^1([0,\infty))$ depending only on r, m_0, L, $\bar{b}$, $\|f^{in}\|_{L_\infty(0,r)}$ and f such that

$$f(t,x) \leq W(t) , \qquad (t,x) \in [0,\infty) \times (0,r) . \tag{8.2.22}$$

(b) If there is $w^{in} > 0$ such that

$$f^{in}(x) \geq w^{in} , \qquad x \in (0,r) , \tag{8.2.23}$$

then there is a positive function $w \in C^1([0,\infty))$ depending only on r, m_0, L, w^{in} and f such that

$$f(t,x) \geq w(t) > 0 , \qquad (t,x) \in [0,\infty) \times (0,r) . \tag{8.2.24}$$

Proof. For $t \geq 0$, we set

$$F(t) := \int_0^r x^{m_0} f(t,x) \, \mathrm{d}x ,$$

which is well defined and belongs to $C^1([0,\infty))$, according to Proposition 8.2.7.

Proof of (a). Let $W \in C^1([0,\infty))$ be the solution to the ordinary differential equation

$$\frac{dW}{dt}(t) = \left(LF(t) + \bar{b} \right) W(t) , \qquad t \in (0,\infty) , \qquad W(0) = \|f^{in}\|_{L_\infty(0,r)} . \tag{8.2.25}$$

We next define

$$\psi(t,x) := (f(t,x) - W(t))_+ , \qquad \sigma(t,x) := \mathbf{1}_{(0,\infty)}(f(t,x) - W(t)) ,$$

and

$$I(t) := \int_0^r x^{m_0} \sigma(t,x) \, \mathrm{d}x ,$$

for $(t, x) \in [0, \infty) \times (0, r)$. Owing to the continuous differentiability with respect to time of f and W, and the Lipschitz continuity of the positive part, the function ψ belongs to $W^1_{\infty, loc}([0, \infty), L_1((0, r), x^{m_0} dx))$. It then follows from (8.0.1a) and the nonnegativity of k, a, b, f and σ that, for $t \geq 0$,

$$\frac{d}{dt} \int_0^r x^{m_0} \psi(t, x) \, dx = \int_0^r x^{m_0} \sigma(t, x) \left[\partial_t f(t, x) - \frac{dW}{dt}(t) \right] \, dx$$

$$\leq J_1(t) + J_2(t) - I(t) \frac{dW}{dt}(t) , \qquad (8.2.26)$$

where

$$J_1(t) := \frac{1}{2} \int_0^r \int_0^x x^{m_0} k(x - y, y) f(t, x - y) \sigma(t, x) f(t, y) \, dy dx ,$$

$$J_2(t) := \int_0^r \int_x^r x^{m_0} a(y) b(x, y) \sigma(t, x) f(t, y) \, dy dx .$$

On the one hand, the property $\sigma \in [0, 1]$ implies that, for $t \in [0, \infty)$ and $0 < y < x$,

$$f(t, x - y) \sigma(t, x) = (f(t, x - y) - W(t)) \sigma(t, x) + W(t) \sigma(t, x)$$

$$\leq \psi(t, x - y) \sigma(t, x) + W(t) \sigma(t, x)$$

$$\leq \psi(t, x - y) + W(t) \sigma(t, x) ,$$

which gives, together with (8.2.12a),

$$2 J_1(t) \leq \int_0^r \int_y^r x^{m_0} k(x - y, y) \psi(t, x - y) f(t, y) \, dx dy$$

$$+ \int_0^r \int_y^r x^{m_0} k(x - y, y) W(t) \sigma(t, x) f(t, y) \, dx dy$$

$$\leq \int_0^r \int_0^{r - y} (x + y)^{m_0} k(x, y) \psi(t, x) f(t, y) \, dx dy$$

$$+ L W(t) \int_0^r \int_y^r x^{m_0} y^{m_0} \sigma(t, x) f(t, y) \, dx dy$$

$$\leq 2^{m_0} L \int_0^r \int_0^r (\max\{x, y\})^{m_0} (\min\{x, y\})^{m_0} \psi(t, x) f(t, y) \, dx dy$$

$$+ L F(t) I(t) W(t)$$

$$\leq 2 L F(t) \int_0^r x^{m_0} \psi(t, x) \, dx + 2 L F(t) I(t) W(t) . \qquad (8.2.27)$$

On the other hand, as above,

$$f(t, y) \sigma(t, x) \leq \psi(t, y) + W(t) \sigma(t, x) , \qquad y \in (0, r) ,$$

and we infer from (8.2.12a), (8.2.21), the nonnegativity of σ, and Fubini's theorem that, for $t \in [0, \infty)$,

$$J_2(t) \leq \int_0^r \int_x^r x^{m_0} a(y) b(x, y) \psi(t, y) \, dy dx + W(t) \int_0^r \int_x^r x^{m_0} \sigma(t, x) a(y) b(x, y) \, dy dx$$

$$\leq \int_0^r a(y) \psi(t, y) n_{m_0}(y) \, dy + \bar{b} W(t) \int_0^r x^{m_0} \sigma(t, x) \, dx$$

$$\leq L^2 \int_0^r y^{m_0} \psi(t,y) \, \mathrm{d}y + \bar{b} I(t) W(t) \ . \tag{8.2.28}$$

Combining (8.2.25), (8.2.26), (8.2.27), and (8.2.28) leads us to

$$\frac{d}{dt} \int_0^r x^{m_0} \psi(t,x) \, \mathrm{d}x \leq (LF(t) + L^2) \int_0^r x^{m_0} \psi(t,x) \, \mathrm{d}x + I(t) \left[(LF(t) + \bar{b}) W(t) - \frac{dW}{dt}(t) \right]$$

$$= (LF(t) + L^2) \int_0^r x^{m_0} \psi(t,x) \, \mathrm{d}x$$

for $t \in [0, \infty)$. Hence, after integration,

$$\int_0^r x^{m_0} \psi(t,x) \, \mathrm{d}x \leq \left(\int_0^r x^{m_0} \psi(0,x) \, \mathrm{d}x \right) \exp \left\{ L^2 t + L \int_0^t F(s) \, \mathrm{d}s \right\} \ , \qquad t \in [0, \infty) \ .$$

Since $\psi(0,x) = 0$ due to the definition (8.2.25) of $W(0)$, we conclude that $\psi(t,x) = 0$ for a.e. $x \in (0,r)$ and all $t \in [0, \infty)$, from which (8.2.22) readily follows.

Proof of (b). We argue as in the proof of [165, Proposition 5.3] and let $w \in C^1([0, \infty))$ be the solution to the ordinary differential equation

$$\frac{dw}{dt}(t) = -L\left(1 + F(t)\right) w(t) \ , \qquad t \in [0, \infty) \ , \qquad w(0) = w^{in} \ . \tag{8.2.29}$$

Observe that w is well defined on $[0, \infty)$ due to the time continuity of f in $L_1((0,r), x^{m_0}\mathrm{d}x)$ provided by Proposition 8.2.7. Also, it readily follows from (8.2.29) and the positivity of w^{in} that $w(t) > 0$ for each $t > 0$.

Now, we infer from (8.0.1a), Proposition 8.2.7, and the nonnegativity of k, a, b, and f that, for $(t,x) \in (0, \infty) \times (0,r)$,

$$\partial_t (w - f)(t,x) \leq \frac{dw}{dt}(t) + \left[\int_0^r k(x,y) f(t,y) \, \mathrm{d}y + a(x) \right] f(t,x) \ .$$

It then follows from (8.2.12a) and (8.2.29) that

$$\partial_t (w - f)(t,x) \leq \frac{dw}{dt}(t) + L\left(F(t) + 1\right) f(t,x)$$

$$\leq -L \left(\int_0^r y^{m_0} f(t,y) \, \mathrm{d}y + 1 \right) (w - f)(t,x) \ .$$

Hence, after integration,

$$(w - f)(t,x) \leq \left(w^{in} - f^{in}(x) \right) \exp \left\{ -Lt - L \int_0^t F(s) \, \mathrm{d}s \right\} \ .$$

Owing to (8.2.23) and (8.2.29), the right-hand side of the previous inequality is nonpositive, from which we conclude that $f(t,x) \geq w(t)$ for $(t,x) \in [0, \infty) \times (0,r)$. □

8.2.1.2 Weak Stability

This section is devoted to the analysis of the behaviour of sequences of solutions to C-F equations and is not only one of the building blocks of the existence proofs presented later in this chapter but also of the construction of self-similar or stationary solutions performed in Chapter 10 by a dynamical approach.

Theorem 8.2.11. *Let $(k_j)_{j\geq 1}$ and $(a_j)_{j\geq 1}$ be two sequences of coagulation kernels and overall fragmentation rates satisfying*

$$\lim_{j\to\infty} k_j(x,y) = k(x,y) , \quad \lim_{j\to\infty} a_j(x) = a(x) \quad \text{for a.e. } (x,y) \in (0,\infty)^2 , \qquad (8.2.30)$$

for some coagulation kernel k and overall fragmentation rate a. Assume further that, for each $r > 1$, there is $K_r > 0$, and $A_r > 0$ such that

$$k_j(x,y) + k(x,y) \leq K_r , \qquad (x,y) \in (1/r,r)^2 , \quad j \geq 1 , \qquad (8.2.31)$$

$$a_j(x) + a(x) \leq A_r , \qquad x \in (0,r) , \quad j \geq 1 . \qquad (8.2.32)$$

Let b be a daughter distribution function satisfying

$$\int_0^y x b(x,y) \, \mathrm{d}x = y , \qquad n_{m_0}(y) = \int_0^y x^{m_0} b(x,y) \, \mathrm{d}x \leq \kappa_{m_0,1} y^{m_0} , \qquad y \in (0,\infty) , \qquad (8.2.33)$$

for some $m_0 \in [0,1]$ and $\kappa_{m_0,1} > 0$.

Next, let $T_0 > 0$ and assume that, for each $j \geq 1$, there is a weak solution f_j on $[0,T_0]$ to the C-F equation (8.0.1), with coefficients (k_j, a_j, b), satisfying

$$M_1(f_j(t)) \leq \mu_1 , \qquad t \in [0,T_0] , \quad j \geq 1 , \qquad (8.2.34)$$

for some $\mu_1 > 0$. Assume further that there is $f \in C([0,T_0), X_{m_0,w})$ such that

$$f_j \longrightarrow f \quad in \quad C([0,T], X_{m_0,w}) \qquad (8.2.35)$$

for all $T \in (0,T_0)$. Assume finally that, for each $r > 0$ and $T \in (0,T_0)$,

$$\omega(r,T) := \sup_{j\geq 1}\left\{ \int_0^T \int_r^\infty x^{m_0} a_j(x) f_j(t,x) \, \mathrm{d}x\mathrm{d}t \right\}$$

$$+ \sup_{j\geq 1}\left\{ \int_0^T \int_r^\infty \int_0^\infty k_j(x,y) f_j(t,x) f_j(t,y) \, \mathrm{d}y\mathrm{d}x\mathrm{d}t \right\} \qquad (8.2.36)$$

$$+ \sup_{j\geq 1}\left\{ \int_0^T \int_0^{1/r} \int_0^\infty x^{m_0} k_j(x,y) f_j(t,x) f_j(t,y) \, \mathrm{d}y\mathrm{d}x\mathrm{d}t \right\}$$

is finite and satisfies

$$\lim_{r\to\infty} \omega(r,T) = 0 . \qquad (8.2.37)$$

Then f is a weak solution on $[0,T_0)$ to the C-F equation (8.0.1) with coefficients (k,a,b).

Proof. Since $m_0 \in [0,1]$, it readily follows from (8.2.34) and (8.2.35) that, for all $r > 1$ and $t \in [0,T_0)$,

$$\int_0^r x f(t,x) \, \mathrm{d}x = \lim_{j\to\infty} \int_0^r x f_j(t,x) \, \mathrm{d}x \leq \mu_1 ,$$

which implies, together with Fatou's lemma, that $M_1(f(t)) \leq \mu_1$. In particular, $f \in L_\infty((0,T_0), X_1)$.

Consider next $T \in (0,T_0)$ and $1 < r < R$. On the one hand, by (8.2.36),

$$\int_0^T \int_r^R x^{m_0} a_j(x) f_j(t,x) \, \mathrm{d}x\mathrm{d}t \leq \omega(r,T) , \qquad j \geq 1 . \qquad (8.2.38)$$

On the other hand, the sequence $(a_j \mathbf{1}_{(r,R)})_{j \geq 1}$ is bounded from above by A_R according to (8.2.32) and is nonnegative. In addition, it converges almost everywhere in $(0, T) \times (0, \infty)$ to $a\mathbf{1}_{(r,R)}$ by (8.2.30). Owing to the convergence (8.2.35), we are in a position to apply Proposition 7.1.12 and conclude that $(a_j f_j \mathbf{1}_{(r,R)})_{j \geq 1}$ converges weakly to $af\mathbf{1}_{(r,R)}$ in X_{m_0} as $j \to \infty$. We then let $j \to \infty$ in (8.2.38) to obtain that

$$\int_0^T \int_r^R x^{m_0} a(x) f(t,x) \, \mathrm{d}x \mathrm{d}t \leq \omega(r,T) \ .$$

We use again Fatou's lemma to let $R \to \infty$ in the previous inequality and end up with

$$\int_0^T \int_r^\infty x^{m_0} a(x) f(t,x) \, \mathrm{d}x \mathrm{d}t \leq \omega(r,T) \ . \tag{8.2.39}$$

Consider again $T \in (0, T_0)$ and $1 < r < R$. By (8.2.36),

$$\int_0^T \int_r^R \int_{1/R}^R k_j(x,y) f_j(t,x) f_j(t,y) \, \mathrm{d}y \mathrm{d}x \mathrm{d}t \leq \omega(r,T) \ , \qquad j \geq 1 \ . \tag{8.2.40}$$

Since

$$\frac{k_j(x,y) \mathbf{1}_{(r,R) \times (1/R,R)}(x,y)}{(xy)^{m_0}} \leq K_R \left(\frac{R}{r} \right)^{m_0} \ , \qquad j \geq 1 \ ,$$

by (8.2.31) and

$$\lim_{j \to \infty} \frac{k_j(x,y) \mathbf{1}_{(r,R) \times (1/R,R)}(x,y)}{(xy)^{m_0}} = \frac{k(x,y) \mathbf{1}_{(r,R) \times (1/R,R)}(x,y)}{(xy)^{m_0}}$$

for almost every $(x,y) \in (0, \infty)^2$ by (8.2.30), it follows from (8.2.35) and Proposition 7.1.12 that

$$\lim_{j \to \infty} \int_0^T \int_r^R \int_{1/R}^R k_j(x,y) f_j(t,x) f_j(t,y) \, \mathrm{d}y \mathrm{d}x \mathrm{d}t$$
$$= \int_0^T \int_r^R \int_{1/R}^R k(x,y) f(t,x) f(t,y) \, \mathrm{d}y \mathrm{d}x \mathrm{d}t \ .$$

Therefore, letting $j \to \infty$ in (8.2.40) leads us to

$$\int_0^T \int_r^R \int_{1/R}^R k(x,y) f(t,x) f(t,y) \, \mathrm{d}y \mathrm{d}x \mathrm{d}t \leq \omega(r,T)$$

and another use of Fatou's lemma gives

$$\int_0^T \int_r^\infty \int_0^\infty k(x,y) f(t,x) f(t,y) \, \mathrm{d}y \mathrm{d}x \mathrm{d}t \leq \omega(r,T) \ , \tag{8.2.41}$$

after letting $R \to \infty$. Similarly, (8.2.35) and (8.2.36) imply that

$$\int_0^T \int_0^{1/r} \int_0^\infty x^{m_0} k(x,y) f(t,x) f(t,y) \, \mathrm{d}y \mathrm{d}x \mathrm{d}t \leq \omega(r,T) \ . \tag{8.2.42}$$

After this preparation, we are ready to take the limit as $j \to \infty$ of the weak formulation of the C-F equation satisfied by f_j. We denote the coagulation and fragmentation terms associated to the coefficients (k_j, a_j, b) by $\mathcal{C}_j$ and $\mathcal{F}_j$, respectively. Let $T \in (0, T_0)$ and $\vartheta \in \Theta^{m_0}$. For $r > 1$ and $j \geq 1$,

$$2\left| \int_0^T \int_0^\infty [\mathcal{C}_j f_j - \mathcal{C}f](t,x)\vartheta(x) \, dxdt \right|$$

$$= \left| \int_0^T \int_0^\infty \int_0^\infty \chi_\vartheta(x,y) \left(k_j(x,y) f_j(t,x) f_j(t,y) - k(x,y) f(t,x) f(t,y) \right) \, dxdydt \right|$$

$$\leq P_1(r,j) + 2P_2(r,j) + 2P_3(r,j) \, ,$$

where

$$P_1(r,j) := \left| \int_0^T \int_{1/r}^r \int_{1/r}^r \chi_\vartheta(x,y) \left(k_j(x,y) f_j(t,x) f_j(t,y) - k(x,y) f(t,x) f(t,y) \right) \, dxdydt \right| \, ,$$

$$P_2(r,j) := \int_0^T \int_r^\infty \int_0^\infty |\chi_\vartheta(x,y)| \left(k_j(x,y) f_j(t,x) f_j(t,y) + k(x,y) f(t,x) f(t,y) \right) \, dxdydt \, ,$$

$$P_3(r,j) := \int_0^T \int_0^{1/r} \int_0^\infty |\chi_\vartheta(x,y)| \left(k_j(x,y) f_j(t,x) f_j(t,y) + k(x,y) f(t,x) f(t,y) \right) \, dxdydt \, .$$

Since $(k_j \mathbf{1}_{(1/r,r) \times (1/r,r)})_{j \geq 1}$ is bounded by (8.2.31) and converges almost everywhere in $(0,T) \times (0,\infty)^2$ to $k \mathbf{1}_{(1/r,r) \times (1/r,r)}$ by (8.2.30), it follows from (8.2.35) that we may apply Proposition 7.1.12 to pass to the limit as $j \to \infty$ in $P_1(r,j)$ and deduce that

$$\lim_{j \to \infty} P_1(r,j) = 0 \, .$$

Next, recalling the definition (8.2.36), we infer from (8.2.41) that

$$P_2(r,j) \leq 3\|\vartheta\|_\infty \int_0^T \int_r^\infty \int_0^\infty \left(k_j(x,y) f_j(t,x) f_j(t,y) + k(x,y) f(t,x) f(t,y) \right) \, dxdydt \, ,$$

$$\leq 6\|\vartheta\|_{C^{0,m_0}} \omega(r,T) \, ,$$

while (8.2.42) and Lemma 8.2.4 ensure that

$$P_3(r,j) \leq 2\|\vartheta\|_{C^{0,m_0}} \int_0^T \int_0^{1/r} \int_0^\infty (\min\{x,y\})^{m_0} k_j(x,y) f_j(t,x) f_j(t,y) \, dxdydt$$

$$+ 2\|\vartheta\|_{C^{0,m_0}} \int_0^T \int_0^{1/r} \int_0^\infty (\min\{x,y\})^{m_0} k(x,y) f(t,x) f(t,y) \, dxdydt$$

$$\leq 4\|\vartheta\|_{C^{0,m_0}} \omega(r,T) \, .$$

Gathering the previously obtained information leads us to

$$\limsup_{j \to \infty} \left| \int_0^T \int_0^\infty [\mathcal{C}_j f_j - \mathcal{C}f](t,x)\vartheta(x) \, dxdt \right| \leq 10\|\vartheta\|_{C^{0,m_0}} \omega(r,T) \, .$$

We then let $r \to \infty$ and deduce from (8.2.37) that

$$\lim_{j \to \infty} \int_0^T \int_0^\infty \mathcal{C}_j f_j(t,x)\vartheta(x) \, dxdt = \int_0^T \int_0^\infty \mathcal{C}f(t,x)\vartheta(x) \, dxdt \, . \tag{8.2.43}$$

Next, for $r > 1$ and $j \geq 1$,

$$\left| \int_0^T \int_0^\infty (\mathscr{F}_j f_j - \mathscr{F} f)(t, x) \vartheta(x) \, \mathrm{d}x \mathrm{d}t \right|$$

$$= \left| \int_0^T \int_0^\infty N_\vartheta(x) \left(a_j(x) f_j(t, x) - a(x) f(t, x) \right) \, \mathrm{d}x \mathrm{d}t \right|$$

$$\leq \left| \int_0^T \int_0^r N_\vartheta(x) \left(a_j(x) f_j(t, x) - a(x) f(t, x) \right) \, \mathrm{d}x \mathrm{d}t \right|$$

$$+ \int_0^T \int_r^\infty |N_\vartheta(x)| \left(a_j(x) f_j(t, x) + a(x) f(t, x) \right) \, \mathrm{d}x \mathrm{d}t \ .$$

Owing to (8.2.33) and Lemma 8.2.4,

$$|N_\vartheta(x)| \leq \|\vartheta\|_{C^{0,m_0}} (x^{m_0} + n_{m_0}(x)) \leq \|\vartheta\|_{C^{0,m_0}} (1 + \kappa_{m_0,1}) x^{m_0} \ , \qquad x \in (0, \infty) \ . \quad (8.2.44)$$

It then follows from (8.2.36), (8.2.39), and (8.2.44) that, for any $r > 1$,

$$\left| \int_0^T \int_0^\infty (\mathscr{F}_j f_j - \mathscr{F} f)(t, x) \vartheta(x) \, \mathrm{d}x \mathrm{d}t \right| \leq \left| \int_0^T \int_0^r N_\vartheta(x) \left(a_j(x) f_j(t, x) - a(x) f(t, x) \right) \, \mathrm{d}x \mathrm{d}t \right|$$

$$+ 2(1 + \kappa_{m_0,1}) \|\vartheta\|_{C^{0,m_0}} \omega(r, T) \ . \quad (8.2.45)$$

According to (8.2.30), (8.2.32), and (8.2.44),

$$x^{-m_0} N_\vartheta(x) a_j(x) \mathbf{1}_{(0,r)}(x) \leq A_r (1 + \kappa_{m_0,1}) \|\vartheta\|_{C^{0,m_0}} \ , \qquad j \geq 1 \ ,$$

and

$$\lim_{j \to \infty} x^{-m_0} N_\vartheta(x) a_j(x) \mathbf{1}_{(0,r)}(x) = x^{-m_0} N_\vartheta(x) a(x) \mathbf{1}_{(0,r)}(x)$$

for almost every $x \in (0, \infty)$. We then deduce from (8.2.35) and Proposition 7.1.12 that the first term of the right-hand side of (8.2.45) vanishes as $j \to \infty$. Therefore,

$$\limsup_{j \to \infty} \left| \int_0^T \int_0^\infty (\mathscr{F}_j f_j - \mathscr{F} f)(t, x) \vartheta(x) \, \mathrm{d}x \mathrm{d}t \right| \leq 2(1 + \kappa_{m_0,1}) \|\vartheta\|_{C^{0,m_0}} \omega(r, T) \ .$$

Since $r > 1$ is arbitrary, we can let $r \to \infty$ and use (8.2.37) to conclude that

$$\lim_{j \to \infty} \int_0^T \int_0^\infty \mathscr{F}_j f_j(t, x) \vartheta(x) \, \mathrm{d}x \mathrm{d}t = \int_0^T \int_0^\infty \mathscr{F} f(t, x) \vartheta(x) \, \mathrm{d}x \mathrm{d}t \ . \quad (8.2.46)$$

Owing to (8.2.35), (8.2.43), and (8.2.46), we are in a position to pass to the limit as $j \to \infty$ in the weak formulation (8.2.4d) of the C-F equation satisfied by f_j (with coefficients (k_j, a_j, b)) and obtain that f is a weak solution on $[0, T_0)$ to the C-F equation (8.0.1) with coefficients (k, a, b). $\qquad \square$

8.2.1.3 Lower Bounds for Coagulation

We begin by establishing a lower bound on the contribution made by coagulation to small sizes in moments of order less than one.

Lemma 8.2.12. *Assume that there is $\lambda \in \mathbb{R}$, and $K_1 > 0$ such that the coagulation kernel satisfies*

$$k(x,y) \geq K_1(xy)^{\lambda/2} , \qquad (x,y) \in (0,1)^2 . \tag{8.2.47}$$

For any $m \in (0,1)$ and $\mu > (m + \lambda)/2$, there exists $C(\lambda, m, \mu) > 0$ such that

$$\left(\int_0^1 x^\mu g(x) \, \mathrm{d}x \right)^2 \leq \frac{C^2}{2K_1} \int_0^1 \int_0^1 k(x,y) \left[x^m + y^m - (x+y)^m \right] g(x)g(y) \, \mathrm{d}y\mathrm{d}x$$

for all nonnegative functions $g \in L_1((0,1), x^m \mathrm{d}x)$.

Proof. We follow the proof of Step 5 in [118, Lemma 3.1] and assume without loss of generality that

$$J := \frac{1}{2} \int_0^1 \int_0^1 k(x,y) \left[x^m + y^m - (x+y)^m \right] g(x)g(y) \, \mathrm{d}y\mathrm{d}x$$

is finite. Since

$$x^m + y^m - (x+y)^m = x \left[x^{m-1} - (x+y)^{m-1} \right] + y \left[y^{m-1} - (x+y)^{m-1} \right] \tag{8.2.48}$$

for $(x,y) \in (0,1)^2$, the integral term J can be written as

$$J = \int_0^1 \int_0^1 xk(x,y) \left[x^{m-1} - (x+y)^{m-1} \right] g(x)g(y) \, \mathrm{d}y\mathrm{d}x .$$

We next use (8.2.47) and the inequality

$$x^{m-1} - (x+y)^{m-1} = (1-m) \int_x^{x+y} z^{m-2} \, \mathrm{d}z \geq (1-m)y(x+y)^{m-2} \tag{8.2.49}$$

for $(x,y) \in (0,1)^2$ to estimate J from below and obtain

$$J \geq (1-m)K_1 \int_0^1 \int_0^1 (x+y)^{m-2}(xy)^{(2+\lambda)/2} g(x)g(y) \, \mathrm{d}y\mathrm{d}x .$$

To proceed further, set $\zeta := 2/(2\mu - m - \lambda) > 0$ and define

$$x_i := i^{-\zeta} \quad \text{and} \quad J_i := \int_{x_{i+1}}^{x_i} x^{(2+\lambda)/2} g(x) \, \mathrm{d}x \quad \text{for} \quad i \geq 1 .$$

Then $(0,1) = \bigcup (x_{i+1}, x_i)$ and

$$J \geq (1-m)K_1 \sum_{i=1}^\infty \int_{x_{i+1}}^{x_i} \int_{x_{i+1}}^{x_i} (x+y)^{m-2}(xy)^{(2+\lambda)/2} g(x)g(y) \, \mathrm{d}y\mathrm{d}x$$

$$\geq (1-m)2^{m-2}K_1 \sum_{i=1}^\infty x_i^{m-2} J_i^2 . \tag{8.2.50}$$

Case 1: $\mu \leq (2+\lambda)/2$. We deduce from the Cauchy–Schwarz inequality that

$$\int_0^1 x^\mu g(x) \, \mathrm{d}x = \sum_{i=1}^\infty \int_{x_{i+1}}^{x_i} x^\mu g(x) \, \mathrm{d}x \leq \sum_{i=1}^\infty x_{i+1}^{(2\mu-2-\lambda)/2} J_i$$

$$\le \left(\sum_{i=1}^{\infty} x_{i+1}^{2\mu-2-\lambda} x_i^{2-m} \right)^{1/2} \left(\sum_{i=1}^{\infty} x_i^{m-2} J_i^2 \right)^{1/2} ,$$

hence

$$\left(\int_0^1 x^\mu g(x) \, dx \right)^2 \le \left(\sum_{i=1}^{\infty} x_{i+1}^{2\mu-2-\lambda} x_i^{2-m} \right) \left(\sum_{i=1}^{\infty} x_i^{m-2} J_i^2 \right) . \tag{8.2.51}$$

Thanks to the choice $\zeta = 2/(2\mu - m - \lambda)$, we have

$$x_{i+1}^{2\mu-2-\lambda} x_i^{2-m} \le \frac{2^{\zeta(\lambda+2-2\mu)}}{i^2} , \qquad i \ge 1 ,$$

so that the first sum in the right-hand side of (8.2.51) is finite. Lemma 8.2.12 then readily follows from (8.2.50) and (8.2.51).

Case 2: $(2+\lambda)/2 < \mu$. Using again the Cauchy–Schwarz inequality we obtain

$$\int_0^1 x^\mu g(x) \, dx = \sum_{i=1}^{\infty} \int_{x_{i+1}}^{x_i} x^\mu g(x) \, dx \le \sum_{i=1}^{\infty} x_i^{(2\mu-2-\lambda)/2} J_i$$

$$\le \left(\sum_{i=1}^{\infty} \frac{1}{i^2} \right)^{1/2} \left(\sum_{i=1}^{\infty} x_i^{m-2} J_i^2 \right)^{1/2} .$$

Combining the above inequality and (8.2.50) completes the proof. □

Remark 8.2.13. *If $\lambda < 1$ and $m \in (\lambda, 1)$, then it is possible to take $\mu = m$ in Lemma 8.2.12.*

For superlinear coagulation kernels we next derive a lower bound for the contribution of the coagulation term to moments of order close to one, but now for large sizes. Such estimates were first observed in [112, Theorem 2.2], the proof given below being slightly different.

Lemma 8.2.14. *Assume that there is $\lambda \ge 1$, and $K_1 > 0$ such that the coagulation kernel satisfies*

$$k(x,y) \ge K_1 (xy)^{\lambda/2} , \qquad (x,y) \in (1,\infty)^2 . \tag{8.2.52}$$

For any $m \in (0,1)$ and $\mu < (m+\lambda)/2$, there exists $C(\lambda, m, \mu) > 0$ such that

$$\left(\int_1^\infty x^\mu g(x) \, dx \right)^2$$

$$\le \frac{C^2}{2K_1} \int_1^\infty \int_1^\infty k(x,y) \left[x^m + y^m - (x+y)^m \right] g(x)g(y) \, dy dx$$

for all nonnegative functions $g \in L_1((1,\infty), x^\mu dx)$.

Proof. As in the proof of Lemma 8.2.12 we deduce from (8.2.48), (8.2.49), and (8.2.52) that

$$J := \frac{1}{2} \int_1^\infty \int_1^\infty k(x,y) \left[x^m + y^m - (x+y)^m \right] g(x)g(y) \, dy dx$$

$$= \int_1^\infty \int_1^\infty x k(x,y) \left[x^{m-1} - (x+y)^{m-1} \right] g(x)g(y) \, dy dx$$

$$\ge (1-m) K_1 \int_1^\infty \int_1^\infty (x+y)^{m-2} (xy)^{(2+\lambda)/2} g(x)g(y) \, dy dx .$$

Introducing

$$x_i := i^\zeta \quad \text{and} \quad J_i := \int_{x_i}^{x_{i+1}} x^{(2+\lambda)/2} g(x) \, \mathrm{d}x \quad \text{for} \quad i \geq 1 \, ,$$

with $\zeta := 2/(\lambda + m - 2\mu) > 0$, we note that $(1, \infty) = \bigcup (x_i, x_{i+1})$ and

$$J \geq (1-m) K_1 \sum_{i=1}^{\infty} \int_{x_i}^{x_{i+1}} \int_{x_i}^{x_{i+1}} (x+y)^{m-2} (xy)^{(2+\lambda)/2} g(x)g(y) \, \mathrm{d}y \mathrm{d}x$$

$$\geq (1-m) 2^{m-2} K_1 \sum_{i=1}^{\infty} x_{i+1}^{m-2} J_i^2 \, . \tag{8.2.53}$$

We next infer from (8.2.53) and the Cauchy–Schwarz inequality that

$$\int_1^\infty x^\mu g(x) \, \mathrm{d}x = \sum_{i=1}^{\infty} \int_{x_i}^{x_{i+1}} x^\mu g(x) \, \mathrm{d}x \leq \sum_{i=1}^{\infty} x_i^{(2\mu-2-\lambda)/2} J_i$$

$$\leq \left(\sum_{i=1}^{\infty} x_i^{2\mu-2-\lambda} x_{i+1}^{2-m} \right)^{1/2} \left(\sum_{i=1}^{\infty} x_{i+1}^{m-2} J_i^2 \right)^{1/2}$$

$$\leq 2^{\zeta(2-m)/2} \left(\sum_{i=1}^{\infty} \frac{1}{i^2} \right)^{1/2} \left(\frac{2^{2-m} J}{(1-m) K_1} \right)^{1/2} \, ,$$

and the proof of Lemma 8.2.14 is complete. □

8.2.1.4 Upper Bounds for Coagulation: Moment Estimates

We begin with an estimate of the contribution of the coagulation term to the growth of the moment of order m when the coagulation kernel grows algebraically.

Lemma 8.2.15. *Assume that there is $\lambda \geq 0$, $\alpha \in [\lambda/2, \lambda]$, and $K_1 > 0$ such that*

$$k(x,y) \leq K_1 (x^\alpha y^{\lambda-\alpha} + x^{\lambda-\alpha} y^\alpha) \, , \qquad (x,y) \in (0,\infty)^2 \, . \tag{8.2.54}$$

For $m \geq 2 + \alpha - \lambda$, there is $C(m) > 0$ such that

$$\frac{1}{2} \int_0^\infty \int_0^\infty k(x,y) \chi_m(x,y) g(x)g(y) \, \mathrm{d}y \mathrm{d}x \leq C(m) K_1 M_1(g) M_{m+\lambda-1}(g)$$

for all functions $g \in X_{1,+} \cap X_{m+\lambda-1}$, the function χ_m being defined in (8.2.2).

Proof. It follows from (8.2.54) and Lemma 7.4.2 that

$$J := \frac{1}{2} \int_0^\infty \int_0^\infty k(x,y) \chi_m(x,y) g(x)g(y) \, \mathrm{d}y \mathrm{d}x$$

$$\leq K_1 C_m \int_0^\infty \int_0^\infty x^\alpha y^{\lambda-\alpha} (xy^{m-1} + x^{m-1}y) g(x)g(y) \, \mathrm{d}y \mathrm{d}x$$

$$\leq K_1 C_m \left[M_{1+\alpha}(g) M_{m+\lambda-1-\alpha}(g) + M_{m+\alpha-1}(g) M_{1+\lambda-\alpha}(g) \right] \, .$$

Since $m \geq 2 + \alpha - \lambda \geq 2 - \alpha$, it follows from Hölder's inequality that

$$M_{1+\alpha}(g) \leq M_{m+\lambda-1}(g)^{\alpha/(m+\lambda-2)} M_1(g)^{(m+\lambda-\alpha-2)/(m+\lambda-2)} \, ,$$

$$M_{m+\lambda-1-\alpha}(g) \leq M_{m+\lambda-1}(g)^{(m+\lambda-\alpha-2)/(m+\lambda-2)} M_1(g)^{\alpha/(m+\lambda-2)} \, ,$$

$$M_{m+\alpha-1}(g) \leq M_{m+\lambda-1}(g)^{(m+\alpha-2)/(m+\lambda-2)} M_1(g)^{(\lambda-\alpha)/(m+\lambda-2)} ,$$
$$M_{1+\lambda-\alpha}(g) \leq M_{m+\lambda-1}(g)^{(\lambda-\alpha)/(m+\lambda-2)} M_1(g)^{(m+\alpha-2)/(m+\lambda-2)} .$$

Consequently, $J \leq 2K_1 C_m M_1(g) M_{m+\lambda-1}(g)$ and the proof of Lemma 8.2.15 is complete.
□

We next turn to sublinear coagulation kernels and uncover an interesting property of the contribution of the coagulation term to the growth of generalised moments associated with functions in C_{VP}, the latter set being defined in Definition 7.1.8 [159, 167].

Lemma 8.2.16. *Assume that the coagulation kernel k is sublinear; that is, there is $K_1 > 0$ such that*

$$k(x,y) \leq K_1(1+x+y) , \qquad (x,y) \in (0,\infty)^2 . \tag{8.2.55}$$

If $\psi \in C_{VP}$ then

$$\frac{1}{2} \int_0^\infty \int_0^\infty k(x,y) \chi_\psi(x,y) g(x) g(y) \, dydx$$
$$\leq 2K_1 M_1(g) \left(\psi''(0) M_1(g) + \int_0^\infty \psi(x) g(x) \, dx \right)$$

for all functions $g \in X_{1,+} \cap L_1((0,\infty), \psi(x)dx)$.

Proof. Owing to the concavity of ψ',

$$\chi_\psi(x,y) = \int_0^x \int_0^y \psi''(x_* + y_*) \, dy_* dx_* \leq \psi''(0) xy , \qquad (x,y) \in (0,\infty)^2 ,$$

while Proposition 7.1.9 (e) asserts that

$$(x+y)\chi_\psi(x,y) \leq 2 \left(x\psi(y) + y\psi(x) \right) , \qquad (x,y) \in (0,\infty)^2 .$$

Consequently, it follows from (8.2.55) that

$$\frac{1}{2} \int_0^\infty \int_0^\infty k(x,y) \chi_\psi(x,y) g(x) g(y) \, dydx$$
$$\leq \frac{K_1}{2} \int_0^\infty \int_0^\infty \left(\chi_\psi(x,y) + (x+y)\chi_\psi(x,y) \right) g(x) g(y) \, dydx$$
$$\leq K_1 \int_0^\infty \int_0^\infty \left(\psi''(0) xy + x\psi(y) + y\psi(x) \right) g(x) g(y) \, dydx$$
$$\leq K_1 \psi''(0) M_1^2(g) + 2K_1 M_1(g) \int_0^\infty \psi(x) g(x) \, dx ,$$

as claimed.
□

8.2.1.5 Upper Bounds for Coagulation: L_p-estimates

We next turn to different estimates which provide a control on L_p-norms and involve monotonicity assumptions on the coagulation kernel instead of growth conditions such as (8.2.54) or (8.2.55). The importance of monotonicity conditions to the derivation of estimates on L_p-norms is noticed in [54, 103] and the cornerstone of the analysis is carried out in [165, 169, 203].

Throughout this section, Φ is a nonnegative and nondecreasing convex function in $C^1([0,\infty))$ and we set

$$\Psi(r) := r\Phi'(r) - \Phi(r) , \qquad r \geq 0 .$$

Lemma 8.2.17. *Assume that there is a subadditive function $\ell : (0, \infty) \to (0, \infty)$ such that the coagulation kernel k satisfies the monotonicity condition*

$$\ell(x - y)k(x - y, y) \leq \ell(x)k(x, y) , \qquad 0 < y < x . \tag{8.2.56}$$

If g is a nonnegative measurable function on $(0, \infty)$ such that $(x, y) \mapsto \ell(x)k(x, y)\Psi(g(x))g(y)$ belongs to $L_1((0, \infty)^2)$, then

$$\int_0^\infty \ell(x)\Phi'(g(x))\mathcal{C}g(x) \, dx \leq - \int_0^\infty \int_x^\infty \ell(x)k(x, y)\Psi(g(x))g(y) \, dydx .$$

Proof. It follows from the convexity of Φ that, for $(r, s) \in (0, \infty)^2$,

$$r(\Phi'(s) - \Phi'(r)) = (r - s)\Phi'(s) - r\Phi'(r) + s\Phi'(s)$$
$$\leq \Phi(r) - \Phi(s) - r\Phi'(r) + s\Phi'(s) ,$$

so that

$$r(\Phi'(s) - \Phi'(r)) \leq \Psi(s) - \Psi(r) , \qquad (r, s) \in (0, \infty)^2 . \tag{8.2.57}$$

Owing to (8.2.57) and the subadditivity of ℓ, which means that $\ell(x + y) \leq \ell(x) + \ell(y)$ for $(x, y) \in (0, \infty)^2$, we obtain

$$[\ell(x + y)\Phi'(g(x + y)) - \ell(x)\Phi'(g(x)) - \ell(y)\Phi'(g(y))] \, g(x)g(y)$$
$$\leq \ell(x) [\Phi'(g(x + y)) - \Phi'(g(x))] \, g(x)g(y) + \ell(y) [\Phi'(g(x + y)) - \Phi'(g(y))] \, g(y)g(x)$$
$$\leq \ell(x) [\Psi(g(x + y)) - \Psi(g(x))] \, g(y) + \ell(y) [\Psi(g(x + y)) - \Psi(g(y))] \, g(x) .$$

Consequently,

$$J := \int_0^\infty \ell(x)\Phi'(g(x))\mathcal{C}g(x) \, dx$$
$$= \frac{1}{2} \int_0^\infty \int_0^\infty k(x, y)\ell(x + y)\Phi'(g(x + y))g(x)g(y) \, dxdy$$
$$- \frac{1}{2} \int_0^\infty \int_0^\infty k(x, y) [\ell(x)\Phi'(g(x)) + \ell(y)\Phi'(g(y))] \, g(x)g(y) \, dxdy$$
$$\leq \int_0^\infty \int_0^\infty \ell(x)k(x, y) [\Psi(g(x + y)) - \Psi(g(x))] \, g(y) \, dxdy$$
$$= \int_0^\infty \int_y^\infty \ell(x - y)k(x - y, y)\Psi(g(x))g(y) \, dxdy$$
$$- \int_0^\infty \int_0^\infty \ell(x)k(x, y)\Psi(g(x))g(y) \, dxdy .$$

We then deduce from the monotonicity assumption (8.2.56) that

$$J \leq - \int_0^\infty \int_0^y \ell(x)k(x, y)\Psi(g(x))g(y) \, dxdy$$
$$= - \int_0^\infty \int_x^\infty \ell(x)k(x, y)\Psi(g(x))g(y) \, dydx ,$$

which is the stated inequality. $\qquad \square$

Two variants of Lemma 8.2.17 are available when ℓ is either constant or the identity. Let us begin with the case $\ell \equiv 1$ and report the following result which provides an improved estimate [165, Lemma 3.5].

Lemma 8.2.18. *Assume that the coagulation kernel k satisfies the monotonicity condition (8.2.56) with $\ell \equiv 1$. If g is a nonnegative measurable function on $(0, \infty)$ such that $(x, y) \mapsto k(x, y)\Psi(g(x))g(y)$ belongs to $L_1((0, \infty)^2)$, then*

$$\int_0^\infty \Phi'(g(x))\mathcal{C}g(x) \, dx \leq -\frac{1}{2}\int_0^\infty \int_x^\infty k(x, y)\Psi(g(x))g(y) \, dy dx$$
$$-\frac{1}{2}\int_0^\infty \int_0^\infty k(x, y)\Phi'(g(x))g(x)g(y) \, dy dx \ .$$

Proof. It follows from (8.2.57) that

$$J := \int_0^\infty \Phi'(g(x))\mathcal{C}g(x) \, dx$$
$$= \frac{1}{2}\int_0^\infty \int_0^\infty k(x, y)\Phi'(g(x+y))g(x)g(y) \, dx dy$$
$$- \int_0^\infty \int_0^\infty k(x, y)\Phi'(g(x))g(x)g(y) \, dx dy$$
$$\leq \frac{1}{2}\int_0^\infty \int_0^\infty k(x, y)\left[\Psi(g(x+y)) - \Psi(g(x))\right]g(y) \, dx dy$$
$$- \frac{1}{2}\int_0^\infty \int_0^\infty k(x, y)\Phi'(g(x))g(x)g(y) \, dx dy$$
$$= \frac{1}{2}\int_0^\infty \int_y^\infty k(x-y, y)\Psi(g(x))g(y) \, dx dy$$
$$- \frac{1}{2}\int_0^\infty \int_0^\infty k(x, y)\left[\Psi(g(x)) + \Phi'(g(x))g(x)\right]g(y) \, dx dy \ .$$

We now use the monotonicity condition (8.2.56) (with $\ell \equiv 1$) to complete the proof of Lemma 8.2.18. $\qquad\square$

When k satisfies (8.2.56) with $\ell = $ id, a weaker version of Lemma 8.2.18 is available [203].

Lemma 8.2.19. *Assume that the coagulation kernel satisfies the monotonicity condition (8.2.56) with $\ell = $ id. If g is a nonnegative measurable function on $(0, \infty)$ such that $(x, y) \mapsto k(x, y)\Phi(g(x))g(y)$ belongs to $L_1((0, \infty)^2)$, then*

$$\int_0^\infty \Phi'(g(x))\mathcal{C}g(x) \, dx$$
$$\leq -\int_0^\infty \int_0^\infty k(x, y)\Phi(g(x))g(y)\mathrm{sign}_+(g(y) - g(x)) \, dy dx \ ,$$

where $\mathrm{sign}_+(r) = 1$ for $r > 0$ and $\mathrm{sign}_+(r) = 0$ for $r \leq 0$.

Proof. We first observe that the monotonicity condition (8.2.56) with $\ell = $ id ensures that

$$\begin{cases} xk(x, y) \leq (x+y)k(y, x+y) \\[2mm] yk(x, y) \leq (x+y)k(x, x+y) \ , \end{cases} \qquad (x, y) \in (0, \infty)^2 \ .$$

Summing the above inequalities and dividing by $x + y$ lead us to

$$k(x, y) \leq k(x, x+y) + k(y, x+y) \ , \qquad (x, y) \in (0, \infty)^2 \ . \qquad (8.2.58)$$

We next split $(0, \infty)^2$ into three sets defined by

$$\mathcal{T}_1 := \left\{ (x, y) \in (0, \infty)^2 \ : \ g(x) > g(y) \right\} ,$$
$$\mathcal{D} := \left\{ (x, y) \in (0, \infty)^2 \ : \ g(x) = g(y) \right\} ,$$
$$\mathcal{T}_2 := \left\{ (x, y) \in (0, \infty)^2 \ : \ g(x) < g(y) \right\} ,$$

and note that

$$(x, y) \in \mathcal{T}_2 \quad \text{if and only if} \quad (y, x) \in \mathcal{T}_1 \ . \tag{8.2.59}$$

This property, along with Fubini's theorem, the convexity of Φ, and (8.2.59), gives

$$
\begin{aligned}
J_1 &:= \frac{1}{2} \int_0^\infty \Phi'(g(x)) \int_0^x k(x - y, y) g(x - y) g(y) \, \mathrm{d}y \mathrm{d}x \\
&= \frac{1}{2} \int_0^\infty \int_0^\infty k(x, y) \Phi'(g(x + y)) g(x) g(y) \, \mathrm{d}y \mathrm{d}x \\
&= \frac{1}{2} \int_{\mathcal{T}_1} k(x, y) \Phi'(g(x + y)) g(x) g(y) \, \mathrm{d}y \mathrm{d}x \\
&\quad + \frac{1}{2} \int_{\mathcal{D}} k(x, y) \Phi'(g(x + y)) g(x) g(y) \, \mathrm{d}y \mathrm{d}x \\
&\quad + \frac{1}{2} \int_{\mathcal{T}_2} k(x, y) \Phi'(g(x + y)) g(x) g(y) \, \mathrm{d}y \mathrm{d}x \\
&\leq \int_{\mathcal{T}_1} k(x, y) \Phi'(g(x + y)) g(x) g(y) \, \mathrm{d}y \mathrm{d}x \\
&\quad + \frac{1}{2} \int_{\mathcal{D}} k(x, y) \Phi'(g(x + y)) g(x) g(y) \, \mathrm{d}y \mathrm{d}x \\
&\leq \int_{\mathcal{T}_1} k(x, y) \left[\Psi(g(x + y)) + \Phi(g(x)) \right] g(y) \, \mathrm{d}y \mathrm{d}x \\
&\quad + \frac{1}{2} \int_{\mathcal{D}} k(x, y) \left[\Psi(g(x + y)) + \Phi(g(x)) \right] g(y) \, \mathrm{d}y \mathrm{d}x \ .
\end{aligned}
$$

We now infer from (8.2.58) and (8.2.59) that

$$
\begin{aligned}
\int_{\mathcal{T}_1} k(x, y) \Psi(g(x + y)) g(y) \, \mathrm{d}y \mathrm{d}x &\leq \int_{\mathcal{T}_1} k(x, x + y) \Psi(g(x + y)) g(y) \, \mathrm{d}y \mathrm{d}x \\
&\quad + \int_{\mathcal{T}_1} k(y, x + y) \Psi(g(x + y)) g(y) \, \mathrm{d}y \mathrm{d}x \\
&\leq \int_{\mathcal{T}_1 \cup \mathcal{T}_2} k(x, x + y) \Psi(g(x + y)) \min\{g(x), g(y)\} \, \mathrm{d}y \mathrm{d}x \ ,
\end{aligned}
$$

and, similarly,

$$
\int_{\mathcal{D}} k(x, y) \Psi(g(x + y)) g(y) \, \mathrm{d}y \mathrm{d}x \leq 2 \int_{\mathcal{D}} k(x, x + y) \Psi(g(x + y)) g(y) \, \mathrm{d}y \mathrm{d}x \ .
$$

Combining the above estimates, we end up with

$$
\begin{aligned}
J_1 &\leq \int_0^\infty \int_0^\infty k(x, x + y) \Psi(g(x + y)) \min\{g(x), g(y)\} \, \mathrm{d}y \mathrm{d}x \\
&\quad + \int_{\mathcal{T}_1 \cup \mathcal{D}} k(x, y) \Phi(g(x)) g(y) \, \mathrm{d}y \mathrm{d}x \ .
\end{aligned}
$$

Also, owing to the symmetry of k,

$$J_2 := \int_0^\infty \int_0^\infty k(x,y)\Phi'(g(x))g(x)g(y) \ \mathrm{d}y\mathrm{d}x$$

$$= \int_0^\infty \int_0^\infty k(x,y) \left[\Psi(g(x)) + \Phi(g(x)) \right] g(y) \ \mathrm{d}y\mathrm{d}x$$

$$= \int_0^\infty \int_0^\infty k(x,y) \left[\Psi(g(y))g(x) + \Phi(g(x))g(y) \right] \ \mathrm{d}y\mathrm{d}x \ .$$

Finally, using the previous inequalities, we obtain

$$J := \int_0^\infty \Phi'(g(x))\mathcal{C}g(x) \ \mathrm{d}x = J_1 - J_2$$

$$\leq \int_0^\infty \int_x^\infty k(x,y)\Psi(g(y)) \min\{g(x), g(y-x)\} \ \mathrm{d}y\mathrm{d}x$$

$$+ \int_{\mathcal{I}_1 \cup \mathcal{D}} k(x,y)\Phi(g(x))g(y) \ \mathrm{d}y\mathrm{d}x$$

$$- \int_0^\infty \int_0^\infty k(x,y) \left[\Phi(g(x))g(y) + \Psi(g(y))g(x) \right] \ \mathrm{d}y\mathrm{d}x$$

$$\leq - \int_{\mathcal{I}_2} k(x,y)\Phi(g(x))g(y) \ \mathrm{d}y\mathrm{d}x \ ,$$

which completes the proof. □

8.2.1.6 Positivity

From a physical point of view, the coagulation mechanism creates particles with larger sizes than those already in the system and thus expands the support of the distribution function f to the right. On the contrary, particles of smaller sizes are formed during breakup events and so fragmentation expands the support to the left. As we shall see below, this expansion takes place at an infinite speed if both the coagulation kernel k and the overall fragmentation rate a are positive. This issue is investigated in [66, 89] for the discrete coagulation equation and in [197] and [113, Lemma 6.1] for the continuous C-F equation.

Proposition 8.2.20. *Consider* $f^{in} \in X_{0,1,+}$ *such that* $f^{in} \not\equiv 0$. *Let* f *be a weak solution to the C-F equation on* $[0,\infty)$ *in the sense of Definition 8.2.1 and define*

$$q(s,x) := a(x) + \int_0^\infty k(x,y)f(s,y) \ \mathrm{d}y \ ,$$

$$I(s,x) := \frac{1}{2} \int_0^x k(x-y,y)f(s,y)f(s,x-y) \ \mathrm{d}y$$

$$+ \int_x^\infty a(y)b(x,y)f(s,y) \ \mathrm{d}y \ ,$$

for $(s,x) \in (0,\infty)^2$. *Assume further that, for all* $R > 0$ *and* $t > 0$,

$$q \in L_1((0,t), L_\infty(0,R)) \ , \qquad I \in L_1((0,t) \times (0,R)) \ . \tag{8.2.60}$$

(a) If $k > 0$ *a.e. in* $(0,\infty)^2$, *then*

$$\int_R^\infty f(t,x) \ \mathrm{d}x > 0 \quad \text{for all} \ R > 0 \ \text{and} \ t > 0 \ .$$

(b) If $a > 0$ a.e. in $(0, \infty)$ and, for all $r > 0$,

$$\int_0^r x b(x, y) \; \mathrm{d}x > 0 , \qquad 0 < r < y , \tag{8.2.61}$$

then

$$\int_0^R f(t, x) \; \mathrm{d}x > 0 \quad \text{for all} \quad R > 0 \quad \text{and} \quad t > 0 .$$

Proof. *(a)* Assume for contradiction that there exists $t_0 > 0$ such that

$$R_0 := \inf \left\{ R \geq 0 , \quad \int_R^\infty f(t_0, x) \; \mathrm{d}x = 0 \right\} < \infty . \tag{8.2.62}$$

Owing to the regularity properties (8.2.60), we infer from (8.0.1) and Fubini's theorem that

$$\int_0^\infty [f(t, x) - f^{in}(x)] \vartheta(x) \; \mathrm{d}x = \int_0^t \int_0^\infty \vartheta(x) \left[-q(s, x) f(s, x) + I(s, x) \right] \; \mathrm{d}x \mathrm{d}s$$

for $t > 0$ and $\vartheta \in L_\infty(0, \infty)$. Then, for all $t > 0$,

$$\partial_t f(t, x) = -q(t, x) f(t, x) + I(t, x) \quad \text{a.e. in} \quad (0, \infty) . \tag{8.2.63}$$

Introducing

$$Q(s, x) := \int_0^s q(\tau, x) \; \mathrm{d}\tau , \qquad (s, x) \in (0, \infty)^2 ,$$

it follows from (8.2.60) and (8.2.63) that, for $R > R_0$,

$$\int_{R_0}^R \left(e^{Q(t_0, x)} f(t_0, x) - f^{in}(x) \right) \; \mathrm{d}x = \int_0^{t_0} \int_{R_0}^R \partial_s \left(e^{Q(s, x)} f(s, x) \right) \; \mathrm{d}x \mathrm{d}s$$

$$= \int_0^{t_0} \int_{R_0}^R e^{Q(s, x)} \left(\partial_s f(s, x) + q(s, x) f(s, x) \right) \; \mathrm{d}x \mathrm{d}s ,$$

hence

$$\int_{R_0}^R e^{Q(t_0, x)} f(t_0, x) \; \mathrm{d}x = \int_{R_0}^R f^{in}(x) \; \mathrm{d}x + \int_0^{t_0} \int_{R_0}^R e^{Q(s, x)} I(s, x) \; \mathrm{d}x \mathrm{d}s .$$

Owing to the definition (8.2.62) of R_0, the left-hand side of the above equality vanishes and, since $R > R_0$ is arbitrary, we deduce from the positivity of e^Q and the nonnegativity of f^{in} and I that

$$f^{in} = 0 \quad \text{a.e. in} \quad (R_0, \infty) , \tag{8.2.64}$$
$$I = 0 \quad \text{a.e. in} \quad (0, t_0) \times (R_0, \infty) . \tag{8.2.65}$$

Now, since $k > 0$ a.e. in $(0, \infty)^2$, it follows from (8.2.65) that

$$f(s, y) \; f(s, x - y) \; \mathbf{1}_{(0, x)}(y) = 0 \quad \text{a.e. in} \quad (0, t_0) \times (R_0, \infty) \times (0, \infty) .$$

Consequently, using Fubini's theorem,

$$0 = \int_0^{t_0} \int_{R_0}^\infty \int_0^x f(s, y) \; f(s, x - y) \; \mathrm{d}y \mathrm{d}x \mathrm{d}s$$

$$= \int_0^{t_0} \int_{R_0}^\infty \int_0^{R_0} f(s, y) \; f(s, x - y) \; \mathrm{d}y \mathrm{d}x \mathrm{d}s$$

$$+ \int_0^{t_0} \int_{R_0}^{\infty} \int_y^{\infty} f(s,y) \, f(s, x-y) \, \mathrm{d}x\mathrm{d}y\mathrm{d}s$$

$$= \int_0^{t_0} \int_0^{R_0} \int_{R_0-y}^{\infty} f(s,y) \, f(s,x) \, \mathrm{d}x\mathrm{d}y\mathrm{d}s$$

$$+ \int_0^{t_0} \int_{R_0}^{\infty} \int_0^{\infty} f(s,y) \, f(s,x) \, \mathrm{d}x\mathrm{d}y\mathrm{d}s \,,$$

from which we readily deduce that

$$\int_0^{t_0} \int_0^{R_0} \int_{R_0-y}^{\infty} f(s,x) \, f(s,y) \, \mathrm{d}x\mathrm{d}y\mathrm{d}s = 0 \,, \tag{8.2.66}$$

$$\int_0^{t_0} \left(\int_{R_0}^{\infty} f(s,x) \, \mathrm{d}x \right) \left(\int_0^{\infty} f(s,y) \, \mathrm{d}y \right) \, \mathrm{d}s = 0 \,. \tag{8.2.67}$$

A first consequence of (8.2.67) is that

$$\int_0^{t_0} \left(\int_{R_0}^{\infty} f(s,x) \, \mathrm{d}x \right)^2 \, \mathrm{d}s = 0 \,,$$

and thus, owing to the time continuity of f in $X_{0,w}$,

$$\int_{R_0}^{\infty} f(s,x) \, \mathrm{d}x = 0 \,, \qquad s \in (0, t_0) \,. \tag{8.2.68}$$

Next, since $R_0 - y \le R_0/2$ for $y \in (R_0/2, R_0)$,

$$\int_0^{t_0} \left(\int_{R_0/2}^{R_0} f(s,x) \, \mathrm{d}x \right)^2 \, \mathrm{d}s \le \int_0^{t_0} \int_{R_0/2}^{R_0} \int_{R_0-y}^{R_0} f(s,x) \, f(s,y) \, \mathrm{d}x\mathrm{d}y\mathrm{d}s \,,$$

and it follows from (8.2.66) that

$$\int_{R_0/2}^{R_0} f(s,x) \, \mathrm{d}x = 0 \,, \qquad s \in (0, t_0) \,,$$

which gives, together with (8.2.68),

$$\int_{R_0/2}^{\infty} f(s,x) \, \mathrm{d}x = 0 \,, \qquad s \in (0, t_0) \,.$$

Recalling the definition (8.2.62) of R_0 we conclude that $R_0 = 0$. Then $f^{in} \equiv 0$ according to (8.2.64), which is clearly a contradiction.

(b) Assume for contradiction that there exists $t_0 > 0$ such that

$$R_0 := \sup \left\{ R \ge 0 \,, \quad \int_0^R f(t_0, x) \, \mathrm{d}x = 0 \right\} > 0 \,. \tag{8.2.69}$$

Arguing as in the proof of (a) (but integrating over $(0, R_0)$ instead of (R_0, R)), we deduce from (8.2.63) and (8.2.69) that

$$f^{in} = 0 \quad \text{a.e. in} \quad (0, R_0) \,, \tag{8.2.70}$$

$$I = 0 \quad \text{a.e. in} \quad (0, t_0) \times (0, R_0) \,. \tag{8.2.71}$$

Owing to the positivity of a, we infer from (8.2.71) that

$$b(x,y)f(s,y)\mathbf{1}_{(x,\infty)}(y) = 0 \quad \text{a.e. in} \quad (0,t_0) \times (0,R_0) \times (0,\infty) \ .$$

In particular, using Fubini's theorem and (8.2.69),

$$0 = \int_0^{t_0} \int_0^{R_0} \int_x^{\infty} xb(x,y)f(s,y) \ dydxds = \int_0^{t_0} \int_0^{R_0} \int_{R_0}^{\infty} xb(x,y)f(s,y) \ dydxds$$

$$= \int_0^{t_0} \int_{R_0}^{\infty} \left(\int_0^{R_0} xb(x,y) \ dx \right) f(s,y) \ dyds \ .$$

It then readily follows from (8.2.61) and the above equality that $f(s,y) = 0$ for almost all $y \in (R_0,\infty)$ and all $s \in (0,t_0)$. The time continuity of f in $X_{0,w}$ further implies that $f(t_0,y) = 0$ for almost every $y \in (R_0,\infty)$, which contradicts the definition (8.2.69) of R_0 and completes the proof.   □

8.2.2   Mass-Conserving Solutions

We begin our study of the existence of mass-conserving weak solutions to the C-F equation (8.0.1) with unbounded coefficients by examining the cases when either the growth of k is moderate, or a grows faster than k. From a physical point of view, the coagulation kernel k is not expected to grow faster than quadratically and we thus assume that there is $K_0 > 0$ such that

$$0 \le k(x,y) = k(y,x) \le K_0(1+x)(1+y) \ , \qquad (x,y) \in (0,\infty)^2 \ . \tag{8.2.72a}$$

We also exclude overall fragmentation rates featuring a singularity for small sizes and thus require that for any $r > 0$ there is $A_r > 0$ such that

$$a(x) \le A_r \ , \qquad x \in (0,r) \ . \tag{8.2.72b}$$

Finally, we always assume that there is no loss of matter during fragmentation events; that is,

$$\int_0^y xb(x,y) \ dx = y \ , \qquad y \in (0,\infty) \ . \tag{8.2.72c}$$

8.2.2.1   Coagulation Kernel with Linear Growth

Let us first consider the case of a coagulation kernel that grows only moderately; that is, there is $K_1 > 0$ such that

$$k(x,y) \le K_1(1+x+y) \ , \qquad (x,y) \in (0,\infty)^2 \ . \tag{8.2.73}$$

With regard to the fragmentation coefficients, we assume that there is $m_0 \in [0,1)$, $\bar{a}_0 > 0$, and $p_0 \in (1,2)$ such that the overall fragmentation rate a satisfies

$$a(y) \le \bar{a}_0 y^{p_0-1} \ , \qquad y \in (0,1) \ , \tag{8.2.74}$$

and, for any $p \in [1,p_0]$ there is a positive constant $\kappa_{m_0,p}$ such that the daughter distribution function b satisfies

$$\int_0^y x^{m_0} b(x,y)^p \ dx \le \kappa_{m_0,p} y^{m_0+1-p} \ , \qquad y \in (0,\infty) \ . \tag{8.2.75}$$

We finally assume that there is $\delta_2 > 0$ such that

$$N_2(y) := y^2 - \int_0^y x^2 b(x,y) \, \mathrm{d}x \geq \delta_2 y^2 \,, \qquad y \in (0,\infty) \,. \tag{8.2.76}$$

Observe that no control is required on the growth of a for large sizes.

Remark 8.2.21. *We point out that (8.2.76) is a special case of (8.1.28) ensuring that the fragmentation operator generates an analytic semigroup in $X_{0,2}$. However, in contrast to Section 8.1.2, we do not impose here any restriction on the growth of the overall fragmentation rate a either at infinity or in relation to the growth rate of the coagulation kernel k. On the other hand, (8.2.74) is more restrictive than (8.1.27) for small sizes as it requires $a(y) \to 0$ as $y \to 0$.*

Example 8.2.22. For illustrative purposes, a model example to be kept in mind throughout this chapter is the following: the coagulation kernel k, the overall fragmentation rate a, and the daughter distribution function b are given by

$$k(x,y) = x^\alpha y^\beta + x^\beta y^\alpha \,, \qquad a(x) = x^\gamma \,, \qquad (x,y) \in (0,\infty)^2 \,, \tag{8.2.77}$$

and

$$b(x,y) = b_\nu(x,y) := (\nu+2)\frac{x^\nu}{y^{\nu+1}} \,, \qquad 0 < x < y \,, \tag{8.2.78}$$

and the parameters α, β, γ, and ν satisfy

$$0 \leq \alpha \leq \beta \leq 1 \,, \qquad \gamma > 0 \,, \qquad \nu > -2 \,. \tag{8.2.79}$$

Clearly, k, a and b satisfy (8.2.72) and (8.2.76) (with $\delta_2 = 1/(\nu+3)$). The growth condition (8.2.73) holds true provided $\alpha+\beta \leq 1$ while any choice of $m_0 \in (-1-\nu, 1) \cap [0,1)$ guarantees the validity of (8.2.74) and (8.2.75) for $p_0 \in (1, \gamma+1)$ sufficiently close to one. On the one hand, observe that $m_0 > 0$ when $\nu \in (-2, -1]$. On the other hand, this choice of coagulation and fragmentation coefficients fits perfectly in the framework developed in Section 8.1.2 when $\gamma > \alpha + \beta$ and $\nu \in (-1, 0]$, which gives the existence and uniqueness of a global classical mass-conserving solution to the C-F equation (8.0.1) in that case. The analysis developed in Section 8.2.2.2 below is thus of interest mainly for the case $\gamma \leq \alpha + \beta$ and $\nu \in (-1, 0]$, more singular daughter distribution functions b_ν corresponding to $\nu \in (-2, -1]$ being considered later in Section 8.2.2.3. $\diamond$

8.2.2.2 Coagulation Kernel with Linear Growth and Integrable Daughter Distribution Function b

In this section we assume further that $m_0 = 0$ in (8.2.75), which implies in particular that

$$n_0(y) = \int_0^y b(x,y) \, \mathrm{d}x \leq \kappa_{0,1} \,, \qquad y \in (0,\infty) \,. \tag{8.2.80}$$

Theorem 8.2.23. *Assume that the coagulation and fragmentation coefficients k, a and b satisfy (8.2.72), (8.2.73), (8.2.74), (8.2.75) with $m_0 = 0$, and (8.2.76). Given $f^{in} \in X_{0,1,+}$, there is at least one mass-conserving weak solution f to the C-F equation (8.0.1) on $[0,\infty)$.*
Furthermore, if $f^{in} \in X_m$ for some $m > 1$, then $f \in L_\infty((0,T), X_m)$ for all $T > 0$.

Theorem 8.2.23 applies in particular to Example 8.2.22 when $\alpha + \beta \leq 1$, $\gamma > 0$, and $\nu > -1$. Several results in the spirit of Theorem 8.2.23 are already available in the literature

[109, 167, 197, 254] and the main contribution of Theorem 8.2.23 is to include multiple fragmentation. We also refer to [14, 160] for similar results for the discrete C-F equations.

Several steps are needed to prove Theorem 8.2.23 and we start by collecting estimates which eventually provide sequential weak compactness in X_1 in a pointwise manner in time. Since the ultimate goal is to consider a sequence of solutions to C-F equations approximating the one we are interested in, we pay special attention to the dependence of the various constants appearing below on the parameters involved in the assumptions on k, a and b. We first study the behaviour for large sizes.

Lemma 8.2.24. *Assume that the coagulation and fragmentation coefficients k, a and b satisfy (8.2.72), (8.2.73) and (8.2.76). Assume further that there is $r_0 > 1$ such that*

$$k(x,y) = 0 \quad \text{if} \quad x + y > r_0 \quad \text{and} \quad a(x) = 0 \quad \text{if} \quad x > r_0 . \qquad (8.2.81)$$

Consider next $f^{in} \in L_1(0, r_0)_+$ and let $f \in C^1([0, \infty), L_1(0, r_0))$ be the corresponding solution to (8.0.1) given by Proposition 8.2.7. Assume further that there is $\psi \in C_{VP}$ such that $f^{in} \in L_1((0, r_0), \psi(x)dx)$ and let $C_0 > 0$ and $C_1 > 0$ be such that

$$\int_0^{r_0} (1 + x) f^{in}(x) \, \mathrm{d}x \le C_0 , \qquad (8.2.82)$$

$$\int_0^{r_0} \psi(x) f^{in}(x) \, \mathrm{d}x \le C_1 . \qquad (8.2.83)$$

Given $T > 0$, there is $C_2(T) > 0$ depending only on K_1 in (8.2.73), δ_2 in (8.2.76), C_0 in (8.2.82), (C_1, ψ) in (8.2.83), and T such that

$$\int_0^{r_0} \psi(x) f(t, x) \, \mathrm{d}x \le C_2(T) , \qquad t \in [0, T] , \qquad (8.2.84a)$$

$$\int_0^T \int_0^{r_0} a(y) \left(y\psi'(y) - \psi(y) \right) f(s, y) \, \mathrm{d}y \mathrm{d}s \le C_2(T) . \qquad (8.2.84b)$$

Furthermore,

$$\int_0^{r_0} x f(t, x) \, \mathrm{d}x = \int_0^{r_0} x f^{in}(x) \, \mathrm{d}x , \qquad t \in [0, T] . \qquad (8.2.84c)$$

Remark 8.2.25. *Since ψ is to be chosen later as a superlinear function, observe that Lemma 8.2.24 not only provides the estimate (8.2.84a) on f for large sizes but also the estimate (8.2.84b) which eventually allows us to control the fragmentation term whatever the growth of a. It is worth pointing out at this stage that the assumptions (8.2.73) and (8.2.76) are used only in the proof of Lemma 8.2.24, and that it is solely the estimates (8.2.84a) and (8.2.84b) that are used in the subsequent analysis.*

Proof of Lemma 8.2.24. We extend f to $[0, \infty) \times (0, \infty)$ by setting $f(t, x) = 0$ for $(t, x) \in [0, \infty) \times (r_0, \infty)$ and recall that Proposition 8.2.7 guarantees that (8.2.84c) holds true. Consequently, thanks to (8.2.82),

$$M_1(f(t)) = \int_0^{r_0} x f(t, x) \, \mathrm{d}x \le C_0 , \qquad t \ge 0 . \qquad (8.2.85)$$

Consider next $T > 0$ and $t \in [0, T]$. On the one hand, we infer from (8.2.73), (8.2.85) and Lemma 8.2.16 that

$$\int_0^{r_0} \mathcal{C} f(t, x) \psi(x) \, \mathrm{d}x \le 2K_1 M_1(f(t)) \left[\psi''(0) M_1(f(t)) + \int_0^{r_0} \psi(x) f(t, x) \, \mathrm{d}x \right]$$

$$\le 2K_1 \left(\psi''(0)C_0^2 + C_0 \int_0^{r_0} \psi(x)f(t,x) \, \mathrm{d}x \right) .$$

On the other hand, (8.2.72c), (8.2.76), the convexity of ψ, and the concavity of $\psi_1 : x \mapsto \psi(x)/x$, see Proposition 7.1.9 (c), ensure that

$$N_\psi(x) = y\psi_1(y) - \int_0^y x\psi_1(x)b(x,y) \, \mathrm{d}x = \int_0^y x(\psi_1(y) - \psi_1(x))b(x,y) \, \mathrm{d}x$$

$$\ge \int_0^y x\psi_1'(y)(y-x)b(x,y) \, \mathrm{d}x = (y\psi'(y) - \psi(y))\frac{N_2(y)}{y^2}$$

$$\ge \delta_2(y\psi'(y) - \psi(y)) , \qquad y \in (0,\infty) .$$

Consequently,

$$\int_0^{r_0} \mathcal{F}f(t,x)\psi(x) \, \mathrm{d}x = -\int_0^{r_0} a(y)f(t,y)N_\psi(y) \, \mathrm{d}y$$

$$\le -\delta_2 \int_0^{r_0} a(y)(y\psi'(y) - \psi(y))f(t,y) \, \mathrm{d}y .$$

Thanks to the previous estimates, we deduce from (8.2.13) and the convexity of ψ that

$$X(t) := \int_0^{r_0} \psi(x)f(t,x) \, \mathrm{d}x$$

$$+ \delta_2 \int_0^t \int_0^{r_0} a(y)(y\psi'(y) - \psi(y))f(s,y) \, \mathrm{d}y\mathrm{d}s \ge 0 , \qquad t \in [0,T] ,$$

satisfies

$$\frac{dX}{dt}(t) := \frac{d}{dt}\int_0^{r_0} \psi(x)f(t,x) \, \mathrm{d}x + \delta_2 \int_0^{r_0} a(y)(y\psi'(y) - \psi(y))f(t,y) \, \mathrm{d}y$$

$$\le 2C_0K_1 \left[\psi''(0)C_0 + \int_0^{r_0} \psi(x)f(t,x) \, \mathrm{d}x \right]$$

$$\le 2C_0^2 K_1\psi''(0) + 2C_0K_1X(t) .$$

We integrate the differential inequality for X and use (8.2.83) to find

$$X(t) \le X(0)e^{2C_0K_1t} + C_0\psi''(0) \left(e^{2C_0K_1t} - 1 \right)$$

$$\le \left(\int_0^{r_0} \psi(x)f^{in}(x) \, \mathrm{d}x + C_0\psi''(0) \right) e^{2C_0K_1t} ,$$

and the proof of Lemma 8.2.24 is complete. □

In the same vein, the temporal growth of moments of order m greater than one may be controlled. Since we aim to involve no moments of smaller orders in the later analysis, we restrict our attention here to the case $m > 1$, and proceed differently according to whether $m \ge 2$ or $m \in (1,2)$.

Lemma 8.2.26. *Assume that the coagulation and fragmentation coefficients k, a and b satisfy (8.2.72), (8.2.73), and (8.2.81) for some $r_0 > 1$. Consider next $f^{in} \in L_1(0,r_0)_+$ satisfying (8.2.82) for some $C_0 > 0$ and let $f \in C^1([0,\infty), L_1(0,r_0))$ be the corresponding solution to (8.0.1) given by Proposition 8.2.7. Then, for $m > 1$ and $T > 0$, there is a positive constant $C(T,m)$, depending only on K_1 in (8.2.73), C_0 in (8.2.82), m and T such that*

$$\int_0^{r_0} x^m f(t,x) \, \mathrm{d}x \le C(T,m) \left(1 + \int_0^{r_0} x^m f^{in}(x) \, \mathrm{d}x \right) , \qquad t \in [0,T] .$$

Proof. We first consider the case $m \geq 2$. Let $T > 0$ and $t \in [0, T]$. Since $N_m \geq 0$ for $m > 1$, we infer from (8.2.13) with $\vartheta(x) = x^m$, $x \in (0, \infty)$, and (8.2.73) that

$$\frac{d}{dt} \int_0^{r_0} x^m f(t, x) \, dx \leq \frac{K_1}{2} \int_0^{r_0} \int_0^{r_0} (1 + x + y) \chi_m(x, y) f(t, x) f(t, y) \, dy dx .$$

It follows from Lemma 7.4.2 and Lemma 7.4.4 that

$$\chi_m(x, y) \leq C(m) \left(x^{m-1} y + x y^{m-1} \right) , \qquad (x, y) \in (0, \infty)^2 ,$$
$$(x + y) \chi_m(x, y) \leq C(m) \left(x^m y + x y^m \right) , \qquad (x, y) \in (0, \infty)^2 .$$

Consequently,

$$\frac{d}{dt} \int_0^{r_0} x^m f(t, x) \, dx \leq K_1 C(m) \left(\int_0^{r_0} x f(t, x) \, dx \right) \left(\int_0^{r_0} y^{m-1} f(t, y) \, dy \right)$$
$$+ K_1 C(m) \left(\int_0^{r_0} x f(t, x) \, dx \right) \left(\int_0^{r_0} y^m f(t, y) \, dy \right) .$$

Thanks to (8.2.14), (8.2.82), and Young's inequality, we further obtain

$$\frac{d}{dt} \int_0^{r_0} x^m f(t, x) \, dx \leq C_0 K_1 C(m) \int_0^{r_0} \left(\frac{m-2}{m-1} y^m + \frac{1}{m-1} y \right) f(t, y) \, dy$$
$$+ C_0 K_1 C(m) \int_0^{r_0} y^m f(t, y) \, dy$$
$$\leq C_0^2 K_1 C(m) + 2 C_0 K_1 C(m) \int_0^{r_0} y^m f(t, y) \, dy .$$

After integration with respect to time, we end up with

$$\int_0^{r_0} x^m f(t, x) \, dx \leq \left(\int_0^{r_0} x^m f^{in}(x) \, dx + \frac{C_0}{2} \right) e^{2 C_0 K_1 C(m) t} ,$$

which proves Lemma 8.2.26 for $m \geq 2$.

We now consider $m \in (1, 2)$ and set $\vartheta(x) = \min\{x^2, x^m\}$, $x \in (0, \infty)$. Since $N_\vartheta \geq 0$, we infer from Proposition 8.2.7 that

$$\frac{d}{dt} \int_0^{r_0} \vartheta(x) f(t, x) \, dx \leq \frac{K_1}{2} \int_0^{r_0} \int_0^{r_0} (1 + x + y) \chi_\vartheta(x, y) f(t, x) f(t, y) \, dy dx .$$

To estimate $(1 + x + y) \chi_\vartheta(x, y)$, we split the analysis into four cases.
– *Case 1.* If $(x, y) \in (0, 1)^2$ satisfy $x + y \in (0, 1)$, then

$$(1 + x + y) \chi_\vartheta(x, y) \leq 2 \chi_2(x, y) \leq 4xy .$$

– *Case 2.* If $(x, y) \in (0, 1)^2$ satisfy $x + y \geq 1$, then

$$(1 + x + y) \chi_\vartheta(x, y) \leq 3 \left[(x + y)^m - x^2 - y^2 \right] \leq 3 \chi_2(x, y) \leq 6xy .$$

– *Case 3.* If $\min\{x, y\} \in (0, 1)$ and $\max\{x, y\} \geq 1$, then, by Lemma 7.4.4,

$$(1 + x + y) \chi_\vartheta(x, y) = (1 + x + y) \left[(x + y)^m - \max\{x, y\}^m - \min\{x, y\}^2 \right]$$
$$= (1 + x + y) \chi_m(x, y) + (1 + x + y) \left[\min\{x, y\}^m - \min\{x, y\}^2 \right]$$
$$\leq 2(x + y) \chi_m(x, y) + 3 \max\{x, y\} \min\{x, y\}^m$$

$$\leq C(m)(x^m y + xy^m) \ .$$

– *Case 4.* Finally, if $x \geq 1$ and $y \geq 1$, then it follows from Lemma 7.4.4 that

$$(1 + x + y)\chi_\vartheta(x,y) = (1 + x + y)\chi_m(x,y) \leq 2(x + y)\chi_m(x,y)$$
$$\leq C(m)(x^m y + xy^m) \ .$$

Summarising, we have shown that

$$\frac{d}{dt} \int_0^{r_0} \vartheta(x) f(t,x) \ \mathrm{d}x \leq K_1 C(m) \int_0^{r_0} \int_0^{r_0} (x^m y + xy^m + xy) f(t,x) f(t,y) \ \mathrm{d}y\mathrm{d}x \ ,$$

and we deduce from (8.2.14) and (8.2.82) that

$$\frac{d}{dt} \int_0^{r_0} \vartheta(x) f(t,x) \ \mathrm{d}x \leq C_0 K_1 C(m) \left(C_0 + 2 \int_0^{r_0} x^m f(t,x) \ \mathrm{d}x \right)$$

$$\leq C_0 K_1 C(m) \left(C_0 + 2 \int_0^1 x^m f(t,x) \ \mathrm{d}x \right)$$

$$+ 2 C_0 K_1 C(m) \int_1^{r_0} x^m f(t,x) \ \mathrm{d}x$$

$$\leq C_0 K_1 C(m) \left(3 C_0 + 2 \int_0^{r_0} \vartheta(x) f(t,x) \ \mathrm{d}x \right) \ .$$

After integration with respect to time, we find

$$\int_0^{r_0} \vartheta(x) f(t,x) \ \mathrm{d}x \leq \left(\int_0^{r_0} x^m f^{in}(x) \ \mathrm{d}x + \frac{3C_0}{2} \right) e^{2 C_0 K_1 C(m) t} \ .$$

Since

$$\int_0^{r_0} x^m f(t,x) \ \mathrm{d}x \leq \int_0^1 x f(t,x) \ \mathrm{d}x + \int_1^{r_0} x^m f(t,x) \ \mathrm{d}x \ ,$$

Lemma 8.2.26 for $m \in (1,2)$ is now a straightforward consequence of (8.2.14), (8.2.82) and the previous two inequalities. $\qquad\square$

We next turn to the behaviour for small sizes which is governed by fragmentation. Here we take advantage of the moderate strength of the breakup mechanism that, according to (8.2.80), only produces a bounded number of fragments.

Lemma 8.2.27. *Assume that the coagulation and fragmentation coefficients k, a and b satisfy (8.2.72), (8.2.75) with $m_0 = 0$, and (8.2.81) for some $r_0 > 1$. Consider next $f^{in} \in L_1(0, r_0)_+$ satisfying (8.2.82) for some $C_0 > 0$ and let $f \in C^1([0,\infty), L_1(0, r_0))$ be the corresponding solution to (8.0.1) given by Proposition 8.2.7. Assume further that for any $T > 0$, there is $C_3(T) > 0$ such that*

$$\int_0^T P(t) \ \mathrm{d}t \leq C_3(T) \ , \qquad P(t) := \int_1^{r_0} a(y) f(t,y) \ \mathrm{d}y \ , \qquad t \in [0,T] \ . \qquad (8.2.86)$$

Given $T > 0$, there is $C_4(T) > 0$ depending only on A_1 in (8.2.72b), $\kappa_{0,1}$ in (8.2.80), C_0 in (8.2.82), $C_3(T)$ in (8.2.86), and T such that

$$\int_0^{r_0} f(t,x) \ \mathrm{d}x \leq C_4(T) \ , \qquad t \in [0,T] \ .$$

Proof. Let $T > 0$ and $t \in [0, T]$. Obviously,

$$\int_0^{r_0} \mathcal{C}f(t, x) \, \mathrm{d}x \leq 0 ,$$

while (8.2.72b), (8.2.80) and (8.2.81) yield

$$\int_0^{r_0} \mathcal{F}f(t, x) \, \mathrm{d}x = \int_0^{r_0} (n_0(y) - 1)a(y)f(t, y) \, \mathrm{d}y \leq \kappa_{0,1} \int_0^{r_0} a(y)f(t, y) \, \mathrm{d}y$$

$$\leq A_1 \kappa_{0,1} \int_0^1 f(t, y) \, \mathrm{d}y + \kappa_{0,1} P(t) .$$

It then follows from (8.2.13) that

$$\frac{d}{dt} \int_0^{r_0} f(t, x) \, \mathrm{d}x \leq \kappa_{0,1} P(t) + A_1 \kappa_{0,1} \int_0^{r_0} f(t, y) \, \mathrm{d}y ,$$

hence

$$\int_0^{r_0} f(t, x) \, \mathrm{d}x \leq \left(\int_0^{r_0} f^{in}(x) \, \mathrm{d}x + \kappa_{0,1} C_3(T) \right) e^{A_1 \kappa_{0,1} t}$$

after integrating with respect to time and using (8.2.82) and (8.2.86). □

The next result is devoted to uniform integrability properties.

Lemma 8.2.28. *Assume that the coagulation and fragmentation coefficients k, a and b satisfy (8.2.72), (8.2.74), (8.2.75) with $m_0 = 0$, and (8.2.81) for some $r_0 > 1$. Consider next $f^{in} \in L_1(0, r_0)_+$ satisfying (8.2.82) for some $C_0 > 0$ and let $f \in C^1([0, \infty), L_1(0, r_0))$ be the corresponding solution to (8.0.1) given by Proposition 8.2.7. Assume further that, for each $T > 0$, there is $C_3(T) > 0$ such that f satisfies (8.2.86). Assume finally that there is $\Phi \in \mathcal{C}_{VP}$, and $C_5 > 0$ such that*

$$\int_0^{r_0} \Phi(f^{in}(x)) \, \mathrm{d}x \leq C_5 . \tag{8.2.87}$$

Then, for any $T > 0$ and $R \in (1, r_0)$ there is $C_6(T, R)$ depending only on (K_0, A_1) in (8.2.72), $\bar{a}_0$ in (8.2.74), $(p_0, \kappa_{0,p_0}, \kappa_{0,1})$ in (8.2.75), C_0 in (8.2.82), $C_3(T)$ in (8.2.86), (Φ, C_5) in (8.2.87), R and T such that

$$\int_0^R \Phi(f(t, x)) \, \mathrm{d}x \leq C_6(T, R) , \qquad t \in [0, T] .$$

Proof. Let $T > 0$. We first infer from (8.2.14) and Lemma 8.2.27 that

$$\int_0^{r_0} (1 + x)f(t, x) \, \mathrm{d}x = \int_0^{r_0} f(t, x) \, \mathrm{d}x + \int_0^{r_0} x f^{in}(x) \, \mathrm{d}x$$

$$\leq C_7(T) := C_0 + C_4(T) , \qquad t \in [0, T] . \tag{8.2.88}$$

Next, since $\Phi \in \mathcal{C}_{VP}$, it follows from (8.2.72b), (8.2.75), (8.2.87) and Proposition 7.1.9 (a) and (g) that we may apply Proposition 8.2.8 to deduce that $\Phi(f(t))$ belongs to $L_1(0, r_0)$ for all $t \in [0, T]$. Since f, k and Φ' are nonnegative, we infer from the convexity of Φ and Proposition 7.1.9 (a) that

$$J_1(t, R) := \int_0^R \mathcal{C}f(t, x)\Phi'(f(t, x)) \, \mathrm{d}x$$

$$\leq \frac{1}{2} \int_0^R \int_0^x k(x-y,y) f(t,x-y) f(t,y) \Phi'(f(t,x)) \, dy dx$$

$$\leq \frac{1}{2} \int_0^R \int_0^x k(x-y,y) f(t,y) \Phi(f(t,x-y)) \, dy dx$$

$$+ \frac{1}{2} \int_0^R \int_0^x k(x-y,y) f(t,y) \left[\Phi'(f(t,x)) f(t,x) - \Phi(f(t,x)) \right] \, dy dx$$

$$\leq \frac{1}{2} \int_0^R \int_y^R k(x-y,y) \Phi(f(t,x-y)) f(t,y) \, dx dy$$

$$+ \frac{1}{2} \int_0^R \int_0^x k(x-y,y) \Phi(f(t,x)) f(t,y) \, dy dx$$

$$\leq \frac{1}{2} \int_0^R \int_0^{R-y} k(x,y) f(t,y) \Phi(f(t,x)) \, dx dy$$

$$+ \frac{1}{2} \int_0^R \int_0^x k(x-y,y) f(t,y) \Phi(f(t,x)) \, dy dx .$$

It now follows from (8.2.72a) and (8.2.88) that

$$J_1(t,R) \leq K_0(1+R)^2 \int_0^R \int_0^R f(t,y) \Phi(f(t,x)) \, dx dy$$

$$\leq C_8(T,R) \int_0^R \Phi(f(t,x)) \, dx , \qquad (8.2.89)$$

with $C_8(T,R) := K_0(1+R)^2 C_7(T)$.

We use again the convexity of Φ, (8.2.81), and Proposition 7.1.9 (a) along with the nonnegativity of f, a, b and Φ' to obtain that

$$J_2(t,R) := \int_0^R \mathscr{F} f(t,x) \Phi'(f(t,x)) \, dx$$

$$\leq \int_0^R \int_x^{r_0} a(y) b(x,y) f(t,y) \Phi'(f(t,x)) \, dy dx$$

$$\leq \int_0^R \int_x^{r_0} a(y) f(t,y) \left[\Phi(b(x,y)) + \Phi'(f(t,x)) f(t,x) - \Phi(f(t,x)) \right] \, dy dx$$

$$\leq \int_0^{r_0} a(y) f(t,y) \int_0^{\min\{y,R\}} \Phi(b(x,y)) \, dx dy$$

$$+ \int_0^R \int_x^{r_0} a(y) f(t,y) \Phi(f(t,x)) \, dy dx .$$

Thanks to (8.2.75), with $m_0 = 0$ and $p = p_0$, and Proposition 7.1.9 (g), we further obtain

$$J_2(t,R) \leq \sup_{r>0}\{\Phi(r) r^{-p_0}\} \int_0^{r_0} a(y) f(t,y) \int_0^y b(x,y)^{p_0} \, dx dy$$

$$+ \left(\int_0^{r_0} a(y) f(t,y) \, dy \right) \int_0^R \Phi(f(t,x)) \, dx$$

$$\leq \kappa_{0,p_0} \sup_{r>0}\{\Phi(r) r^{-p_0}\} \int_0^{r_0} a(y) y^{1-p_0} f(t,y) \, dy$$

$$+ \left(\int_0^{r_0} a(y) f(t,y) \, dy \right) \int_0^R \Phi(f(t,x)) \, dx .$$

Since

$$\int_0^{r_0} a(y) y^{1-p_0} f(t,y) \, dy \le \bar{a}_0 \int_0^1 f(t,y) \, dy + P(t) \le \bar{a}_0 C_7(T) + P(t) \,,$$

$$\int_0^{r_0} a(y) f(t,y) \, dy \le \int_0^1 a(y) f(t,y) \, dy + P(t) \le A_1 C_7(T) + P(t) \,,$$

by (8.2.72b), (8.2.74), and (8.2.88), we end up with

$$J_2(t,R) \le \kappa_{0,p_0} \sup_{r>0}\{\Phi(r) r^{-p_0}\} \left(\bar{a}_0 C_7(T) + P(t)\right)$$

$$+ \left(A_1 C_7(T) + P(t)\right) \int_0^R \Phi(f(t,x)) \, dx \,. \tag{8.2.90}$$

We now infer from (8.0.1), (8.2.89), and (8.2.90) that

$$\frac{d}{dt} \int_0^R \Phi(f(t,x)) \, dx \le [C_8(T,R) + A_1 C_7(T) + P(t)] \int_0^R \Phi(f(t,x)) \, dx$$

$$+ \kappa_{0,p_0} \sup_{r>0}\{\Phi(r) r^{-p_0}\} \left(\bar{a}_0 C_7(T) + P(t)\right) \,,$$

hence, after integration with respect to time,

$$\int_0^R \Phi(f(t,x)) \, dx \le \left[\int_0^R \Phi(f^{in}(x)) \, dx + \int_0^t Z(s) e^{-Y(s)} \, ds\right] e^{Y(t)} \,, \qquad t \in [0,T] \,,$$

where

$$Y(t) := \int_0^t \left(C_8(T,R) + A_1 C_7(T) + P(t)\right) \, ds \,,$$

$$Z(t) := \kappa_{0,p_0} \sup_{r>0}\{\Phi(r) r^{-p_0}\} \left(\bar{a}_0 C_7(T) + P(t)\right) \,.$$

We finally use (8.2.86) and (8.2.87) to complete the proof.   □

As we shall see below in the proof of Theorem 8.2.23, combining Lemma 8.2.24, Lemma 8.2.27 and Lemma 8.2.28 guarantees that, given $T > 0$, solutions to the C-F equation lie in a relatively weakly sequentially compact subset of X_0 for each $t \in [0,T]$. We now turn to compactness with respect to the time variable which will be deduced from the following equicontinuity result.

Lemma 8.2.29. *Assume that the coagulation and fragmentation coefficients k, a and b satisfy (8.2.72), (8.2.75) with $m_0 = 0$, and (8.2.81) for some $r_0 > 1$. Consider next $f^{in} \in L_1(0,r_0)_+$ satisfying (8.2.82) for some $C_0 > 0$, and let $f \in C^1([0,\infty), L_1(0,r_0))$ be the corresponding solution to (8.0.1) given by Proposition 8.2.7. Assume further that, for each $T > 0$, there is $C_3(T) > 0$ such that f satisfies (8.2.86). For each $T > 0$ there is $C_9(T) > 0$ depending only on A_1 in (8.2.72b), $\kappa_{0,1}$ in (8.2.80), C_0 in (8.2.82), $C_3(T)$ in (8.2.86), and T such that, for $t_1 \in [0,T]$ and $t_2 \in [t_1, T]$,*

$$\int_0^{r_0} |f(t_2,x) - f(t_1,x)| \, dx \le C_9(T)(t_2 - t_1) + (1 + \kappa_{0,1}) \int_{t_1}^{t_2} \int_1^{r_0} a(x) f(s,x) \, dx \, ds \,.$$

Proof. Let $T > 0$. We infer from (8.2.14) and Lemma 8.2.27 that

$$\int_0^{r_0} (1+x)f(t,x)\ \mathrm{d}x = \int_0^{r_0} f(t,x)\ \mathrm{d}x + \int_0^{r_0} x f^{in}(x)\ \mathrm{d}x$$
$$\leq C_7(T) = C_0 + C_4(T)\ , \qquad t \in [0,T]\ . \tag{8.2.91}$$

Next, let $t \in [0,T]$. On the one hand, it follows from (8.2.72a) and (8.2.91) that

$$\left| \int_0^{r_0} \mathcal{C}f(t,x)\ \mathrm{d}x \right| \leq \frac{K_0}{2} \int_0^{r_0} \int_0^x (1+x)(1+y)f(t,x)f(t,y)\ \mathrm{d}y\mathrm{d}x$$
$$+ K_0 \int_0^{r_0} \int_0^{r_0} (1+x)(1+y)f(t,x)f(t,y)\ \mathrm{d}y\mathrm{d}x$$
$$\leq 2K_0 C_7(T)^2\ .$$

On the other hand, we infer from (8.2.72b), (8.2.80) and (8.2.91) that

$$\left| \int_0^{r_0} \mathcal{F}f(t,x)\ \mathrm{d}x \right| \leq \int_0^{r_0} a(x)f(t,x)\ \mathrm{d}x + \int_0^{r_0} \int_x^{r_0} a(y)b(x,y)f(t,y)\ \mathrm{d}y\mathrm{d}x$$
$$\leq \int_0^{r_0} a(x)(1+n_0(x))f(t,x)\ \mathrm{d}x$$
$$\leq (1+\kappa_{0,1}) \left(A_1 \int_0^1 f(t,x)\ \mathrm{d}x + \int_1^{r_0} a(x)f(t,x)\ \mathrm{d}x \right)$$
$$\leq (1+\kappa_{0,1}) \left(A_1 C_7(T) + \int_1^{r_0} a(x)f(t,x)\ \mathrm{d}x \right)\ .$$

Consequently, by (8.0.1),

$$\int_0^{r_0} |\partial_t f(t,x)|\ \mathrm{d}x \leq C_9(T) + (1+\kappa_{0,1}) \int_1^{r_0} a(x)f(t,x)\ \mathrm{d}x\ ,$$

with $C_9(T) := 2K_0 C_7(T)^2 + A_1(1+\kappa_{0,1})C_7(T)$. Now, for $0 \leq t_1 \leq t_2 \leq T$, it follows from the above inequality and Fubini's theorem that

$$\int_0^{r_0} |f(t_2,x) - f(t_1,x)|\ \mathrm{d}x \leq \int_0^{r_0} \int_{t_1}^{t_2} |\partial_t f(t,x)|\ \mathrm{d}t\mathrm{d}x$$
$$\leq C_9(T)(t_2-t_1) + (1+\kappa_{0,1}) \int_{t_1}^{t_2} \int_1^{r_0} a(x)f(t,x)\ \mathrm{d}x\mathrm{d}t\ ,$$

and the proof of Lemma 8.2.29 is complete. $\qquad\square$

The results derived so far provide the tools required for the proof of Theorem 8.2.23 which we perform now.

Proof of Theorem 8.2.23. Let k, a and b be coagulation and fragmentation coefficients satisfying (8.2.72), (8.2.73), (8.2.74), (8.2.75) with $m_0 = 0$, and (8.2.76). Also let $f^{in} \in X_{0,1,+}$ and put

$$\varrho := M_1(f^{in}) \quad \text{and} \quad C_0 := M_0(f^{in}) + \varrho\ . \tag{8.2.92}$$

Then, $f^{in} \in L_1(0,\infty)$ and $x \mapsto x \in L_1((0,\infty), f^{in}(x)\mathrm{d}x)$, so that, according to the de la Vallée-Poussin theorem (Theorem 7.1.6), there is $\Phi \in \mathcal{C}_{VP,\infty}$, and $\psi \in \mathcal{C}_{VP,\infty}$ such that

$$\Phi(f^{in}) \in L_1(0,\infty) \quad \text{and} \quad M_\psi(f^{in}) < \infty\ . \tag{8.2.93}$$

We begin the proof with the construction of a sequence of approximations which is done as follows: given a positive integer $j \geq 1$ and $(x, y) \in (0, \infty)^2$, we define

$$k_j(x, y) := k(x, y) \mathbf{1}_{(0,j)}(x + y) , \qquad a_j(x) := a(x) \mathbf{1}_{(0,j)}(x) , \qquad (8.2.94)$$

and

$$f_j^{in}(x) := f^{in}(x) \mathbf{1}_{(0,j)}(x) . \qquad (8.2.95)$$

Then f_j^{in} is a nonnegative function in $L_1(0, j)$ and k_j and a_j satisfy (8.2.81) with $r_0 = j$, so that the C-F equation (8.0.1) with coefficients (k_j, a_j, b) has a unique solution $f_j \in C^1([0, \infty), L_1(0, j))$ by Proposition 8.2.7. We extend f_j to $[0, \infty) \times (0, \infty)$ by setting $f_j(t, x) = 0$ for $(t, x) \in [0, \infty) \times (j, \infty)$. Then, still by Proposition 8.2.7,

$$M_1(f_j(t)) = \int_0^j x f_j(t, x) \, dx = \int_0^j x f_j^{in}(x) \, dx = \int_0^j x f^{in}(x) \, dx \leq \varrho , \qquad t \geq 0 , \quad (8.2.96)$$

the last inequality being a consequence of (8.2.92). Also, $k_j \leq k$ and $a_j \leq a$, and, since k, a and b satisfy (8.2.72), (8.2.73), (8.2.74), (8.2.75) with $m_0 = 0$, and (8.2.76), it readily follows that (k_j, a_j, b) also satisfies these assumptions uniformly with respect to $j \geq 1$. Similarly, $f_j^{in} \leq f^{in}$, (8.2.93) and the monotonicity of Φ and ψ imply that

$$\int_0^j \psi(x) f_j^{in}(x) \, dx \leq C_1 := M_\psi(f^{in}) \quad \text{and} \quad \int_0^j \Phi(f_j^{in}(x)) \, dx \leq C_5 := \|\Phi(f^{in})\|_1 .$$
$$(8.2.97)$$

Step 1: Estimates. We are now in a position to apply Lemma 8.2.24 and deduce that, for each $T > 0$ and $j \geq 1$,

$$M_\psi(f_j(t)) \leq C_2(T) , \qquad t \in [0, T] , \qquad (8.2.98)$$

and

$$\int_0^T \int_0^\infty a(y) (y\psi'(y) - \psi(y)) f_j(s, y) \, dy ds \leq C_2(T) . \qquad (8.2.99)$$

Since $y\psi'(y) - \psi(y) \geq 0$ for $y > 0$ by Proposition 7.1.9 (a), and

$$y\psi'(y) - \psi(y) \geq \frac{[\psi'(y)]^2}{2\psi''(0)} \geq \frac{[\psi'(y_0)]^2}{2\psi''(0)} , \qquad y > y_0 , \quad y_0 > 0 ,$$

by Proposition 7.1.9 (e), the following estimate is a straightforward consequence of (8.2.99): for any $T > 0$ and $j \geq 1$,

$$\int_0^T \int_R^\infty a(y) f_j(s, y) \, dy ds \leq \frac{[\psi'(1)]^2}{[\psi'(R)]^2} C_3(T) , \qquad R \in (1, j) , \qquad (8.2.100)$$

with $C_3(T) := 2\psi''(0) C_2(T) / [\psi'(1)]^2$. Now, (8.2.82), (8.2.86) and (8.2.87) are consequences of (8.2.92), (8.2.95), (8.2.100) (with $R = 1$), and (8.2.97), respectively, and, since k_j, a_j and b satisfy (8.2.72), (8.2.74), (8.2.75) with $m_0 = 0$, and (8.2.81) with $r_0 = j$, we may apply Lemmas 8.2.27, 8.2.28 and 8.2.29 to establish that, for $T > 0$ and $R \in (1, j)$,

$$M_0(f_j(t)) = \int_0^j f_j(t, x) \, dx \leq C_4(T) , \qquad t \in [0, T] , \qquad (8.2.101)$$

$$\int_0^R \Phi(f_j(t, x)) \, dx \leq C_6(T, R) , \qquad t \in [0, T] , \qquad (8.2.102)$$

and

$$\|f_j(t_2) - f_j(t_1)\|_1 = \int_0^j |f_j(t_2, x) - f_j(t_1, x)| \, dx$$

$$\leq C_9(T)(t_2 - t_1) + (1 + \kappa_{0,1}) \int_{t_1}^{t_2} \int_1^j a(x) f_j(s, x) \, dx ds$$

for $t_1 \in [0, T]$ and $t_2 \in [t_1, T]$. Combining the last inequality with (8.2.74), (8.2.100) and (8.2.101) gives, for $T > 0$, $t_1 \in [0, T]$, $t_2 \in [t_1, T]$ and $R \in (1, j)$,

$$\|f_j(t_2) - f_j(t_1)\|_1 \leq C_9(T)(t_2 - t_1) + (1 + \kappa_{0,1}) A_R \int_{t_1}^{t_2} \int_1^R f_j(s, x) \, dx ds$$

$$+ (1 + \kappa_{0,1}) \int_{t_1}^{t_2} \int_R^j a(x) f_j(s, x) \, dx ds$$

$$\leq [C_9(T) + A_R(1 + \kappa_{0,1}) C_4(T)] \, (t_2 - t_1) + (1 + \kappa_{0,1}) \frac{[\psi'(1)]^2}{[\psi'(R)]^2} C_3(T) \ .$$

The previous inequality being valid for all $R \in (1, j)$, we end up with

$$\|f_j(t_2) - f_j(t_1)\|_1 \leq C_{10}(T) \omega_1(t_2 - t_1) \ , \qquad 0 \leq t_1 \leq t_2 \leq T \ , \tag{8.2.103}$$

where

$$\omega_1(s) := s + \inf_{R>1} \left\{ A_R s + \frac{1}{[\psi'(R)]^2} \right\} \ , \qquad s \in [0, T] \ , \tag{8.2.104}$$

and

$$C_{10}(T) := C_9(T) + (1 + \kappa_{0,1}) C_4(T) + (1 + \kappa_{0,1})[\psi'(1)]^2 C_3(T) \ .$$

We claim that the property $\psi \in \mathcal{C}_{VP,\infty}$ implies that

$$\lim_{s \to 0} \omega_1(s) = 0 \ . \tag{8.2.105}$$

Indeed, it follows from (8.2.104) that, for all $R > 1$,

$$\limsup_{s \to 0} \omega_1(s) \leq \frac{1}{[\psi'(R)]^2} \ .$$

Since $\psi'(R) \to \infty$ as $R \to \infty$, we may let $R \to \infty$ in the above inequality and obtain (8.2.105).

Step 2: Compactness. Now, fix $T > 0$ and define

$$\mathcal{E}(T) := \{f_j(t) \ : \ t \in [0, T] \ , \ j \geq 1\} \ ,$$

If E is a measurable subset of $(0, \infty)$ with finite measure and $R > 1$, then it follows from (8.2.96), (8.2.102) and the monotonicity of $r \mapsto \Phi(r)/r$ (Proposition 7.1.9 (c)) that, for all $t \in [0, T]$, $j \geq 1$, $R > 1$, and $K > 1$,

$$\int_E f_j(t, x) \, dx \leq \int_{E \cap (0, R)} f_j(t, x) \, dx + \int_R^\infty f_j(t, x) \, dx$$

$$\leq \int_{E \cap (0, R)} f_j(t, x) \mathbf{1}_{(0, K)}(f_j(t, x)) \, dx$$

$$+ \int_{E \cap (0, R)} f_j(t, x) \mathbf{1}_{[K, \infty)}(f_j(t, x)) \, dx + \frac{\varrho}{R}$$

$$\leq K|E| + \frac{K}{\Phi(K)} \int_0^R \Phi(f_j(t,x)) \, \mathrm{d}x + \frac{\varrho}{R}$$

$$\leq K|E| + \frac{KC_6(T,R)}{\Phi(K)} + \frac{\varrho}{R} \ .$$

Therefore the modulus of uniform integrability $\eta\{\mathcal{E}(T)\}$ of $\mathcal{E}(T)$ in X_0, see Definition 7.1.2, satisfies

$$\eta\{\mathcal{E}(T)\} \leq \frac{KC_6(T,R)}{\Phi(K)} + \frac{\varrho}{R}$$

for all $K > 1$ and $R > 1$. Since $\Phi \in \mathcal{C}_{VP,\infty}$, we can let $K \to \infty$ in the previous inequality to obtain that $\eta\{\mathcal{E}(T)\} \leq \varrho/R$ for all $R > 1$. We then let $R \to \infty$ to conclude that

$$\eta\{\mathcal{E}(T)\} = 0 \ . \tag{8.2.106}$$

Moreover, by (8.2.96),

$$\int_R^\infty f_j(t,x) \, \mathrm{d}x \leq \frac{\varrho}{R}$$

for $j \geq 1$, $t \in [0,T]$, and $R > 0$, so that

$$\lim_{R \to \infty} \sup_{g \in \mathcal{E}(T)} \left\{ \int_R^\infty g(x) \, \mathrm{d}x \right\} = 0 \ .$$

In particular, $\mathcal{E}(T)$ is a relatively sequentially weakly compact subset of X_0 according to the Dunford–Pettis theorem (Theorem 7.1.3) while (8.2.105) and (8.2.106) imply that the sequence $(f_j)_{j\geq 1}$ is strongly and thus also weakly equicontinuous in X_0 at each $t \in [0,T]$ in the sense of Definition 7.1.15. We are then in a position to apply the version of the Arzelà–Ascoli theorem recalled in Theorem 7.1.16 to deduce that $(f_j)_{j\geq 1}$ is relatively sequentially compact in $C([0,T], X_{0,w})$. Therefore, since T is arbitrary, a diagonal process ensures the existence of a subsequence of $(f_j)_{j\geq 1}$ (not relabelled) and $f \in C([0,\infty), X_{0,w})$ such that

$$f_j \longrightarrow f \ \text{ in } \ C([0,T], X_{0,w}) \ \text{ for all } \ T > 0 \ . \tag{8.2.107}$$

A first consequence of (8.2.107) and the nonnegativity of each f_j is that $f(t)$ is a nonnegative function in X_0 for all $t > 0$.

We now exploit (8.2.98) and the fact that $\psi \in \mathcal{C}_{VP,\infty}$ to improve (8.2.107). To this end, let $T > 0$ and observe that (8.2.98), (8.2.107) and the nonnegativity of ψ ensure that, for all $t \in [0,T]$ and $R > 1$,

$$\int_0^R \psi(x) f(t,x) \, \mathrm{d}x = \lim_{j \to \infty} \int_0^R \psi(x) f_j(t,x) \, \mathrm{d}x \leq C_2(T) \ ,$$

hence

$$M_\psi(f(t)) \leq \liminf_{R \to \infty} \int_0^R \psi(x) f(t,x) \, \mathrm{d}x \leq C_2(T) \ , \qquad t \in [0,T] \ , \tag{8.2.108}$$

the first inequality used above being a consequence of Fatou's lemma. We next consider $\vartheta \in L_\infty(0,\infty)$, $j \geq 1$, $t \in [0,T]$ and $R > 0$. We infer from (8.2.98), (8.2.108) and the monotonicity of $x \mapsto \psi(x)/x$ (Proposition 7.1.9 (c)) that

$$\left| \int_0^\infty x \vartheta(x) (f_j - f)(t,x) \, \mathrm{d}x \right|$$

$$\le \left| \int_0^R x\vartheta(x)(f_j - f)(t,x)\,\mathrm{d}x \right| + \|\vartheta\|_\infty \int_R^\infty x(f + f_j)(t,x)\,\mathrm{d}x$$

$$\le \left| \int_0^R x\vartheta(x)(f_j - f)(t,x)\,\mathrm{d}x \right| + \frac{R}{\psi(R)}\|\vartheta\|_\infty \int_R^\infty \psi(x)(f + f_j)(t,x)\,\mathrm{d}x$$

$$\le \left| \int_0^R x\vartheta(x)(f_j - f)(t,x)\,\mathrm{d}x \right| + \frac{2RC_2(T)}{\psi(R)}\|\vartheta\|_\infty \ .$$

Therefore,

$$\sup_{t\in[0,T]} \left| \int_0^\infty x\vartheta(x)(f_j - f)(t,x)\,\mathrm{d}x \right| \le \sup_{t\in[0,T]} \left| \int_0^R x\vartheta(x)(f_j - f)(t,x)\,\mathrm{d}x \right|$$
$$+ \frac{2RC_2(T)}{\psi(R)}\|\vartheta\|_\infty \ .$$

Owing to (8.2.107), we may let $j \to \infty$ in the previous inequality and obtain

$$\limsup_{j\to\infty} \sup_{t\in[0,T]} \left| \int_0^\infty x\vartheta(x)(f_j - f)(t,x)\,\mathrm{d}x \right| \le \frac{2RC_2(T)}{\psi(R)}\|\vartheta\|_\infty \ .$$

Since $\psi \in \mathcal{C}_{VP,\infty}$, letting $R \to \infty$ gives

$$\lim_{j\to\infty} \sup_{t\in[0,T]} \left| \int_0^\infty x\vartheta(x)(f_j - f)(t,x)\,\mathrm{d}x \right| = 0 \ ,$$

thereby establishing that

$$f_j \longrightarrow f \quad \text{in} \ \ C([0,T], X_{0,w} \cap X_{1,w}) \ \ \text{for all} \ \ T > 0 \ . \tag{8.2.109}$$

Step 3: Mass conservation. Since $f^{in} \in X_1$, a straightforward consequence of (8.2.95) is that

$$\lim_{j\to\infty} M_1(f_j^{in}) = \varrho = M_1(f^{in}) \ ,$$

and combining this with (8.2.96) and (8.2.109) gives

$$M_1(f(t)) = \lim_{j\to\infty} M_1(f_j(t)) = \lim_{j\to\infty} M_1(f_j^{in}) = \varrho \ , \qquad t \ge 0 \ . \tag{8.2.110}$$

Step 4: Limit equation. It remains to verify that the limit f of the sequence $(f_j)_{j\ge1}$ obtained in (8.2.107) is a weak solution to the C-F equation (8.0.1) with coefficients (k,a,b). To this end, we aim to apply the weak stability result (Theorem 8.2.11) and check that the conditions to use it are met. The almost everywhere convergence (8.2.30) as $j \to \infty$ of $(k_j)_{j\ge1}$ and $(a_j)_{j\ge1}$ to k and a, respectively, readily follows from (8.2.94), while the local boundedness (8.2.31) and (8.2.32) of (k_j, k, a_j, a) and the properties (8.2.33) of b are given by (8.2.72), (8.2.80) and (8.2.94). Moreover, for any $T > 0$, (8.2.96) and (8.2.107) ensure the boundedness (8.2.34) of $(M_1(f_j))_{j\ge1}$ and the required weak convergence (8.2.35) (with $m_0 = 0$) of $(f_j)_{j\ge1}$ to f. Finally, since $\psi \in \mathcal{C}_{VP,\infty}$, we deduce from (8.2.100) (with $R = r$) that, for $T > 0$,

$$\limsup_{r\to\infty} \left\{ \int_0^T \int_r^\infty a_j(x)f_j(t,x)\,\mathrm{d}x\mathrm{d}t \right\} \le \lim_{r\to\infty} \frac{2\psi''(0)}{[\psi'(r)]^2}C_2(T) = 0 \ . \tag{8.2.111}$$

We also infer from (8.2.72a), (8.2.96), (8.2.98), (8.2.101) and Proposition 7.1.9 (c) that, for $T > 0$ and $r > 1$,

$$\sup_{j \geq 1} \left\{ \int_0^T \int_r^\infty \int_0^\infty k_j(x,y) f_j(t,x) f_j(t,y) \, dy dx dt \right\}$$

$$\leq K_0 \sup_{j \geq 1} \left\{ \int_0^T [M_0(f_j(t)) + M_1(f_j(t))] \left(\int_r^\infty (1+x) f_j(t,x) \, dx \right) dt \right\}$$

$$\leq 2 K_0 (\varrho + C_4(T)) \frac{r}{\psi(r)} \sup_{j \geq 1} \left\{ \int_0^T M_\psi(f_j(t)) \, dt \right\}$$

$$\leq 2 K_0 T C_2(T)(\varrho + C_4(T)) \frac{r}{\psi(r)} .$$

Since $\psi \in \mathcal{C}_{VP,\infty}$, letting $r \to \infty$ in the previous inequality gives

$$\lim_{r \to \infty} \sup_{j \geq 1} \left\{ \int_0^T \int_r^\infty \int_0^\infty k_j(x,y) f_j(t,x) f_j(t,y) \, dy dx dt \right\} = 0 \tag{8.2.112}$$

for all $T > 0$. Similarly, by (8.2.72a), (8.2.96) and (8.2.101), one has, for $T > 0$ and $r > 1$,

$$\sup_{j \geq 1} \left\{ \int_0^T \int_0^{1/r} \int_0^\infty k_j(x,y) f_j(t,x) f_j(t,y) \, dy dx dt \right\}$$

$$\leq K_0 \sup_{j \geq 1} \left\{ \int_0^T [M_0(f_j(t)) + M_1(f_j(t))] \left(\int_0^{1/r} (1+x) f_j(t,x) \, dx \right) dt \right\}$$

$$\leq 2 K_0 (\varrho + C_4(T)) \sup_{j \geq 1} \left\{ \int_0^T \int_0^{1/r} f_j(t,x) \, dx dt \right\} ,$$

and the right-hand side of the previous inequality converges to zero as $r \to \infty$ according to (8.2.106). Therefore,

$$\lim_{r \to \infty} \sup_{j \geq 1} \left\{ \int_0^T \int_0^{1/r} \int_0^\infty k_j(x,y) f_j(t,x) f_j(t,y) \, dy dx dt \right\} = 0 \tag{8.2.113}$$

for all $T > 0$. Collecting (8.2.111), (8.2.112) and (8.2.113) gives (8.2.37), so that all the assumptions required to apply Theorem 8.2.11 are satisfied. We thus conclude that f is a weak solution to the C-F equation (8.0.1) on $[0,\infty)$ with coefficients (k,a,b), thereby completing the proof of the first part of Theorem 8.2.23.

The last statement of Theorem 8.2.23 is a straightforward consequence of the convergence (8.2.107), Lemma 8.2.26 and Fatou's lemma. □

A consequence of Theorem 8.2.23 is the propagation of (algebraic) moments along time evolution for sublinear coagulation kernels in the sense that, for each $m > 1$, $f(t)$ belongs to X_m for all $t > 0$ as soon as $f^{in} \in X_m$. This property does not extend at all to exponential moments, at least in the absence of fragmentation. Indeed, as observed in [104], when $k \equiv 2$ and $a \equiv 0$, an elementary computation gives

$$\frac{d}{dt} \int_0^\infty (e^{\xi x} - 1) f(t,x) \, dx = \left(\int_0^\infty (e^{\xi x} - 1) f(t,x) \, dx \right)^2 ,$$

for all $\xi > 0$, from which we readily deduce that

$$\lim_{t \to T(\xi)} \int_0^\infty (e^{\xi x} - 1) f(t, x) \, \mathrm{d}x = \infty \,, \qquad T(\xi) := \left(\int_0^\infty (e^{\xi x} - 1) f^{in}(x) \, \mathrm{d}x \right)^{-1} .$$

Therefore, if $f^{in} \in L_1((0, \infty), (e^{\xi x} - 1)\mathrm{d}x)$ for some $\xi > 0$, then f blows up in finite time in this space. Whether fragmentation prevents this phenomenon from occurring seems not to have been studied.

8.2.2.3 Coagulation Kernel with Linear Growth and Non-integrable Daughter Distribution Function b

Let us now investigate the case where the number of fragments $n_0(y)$, defined in (8.2.80), resulting from the breakage of a particle of size $y > 0$ is infinite for all $y > 0$. Specifically, we assume that the coagulation and fragmentation coefficients k, a and b still satisfy (8.2.72), (8.2.73), (8.2.74), (8.2.75) and (8.2.76), but with $m_0 \in (0, 1)$. We further assume that both k and a vanish in a suitable way as $(x, y) \to (0, 0)$: for all $R > 0$, there is $L_R > 0$ such that

$$k(x, y) \le L_R \left(\min\{x, y\} \right)^{m_0} \,, \qquad (x, y) \in (0, R)^2 \,, \tag{8.2.114a}$$

and, moreover, the exponent p_0 in (8.2.74) and (8.2.75) satisfies

$$p_0 \ge 1 + m_0 \,. \tag{8.2.114b}$$

Theorem 8.2.30. *Assume that the coagulation and fragmentation coefficients k, a and b satisfy (8.2.72), (8.2.73), (8.2.74), (8.2.75) with $m_0 \in (0, 1)$, (8.2.76), and (8.2.114). Let $f^{in} \in X_{m_0,+} \cap X_1$ be such that*

$$\int_0^\infty \psi(x) f^{in}(x) \, \mathrm{d}x < \infty \,, \tag{8.2.115}$$

where $\psi \in \mathcal{C}_{VP,\infty}$ satisfies

$$\lim_{x \to \infty} \frac{x\psi'(x) - \psi(x)}{x^{m_0}} = \infty \quad and \quad \delta_0 := \inf_{x \in (1,\infty)} \left\{ \frac{x\psi'(x) - \psi(x)}{x^{m_0}} \right\} > 0 \,. \tag{8.2.116}$$

Then there is at least one mass-conserving weak solution f to the C-F equation (8.0.1) on $[0, \infty)$. Furthermore, if $f^{in} \in X_m$ for some $m > 1$, then $f \in L_\infty((0, T), X_m)$.

Theorem 8.2.30 applies in particular to the coagulation and fragmentation coefficients described in Example 8.2.22 when $\nu \in (-2, -1]$ and $\alpha > -\nu - 1$. Indeed, the assumptions required to apply Theorem 8.2.30 are satisfied for any $m_0 \in (-\nu - 1, \alpha)$ and $p_0 \in (1, (1 + m_0)/|\nu|)$ such that $p_0 \le 1 + \gamma$. Observe that the constraint $\alpha > -\nu - 1$ implies $\alpha > 0$, so that the constant coagulation kernel is not covered by Theorem 8.2.30. Theorem 8.2.30 actually extends the existence result obtained in [162] for $b = b_\nu$ and $\nu \in (-2, -1]$, see Example 8.2.22, to a broader class of coagulation and fragmentation coefficients featuring non-integrable daughter distribution functions b. A further comment is that, unlike Theorem 8.2.23, not all initial conditions in $X_{m_0,+} \cap X_1$ are included in Theorem 8.2.30, due to the additional assumption (8.2.115). Still, it covers a large class of initial conditions in $X_{m_0,+} \cap X_1$, according to Corollary 8.2.36 below.

Remark 8.2.31. *Observe that assumption (8.2.114a) requires the coagulation kernel k to vanish as $(x, y) \to (0, 0)$, a constraint which is not required when $m_0 = 0$. This assumption, in particular, makes Theorem 8.2.30 not applicable to coagulation kernels k behaving as a positive constant for small sizes.*

The proof of Theorem 8.2.30 looks similar to that of Theorem 8.2.23 performed in the previous section. This is true as far as the behaviour for large sizes is concerned but the analysis for small sizes requires several modifications to overcome the lack of integrability of b. We thus begin the proof of Theorem 8.2.30 by establishing a control on the behaviour for large sizes. Observing that the property $f^{in} \in X_0$ is not used in the proof of Lemma 8.2.24, its outcome remains valid and provides the following information.

Lemma 8.2.32. *Assume that the coagulation and fragmentation coefficients k, a and b satisfy (8.2.72), (8.2.73) and (8.2.76). Assume further that there is $r_0 > 1$ such that*

$$k(x, y) = 0 \quad \text{if} \quad x + y > r_0 \quad \text{and} \quad a(x) = 0 \quad \text{if} \quad x > r_0 . \tag{8.2.117}$$

Consider next $f^{in} \in L_1((0, r_0), x^{m_0}\mathrm{d}x)_+$, and let $f \in C^1([0, \infty), L_1((0, r_0), x^{m_0}\mathrm{d}x))$ be the corresponding solution to (8.0.1) given by Proposition 8.2.7. Assume further that there is $\psi \in C_{VP}$ such that $f^{in} \in L_1((0, r_0), \psi(x)\mathrm{d}x)$ and let $C_0 > 0$ and $C_1 > 0$ be such that

$$\int_0^{r_0} (x^{m_0} + x) f^{in}(x) \, \mathrm{d}x \le C_0 , \tag{8.2.118}$$

$$\int_0^{r_0} \psi(x) f^{in}(x) \, \mathrm{d}x \le C_1 . \tag{8.2.119}$$

Given $T > 0$, there is $C_2(T) > 0$ depending only on K_1 in (8.2.73), δ_2 in (8.2.76), C_0 in (8.2.118), (C_1, ψ) in (8.2.119), and T such that

$$\int_0^{r_0} \psi(x) f(t, x) \, \mathrm{d}x \le C_2(T) , \qquad t \in [0, T] , \tag{8.2.120a}$$

$$\int_0^T \int_0^{r_0} a(y) \left(y\psi'(y) - \psi(y) \right) f(s, y) \, \mathrm{d}y\mathrm{d}s \le C_2(T) , \tag{8.2.120b}$$

as well as

$$\int_0^{r_0} x f(t, x) \, \mathrm{d}x = \int_0^{r_0} x f^{in}(x) \, \mathrm{d}x , \qquad t \in [0, T] . \tag{8.2.120c}$$

The next step is the analysis of the behaviour for small sizes which, as usual, is governed by fragmentation. Here we have to cope with the formation of an infinite number of particles during each fragmentation event and, for this purpose, we establish the following analogue of Lemma 8.2.27.

Lemma 8.2.33. *Assume that the coagulation and fragmentation coefficients k, a and b satisfy (8.2.72), (8.2.75) with $m_0 \in (0, 1)$, and (8.2.117) for some $r_0 > 1$. Consider next $f^{in} \in L_1((0, r_0), x^{m_0}\mathrm{d}x)_+$ satisfying (8.2.118) for some $C_0 > 0$ and let $f \in C^1([0, \infty), L_1((0, r_0), x^{m_0}\mathrm{d}x))$ be the corresponding solution to (8.0.1) given by Proposition 8.2.7. Assume further that, for all $T > 0$, there is $C_3(T) > 0$ such that*

$$\int_0^T P(t) \, \mathrm{d}t \le C_3(T) , \qquad P(t) := \int_1^{r_0} y^{m_0} a(y) f(t, y) \, \mathrm{d}y , \qquad t \in [0, T] . \tag{8.2.121}$$

Given $T > 0$, there is $C_4(T) > 0$ depending only on A_1 in (8.2.72b), $\kappa_{m_0,1}$ in (8.2.75), C_0 in (8.2.118), $C_3(T)$ in (8.2.121), and T such that

$$\int_0^{r_0} x^{m_0} f(t, x) \, \mathrm{d}x \le C_4(T) , \qquad t \in [0, T] .$$

Proof. Setting $\vartheta(x) = \max\{x, x^{m_0}\}$ for $x \in (0, \infty)$, we note that

$$
\chi_\vartheta(x, y) = \begin{cases}
(x+y)^{m_0} - x^{m_0} - y^{m_0} \leq 0 & \text{for} \quad (x, y) \in (0,1)^2 , \ x + y \in (0,1) , \\
\begin{aligned} x^{m_0}\left(x^{1-m_0} - 1\right) \\ + y^{m_0}\left(1 - y^{1-m_0}\right) \leq 0 \end{aligned} & \text{for} \quad (x, y) \in (0,1)^2 , \ x + y \geq 1 , \\
x^{m_0}\left(x^{1-m_0} - 1\right) \leq 0 & \text{for} \quad (x, y) \in (0,1) \times [1, \infty) , \\
y^{m_0}\left(y^{1-m_0} - 1\right) \leq 0 & \text{for} \quad (x, y) \in [1, \infty) \times (0,1) , \\
0 & \text{for} \quad (x, y) \in [1, \infty)^2 ,
\end{cases}
$$

while (8.2.72c) and (8.2.75) yield

$$
-N_\vartheta(y) = -y^{m_0} + \int_0^y x^{m_0} b(x, y) \, \mathrm{d}x \leq \kappa_{m_0,1} y^{m_0} \quad \text{for } y \in (0,1) ,
$$

$$
-N_\vartheta(y) = -y + \int_0^1 x^{m_0} b(x, y) \, \mathrm{d}x + \int_1^y x b(x, y) \, \mathrm{d}x \leq \int_0^1 x^{m_0} b(x, y) \, \mathrm{d}x
$$

$$
\leq n_{m_0}(y) \leq \kappa_{m_0,1} y^{m_0} \quad \text{for } y \in [1, \infty) .
$$

Combining the previous inequalities with (8.2.13) and (8.2.72b) gives, for $t \geq 0$,

$$
\frac{d}{dt} \int_0^{r_0} \vartheta(x) f(t, x) \, \mathrm{d}x \leq \int_0^{r_0} \mathscr{F} f(t, x) \vartheta(x) \, \mathrm{d}x
$$

$$
\leq \kappa_{m_0,1} \left[\int_0^1 a(x) x^{m_0} f(t, x) \, \mathrm{d}x + P(t) \right]
$$

$$
\leq \kappa_{m_0,1} \left[A_1 \int_0^1 \vartheta(x) f(t, x) \, \mathrm{d}x + P(t) \right]
$$

$$
\leq (1 + A_1)\kappa_{m_0,1} \left[\int_0^{r_0} \vartheta(x) f(t, x) \, \mathrm{d}x + P(t) \right] .
$$

We now integrate with respect to time and use (8.2.118) and (8.2.121) to conclude that, for $t \geq 0$,

$$
\int_0^{r_0} \vartheta(x) f(t, x) \, \mathrm{d}x \leq \left(\int_0^{r_0} \vartheta(x) f^{in}(x) \, \mathrm{d}x + (1 + A_1)\kappa_{m_0,1} C_3(T) \right) e^{(1+A_1)\kappa_{m_0,1} t}
$$

$$
\leq (C_0 + (1 + A_1)\kappa_{m_0,1} C_3(T)) \, e^{(1+A_1)\kappa_{m_0,1} t} .
$$

Recalling the definition of ϑ completes the proof. $\qquad\square$

We next turn to the uniform integrability which is studied as in Lemma 8.2.28. Since small sizes come into play here as well, the proof of Lemma 8.2.28 has to be suitably adapted.

Lemma 8.2.34. *Assume that the coagulation and fragmentation coefficients k, a and b satisfy (8.2.72), (8.2.74), (8.2.75) with $m_0 \in (0, 1)$, (8.2.114), and (8.2.117) for some $r_0 > 1$. Consider next $f^{in} \in L_1((0, r_0), x^{m_0}\mathrm{d}x)_+$ satisfying (8.2.118) for some $C_0 > 0$ and let $f \in C^1([0, \infty), L_1((0, r_0), x^{m_0}\mathrm{d}x))$ be the corresponding solution to (8.0.1) given by Proposition 8.2.7. Assume further that, for all $T > 0$, there is $C_3(T)$ such that f satisfies (8.2.121) and that there is $\Phi \in \mathcal{C}_{VP}$, and $C_5 > 0$ such that*

$$
\int_0^{r_0} x^{m_0} \Phi(f^{in}(x)) \, \mathrm{d}x \leq C_5 . \tag{8.2.122}
$$

Given $T > 0$ and $R \in (1, r_0)$, there is $C_6(T, R) > 0$ depending only on A_1 in (8.2.72b), $\bar{a}_0$

in (8.2.74), κ_{m_0,p_0} *in* (8.2.75), L_R *in* (8.2.114), C_0 *in* (8.2.118), $C_3(T) > 0$ *in* (8.2.121), Φ *in* (8.2.122), R *and* T *such that*

$$\int_0^R x^{m_0} \Phi(f(t,x)) \, dx \le C_6(T,R) , \qquad t \in [0,T] .$$

Proof. Let $T > 0$ and $R \in (1, r_0)$. We first analyse the contribution of the coagulation term and deduce, from the nonnegativity of Φ', that for $t \in [0,T]$,

$$\int_0^R x^{m_0} \mathcal{C} f(t,x) \Phi'(f(t,x)) \, dx \le \frac{1}{2} J_1(t,R) ,$$

where

$$J_1(t,R) := \int_0^R \int_0^x x^{m_0} k(x-y,y) \Phi'(f(t,x)) f(t,x-y) f(t,y) \, dy dx .$$

Owing to the subadditivity of $x \mapsto x^{m_0}$, the symmetry of k, the convexity of Φ, and Proposition 7.1.9 (b),

$$J_1(t,R) \le \int_0^R \int_0^x ((x-y)^{m_0} + y^{m_0}) \, k(x-y,y) \Phi'(f(t,x)) f(t,x-y) f(t,y) \, dy dx$$

$$\le 2 \int_0^R \int_0^x (x-y)^{m_0} k(x-y,y) \Phi'(f(t,x)) f(t,x-y) f(t,y) \, dy dx$$

$$\le 2 \int_0^R \int_0^x (x-y)^{m_0} k(x-y,y) \left[\Phi(f(t,x)) + \Phi(f(t,x-y)) \right] f(t,y) \, dy dx .$$

We next infer from (8.2.114) and Lemma 8.2.33 that

$$J_1(t,R) \le 2 L_R \int_0^R \int_y^R (x-y)^{m_0} y^{m_0} \Phi(f(t,x-y)) f(t,y) \, dx dy$$

$$+ 2 L_R \int_0^R \int_0^x x^{m_0} y^{m_0} \Phi(f(t,x)) f(t,y) \, dy dx$$

$$\le 4 L_R \left(\int_0^R x^{m_0} \Phi(f(t,x)) \, dx \right) \left(\int_0^R y^{m_0} f(t,y) \, dy \right)$$

$$\le 4 L_R C_4(T) \int_0^R x^{m_0} \Phi(f(t,x)) \, dx .$$

Consequently, for $t \in [0,T]$,

$$\int_0^R x^{m_0} \mathcal{C} f(t,x) \Phi'(f(t,x)) \, dx \le 2 L_R C_4(T) \int_0^R x^{m_0} \Phi(f(t,x)) \, dx . \qquad (8.2.123)$$

We use again the monotonicity and convexity of Φ as well as Proposition 7.1.9 (b) to estimate the contribution of the fragmentation term and obtain that, for $t \in [0,T]$,

$$J_2(t,R) := \int_0^R x^{m_0} \mathcal{F} f(t,x) \Phi'(f(t,x)) \, dx$$

$$\le \int_0^R x^{m_0} \Phi'(f(t,x)) \int_x^{r_0} a(y) b(x,y) f(t,y) \, dy dx$$

$$\le \int_0^R \int_x^{r_0} x^{m_0} a(y) f(t,y) \left[\Phi(b(x,y)) + \Phi(f(t,x)) \right] \, dy dx$$

$$= J_3(t, R) + J_4(t, R) \ ,$$

where

$$J_3(t, R) := \int_0^{r_0} a(y) f(t, y) \int_0^{\min\{y, R\}} x^{m_0} \Phi(b(x, y)) \ \mathrm{d}x \mathrm{d}y \ ,$$

$$J_4(t, R) := \int_0^{r_0} a(y) f(t, y) \int_0^{\min\{y, R\}} x^{m_0} \Phi(f(t, x)) \ \mathrm{d}x \mathrm{d}y \ .$$

On the one hand, by Proposition 7.1.9 (g),

$$\Lambda := \sup_{r > 0} \left\{ \frac{\Phi(r)}{r^{p_0}} \right\} < \infty \ ,$$

and we infer from (8.2.74), (8.2.75) and Lemma 8.2.33 that

$$J_3(t, R) \le \Lambda \int_0^{r_0} a(y) f(t, y) \int_0^{\min\{y, R\}} x^{m_0} b(x, y)^{p_0} \ \mathrm{d}x \mathrm{d}y$$

$$\le \kappa_{m_0, p_0} \Lambda \int_0^{r_0} y^{m_0 + 1 - p_0} a(y) f(t, y) \ \mathrm{d}y$$

$$\le \kappa_{m_0, p_0} \Lambda \bar{a}_0 \int_0^1 y^{m_0} f(t, y) \ \mathrm{d}y$$

$$+ \kappa_{m_0, p_0} \Lambda \int_1^{r_0} y^{m_0} a(y) f(t, y) \ \mathrm{d}y$$

$$\le \kappa_{m_0, p_0} \Lambda \left(\bar{a}_0 C_4(T) + P(t) \right) \ ,$$

recalling that P is defined in (8.2.121). On the other hand, it follows from (8.2.74), (8.2.114b) and Lemma 8.2.33 that

$$J_4(t, R) \le \bar{a}_0 \left(\int_0^1 y^{p_0 - 1} f(t, y) \ \mathrm{d}y \right) \int_0^R x^{m_0} \Phi(f(t, x)) \ \mathrm{d}x$$

$$+ \left(\int_1^{r_0} a(y) f(t, y) \ \mathrm{d}y \right) \int_0^R x^{m_0} \Phi(f(t, x)) \ \mathrm{d}x$$

$$\le \bar{a}_0 \left(\int_0^1 y^{m_0} f(t, y) \ \mathrm{d}y \right) \int_0^R x^{m_0} \Phi(f(t, x)) \ \mathrm{d}x$$

$$+ \left(\int_1^{r_0} y^{m_0} a(y) f(t, y) \ \mathrm{d}y \right) \int_0^R x^{m_0} \Phi(f(t, x)) \ \mathrm{d}x$$

$$\le [\bar{a}_0 C_4(T) + P(t)] \int_0^R x^{m_0} \Phi(f(t, x)) \ \mathrm{d}x \ .$$

Consequently,

$$J_2(t, R) \le C_7(T) (1 + P(t)) \left[1 + \int_0^R x^{m_0} \Phi(f(t, x)) \ \mathrm{d}x \right] \ , \qquad (8.2.124)$$

with $C_7(T) := (1 + \kappa_{m_0, p_0} \Lambda)(1 + \bar{a}_0 C_4(T))$.

Now, setting $C_8(T, R) := 2 L_R C_4(T) + C_7(T)$, it follows from (8.0.1), (8.2.123) and (8.2.124) that for $t \in [0, T]$,

$$\frac{d}{dt} \int_0^R x^{m_0} \Phi(f(t, x)) \ \mathrm{d}x \le C_8(T, R)(1 + P(t)) \left[1 + \int_0^R x^{m_0} \Phi(f(t, x)) \ \mathrm{d}x \right] \ .$$

Hence, by (8.2.121), (8.2.122) and Gronwall's lemma,

$$1 + \int_0^R x^{m_0} \Phi(f(t,x)) \, dx$$

$$\leq \left(1 + \int_0^R x^{m_0} \Phi(f^{in}) \, dx\right) \exp\left\{C_8(T,R)\left(t + \int_0^t P(s) \, ds\right)\right\}$$

$$\leq (1 + C_5)e^{C_8(T,R)(T+C_3(T))}$$

for $t \in [0,T]$ and the proof of Lemma 8.2.34 is complete. $\qquad\qquad\square$

We finally investigate time equicontinuity properties.

Lemma 8.2.35. *Assume that the coagulation and fragmentation coefficients k, a and b satisfy (8.2.72), (8.2.74), (8.2.75) with $m_0 \in (0,1)$, (8.2.114), and (8.2.117) for some $r_0 > 1$. Consider next $f^{in} \in L_1((0,r_0), x^{m_0}dx)_+$ satisfying (8.2.118) for some $C_0 > 0$ and let $f \in C^1([0,\infty), L_1((0,r_0), x^{m_0}dx))$ be the corresponding solution to (8.0.1) given by Proposition 8.2.7. Assume further that, for all $T > 0$, there is $C_3(T) > 0$ such that f satisfies (8.2.121). Given $T > 0$ and $R \in (1, r_0)$, there is $C_9(T,R) > 0$ depending only on K_0 in (8.2.72a), A_R in (8.2.72b), $\kappa_{m_0,1}$ in (8.2.75), L_R in (8.2.114a), C_0 in (8.2.118), R and T such that*

$$\int_0^R x^{m_0}|f(t_2,x) - f(t_1,x)| \, dx \leq C_9(T,R)(t_2 - t_1)$$

$$+ \kappa_{m_0,1} \int_{t_1}^{t_2} \int_R^{r_0} x^{m_0} a(x)f(t,x) \, dx dt$$

for $t_1 \in [0,T]$ and $t_2 \in [t_1, T]$.

Proof. Let $R \in (1, r_0)$, $T > 0$, and $t \in [0,T]$. It follows from (8.2.14), (8.2.72a), (8.2.114a), (8.2.118), (8.2.120c), Fubini's theorem and Lemma 8.2.33 that

$$\int_0^R x^{m_0}|\mathcal{C}f(t,x)| \, dx \leq \frac{1}{2}\int_0^R \int_0^{R-y} (x+y)^{m_0}k(x,y)f(t,x)f(t,y) \, dxdy$$

$$+ \int_0^R \int_0^{r_0} x^{m_0}k(x,y)f(t,x)f(t,y) \, dydx$$

$$\leq L_R \int_0^R \int_0^{R-y} \min\{x,y\}^{m_0}\max\{x,y\}^{m_0}f(t,x)f(t,y) \, dxdy$$

$$+ L_R \int_0^R \int_0^R x^{m_0}\min\{x,y\}^{m_0}f(t,x)f(t,y) \, dydx$$

$$+ K_0 \int_0^R \int_R^{r_0} x^{m_0}(1+x)(1+y)f(t,x)f(t,y) \, dydx$$

$$\leq 2L_R\left(\int_0^R x^{m_0}f(t,x) \, dx\right)^2$$

$$+ 2K_0(1+R)\left(\int_0^R x^{m_0}f(t,x) \, dx\right)\left(\int_R^{r_0} yf(t,y) \, dy\right)$$

$$\leq 2\left(L_R + K_0(1+R)\right)\left[C_4(T)^2 + C_0C_4(T)\right].$$

We next infer from (8.2.72b), (8.2.75) and Lemma 8.2.33 that

$$
\int_0^R x^{m_0} |\mathcal{F}f(t,x)| \ \mathrm{d}x \le \int_0^R y^{m_0} a(y) f(t,y) \ \mathrm{d}y
$$

$$
+ \int_0^{r_0} a(y) f(t,y) \int_0^{\min\{y,R\}} x^{m_0} b(x,y) \ \mathrm{d}x \mathrm{d}y
$$

$$
\le A_R C_4(T) + \kappa_{m_0,1} \int_0^{r_0} y^{m_0} a(y) f(t,y) \ \mathrm{d}y
$$

$$
\le (1 + \kappa_{m_0,1}) A_R C_4(T) + \kappa_{m_0,1} \int_R^{r_0} y^{m_0} a(y) f(t,y) \ \mathrm{d}y \ .
$$

Combining the previous estimates with (8.0.1) readily gives Lemma 8.2.35. $\qquad\square$

Proof of Theorem 8.2.30. Thanks to the previously established lemmas, the proof of Theorem 8.2.30 is similar to that of Theorem 8.2.23, the last statement of Theorem 8.2.30 being a consequence of Lemma 8.2.26 which is valid in the present case as well. The main difference is that the control on

$$
\int_0^T \int_R^{r_0} a(y) y^{m_0} f(t,y) \ \mathrm{d}y \mathrm{d}t
$$

(see (8.2.100)) now has to be derived from (8.2.120b) with the help of the assumption (8.2.116), instead of using Proposition 7.1.9 (f). $\qquad\square$

Theorem 8.2.30 provides the existence of at least one mass-conserving (extended) weak solution to (8.0.1) for initial conditions $f^{in} \in X_{m_0,+} \cap X_1$ which satisfy the additional growth conditions (8.2.115)–(8.2.116) for large sizes. Since the latter is not very informative, let us now show that the class of initial conditions to which Theorem 8.2.30 applies is quite large.

Given an integer $p \ge 2$ we define by induction the function $\exp_p$ by

$$
\exp_{p+1} := \exp \circ \exp_p \ , \qquad p \ge 1 \ , \qquad \exp_1 := \exp \ ,
$$

which is increasing from $\mathbb{R}$ onto $(\exp_{p-1}(0), \infty)$. Let $\ln_p := \exp_p^{-1}$ be its inverse which is defined on $(\exp_{p-1}(0), \infty)$ and positive on $(\exp_p(0), \infty)$.

Corollary 8.2.36. *Assume that the coagulation and fragmentation coefficients k, a and b satisfy (8.2.72), (8.2.73), (8.2.74), (8.2.75) with $m_0 \in (0,1)$, (8.2.76), and (8.2.114). Let $f^{in} \in X_{m_0,+} \cap X_1$ be such that*

$$
\int_0^\infty \left[(x+\zeta) \ln_p(x+\zeta) - (x+\zeta) \ln_p(\zeta) - x\zeta \ln_p'(\zeta) \right] f^{in}(x) \ \mathrm{d}x < \infty
$$

for some $\zeta \ge \zeta_p$, where ζ_p is the unique solution in $(\exp_p(0), \infty)$ to

$$
\sum_{i=0}^{p-1} \left(\frac{1}{\ln_p \zeta_p} \right)^i = \frac{3}{2} \ . \tag{8.2.125}
$$

Then there is at least one mass-conserving weak solution f to the C-F equation (8.0.1) on $[0, \infty)$.

Corollary 8.2.36 shows that the additional integrability of f^{in} for large sizes required in Theorem 8.2.30 is rather mild. Corollary 8.2.36 readily follows from Theorem 8.2.30 as soon as we show that the function

$$W_p(x) := (x + \zeta) \ln_p(x + \zeta) - (x + \zeta) \ln_p(\zeta) - x\zeta \ln'_p(\zeta) , \qquad x \in [0, \infty) , \qquad (8.2.126)$$

satisfies (8.2.116) for $\zeta \geq \zeta_p$ and $p \geq 2$. This is the purpose of the next lemma.

Lemma 8.2.37. *Let $p \geq 2$ and $\zeta \geq \zeta_p$. The function W_p defined in (8.2.126) belongs to $\mathcal{C}_{VP,\infty}$ and satisfies (8.2.116) for all $m_0 \in (0, 1)$.*

Proof. The function $\ln_q$ satisfies the recurrence formula

$$\ln_{q+1} = \ln \circ \ln_q \quad \text{in } (\exp_q(0), \infty) , \qquad q \geq 1 , \qquad \ln_1 = \ln .$$

Setting $\ln_0 := \mathrm{id}$, the chain rule gives, for $q \geq 1$,

$$\ln'_q(x) = \prod_{i=0}^{q-1} \frac{1}{\ln_i(x)} , \qquad x \in (\exp_{q-1}(0), \infty) ,$$

and

$$\ln''_q(x) = -\ln'_q(x) \sum_{i=0}^{q-1} \ln'_{i+1}(x) , \qquad x \in (\exp_{q-1}(0), \infty) .$$

Also, since $x \geq \ln x$ for $x \in (0, \infty)$,

$$\ln_{q+1}(x) \leq \ln_q(x) , \qquad x \in (\exp_q(0), \infty) , \qquad q \geq 1 .$$

Thanks to the previous identities,

$$W'_p(x) = \ln_p(x + \zeta) - \ln_p(\zeta) + (x + \zeta) \ln'_p(x + \zeta) - \zeta \ln'_p(\zeta) , \qquad x \in (0, \infty) ,$$

from which we readily deduce that $W_p(0) = W'_p(0) = 0$, while the monotonicity of $\ln_p$ implies $W'_p(x) \geq 0$ for $x \geq 0$. Next,

$$W''_p(x) = 2 \ln'_p(x + \zeta) - (x + \zeta) \ln'_p(x + \zeta) \sum_{i=0}^{p-1} \ln'_{i+1}(x + \zeta) , \qquad x \in (0, \infty) ,$$

and

$$W'''_p(x) = 3 \ln''_p(x + \zeta) - (x + \zeta) \ln''_p(x + \zeta) \sum_{i=0}^{p-1} \ln'_{i+1}(x + \zeta)$$

$$+ (x + \zeta) \ln'_p(x + \zeta) \sum_{i=0}^{p-1} \sum_{j=0}^{i} \ln'_{j+1}(x + \zeta) \ln'_{i+1}(x + \zeta) , \qquad x \in (0, \infty) .$$

Now, for $i \in \{1, \ldots, p-1\}$ and $x \in [0, \infty)$,

$$\ln'_{i+1}(x + \zeta) = \prod_{j=0}^{i} \frac{1}{\ln_j(x + \zeta)} \leq \frac{1}{x + \zeta} \prod_{j=1}^{i} \frac{1}{\ln_j(\zeta)} \leq \frac{1}{x + \zeta} \left(\frac{1}{\ln_p \zeta} \right)^i ,$$

while

$$\ln'_1(x + \zeta) = \frac{1}{x + \zeta} , \qquad x \in [0, \infty) .$$

Since $\ln_q' \geq 0$ and $\ln_q'' \leq 0$ for all integers $q \in \{1, \ldots, p\}$, we obtain

$$W_p'''(x) \leq -\ln_p''(x+\zeta) \left[\sum_{i=0}^{p-1} \left(\frac{1}{\ln_p \zeta} \right)^i - 3 \right]$$

$$+ \ln_p'(x+\zeta) \sum_{i=0}^{p-1} \ln_{i+1}'(x+\zeta) \sum_{j=0}^{i} \left(\frac{1}{\ln_p \zeta} \right)^j$$

$$\leq -\ln_p''(x+\zeta) \left[\sum_{i=0}^{p-1} \left(\frac{1}{\ln_p \zeta} \right)^i - 3 \right]$$

$$+ \ln_p'(x+\zeta) \sum_{i=0}^{p-1} \ln_{i+1}'(x+\zeta) \sum_{j=0}^{p} \left(\frac{1}{\ln_p \zeta} \right)^j$$

$$\leq -\ln_p''(x+\zeta) \left[2 \sum_{i=0}^{p-1} \left(\frac{1}{\ln_p \zeta} \right)^i - 3 \right], \qquad x \in [0, \infty) .$$

Since $\zeta \geq \zeta_p$ and $1/\ln_p$ is decreasing, we infer from (8.2.125) and the previous inequality that $W_p'''(x) \leq 0$ for $x \in [0, \infty)$ from which we deduce that

$$W_p''(x) \geq \lim_{y \to \infty} W_p''(y) = 0 , \qquad x \in [0, \infty) .$$

We have thus shown that W_p is a nonnegative, convex and increasing function with a concave derivative in $[0, \infty)$. In addition,

$$x W_p'(x) - W_p(x) = x \prod_{i=1}^{p-1} \frac{1}{\ln_i(x+\zeta)} - \zeta \ln_p(x+\zeta) + \zeta \ln_p(\zeta) , \qquad x \in [0, \infty) .$$

Consequently, for any $m_0 \in (0, 1)$,

$$\lim_{x \to \infty} \frac{x W_p'(x) - W_p(x)}{x^{m_0}} = \lim_{x \to \infty} x^{1-m_0} \prod_{i=1}^{p-1} \frac{1}{\ln_i(x+\zeta)} = \infty ,$$

and the proof is complete. $\qquad \square$

8.2.2.4 Strong Fragmentation

We now consider the case where the growth of the coagulation kernel k is dominated by that of the overall fragmentation rate a. More precisely, we assume that there is $K_1 > 0$, $a_0 > 0$, and

$$0 \leq \alpha \leq \beta \leq 1 , \qquad \gamma > \lambda - 1 > 0 , \qquad \lambda := \alpha + \beta , \qquad (8.2.127)$$

such that

$$k(x, y) \leq K_1 \left[(1+x)^\alpha (1+y)^\beta + (1+x)^\beta (1+y)^\alpha \right] , \qquad (x, y) \in (0, \infty)^2 , \qquad (8.2.128)$$

and

$$a(x) \geq a_0 x^\gamma , \qquad x \in (0, \infty) . \qquad (8.2.129)$$

We further assume that a and b satisfy (8.2.72), (8.2.74) with $p_0 \in (1, 1+\gamma)$, (8.2.75) with $m_0 = 0$, and (8.2.76), so that b enjoys the property (8.2.80). As we shall see now, these assumptions also guarantee the existence of a mass-conserving weak solution to the C-F equation (8.0.1).

Theorem 8.2.38. *Assume that the coagulation and fragmentation coefficients k, a and b satisfy (8.2.72), (8.2.74), (8.2.75) with $m_0 = 0$, (8.2.76), (8.2.128) and (8.2.129). Given $f^{in} \in X_{0,1,+}$, there is at least one mass-conserving solution f to the C-F equation (8.0.1) satisfying the following property: for all $m > 1$, there is a positive constant $C(m) > 0$ such that*

$$M_m(f(t)) \le C(m) \left(1 + \frac{1}{t} \right)^{(m-1)/\gamma} , \qquad t > 0 . \tag{8.2.130}$$

Theorem 8.2.38 applies in particular to the coagulation and fragmentation coefficients given in Example 8.2.22 when $\gamma > \alpha + \beta - 1$ and $\nu > -1$. It somewhat extends [109, Theorem 1.2.1] to multiple fragmentation, see also [88] for the discrete C-F equation.

Remark 8.2.39. *The assumption $\lambda > 1$ is actually not restrictive. Indeed, if the coagulation kernel k satisfies (8.2.128) for some $\lambda \le 1$, then it satisfies (8.2.73) so that the existence of at least one mass-conserving solution to the C-F equation (8.0.1) follows from Theorem 8.2.23 (and holds true without requiring the lower bound (8.2.129) on a). If a satisfies (8.2.129), this solution can also be shown to satisfy the property (8.2.130) with the help of Lemma 8.2.41 below.*

Remark 8.2.40. *The assumptions (8.2.127) and (8.2.128) on the coagulation kernel differ slightly from (8.1.31) and (8.1.32) which ensure the existence of classical solutions, see Theorem 8.1.2. In particular, a faster growth of k relative to a is allowed. However, a lower bound on a such as (8.2.129) is not needed in (8.1.31) and (8.1.32), due to the assumed intertwining of the growth of k and a. Another difference is that classical solvability requires more regularity on f^{in} than Theorem 8.2.38.*

As in the proof of Theorem 8.2.23 we begin with estimates for large sizes but here we take advantage of the lower bound (8.2.129) on the growth of a.

Lemma 8.2.41. *Assume that the coagulation and fragmentation coefficients k, a and b satisfy (8.2.72), (8.2.76) and (8.2.128). Assume further that there is $r_0 > 1$, a_0, and $\gamma > \lambda - 1$ such that*

$$k(x,y) = 0 \quad \text{if } x + y > r_0 , \tag{8.2.131}$$

and

$$a(x) = 0 \quad \text{for } x > r_0 , \quad a(x) \ge a_0 x^\gamma \quad \text{for } x \in (0, r_0) . \tag{8.2.132}$$

Consider next $f^{in} \in L_1(0, r_0)_+$, and let $f \in C^1([0, \infty), L_1(0, r_0))$ be the corresponding solution to (8.0.1) given by Proposition 8.2.7. Let $C_0 > 0$ be such that

$$\int_0^{r_0} (1 + x) f^{in}(x) \, \mathrm{d}x \le C_0 . \tag{8.2.133}$$

Given $m > 1$, there is a positive constant $C_{11}(m)$ depending only on δ_2 in (8.2.76), (α, λ, K_1) in (8.2.128), (a_0, γ) in (8.2.132), C_0 in (8.2.133), and m such that

$$\int_0^{r_0} x^m f(t, x) \, \mathrm{d}x \le C_{11}(m) \left(1 + \frac{1}{t} \right)^{(m-1)/\gamma} , \qquad t > 0 .$$

Proof. We extend f to $[0, \infty) \times (0, \infty)$ by setting $f(t, x) = 0$ for $x > r_0$ and $t \ge 0$.

Let $m \ge 2$. On the one hand, thanks to the monotonicity of $x \mapsto x^{m-2}$ and (8.2.76),

$$N_m(y) = y^m - \int_0^y x^m b(x, y) \, \mathrm{d}x \ge y^m - \int_0^y x^2 y^{m-2} b(x, y) \, \mathrm{d}x$$

$$\ge y^{m-2} \left(y^2 - \int_0^y x^2 b(x, y) \, \mathrm{d}x \right) = y^{m-2} N_2(y) \ge \delta_2 y^m , \qquad y \in (0, \infty) .$$

It then follows from (8.2.132) that, for $t > 0$,

$$\int_0^{r_0} \mathscr{F} f(t,x) x^m \, \mathrm{d}x = -\int_0^{r_0} a(x) N_m(x) f(t,x) \, \mathrm{d}x \leq -\delta_2 \int_0^{r_0} a(x) x^m f(t,x) \, \mathrm{d}x$$
$$\leq -a_0 \delta_2 M_{m+\gamma}(f(t)) \ . \tag{8.2.134}$$

On the other hand, by Lemma 7.4.2,

$$\chi_m(x,y) \leq C(m) \left(x^{m-1} y + x y^{m-1} \right) \ , \qquad (x,y) \in (0,\infty)^2 \ ,$$

and we infer from (8.2.128) that, for $t > 0$,

$$\int_0^{r_0} \mathcal{C} f(t,x) x^m \, \mathrm{d}x = \frac{1}{2} \int_0^{r_0} \int_0^{r_0} \chi_m(x,y) k(x,y) f(t,x) f(t,y) \, \mathrm{d}y \mathrm{d}x$$
$$\leq \frac{K_1}{2} \int_0^{r_0} \int_0^{r_0} \chi_m(x,y)(1+x)^\alpha (1+y)^\beta f(t,x) f(t,y) \, \mathrm{d}y \mathrm{d}x$$
$$+ \frac{K_1}{2} \int_0^{r_0} \int_0^{r_0} \chi_m(x,y)(1+x)^\beta (1+y)^\alpha f(t,x) f(t,y) \, \mathrm{d}y \mathrm{d}x$$
$$\leq K_1 \int_0^{r_0} \int_0^{r_0} \chi_m(x,y)(1+x)^\alpha (1+y)^\beta f(t,x) f(t,y) \, \mathrm{d}y \mathrm{d}x$$
$$\leq K_1 C(m) \int_0^{r_0} \int_0^{r_0} x^{m-1} y (1+x^\alpha)(1+y^\beta) f(t,x) f(t,y) \, \mathrm{d}y \mathrm{d}x$$
$$+ K_1 C(m) \int_0^{r_0} \int_0^{r_0} x y^{m-1} (1+x^\alpha)(1+y^\beta) f(t,x) f(t,y) \, \mathrm{d}y \mathrm{d}x$$
$$\leq K_1 C(m) \left[M_{m-1}(f(t)) M_1(f(t)) + M_{m+\alpha-1}(f(t)) M_1(f(t)) \right]$$
$$+ K_1 C(m) \left[M_{m-1}(f(t)) M_{\beta+1}(f(t)) + M_{m+\alpha-1}(f(t)) M_{\beta+1}(f(t)) \right]$$
$$+ K_1 C(m) \left[M_{m-1}(f(t)) M_1(f(t)) + M_{m-1}(f(t)) M_{1+\alpha}(f(t)) \right]$$
$$+ K_1 C(m) \left[M_{m+\beta-1}(f(t)) M_1(f(t)) + M_{m+\beta-1}(f(t)) M_{1+\alpha}(f(t)) \right] \ .$$

Since

$$M_1(f(t)) = \int_0^{r_0} x f(t,x) \, \mathrm{d}x = \int_0^{r_0} x f^{in}(x) \, \mathrm{d}x \leq C_0 \ , \qquad t \geq 0 \ , \tag{8.2.135}$$

by (8.2.14) and (8.2.133), we infer from (8.2.135) and Hölder's and Young's inequalities that

$$M_{m-1}(f(t)) M_1(f(t)) \leq C_0^{(m+2\lambda-2)/(m+\lambda-2)} M_{m+\lambda-1}(f(t))^{(m-2)/(m+\lambda-2)}$$
$$\leq \frac{\lambda}{m+\lambda-2} C_0^2 + \frac{m-2}{m+\lambda-2} C_0 M_{m+\lambda-1}(f(t)) \ ,$$

$$M_{m+\alpha-1}(f(t)) M_1(f(t)) \leq C_0^{(m+2\lambda-\alpha-2)/(m+\lambda-2)} M_{m+\lambda-1}(f(t))^{(m+\alpha-2)/(m+\lambda-2)}$$
$$\leq \frac{\beta}{m+\lambda-2} C_0^2 + \frac{m+\alpha-2}{m+\lambda-2} C_0 M_{m+\lambda-1}(f(t)) \ ,$$

$$M_{m-1}(f(t)) M_{\beta+1}(f(t)) \leq C_0^{(m+\lambda+\alpha-2)/(m+\lambda-2)} M_{m+\lambda-1}(f(t))^{(m+\beta-2)/(m+\lambda-2)}$$
$$\leq \frac{\alpha}{m+\lambda-2} C_0^2 + \frac{m+\beta-2}{m+\lambda-2} C_0 M_{m+\lambda-1}(f(t)) \ ,$$

$$M_{m+\alpha-1}(f(t)) M_{\beta+1}(f(t)) \leq C_0 M_{m+\lambda-1}(f(t)) \ ,$$

$$M_{m-1}(f(t))M_{\alpha+1}(f(t)) \leq C_0^{(m+2\lambda-\alpha-2)/(m+\lambda-2)} M_{m+\lambda-1}(f(t))^{(m+\alpha-2)/(m+\lambda-2)}$$

$$\leq \frac{\beta}{m+\lambda-2}C_0^2 + \frac{m+\alpha-2}{m+\lambda-2}C_0 M_{m+\lambda-1}(f(t)) ,$$

$$M_{m+\beta-1}(f(t))M_1(f(t)) \leq C_0^{(m+\lambda+\alpha-2)/(m+\lambda-2)} M_{m+\lambda-1}(f(t))^{(m+\beta-2)/(m+\lambda-2)}$$

$$\leq \frac{\alpha}{m+\lambda-2}C_0^2 + \frac{m+\beta-2}{m+\lambda-2}C_0 M_{m+\lambda-1}(f(t)) ,$$

and

$$M_{m+\beta-1}(f(t))M_{1+\alpha}(f(t)) \leq C_0 M_{m+\lambda-1}(f(t)) .$$

Combining the above estimates leads us to

$$\int_0^{r_0} \mathcal{C}f(t,x)x^m \, \mathrm{d}x \leq C_{12}(m)\left[1 + M_{m+\lambda-1}(f(t))\right]$$

for some positive constant $C_{12}(m) > 0$ depending only on α, β, K_1, C_0 and m. Since $m + \gamma > m + \lambda - 1 > 1$, a further use of (8.2.135) and Hölder's and Young's inequalities yields the existence of a positive constant $C_{13}(m)$ depending only on α, β, K_1, C_0, a_0, γ, δ_2 and m such that

$$\int_0^{r_0} \mathcal{C}f(t,x)x^m \, \mathrm{d}x \leq C_{12}(m)\left[1 + M_1(f(t))^{(\gamma+1-\lambda)/(m+\gamma-1)} M_{m+\gamma}(f(t))^{(m+\lambda-2)/(m+\gamma-1)}\right]$$

$$\leq C_{13}(m) + \frac{a_0\delta_2}{2}M_{m+\gamma}(f(t)) . \tag{8.2.136}$$

We now infer from (8.2.13) and the estimates (8.2.134) and (8.2.136) that

$$\frac{d}{dt}M_m(f(t)) + \frac{a_0\delta_2}{2}M_{m+\gamma}(f(t)) \leq C_{13}(m) , \qquad t \geq 0 .$$

We use once more Hölder's inequality and (8.2.135) to obtain

$$M_m(f(t)) \leq M_1(f(t))^{\gamma/(m+\gamma-1)} M_{m+\gamma}(f(t))^{(m-1)/(m+\gamma-1)}$$

$$\leq C_0^{\gamma/(m+\gamma-1)} M_{m+\gamma}(f(t))^{(m-1)/(m+\gamma-1)} ,$$

and thus

$$\frac{d}{dt}M_m(f(t)) + \frac{a_0\delta_2}{2C_0^{\gamma/(m-1)}}M_m(f(t))^{(m+\gamma-1)/(m-1)} \leq C_{13}(m) , \qquad t \geq 0 . \tag{8.2.137}$$

Introducing

$$Y(t) := \left(Y_1 + \frac{Y_2}{t}\right)^{(m-1)/\gamma} , \qquad t > 0 ,$$

with

$$Y_2 := \frac{2(m-1)}{\gamma a_0 \delta_2}C_0^{\gamma/(m-1)} , \qquad Y_1 = \left(\frac{\gamma Y_2 C_{13}(m)}{m-1}\right)^{\gamma/(m+\gamma-1)} ,$$

an easy computation shows that Y is a supersolution to (8.2.137) such that $Y(t) \to \infty$ as $t \to 0$. The comparison principle then implies that $M_m(f(t)) \leq Y(t)$ for $t > 0$, which proves Lemma 8.2.41 for $m \geq 2$.

Finally, for $m \in (1,2)$, combining Hölder's inequality, Lemma 8.2.41 for $m = 2$, and (8.2.135) gives

$$M_m(f(t)) \leq M_1(f(t))^{2-m} M_2(f(t))^{m-1} \leq C_0^{2-m} C_{11}(2)^{m-1}\left(1 + \frac{1}{t}\right)^{(m-1)/\gamma}$$

for $t > 0$, thereby completing the proof of Lemma 8.2.41.   □

The next step is to show that, though the growth of k allowed by (8.2.128) may be faster than the linear one required by (8.2.73), an analogue of Lemma 8.2.24 may be derived from Lemma 8.2.41.

Lemma 8.2.42. *Assume that the coagulation and fragmentation coefficients k, a and b satisfy (8.2.72), (8.2.75) with $m_0 = 0$, (8.2.76), (8.2.128), as well as (8.2.131) and (8.2.132) for some $r_0 > 0$. Consider next $f^{in} \in L_1(0, r_0)_+$ satisfying (8.2.133) for some $C_0 > 0$, and let $f \in C^1([0, \infty), L_1(0, r_0))$ be the corresponding solution to (8.0.1) given by Proposition 8.2.7. Assume further that there is $\psi \in \mathcal{C}_{VP}$ such that $f^{in} \in L_1((0, r_0), \psi(x)\mathrm{d}x)$ and let $C_1 > 0$ be such that*

$$\int_0^{r_0} \psi(x) f^{in}(x) \, \mathrm{d}x \leq C_1 . \tag{8.2.138}$$

Given $T > 0$, there is $C_{14}(T) > 0$ depending only on δ_2 in (8.2.76), (α, β, K_1) in (8.2.128), C_0 in (8.2.133), (C_1, ψ) in (8.2.138), and T such that

$$\int_0^{r_0} \psi(x) f(t, x) \, \mathrm{d}x \leq C_{14}(T) , \qquad t \in [0, T] , \tag{8.2.139a}$$

$$\int_0^T \int_0^{r_0} a(y) \left(y \psi'(y) - \psi(y) \right) f(s, y) \, \mathrm{d}y \mathrm{d}s \leq C_{14}(T) , \tag{8.2.139b}$$

as well as

$$\int_0^{r_0} x f(t, x) \, \mathrm{d}x = \int_0^{r_0} x f^{in}(x) \, \mathrm{d}x , \qquad t \in [0, T] . \tag{8.2.139c}$$

Proof. We extend f to $[0, \infty) \times (0, \infty)$ by setting $f(t, x) = 0$ for $x > r_0$ and $t \geq 0$. We next recall that Proposition 8.2.7 guarantees that (8.2.139c) holds true and thus, thanks to (8.2.133),

$$\int_0^{r_0} x f(t, x) \, \mathrm{d}x \leq C_0 , \qquad t \geq 0 . \tag{8.2.140}$$

Let $(x, y) \in (0, \infty)^2$. Since

$$\chi_\psi(x, y) = \int_0^x \int_0^y \psi''(x_* + y_*) \, \mathrm{d}y_* \mathrm{d}x_* \in [0, \psi''(0)xy]$$

by the convexity of ψ, the concavity of ψ', and

$$\chi_\psi(x, y) \leq \frac{2x}{x + y} \psi(y) + \frac{2y}{x + y} \psi(x)$$

by Proposition 7.1.9 (e), one has

$$\begin{aligned}
0 \leq{} & \chi_\psi(x, y)(1 + x)^\alpha (1 + y)^\beta \\
\leq{} & \chi_\psi(x, y) \left(1 + x^\alpha + y^\beta + x^\alpha y^\beta \right) \\
\leq{} & \psi''(0)xy + \left(x^\alpha + y^\beta + x^\alpha y^\beta \right) \left(\frac{2x}{x + y} \psi(y) + \frac{2y}{x + y} \psi(x) \right) \\
\leq{} & \psi''(0)xy + 2 \left[\frac{x}{(x + y)^{1-\alpha}} + \frac{x}{(x + y)^{1-\beta}} + \frac{x^{1+\alpha}}{(x + y)^{1-\beta}} \right] \psi(y) \\
& + 2 \left[\frac{y}{(x + y)^{1-\alpha}} + \frac{y}{(x + y)^{1-\beta}} + \frac{y^{1+\beta}}{(x + y)^{1-\alpha}} \right] \psi(x) \\
\leq{} & \psi''(0)xy + 2 \left[\frac{x}{y^{1-\alpha}} + \frac{x}{y^{1-\beta}} + x^\lambda \right] \psi(y)
\end{aligned}$$

$$+ 2 \left[\frac{y}{x^{1-\alpha}} + \frac{y}{x^{1-\beta}} + y^\lambda \right] \psi(x) .$$

Let $t \geq 0$. Thanks to (8.2.128), the nonnegativity of χ_ψ and the above estimate, we find

$$\int_0^{r_0} \mathcal{C} f(t,x) \psi(x) \, \mathrm{d}x \leq \frac{K_1}{2} \int_0^{r_0} \int_0^{r_0} \chi_\psi(x,y)(1+x)^\alpha (1+y)^\beta f(t,x) f(t,y) \, \mathrm{d}y\mathrm{d}x$$

$$+ \frac{K_1}{2} \int_0^{r_0} \int_0^{r_0} \chi_\psi(x,y)(1+x)^\beta (1+y)^\alpha f(t,x) f(t,y) \, \mathrm{d}y\mathrm{d}x$$

$$\leq K_1 \int_0^{r_0} \int_0^{r_0} \chi_\psi(x,y)(1+x)^\alpha (1+y)^\beta f(t,x) f(t,y) \, \mathrm{d}y\mathrm{d}x$$

$$\leq K_1 \psi''(0) M_1(f(t))^2 + 4K_1 M_\lambda(f(t)) M_\psi(f(t))$$

$$+ 4K_1 M_1(f(t)) \int_0^{r_0} \left(\frac{\psi(y)}{y^{1-\alpha}} + \frac{\psi(y)}{y^{1-\beta}} \right) f(t,y) \, \mathrm{d}y ,$$

hence

$$\int_0^{r_0} \mathcal{C} f(t,x) \psi(x) \, \mathrm{d}x \leq K_1 \psi''(0) C_0^2 + 4K_1 M_\lambda(f(t)) M_\psi(f(t))$$

$$+ 4K_1 C_0 \int_0^{r_0} \left(\frac{\psi(y)}{y^{1-\alpha}} + \frac{\psi(y)}{y^{1-\beta}} \right) f(t,y) \, \mathrm{d}y$$

by (8.2.140). Now, observe that, when $\alpha = 1$,

$$\int_0^{r_0} \frac{\psi(y)}{y^{1-\alpha}} f(t,y) \, \mathrm{d}y = M_\psi(f(t)) ,$$

whereas, when $\alpha \in [0,1)$,

$$\int_0^{r_0} \frac{\psi(y)}{y^{1-\alpha}} f(t,y) \, \mathrm{d}y \leq \sup_{y>0} \left\{ \frac{\psi(y)}{y^{2-\alpha}} \right\} M_1(f(t)) ,$$

which is finite by Proposition 7.1.9 (g) since $2 - \alpha \in (1,2]$. A similar result being true when α is replaced by β, we finally obtain that there is a positive constant $C_{15} > 0$ depending only on K_1, α, β, C_0 and ψ such that

$$\int_0^{r_0} \mathcal{C} f(t,x) \psi(x) \, \mathrm{d}x \leq C_{15} \left(1 + M_\psi(f(t)) + M_\lambda(f(t)) M_\psi(f(t)) \right) , \qquad t \geq 0 . \qquad (8.2.141)$$

Next, concerning the fragmentation term, we proceed as in the proof of Lemma 8.2.24 and deduce from (8.2.76) that

$$\int_0^{r_0} \mathcal{F} f(t,x) \psi(x) \, \mathrm{d}x \leq -\delta_2 \int_0^{r_0} a(y)(y\psi'(y) - \psi(y)) f(t,y) \, \mathrm{d}y , \qquad t \geq 0 . \qquad (8.2.142)$$

Now, introducing

$$Y(t) := M_\psi(f(t)) + \delta_2 \int_0^t \int_0^{r_0} a(y)(y\psi'(y) - \psi(y)) f(s,y) \, \mathrm{d}y\mathrm{d}s , \qquad t \geq 0 ,$$

we infer from (8.2.13), (8.2.141) and (8.2.142) that, for $t \geq 0$,

$$\frac{d}{dt} Y(t) \leq C_{15} \left[1 + Y(t) + M_\lambda(f(t)) Y(t) \right] .$$

Integrating with respect to time and using $Y(0) = M_\psi(f^{in})$, we end up with

$$Y(t) \le (Y(0) + C_{15}t) \exp\left(C_{15} \int_0^t [1 + M_\lambda(f(s))] \, ds\right) , \qquad t \ge 0 . \qquad (8.2.143)$$

Since $\lambda \in (1, \gamma + 1)$, it follows from Lemma 8.2.41 that, for $t \ge 0$,

$$\int_0^t M_\lambda(f(s)) \, ds \le C(\lambda) \int_0^t \left(1 + \frac{1}{s}\right)^{(\lambda-1)/\gamma} \, ds \le C_{11}(\lambda) \left(t + \frac{\gamma}{\gamma - \lambda + 1} t^{(\gamma-\lambda+1)/\gamma}\right) .$$

Combining this inequality with (8.2.138) and (8.2.143) ends the proof of Lemma 8.2.42. □

At this point, we recall that the other results used in the proof of Theorem 8.2.23 (Lemma 8.2.27, Lemma 8.2.28, and Lemma 8.2.29) do not rely on the specific growth assumptions on k and a but only on (8.2.72), (8.2.74), (8.2.75) with $m_0 = 0$, and a suitable control on $M_a(f)$ which is provided by (8.2.139b). We are thus in a position to prove Theorem 8.2.38.

Proof of Theorem 8.2.38. As in the proof of Theorem 8.2.23 we put

$$\varrho := M_1(f^{in}) \quad \text{and} \quad C_0 := M_0(f^{in}) + \varrho .$$

Then, $f^{in} \in L_1(0, \infty)$ and $x \mapsto x \in L_1((0, \infty), f^{in}(x)dx)$, so that, according to the de la Vallée-Poussin theorem (Theorem 7.1.6), there is $\Phi \in \mathcal{C}_{VP,\infty}$, and $\psi \in \mathcal{C}_{VP,\infty}$ such that

$$\Phi(f^{in}) \in L_1(0, \infty) \quad \text{and} \quad M_\psi(f^{in}) < \infty .$$

Next, given a positive integer $j \ge 1$ and $(x, y) \in (0, \infty)^2$, we define

$$k_j(x, y) := k(x, y)\mathbf{1}_{(0,j)}(x + y) , \qquad a_j(x) := a(x)\mathbf{1}_{(0,j)}(x) ,$$

and

$$f_j^{in}(x) := f^{in}(x)\mathbf{1}_{(0,j)}(x) .$$

Then f_j^{in} is a nonnegative function in $L_1(0, j)$ and k_j and a_j satisfy (8.2.131) and (8.2.132) with $r_0 = j$, so that the C-F equation (8.0.1) with coefficients (k_j, a_j, b) has a unique solution $f_j \in C^1([0, \infty), L_1(0, j))$ by Proposition 8.2.7. We extend f_j to $[0, \infty) \times (0, \infty)$ by setting $f_j(t, x) = 0$ for $(t, x) \in [0, \infty) \times (j, \infty)$. We first infer from Lemma 8.2.41 that, for all $m > 1$ and $j \ge 1$,

$$M_m(f_j(t)) \le C_{11}(m) \left(1 + \frac{1}{t}\right)^{(m-1)/\gamma} , \qquad t > 0 . \qquad (8.2.144)$$

Moreover, Lemma 8.2.42 ensures the validity of (8.2.139b) which, together with the properties of ψ (see Proposition 7.1.9 (f)) implies that f_j satisfies (8.2.86) with a constant which does not depend on j. The assumptions required to use Lemmas 8.2.27, 8.2.28 and 8.2.29 being satisfied, we argue as in the proof of Theorem 8.2.23 to obtain the existence of a subsequence of $(f_j)_{j\ge1}$ (not relabelled) and a nonnegative function $f \in C([0, \infty), X_{0,w} \cap X_{1,w})$ such that

$$f_j \longrightarrow f \quad \text{in} \quad C([0, T], X_{0,w} \cap X_{1,w}) \qquad (8.2.145)$$

for all $T > 0$. The function f is a mass-conserving weak solution on $[0, \infty)$ to the C-F equation (8.0.1) with coefficients (k, a, b). Finally, it follows from (8.2.144) and (8.2.145) that for all $m > 1$, $t > 0$ and $R > 0$,

$$\int_0^R x^m f(t, x) \, \mathrm{d}x = \lim_{j \to \infty} \int_0^R x^m f_j(t, x) \, \mathrm{d}x \leq C_{11}(m) \left(1 + \frac{1}{t} \right)^{(m-1)/\gamma} .$$

Letting $R \to \infty$ in the previous inequality gives (8.2.130) by Fatou's lemma. $\square$

We finally point out that there are coagulation and fragmentation coefficients which do not satisfy the assumptions of either Theorem 8.2.23 or Theorem 8.2.38 but for which the existence of mass-conserving solutions is nevertheless available [95, 113, 164, 156, 165, 173]. In these papers, the authors either deal with some specific choices of coagulation and fragmentation coefficients which allow one to use the Laplace transform to derive a more tractable equation [95, 173] or follow basically the approach depicted in the proofs of Theorems 8.2.23 and 8.2.38 [113, 164, 156, 165]. We refer to Sections 10.3.2 and 10.3.3.2 for a more precise account of these results.

8.2.3   Weak Solutions

We now turn to situations where we cannot expect the existence of mass-conserving solutions to the C-F equation (8.0.1) on $[0, \infty)$ due to the already mentioned gelation phenomenon. Nevertheless, weak solutions with a nonincreasing total mass may be constructed, as we show now. As before, we assume that there is $K_0 > 0$ such that

$$k(x, y) \leq K_0 (1 + x)(1 + y) , \qquad (x, y) \in (0, \infty)^2 . \qquad (8.2.146a)$$

We also exclude overall fragmentation rates featuring a singularity for small sizes and thus require that for any $r > 0$ there is $A_r > 0$ such that

$$a(x) \leq A_r , \qquad x \in (0, r) . \qquad (8.2.146b)$$

Finally, we always assume that there is no loss of matter during fragmentation events; that is,

$$\int_0^y x b(x, y) \, \mathrm{d}x = y , \qquad y \in (0, \infty) . \qquad (8.2.146c)$$

Next, in sharp contrast with the previous section, here the only expected control for large sizes is the obvious bound on the total mass, and this feature places strong constraints on the growth of k and a. We actually handle simultaneously integrable and non-integrable daughter distribution functions b and assume that there is $m_0 \in [0, 1)$, and $\kappa_{m_0, 1} > 0$ such that

$$n_{m_0}(y) = \int_0^y x^{m_0} b(x, y) \, \mathrm{d}x \leq \kappa_{m_0, 1} y^{m_0} , \qquad y > 0 . \qquad (8.2.147a)$$

We supplement (8.2.147a) with a joint uniform integrability property of a and b; that is, we assume that there exists a positive function $\omega \in C([0, \infty)^2)$ such that for all $R \geq 0$ and $\varepsilon > 0$, $\omega(R, 0) = 0$ and

$$a(y) \int_0^{\min\{y, R\}} x^{m_0} \mathbf{1}_E(x) b(x, y) \, \mathrm{d}x \leq \omega(R, \varepsilon) \left(y^{m_0} + y \right) \qquad (8.2.147b)$$

for all $y > 0$ and all measurable subsets E of $(0, \infty)$ satisfying $|E| \leq \varepsilon$. We also require that k vanishes appropriately as $(x, y) \to (0, 0)$ in the following sense: for any $R > 0$ there is $L_R > 0$ such that

$$k(x, y) \leq L_R \left(\min\{x, y\} \right)^{m_0} , \qquad (x, y) \in (0, R)^2 . \qquad (8.2.147c)$$

Note that, when $m_0 = 0$, the assumption (8.2.146a) obviously implies (8.2.147c).

Finally, a suitable control on the growth of both k and a for large sizes is needed: specifically, we assume that

$$\lim_{\substack{x \to \infty \\ y > 0}} \sup \frac{k(x, y)}{(x^{m_0} + x)(y^{m_0} + y)} = 0 , \tag{8.2.147d}$$

and

$$\lim_{x \to \infty} \frac{a(x) n_{m_0}(x)}{x} = 0 . \tag{8.2.147e}$$

Weak solutions to the C-F equation (8.0.1) exist when the above assumptions are satisfied, as stated now.

Theorem 8.2.43. *Assume that the coagulation and fragmentation coefficients k, a and b satisfy (8.2.146) and (8.2.147). Given $f^{in} \in X_{m_0, +} \cap X_1$, there exists at least one weak solution f to the C-F equation (8.0.1) on $[0, \infty)$ which satisfies*

$$M_1(f(t)) \leq M_1(f^{in}) , \qquad t \geq 0 . \tag{8.2.148}$$

Several results in the spirit of Theorem 8.2.43 are already available in the literature [124, 127, 128, 167, 252]. We also refer to [189, 250, 273] for similar results for the discrete C-F equations. As already pointed out, (8.2.147b) and (8.2.147e) place strong restrictions on the growth of a. This is easily seen in Example 8.2.22 to which Theorem 8.2.43 only applies when $0 \leq \alpha \leq \beta < 1$, $\gamma \in (0, \min\{1, 2 + \nu\})$, and $\alpha > -(1 + \nu)$. Note that it implies $\gamma \in (0, 1)$ when $\nu \geq -1$ whereas only the smaller range $\gamma \in (0, 2 + \nu)$ is allowed for when $\nu \in (-2, -1)$. Nevertheless, that Theorem 8.2.43 applies to non-integrable daughter distribution functions b is a novel contribution.

The proof of Theorem 8.2.43 involves two steps: we first exploit the boundedness (8.2.147a) and the local uniform integrability (8.2.147b) of b, together with (8.2.147c), to obtain sequential weak compactness in X_{m_0} pointwise in time, along with weak equicontinuity with respect to time. We next use the bounds (8.2.147d) and (8.2.147e) on the growth of k and a for large sizes, as well as Theorem 8.2.11, to pass to the limit as the approximation parameter diverges to infinity. To begin with, we study the behaviour for small sizes.

Lemma 8.2.44. *Assume that the coagulation and fragmentation coefficients k, a and b satisfy (8.2.146) and (8.2.147). Assume further that there is $r_0 > 1$ such that*

$$k(x, y) = 0 \quad \text{if } x + y > r_0 \quad \text{and} \quad a(x) = 0 \quad \text{if } x > r_0 . \tag{8.2.149}$$

Consider next $f^{in} \in L_1((0, r_0), x^{m_0} dx)_+$ and let $f \in C^1([0, \infty), L_1(0, r_0))$ be the corresponding solution to the C-F equation (8.0.1) given by Proposition 8.2.7. Besides, let $C_0 > 0$ be such that

$$\int_0^{r_0} (x^{m_0} + x) f^{in}(x) \, dx \leq C_0 . \tag{8.2.150}$$

Given $T > 0$ there is $C_1(T) > 0$ depending only on $\kappa_{m_0, 1}$ in (8.2.147a), C_0 in (8.2.150), and T such that

$$\int_0^{r_0} (x^{m_0} + x) f(t, x) \, dx \leq C_1(T) , \qquad t \in [0, T] .$$

Proof. Setting $\vartheta(x) = \max\{x, x^{m_0}\}$ for $x \in (0, \infty)$, we recall that

$$\chi_\vartheta(x, y) \leq 0 , \qquad (x, y) \in (0, \infty)^2 ,$$

see the proof of Lemma 8.2.33, while (8.2.146c) and (8.2.147a) ensure that

$$-N_\vartheta(y) = -y^{m_0} + \int_0^y x^{m_0} b(x,y)\ dx \le \kappa_{m_0,1} y^{m_0} \quad \text{for } y \in (0,1)\ ,$$

$$-N_\vartheta(y) = -y + \int_0^1 x^{m_0} b(x,y)\ dx + \int_1^y x b(x,y)\ dx \le \int_0^1 x^{m_0} b(x,y)\ dx$$
$$\le n_{m_0}(y) \le \kappa_{m_0,1} y^{m_0} \le \kappa_{m_0,1} y \quad \text{for } y \in [1,\infty)\ .$$

It then follows from (8.2.13) that

$$\frac{d}{dt} \int_0^{r_0} \vartheta(x) f(t,x)\ dx \le \kappa_{m_0,1} \int_0^{r_0} \vartheta(y) f(t,y)\ dy\ , \qquad t \in [0,\infty)\ .$$

Integrating with respect to time and using (8.2.150), we find

$$\int_0^{r_0} \vartheta(x) f(t,x)\ dx \le e^{\kappa_{m_0,1} t} \int_0^{r_0} \vartheta(x) f^{in}(x)\ dx \le C_0 e^{\kappa_{m_0,1} t}\ , \qquad t \in [0,\infty)\ ,$$

which gives Lemma 8.2.44. □

The previous lemma ensures the boundedness of weak solutions in X_{m_0} and we shall now improve this to uniform integrability. In contrast to the proof of Lemmas 8.2.28 and 8.2.34, we do not use here the de la Vallée-Poussin theorem, but the original definition of uniform integrability, see Definition 7.1.2, as in the pioneering work of Stewart [252].

Lemma 8.2.45. *Assume that the coagulation and fragmentation coefficients k, a and b satisfy (8.2.146) and (8.2.147). Assume further that there is $r_0 > 1$ such that (8.2.149) holds true. Consider next $f^{in} \in L_1((0,r_0), x^{m_0} dx)_+$ satisfying (8.2.150) and let $f \in C^1([0,\infty), L_1((0,r_0), x^{m_0} dx))$ be the corresponding solution to the C-F equation (8.0.1) given by Proposition 8.2.7. Given $T > 0$, $R \in (0,r_0)$, and $\varepsilon > 0$, there is $C_2(T,R) > 0$ depending only on $\kappa_{m_0,1}$ in (8.2.147a), L_R in (8.2.147c), C_0 in (8.2.149), R and T such that*

$$\xi(t,R,\varepsilon) := \sup \left\{ \int_E x^{m_0} f(t,x)\ dx\ :\ E \subset (0,R)\ ,\ |E| \le \varepsilon \right\}$$
$$\le C_2(T,R) \left[\sup \left\{ \int_E x^{m_0} f^{in}(x)\ dx\ :\ E \subset (0,r_0)\ ,\ |E| \le \varepsilon \right\} + \omega(R,\varepsilon) \right]\ ,$$

for all $t \in [0,T]$, where ω is defined in (8.2.147b).

Proof. Let $T > 0$, $\varepsilon > 0$, $R \in (0,r_0)$ and $t \in [0,T]$. Given a measurable subset E of $(0,R)$ satisfying $|E| \le \varepsilon$, we infer from (8.2.13) with $\vartheta(x) = x^{m_0} \mathbf{1}_E(x)$, $x \in (0,r_0)$, (8.2.147b), and (8.2.147c) that

$$\frac{d}{dt} \int_0^{r_0} x^{m_0} \mathbf{1}_E(x) f(t,x)\ dx$$
$$\le \frac{1}{2} \int_0^R \int_0^{R-x} k(x,y)(x+y)^{m_0} \mathbf{1}_E(x+y) f(t,x) f(t,y)\ dy dx$$
$$+ \int_0^{r_0} a(y) f(t,y) \int_0^{\min\{y,R\}} x^{m_0} \mathbf{1}_E(x) b(x,y)\ dx dy$$
$$\le L_R \int_0^R \int_0^R \min\{x,y\}^{m_0} \max\{x,y\}^{m_0} \mathbf{1}_E(x+y) f(t,x) f(t,y)\ dy dx$$

$$+ \omega(R, \varepsilon) \int_0^{r_0} (y^{m_0} + y) f(t, y) \, \mathrm{d}y$$

$$\leq L_R \int_0^R x^{m_0} f(t, x) \int_0^R y^{m_0} \mathbf{1}_{-x+E}(y) f(t, y) \, \mathrm{d}y \mathrm{d}x$$

$$+ \omega(R, \varepsilon) \int_0^{r_0} (y^{m_0} + y) f(t, y) \, \mathrm{d}y \ .$$

We now deduce from Lemma 8.2.44 and the properties $-x + E \subset (0, R)$ and $|-x + E| = |E|$ that

$$\frac{d}{dt} \int_0^{r_0} x^{m_0} \mathbf{1}_E(x) f(t, x) \, \mathrm{d}x \leq L_R \xi(t, R, \varepsilon) \int_0^{r_0} x^{m_0} f(t, x) \, \mathrm{d}x + C_1(T) \omega(R, \varepsilon)$$

$$\leq C_1(T) \left[L_R \xi(t, R, \varepsilon) + \omega(R, \varepsilon) \right] \ .$$

Integrating with respect to time gives, for $t \in [0, T]$,

$$\int_0^{r_0} x^{m_0} \mathbf{1}_E(x) f(t, x) \, \mathrm{d}x \leq \int_0^{r_0} x^{m_0} \mathbf{1}_E(x) f^{in}(x) \, \mathrm{d}x$$

$$+ C_1(T) \left[L_R \int_0^t \xi(s, R, \varepsilon) \, \mathrm{d}s + \omega(R, \varepsilon) t \right]$$

$$\leq \xi(0, R, \varepsilon) + C_1(T) \left[L_R \int_0^t \xi(s, R, \varepsilon) \, \mathrm{d}s + T \omega(R, \varepsilon) \right] \ .$$

Taking the supremum with respect to all measurable subsets $E \subset (0, R)$ satisfying $|E| \leq \varepsilon$, we end up with

$$\xi(t, R, \varepsilon) \leq \xi(0, R, \varepsilon) + T C_1(T) \omega(R, \varepsilon) + C_1(T) L_R \int_0^t \xi(s, R, \varepsilon) \, \mathrm{d}s$$

for all $t \in [0, T]$. Using Gronwall's lemma gives

$$\xi(t, R, \varepsilon) \leq \left[\xi(0, R, \varepsilon) + T C_1(T) \omega(R, \varepsilon) \right] e^{C_1(T) L_R t} \ ,$$

and completes the proof of Lemma 8.2.45. □

Let us finally study the time equicontinuity.

Lemma 8.2.46. *Assume that the coagulation and fragmentation coefficients k, a and b satisfy (8.2.146) and (8.2.147). Assume further that there is $r_0 > 1$ such that (8.2.149) holds true. Consider next $f^{in} \in L_1((0, r_0), x^{m_0} \mathrm{d}x)_+$ satisfying (8.2.150) and let $f \in C^1([0, \infty), L_1((0, r_0), x^{m_0} \mathrm{d}x))$ be the corresponding solution to the C-F equation (8.0.1) given by Proposition 8.2.7. Given $T > 0$ and $R \in (1, r_0)$, there is $C_3(T, R) > 0$ depending only on K_0 in (8.2.146a), A_R in (8.2.146b), $\kappa_{m_0,1}$ in (8.2.147a), $\omega(R, R)$ in (8.2.147b), L_R in (8.2.147c), C_0 in (8.2.149), R and T such that*

$$\int_0^R x^{m_0} |f(t_2, x) - f(t_1, x)| \, \mathrm{d}x \leq C_3(T, R)(t_2 - t_1) \ , \qquad 0 \leq t_1 \leq t_2 \leq T \ .$$

Proof. Let $R \in (1, r_0)$, $T > 0$, and $t \in [0, T]$. Owing to (8.2.146a), (8.2.147c), Fubini's theorem and Lemma 8.2.44, we obtain

$$\int_0^R x^{m_0} |\mathcal{C} f(t, x)| \, \mathrm{d}x \leq \frac{1}{2} \int_0^R \int_0^{R-y} (x + y)^{m_0} k(x, y) f(t, x) f(t, y) \, \mathrm{d}x \mathrm{d}y$$

$$+ \int_0^R \int_0^{r_0} x^{m_0} k(x,y) f(t,x) f(t,y) \; \mathrm{d}y \mathrm{d}x$$

$$\leq L_R \int_0^R \int_0^{R-y} \min\{x,y\}^{m_0} \max\{x,y\}^{m_0} f(t,x) f(t,y) \; \mathrm{d}x \mathrm{d}y$$

$$+ L_R \int_0^R \int_0^R x^{m_0} \min\{x,y\}^{m_0} f(t,x) f(t,y) \; \mathrm{d}y \mathrm{d}x$$

$$+ K_0 \int_0^R \int_R^{r_0} x^{m_0} (1+x)(1+y) f(t,x) f(t,y) \; \mathrm{d}y \mathrm{d}x$$

$$\leq 2 L_R \left(\int_0^R x^{m_0} f(t,x) \; \mathrm{d}x \right)^2$$

$$+ 2 K_0 (1+R) \left(\int_0^R x^{m_0} f(t,x) \; \mathrm{d}x \right) \left(\int_R^{r_0} y f(t,y) \; \mathrm{d}y \right)$$

$$\leq 2 \left(L_R + K_0(1+R) \right) C_1(T)^2 \; .$$

Next, it follows from (8.2.146b), (8.2.147b) and Lemma 8.2.44 that

$$\int_0^R x^{m_0} |\mathscr{F} f(t,x)| \; \mathrm{d}x \leq \int_0^R y^{m_0} a(y) f(t,y) \; \mathrm{d}y$$

$$+ \int_0^{r_0} a(y) f(t,y) \int_0^{\min\{y,R\}} x^{m_0} b(x,y) \; \mathrm{d}x \mathrm{d}y$$

$$\leq A_R C_1(T) + \omega(R,R) \int_0^{r_0} (y^{m_0} + y) f(t,y) \; \mathrm{d}y$$

$$\leq C_1(T) \left(A_R + \omega(R,R) \right) \; .$$

Consequently, by the C-F equation (8.0.1),

$$\int_0^R x^{m_0} |\partial_t f(t,x)| \; \mathrm{d}x \leq C_3(T,R) \; , \qquad t \in [0,T] \; ,$$

from which Lemma 8.2.46 follows.   □

We are now in a position to complete the proof of Theorem 8.2.43 and thereby show the existence of weak solutions.

Proof of Theorem 8.2.43. Let k, a and b be coagulation and fragmentation coefficients satisfying (8.2.146) and (8.2.147). Also let $f^{in} \in X_{m_0,+} \cap X_1$ and put

$$\varrho := M_1(f^{in}) \quad \text{and} \quad C_0 := M_{m_0}(f^{in}) + \varrho \; . \tag{8.2.151}$$

As before, we construct a sequence $(k_j, a_j, b)_{j \geq 1}$ of approximations of (k, a, b) as follows. Given a positive integer $j \geq 1$, we define

$$k_j(x,y) := k(x,y) \mathbf{1}_{(0,j)}(x+y) \; , \qquad a_j(x) := a(x) \mathbf{1}_{(0,j)}(x) \; , \tag{8.2.152}$$

for $(x,y) \in (0,\infty)^2$ and

$$f_j^{in}(x) := f^{in}(x) \mathbf{1}_{(0,j)}(x) \; , \qquad x \in (0,\infty) \; . \tag{8.2.153}$$

Clearly (8.2.152) ensures that, for $j \geq 1$, (k_j, a_j, b) satisfies (8.2.146) and (8.2.147) uniformly with respect to j. Furthermore, f_j^{in} is a nonnegative function in $L_1((0,j), x^{m_0} dx)$ and it follows from Proposition 8.2.7 that the C-F equation (8.0.1) with coefficients (k_j, a_j, b) has a unique solution $f_j \in C^1([0,\infty), L_1((0,j), x^{m_0} dx))$. We extend f_j to $[0,\infty) \times (0,\infty)$ by setting $f_j(t,x) = 0$ for $(t,x) \in [0,\infty) \times (j,\infty)$. Also, by (8.2.14),

$$M_1(f_j(t)) = \int_0^j x f_j(t,x) \, dx = \int_0^j x f_j^{in}(x) \, dx = \int_0^j x f^{in}(x) \, dx \le \varrho \,, \qquad t \ge 0 \,. \quad (8.2.154)$$

Step 1: Compactness. We infer from Lemma 8.2.44, Lemma 8.2.45 and Lemma 8.2.46 that, given $T > 0$ and $R > 0$, there is $C_4(T) > 0$, depending only on $\kappa_{m_0,1}$, C_0, and T, and $C_5(T,R)$, depending only on $\kappa_{m_0,1}$, C_0, L_R, K_0, A_R, $\omega(R,R)$, R and T, such that, for all $\varepsilon > 0$, $j \ge 1$, and $(t,s) \in [0,T]^2$,

$$M_{m_0}(f_j(t)) + M_1(f_j(t)) \le C_4(T) \,, \quad (8.2.155)$$

$$\sup \left\{ \int_E x^{m_0} f_j(t,x) \, dx \ : \ E \subset (0,R) \,, \ |E| \le \varepsilon \right\}$$
$$\le C_5(T,R) \left[\eta \left\{ \{f^{in}\}, \varepsilon; X_{m_0} \right\} + \omega(R,\varepsilon) \right] \,, \quad (8.2.156)$$

and

$$\int_0^R x^{m_0} |f_j(t,x)) - f_j(s,x)| \, dx \le C_5(T,R)|t-s| \,, \quad (8.2.157)$$

the quantity $\eta \left\{ \{f^{in}\}, \varepsilon; X_{m_0} \right\}$ being defined in Definition 7.1.2.

Now, let $T > 0$ and $\varepsilon > 0$ and define

$$\mathcal{E}(T) := \{f_j(t) \ : \ t \in [0,T] \,, \ j \ge 1\} \,.$$

If E is a measurable subset of $(0,\infty)$ with $|E| \le \varepsilon$, then it follows from (8.2.155) and (8.2.156) that, for $j \ge 1$, $R > 0$ and $t \in [0,T]$,

$$\int_E x^{m_0} f_j(t,x) \, dx \le \int_{E \cap (0,R)} x^{m_0} f_j(t,x) \, dx + \int_R^\infty x^{m_0} f_j(t,x) \, dx$$
$$\le C_5(T,R) \left[\eta \left\{ \{f^{in}\}, \varepsilon; X_{m_0} \right\} + \omega(R,\varepsilon) \right] + \frac{C_4(T)}{R^{1-m_0}} \,.$$

Therefore,

$$\eta\{\mathcal{E}(T), \varepsilon; X_{m_0}\} \le C_5(T,R) \left[\eta \left\{ \{f^{in}\}, \varepsilon; X_{m_0} \right\} + \omega(R,\varepsilon) \right] + \frac{C_4(T)}{R^{1-m_0}}$$

for all $R > 0$ and $\varepsilon > 0$. We now let $\varepsilon \to 0$ and use the integrability of $x \mapsto x^{m_0} f^{in}(x)$ in $(0,\infty)$ and the property $\omega(R,0) = 0$ to deduce that the modulus of uniform integrability in X_{m_0}, see (7.1.2), satisfies

$$\eta\{\mathcal{E}(T); X_{m_0}\} = \lim_{\varepsilon \to 0} \eta\{\mathcal{E}(T), \varepsilon; X_{m_0}\} \le \frac{C_4(T)}{R^{1-m_0}}$$

for all $R > 0$. Letting $R \to \infty$, we end up with

$$\eta\{\mathcal{E}(T); X_{m_0}\} = 0 \,. \quad (8.2.158)$$

Furthermore, for $j \ge 1$, $t \in [0,T]$ and $R > 0$, we infer from (8.2.155) that

$$\int_R^\infty x^{m_0} f_j(t,x) \, dx \le \frac{C_4(T)}{R^{1-m_0}} \,,$$

so that

$$\lim_{R\to\infty} \sup_{g\in\mathcal{E}(T)} \left\{ \int_R^\infty x^{m_0} g(x)\,\mathrm{d}x \right\} = 0 \ . \tag{8.2.159}$$

In addition, (8.2.155) and (8.2.157) imply that, for all $j \geq 1$, $t_1 \in [0,T]$, $t_2 \in [t_1, T]$ and $R > 0$,

$$\int_0^\infty x^{m_0} |f_j(t_2, x) - f_j(t_1, x)|\,\mathrm{d}x \leq \int_0^R x^{m_0} |f_j(t_2, x) - f_j(t_1, x)|\,\mathrm{d}x$$

$$+ \int_R^\infty x^{m_0} [f_j(t_2, x) + f_j(t_1, x)]\,\mathrm{d}x$$

$$\leq C_5(T, R)(t_2 - t_1) + \frac{2C_4(T)}{R^{1-m_0}} \ ,$$

and thus

$$\int_0^\infty x^{m_0} |f_j(t_2, x) - f_j(t_1, x)|\,\mathrm{d}x \leq \omega_1(T, t_2 - t_1) \ , \qquad 0 \leq t_1 \leq t_2 \leq T \ , \ j \geq 1 \ , \tag{8.2.160}$$

where

$$\omega_1(T, s) := \inf_{R>0} \left\{ C_5(T, R)s + \frac{2C_4(T)}{R^{1-m_0}} \right\} \ , \qquad s \in [0, T] \ .$$

Since

$$\limsup_{s\to 0} \omega_1(T, s) \leq \limsup_{s\to 0} \left(C_5(T, R)s + \frac{2C_4(T)}{R^{1-m_0}} \right) = \frac{2C_4(T)}{R^{1-m_0}}$$

for all $R > 0$, we may pass to the limit as $R \to \infty$ in the previous inequality and conclude that

$$\lim_{s\to 0} \omega_1(T, s) = 0 \ . \tag{8.2.161}$$

Owing to (8.2.160) and (8.2.161), the sequence $(f_j)_{j\geq 1}$ is strongly, and thus also weakly, equicontinuous in X_{m_0} at each $t \in [0,T]$ in the sense of Definition 7.1.15. This property, together with the relative sequential weak compactness of $\mathcal{E}(T)$ in X_{m_0}, which follows from (8.2.158), (8.2.159) and the Dunford–Pettis theorem (Theorem 7.1.3), allows us to apply the version of the Arzelà–Ascoli theorem recalled in Theorem 7.1.16 and conclude that $(f_j)_{j\geq 1}$ is relatively sequentially compact in $C([0,T], X_{m_0,w})$. As T is arbitrary, there is a subsequence of $(f_j)_{j\geq 1}$ (not relabelled), and $f \in C([0,\infty), X_{m_0,w})$ such that

$$f_j \longrightarrow f \quad \text{in } C([0,T], X_{m_0,w}) \text{ for all } T > 0 \ . \tag{8.2.162}$$

Thanks to (8.2.154), (8.2.155), (8.2.162), and the nonnegativity of f_j for each $j \geq 1$, the function $f(t)$ is a nonnegative function in $X_{m_0} \cap X_1$ for all $t > 0$ and it satisfies

$$M_1(f(t)) \leq M_1(f^{in}) \ , \qquad t \geq 0 \ , \tag{8.2.163a}$$

$$M_{m_0}(f(t)) \leq C_4(T) \ , \qquad t \in [0,T] \ , \qquad T > 0 \ . \tag{8.2.163b}$$

Step 2: Limit equation. To complete the proof, we shall pass to the limit as $j \to \infty$ with the help of Theorem 8.2.11 and thus check that the assumptions required for its use are satisfied. The almost everywhere convergence (8.2.30) of (k_j, a_j) to (k, a) is a straightforward consequence of (8.2.152), while the upper bounds (8.2.31) and (8.2.32) follow from (8.2.146a), (8.2.146b) and (8.2.152). That b satisfies (8.2.33) is guaranteed by (8.2.146c) and (8.2.147a) and the boundedness (8.2.34) in X_1 and the convergence (8.2.35) in $C([0, T], X_{m_0,w})$ are provided by (8.2.155) and (8.2.162), respectively. We are left with the control (8.2.36)–(8.2.37) for small and large sizes. Let us start with the fragmentation term. Since

$$x^{m_0} a(x) = x^{m_0-1} a(x) \int_0^x y b(y, x) \, dy \leq a(x) n_{m_0}(x) , \qquad x \in (0, \infty) ,$$

by (8.2.146c), we infer from (8.2.152) and (8.2.155) that, for $T > 0$, $r > 1$ and $j \geq 1$,

$$\int_0^T \int_r^\infty x^{m_0} a_j(x) f_j(t, x) \, dxdt \leq \sup_{x>r} \left\{ \frac{a(x) n_{m_0}(x)}{x} \right\} \int_0^T \int_r^\infty x f_j(t, x) \, dxdt$$

$$\leq T C_4(T) \sup_{x>r} \left\{ \frac{a(x) n_{m_0}(x)}{x} \right\} .$$

Thanks to (8.2.147e), we obtain

$$\lim_{r\to\infty} \sup_{j\geq 1} \int_0^T \int_r^\infty x^{m_0} a_j(x) f_j(t, x) \, dxdt = 0 . \tag{8.2.164}$$

Next it follows from (8.2.152) and (8.2.155) that, for $T > 0$, $r > 1$ and $j \geq 1$,

$$\int_0^T \int_r^\infty \int_0^\infty k_j(x, y) f_j(t, x) f_j(t, y) \, dydxdt$$

$$\leq \sup_{x>r,y>0} \left\{ \frac{k(x, y)}{(x^{m_0} + x)(y^{m_0} + y)} \right\}$$

$$\times \int_0^T \int_r^\infty \int_0^\infty (x^{m_0} + x) f_j(t, x)(y^{m_0} + y) f_j(t, y) \, dydxdt$$

$$\leq 4 T C_4(T)^2 \sup_{x>r,y>0} \left\{ \frac{k(x, y)}{(x^{m_0} + x)(y^{m_0} + y)} \right\} ,$$

and (8.2.147d) implies that

$$\lim_{r\to\infty} \sup_{j\geq 1} \int_0^T \int_r^\infty \int_0^\infty k_j(x, y) f_j(t, x) f_j(t, y) \, dydxdt = 0 . \tag{8.2.165}$$

Finally, by (8.2.147c), (8.2.147d), (8.2.152), (8.2.155) and the symmetry of k,

$$\int_0^T \int_0^{1/r} \int_0^\infty x^{m_0} k_j(x, y) f_j(t, x) f_j(t, y) \, dydxdt$$

$$\leq L_1 \int_0^T \int_0^{1/r} \int_0^1 x^{m_0} y^{m_0} f_j(t, x) f_j(t, y) \, dydxdt$$

$$+ \int_0^T \int_0^{1/r} \int_1^\infty x^{m_0} k(x, y) f_j(t, x) f_j(t, y) \, dydxdt$$

$$\leq L_1 T C_4(T) \eta \left\{ \mathcal{E}(T), \frac{1}{r}; X_{m_0} \right\}$$

$$+ \frac{1}{r^{m_0}} \int_0^T \int_1^\infty \int_0^{1/r} k(x, y) f(t, x) f(t, y) \, dxdydt$$

$$\leq L_1 T C_4(T) \eta \left\{ \mathcal{E}(T), \frac{1}{r}; X_{m_0} \right\}$$

$$+ \frac{4TC_4(T)^2}{r^{m_0}} \sup_{x>1,y>0} \left\{ \frac{k(x,y)}{(x^{m_0}+x)(y^{m_0}+y)} \right\} .$$

Recalling (8.2.158), we deduce from the previous inequality that

$$\limsup_{r\to\infty} \sum_{j\geq 1} \int_0^T \int_0^{1/r} \int_0^\infty x^{m_0} k_j(x,y) f_j(t,x) f_j(t,y) \, dy dx dt = 0 . \qquad (8.2.166)$$

Collecting (8.2.164), (8.2.165) and (8.2.166), we have established (8.2.36) and (8.2.37). It remains to apply Theorem 8.2.11 to conclude that f is a weak solution to the C-F equation (8.0.1) on $[0,\infty)$ with coefficients (k,a,b). □

Theorem 8.2.43 applies in particular to the coagulation and fragmentation coefficients from Example 8.2.22

$$k(x,y) = x^\alpha y^\beta + x^\beta y^\alpha , \quad a(x) = x^\gamma , \quad b(x,y) = b_\nu(x,y) = \frac{\nu+2}{y} \left(\frac{x}{y} \right)^\nu ,$$

when $0 \leq \alpha \leq \beta < 1$, $\gamma \in (0,1)$, and $\alpha + \nu > -1$. Note that α has to be positive if $\nu \in (-2,-1]$; that is, when the daughter distribution function b_ν is not integrable.

We conclude this section with the existence of weak solutions to the C-F equation (8.0.1) for coagulation kernels not satisfying (8.2.146a) but having a "product structure". More precisely, assume that there are nonnegative functions $k_0 \in C([0,\infty))$ and $k_1 \in C([0,\infty)^2)$ and a positive constant K_1 such that

$$k(x,y) = k_0(x)k_0(y) + k_1(x,y) , \qquad (x,y) \in (0,\infty)^2 , \qquad (8.2.167a)$$

and

$$0 \leq k_1(x,y) = k_1(y,x) \leq K_1 k_0(x) k_0(y) , \qquad (x,y) \in (0,\infty)^2 . \qquad (8.2.167b)$$

We next assume the fragmentation to be suitably dominated by k_0 in the following sense: the overall fragmentation rate a and the daughter distribution function b satisfy (8.2.146b) and (8.2.146c), and there is a positive constant $K_2 > 0$, and a nonincreasing function $\omega_2 \in C([0,\infty))$ such that $\omega(x) \to 0$ as $x \to \infty$ and

$$a(x)b(y,x) \leq K_2 \left(1 + \max\{x, k_0(x)\} \right) , \qquad 0 < y < x , \qquad (8.2.167c)$$

$$a(x)n_0(x) \leq \omega_2(x) \max\{x, k_0(x)\} , \qquad x \in (0,\infty) . \qquad (8.2.167d)$$

Observe that no growth condition is required on the function k_0.

Theorem 8.2.47. *[158] Assume that the coagulation and fragmentation coefficients k, a and b satisfy (8.2.146b), (8.2.146c) and (8.2.167). Given $f^{in} \in X_{0,1,+}$, there exists at least one weak solution f to the C-F equation (8.0.1) on $[0,\infty)$ which satisfies*

$$M_1(f(t)) \leq M_1(f^{in}) , \qquad t \geq 0 . \qquad (8.2.168)$$

We refer to [158] for the proof of Theorem 8.2.47 which follows the lines of that of Theorem 8.2.43. The main difference is that the boundedness of the first moment is no longer sufficient to control the behaviour for large sizes as the growth of k_0 is arbitrary. Instead, we take advantage of the product structure (8.2.167a) of the coagulation kernel k and the cornerstone of the proof is the estimate

$$\int_0^t \left[\int_R^\infty k_0(x) f(s,x) \, dx \right]^2 ds \leq 2 \left(\frac{1}{R} + t\omega_2(R) \right) M_1(f^{in}) + \omega_2(R)^2 , \qquad (8.2.169)$$

which is valid for $t > 0$ and $R \geq 1$ and provides the missing information necessary to control the behaviour for large sizes. We provide a sketch of the proof below, bearing in mind that the derivation of (8.2.169) is of course carried out on a C-F equation with truncated coefficients as in the proof of Theorem 8.2.43. Let $t > 0$ and $R \geq 1$. Using $\vartheta(x) = \min\{R, x\}$, $x \in (0, \infty)$, as a test function in (8.2.4d) and noticing that

$$\chi_\vartheta(x, y) \leq 0 , \qquad (x, y) \in (0, \infty)^2 , \qquad \chi_\vartheta(x, y) \leq -R , \qquad (x, y) \in (R, \infty)^2 ,$$
$$N_\vartheta(y) = 0 , \qquad y \in (0, R) , \qquad N_\vartheta(y) \geq -R n_0(y) , \qquad y \in (R, \infty) ,$$

we obtain

$$R \int_0^t \int_R^\infty \int_R^\infty k(x, y) f(s, x) f(s, y) \, dy dx ds$$
$$\leq \int_0^\infty \vartheta(x) f^{in}(x) \, dx + R \int_0^t \int_R^\infty a(y) n_0(y) f(s, y) \, dy ds .$$

Thanks to (8.2.167a) and (8.2.167d), we further obtain

$$\int_0^t \left[\int_R^\infty k_0(x) f(s, x) \, dx \right]^2 ds \leq \frac{M_1(f^{in})}{R} + \omega_2(R) \int_0^t \int_R^\infty (y + k_0(y)) f(s, y) \, dy ds ,$$

hence (8.2.169), after using the boundedness of $M_1(f)$ and the Cauchy–Schwarz inequality. Let us point out that the approximation of the coagulation kernel designed for the proof of Theorem 8.2.47 differs from that given by (8.2.152) which is used in the proof of Theorem 8.2.43. Indeed, it is chosen to be $k_j(x, y) = k(x, y) \mathbf{1}_{(0,j)}(x) \mathbf{1}_{(0,j)}(y)$, $(x, y) \in (0, \infty)^2$, for $j \geq 1$, in order to exploit the product structure and derive the counterpart of (8.2.169) at the level of the approximation.

8.2.4 Singular Coagulation Coefficients

In this section, our aim is to handle coagulation kernels featuring a singularity as $x \to 0$ and/or $y \to 0$, such as Smoluchowski's coagulation kernel

$$k(x, y) = \left(x^{1/3} + y^{1/3} \right) \left(\frac{1}{x^{1/3}} + \frac{1}{y^{1/3}} \right) , \qquad (x, y) \in (0, \infty)^2 ,$$

see [247, 248] and Chapter 2 in Volume I. More precisely, we assume that there is $K_1 > 0$, and $\sigma > 0$ such that

$$0 \leq k(x, y) = k(y, x) \leq K_1 (xy)^{-\sigma} , \qquad (x, y) \in (0, 1) \times (0, 1) , \qquad \text{(8.2.170a)}$$

$$0 \leq k(x, y) = k(y, x) \leq K_1 y x^{-\sigma} , \qquad (x, y) \in (0, 1) \times (1, \infty) , \qquad \text{(8.2.170b)}$$

and

$$0 \leq k(x, y) = k(y, x) \leq K_1 (x + y) , \qquad (x, y) \in [1, \infty)^2 . \qquad \text{(8.2.170c)}$$

Next, a control on the growth of the overall fragmentation rate is required: for any $r > 0$ there is $A_r > 0$ such that

$$a(x) \leq A_r , \qquad x \in (0, r) , \qquad \text{(8.2.171a)}$$

and there is $\bar{A} > 0$ such that

$$a(x) \leq \bar{A} x^{1+2\sigma} , \qquad x \in [1, \infty) . \qquad \text{(8.2.171b)}$$

In addition, the daughter distribution function b is a nonnegative measurable function defined on $\{(x,y) \in (0,\infty)^2 \ : \ x < y\}$ satisfying

$$\int_0^y x b(x,y) \ \mathrm{d}x = y \ , \qquad y \in (0,\infty) \ , \tag{8.2.172a}$$

and there is $p_0 > 1$ satisfying $p_0 - 1 \le \sigma p_0$, and a positive constant $\kappa_1 > 0$ such that

$$y^{2\sigma} \int_0^y x^{-2\sigma} b(x,y) \ \mathrm{d}x + y^{p_0(1+\sigma)-1} \int_0^y x^{-\sigma p_0} b(x,y)^{p_0} \ \mathrm{d}x \le \kappa_1 \tag{8.2.172b}$$

for $y \in (0,\infty)$. Finally, we assume that there is $\delta_2 > 0$ such that

$$N_2(y) = y^2 - \int_0^y x^2 b(x,y) \ \mathrm{d}x \ge \delta_2 y^2 \ , \qquad y \in (0,\infty) \ . \tag{8.2.172c}$$

These assumptions ensure the existence of a mass-conserving weak solution to (8.0.1) on $[0,\infty)$.

Theorem 8.2.48. *Assume that the coagulation and fragmentation coefficients k, a and b satisfy (8.2.170), (8.2.171) and (8.2.172). Given $f^{in} \in X_{-2\sigma,+} \cap X_1$, there exists at least one mass-conserving solution to the C-F equation (8.0.1) on $[0,\infty)$ which additionally belongs to $L_\infty((0,T), X_{-2\sigma})$ for all $T > 0$.*

Remark 8.2.49. *The assumption (8.2.170a) provides only an upper bound on the singularity of k for small sizes and leaves the actual singularity somewhat undetermined. In particular, the three kernels $K(x,y) = (xy)^{-\sigma}$, $K(x,y) = (x+y)^{-2\sigma}$ and $K(x,y) = x^{-\sigma} + y^{-\sigma}$ all satisfy (8.2.170) but blow up in a different way as $x \to 0$ and/or $y \to 0$. As already observed in [39, 86, 87, 113], this weak control on the singularity requires the existence of weak solutions to be investigated in a smaller space, namely $X_{-2\sigma} \cap X_1$ instead of $X_0 \cap X_1$. If more precise information on the singularity of k is available, existence can be obtained in a larger space [111, 113]. We shall come back to this issue in Section 10.3.3.2.*

We begin with an estimate of the moment of order -2σ.

Lemma 8.2.50. *Assume that the coagulation and fragmentation coefficients k, a and b satisfy (8.2.170), (8.2.171) and (8.2.172). Assume further that there is $r_0 > 1$ such that*

$$k(x,y) = 0 \ \text{ if } \ x + y > r_0 \ \text{ or } \ \min\{x,y\} < r_0 \ , \qquad a(x) = 0 \ \text{ if } \ x > r_0 \ . \tag{8.2.173}$$

Consider next $f^{in} \in L_1((0,r_0), (1 + x^{-2\sigma})\mathrm{d}x)_+$ and let $f \in C^1([0,\infty), L_1(0,r_0))$ be the corresponding solution to the C-F equation (8.0.1) given by Proposition 8.2.7. Besides, let $C_0 > 0$ be such that

$$\int_0^{r_0} \left(x^{-2\sigma} + x \right) f^{in}(x) \ \mathrm{d}x \le C_0 \ . \tag{8.2.174}$$

Given $T > 0$ there is $C_1(T) > 0$ depending only on σ, $(A_1, \bar{A})$ in (8.2.171), κ_1 in (8.2.172), C_0 in (8.2.174), and T such that, for all $m \in [-2\sigma, 1]$,

$$\int_0^{r_0} x^m f(t,x) \ \mathrm{d}x \le C_1(T) \ , \qquad t \in [0,T] \ . $$

Proof. We first recall that Proposition 8.2.7 and (8.2.174) guarantee that

$$\int_0^{r_0} x f(t,x) \ \mathrm{d}x = \int_0^{r_0} x f^{in}(x) \ \mathrm{d}x \le C_0 \ , \qquad t \ge 0 \ . \tag{8.2.175}$$

Next, let $t \geq 0$. For $\varepsilon \in (0,1)$ and $x \in (0, \infty)$, we define $\vartheta(x) = (x + \varepsilon)^{-2\sigma}$ and note that $\chi_\vartheta \leq 0$, so that

$$\int_0^{r_0} \mathcal{C}f(t,x)(x + \varepsilon)^{-2\sigma} \, \mathrm{d}x \leq 0 .$$

We next infer from (8.2.171), (8.2.172b), (8.2.175) and the monotonicity of $x \mapsto (x/(x+\varepsilon))^{2\sigma}$ that

$$\int_0^{r_0} \mathcal{F}f(t,x)(x+\varepsilon)^{-2\sigma} \, \mathrm{d}x \leq \int_0^{r_0} a(y)f(t,y) \int_0^y (x+\varepsilon)^{-2\sigma}b(x,y) \, \mathrm{d}x\mathrm{d}y$$

$$\leq \int_0^{r_0} a(y)f(t,y) \left(\frac{y}{y+\varepsilon}\right)^{2\sigma} \int_0^y x^{-2\sigma}b(x,y) \, \mathrm{d}x\mathrm{d}y$$

$$\leq \kappa_1 \int_0^{r_0} a(y)(y+\varepsilon)^{-2\sigma}f(t,y) \, \mathrm{d}y$$

$$\leq \kappa_1 A_1 \int_0^1 (y+\varepsilon)^{-2\sigma}f(t,y) \, \mathrm{d}y + \kappa_1 \bar{A} \int_1^{r_0} yf(t,y) \, \mathrm{d}y$$

$$\leq \kappa_1 A_1 \int_0^{r_0} (y+\varepsilon)^{-2\sigma}f(t,y) \, \mathrm{d}y + C_0\kappa_1\bar{A} .$$

It then follows from (8.2.13) and the previous two estimates that

$$\frac{d}{dt}\int_0^{r_0} (y+\varepsilon)^{-2\sigma}f(t,y) \, \mathrm{d}y \leq \kappa_1 A_1 \int_0^{r_0}(y+\varepsilon)^{-2\sigma}f(t,y) \, \mathrm{d}y + C_0\kappa_1\bar{A} , \qquad t \geq 0 ,$$

and thus, after integration with respect to time,

$$\int_0^{r_0} (y+\varepsilon)^{-2\sigma}f(t,y) \, \mathrm{d}y \leq e^{\kappa_1 A_1 t}\left[\int_0^{r_0}(y+\varepsilon)^{-2\sigma}f^{in}(y) \, \mathrm{d}y + C_0\frac{\bar{A}}{A_1}\right] , \qquad t \geq 0 .$$

Owing to (8.2.174), the right-hand side of the above inequality is bounded independently of ε, so that we may let $\varepsilon \to 0$ and deduce from Fatou's lemma that $f(t) \in L_1((0,r_0), x^{-2\sigma}\mathrm{d}x)$ and

$$\int_0^{r_0} y^{-2\sigma}f(t,y) \, \mathrm{d}y \leq C_0 e^{\kappa_1 A_1 t}\left(1 + \frac{\bar{A}}{A_1}\right) , \qquad t \geq 0 . \tag{8.2.176}$$

Finally, let $m \in [-2\sigma, 1]$. Thanks to the elementary inequality

$$x^m \leq x^{-2\sigma} + x , \qquad x \in (0, \infty) ,$$

the statement of Lemma 8.2.50 follows at once from (8.2.175) and (8.2.176). $\qquad \square$

We next turn to the control of f for large sizes and proceed as in the proof of Lemma 8.2.24, adapting the argument to cope with the possible unboundedness of k for small sizes.

Lemma 8.2.51. *Assume that the coagulation and fragmentation coefficients k, a and b satisfy (8.2.170), (8.2.171) and (8.2.172). Assume further that there is $r_0 > 1$ such that (8.2.173) holds true. Consider next $f^{in} \in L_1((0,r_0), (1+x^{-2\sigma})\mathrm{d}x)_+$ satisfying (8.2.174) and let $f \in C^1([0,\infty), L_1(0,r_0))$ be the corresponding solution to the C-F equation (8.0.1) given by Proposition 8.2.7. Assume moreover that $f^{in} \in L_1((0,r_0), \psi(x)\mathrm{d}x)$ for some $\psi \in \mathcal{C}_{VP}$ and let $C_0' > 0$ be such that*

$$\int_0^{r_0} \psi(x)f^{in}(x) \, \mathrm{d}x \leq C_0' . \tag{8.2.177}$$

Given $T > 0$ there is $C_2(T) > 0$ depending only on σ, K_1 in (8.2.170), $(A_1, \bar{A})$ in (8.2.171), κ_1 in (8.2.172), C_0 in (8.2.174), (ψ, C_0') in (8.2.177), and T such that

$$\int_0^{r_0} \psi(x) f(t,x) \, dx \le C_2(T) \,, \qquad t \in [0,T] \,, \tag{8.2.178}$$

$$\int_0^T \int_0^{r_0} a(y)(y\psi'(y) - \psi(y)) f(s,y) \, dy ds \le C_2(T) \,. \tag{8.2.179}$$

Proof. Let $T > 0$ and $t \in [0,T]$. To estimate the contribution of the coagulation term to the evolution of the moment of $f(t)$ associated to the weight ψ we split it into three contributions:

$$\int_0^{r_0} \mathcal{C} f(t,x)\psi(x) \, dx = J_1(t) + J_2(t) + J_3(t) \,,$$

with

$$J_1(t) := \frac{1}{2} \int_0^1 \int_0^1 k(x,y)\chi_\psi(x,y) f(t,x) f(t,y) \, dy dx$$

$$J_2(t) := \int_0^1 \int_1^{r_0} k(x,y)\chi_\psi(x,y) f(t,x) f(t,y) \, dy dx$$

$$J_3(t) := \frac{1}{2} \int_1^{r_0} \int_1^{r_0} k(x,y)\chi_\psi(x,y) f(t,x) f(t,y) \, dy dx \,.$$

First, owing to the concavity of ψ',

$$\chi_\psi(x,y) = \int_0^x \int_0^y \psi''(x_* + y_*) \, dy_* dx_* \le \psi''(0)xy \,, \qquad (x,y) \in (0,\infty)^2 \,,$$

and we infer from (8.2.170a) and Lemma 8.2.50 (with $m = 1 - \sigma \in [-2\sigma, 1]$) that

$$J_1(t) \le K_1 \psi''(0) \int_0^1 \int_0^1 (xy)^{1-\sigma} f(t,x) f(t,y) \, dy dx$$

$$\le K_1 \psi''(0) \left(\int_0^{r_0} x^{1-\sigma} f(t,x) \, dx \right)^2 \le K_1 \psi''(0) C_1(T)^2 \,.$$

Next, on using once more the concavity of ψ' as well as Proposition 7.1.9 (a), we obtain

$$\chi_\psi(x,y) \le \int_0^x \int_0^y \psi''(y_*) \, dy_* dx_* \le x\psi'(y) \le 2x\frac{\psi(y)}{y} \,, \qquad (x,y) \in (0,\infty)^2 \,.$$

Therefore, by (8.2.170b) and Lemma 8.2.50 (with $m = 1 - \sigma$),

$$J_2(t) \le 2K_1 \int_0^1 \int_1^{r_0} x^{1-\sigma} \psi(y) f(t,x) f(t,y) \, dy dx$$

$$\le 2K_1 C_1(T) \int_0^{r_0} \psi(y) f(t,y) \, dy \,.$$

Finally it follows from (8.2.170c), (8.2.175) and Proposition 7.1.9 (e) that

$$J_3(t) \le K_1 \int_1^{r_0} \int_1^{r_0} [x\psi(y) + y\psi(x)] f(t,x) f(t,y) \, dy dx$$

$$\le 2C_0 K_1 \int_0^{r_0} \psi(x) f(t,x) \, dx \,.$$

Collecting the previous estimates, we end up with

$$\int_0^{r_0} \mathcal{C}f(t,x)\psi(x)\,\mathrm{d}x \le K_1\psi''(0)C_1(T)^2$$

$$+ 2K_1[C_0 + C_1(T)]\int_0^{r_0} \psi(x)f(t,x)\,\mathrm{d}x\ . \tag{8.2.180}$$

Concerning the fragmentation term, we proceed as in the proof of Lemma 8.2.24 and use (8.2.172) and the concavity of $x \mapsto \psi(x)/x$, established in Proposition 7.1.9 (c), to obtain

$$\int_0^{r_0} \mathcal{F}f(t,x)\psi(x)\,\mathrm{d}x \le -\delta_2 \int_0^{r_0} a(y)(y\psi'(y) - \psi(y))f(t,y)\,\mathrm{d}y\ . \tag{8.2.181}$$

Introducing

$$P(t) := \int_0^{r_0} \psi(x)f(t,x)\,\mathrm{d}x + \delta_2 \int_0^t \int_0^{r_0} a(y)(y\psi'(y) - \psi(y))f(s,y)\,\mathrm{d}y\mathrm{d}s$$

for $t \in [0,T]$, we infer from (8.2.13), (8.2.180) and (8.2.181) that, for $t \in [0,T]$,

$$\frac{dP}{dt}(t) \le K_1C_3(T)(1 + P(t))\ ,$$

with $C_3(T) := \max\{\psi''(0)C_1(T)^2, 2[C_0 + C_1(T)]\}$. Integrating the previous differential inequality, and using (8.2.177), gives

$$P(t) \le e^{K_1C_3(T)t}\left(\int_0^{r_0} \psi(x)f^{in}(x)\,\mathrm{d}x + 1\right)\ ,$$

and the proof of Lemma 8.2.51 is complete. $\qquad\square$

We now deal with the uniform integrability issue. As already observed in [39, 86, 87], the possible singularity of k requires using a suitable weight, though the proof turns out to be quite close to that of the same result for weak solutions, see Lemma 8.2.45.

Lemma 8.2.52. *Assume that the coagulation and fragmentation coefficients k, a and b satisfy (8.2.170), (8.2.171) and (8.2.172). Assume further that there is $r_0 > 1$ such that (8.2.173) holds true. Consider next $f^{in} \in L_1((0,r_0),(1+x^{-2\sigma})\mathrm{d}x)_+$ satisfying (8.2.174) and let $f \in C^1([0,\infty), L_1(0,r_0))$ be the corresponding solution to the C-F equation (8.0.1) given by Proposition 8.2.7. Given $T > 0$, $R \in (0,r_0)$, and $\varepsilon > 0$, there is $C_4(T,R) > 0$ depending only on σ, K_1 in (8.2.170), A_1 in (8.2.171), (p_0,κ_1) in (8.2.172), C_0 in (8.2.175), R and T such that*

$$\xi(t,R,\varepsilon) := \sup\left\{\int_E x^{-\sigma}f(t,x)\,\mathrm{d}x\ :\ E \subset (0,R)\ ,\ |E| \le \varepsilon\right\}$$

$$\le C_4(T,R)\sup\left\{\int_E x^{-\sigma}f^{in}(x)\,\mathrm{d}x\ :\ E \subset (0,r_0)\ ,\ |E| \le \varepsilon\right\}$$

$$+ \varepsilon^{(p_0-1)/p_0}C_4(T,R)\left[1 + \int_0^T \int_1^{r_0} a(y)f(s,y)\,\mathrm{d}y\mathrm{d}s\right]\ ,$$

for all $t \in [0,T]$.

Proof. Let $T > 0$, $\varepsilon > 0$, $R \in (0, r_0)$, $t \in [0, T]$, and let E be a measurable subset of $(0, R)$ satisfying $|E| \leq \varepsilon$. We note that $k(x, y) \leq 2K_1 R^{1+2\sigma}(xy)^{-\sigma}$ for $(x, y) \in (0, R)^2$ by (8.2.170) and infer from (8.2.13) (with $\vartheta(x) = x^{-\sigma}\mathbf{1}_E(x)$), (8.2.172b), and Hölder's inequality that

$$\frac{d}{dt} \int_0^{r_0} x^{-\sigma}\mathbf{1}_E(x)f(t, x) \, dx$$

$$\leq \frac{1}{2} \int_0^R \int_0^{R-x} k(x, y)(x + y)^{-\sigma}\mathbf{1}_E(x + y)f(t, x)f(t, y) \, dydx$$

$$+ \int_0^{r_0} a(y)f(t, y) \int_0^{\min\{y, R\}} x^{-\sigma}\mathbf{1}_E(x)b(x, y) \, dxdy$$

$$\leq K_1 R^{1+2\sigma} \int_0^R x^{-2\sigma}f(t, x) \int_0^R y^{-\sigma}\mathbf{1}_{-x+E}(y)f(t, y) \, dydx$$

$$+ \int_0^{r_0} |E|^{(p_0-1)/p_0} \left(\int_0^y x^{-p_0\sigma}b(x, y)^{p_0} \, dx \right)^{1/p_0} a(y)f(t, y) \, dy \,.$$

Since $-x + E \subset (0, R)$ and $|-x + E| = |E| \leq \varepsilon$, we further obtain

$$\frac{d}{dt} \int_0^{r_0} x^{-\sigma}\mathbf{1}_E(x)f(t, x) \, dx \leq K_1 R^{1+2\sigma}\xi(t, R, \varepsilon) \int_0^R x^{-2\sigma}f(t, x) \, dx$$

$$+ \kappa_1^{1/p_0}\varepsilon^{(p_0-1)/p_0} \int_0^{r_0} a(y)y^{(1-p_0(1+\sigma))/p_0}f(t, y) \, dy \,.$$

Now, as $1 - p_0(1 + \sigma) \leq 0$ and

$$\frac{1 - p_0(1 + \sigma)}{p_0} = -2\sigma + \frac{\sigma p_0 + 1 - p_0}{p_0} \geq -2\sigma \,,$$

it follows that

$$\int_0^{r_0} a(y)y^{(1-p_0(1+\sigma))/p_0}f(t, y) \, dy \leq \int_0^1 a(y)y^{-2\sigma}f(t, y) \, dy + \int_1^{r_0} a(y)f(t, y) \, dy \,,$$

and, on using (8.2.171a) and Lemma 8.2.50, we obtain

$$\frac{d}{dt} \int_0^{r_0} x^{-\sigma}\mathbf{1}_E(x)f(t, x) \, dx$$

$$\leq K_1 C_1(T)R^{1+2\sigma}\xi(t, R, \varepsilon)$$

$$+ \kappa_1^{1/p_0}\varepsilon^{(p_0-1)/p_0} \left[A_1 \int_0^1 y^{-2\sigma}f(t, y) \, dy + \int_1^{r_0} a(y)f(t, y) \, dy \right]$$

$$\leq K_1 C_1(T)R^{1+2\sigma}\xi(t, R, \varepsilon)$$

$$+ \kappa_1^{1/p_0}\varepsilon^{(p_0-1)/p_0} \left[A_1 C_1(T) + \int_1^{r_0} a(y)f(t, y) \, dy \right] \,.$$

Integrating with respect to time gives, for $t \in [0, T]$,

$$\int_0^{r_0} x^{-\sigma}\mathbf{1}_E(x)f(t, x) \, dx \leq \int_0^{r_0} x^{-\sigma}\mathbf{1}_E(x)f^{in}(x) \, dx + K_1 C_1(T)R^{1+2\sigma} \int_0^t \xi(s, R, \varepsilon) \, ds$$

$$+ \kappa_1^{1/p_0}\varepsilon^{(p_0-1)/p_0} \left[A_1 C_1(T)t + \int_0^t \int_1^{r_0} a(y)f(s, y) \, dyds \right]$$

$$\leq \xi(0, R, \varepsilon) + K_1 C_1(T) R^{1+2\sigma} \int_0^t \xi(s, R, \varepsilon) \, ds$$

$$+ \kappa_1^{1/p_0} \varepsilon^{(p_0-1)/p_0} \left[A_1 T C_1(T) + \int_0^T \int_1^{r_0} a(y) f(s, y) \, dyds \right] .$$

Taking the supremum with respect to all measurable subsets $E \subset (0, R)$ satisfying $|E| \leq \varepsilon$, we end up with

$$\xi(t, R, \varepsilon) \leq \xi(0, R, \varepsilon) + K_1 C_1(T) R^{1+2\sigma} \int_0^t \xi(s, R, \varepsilon) \, ds$$

$$+ \kappa_1^{1/p_0} \varepsilon^{(p_0-1)/p_0} \left[A_1 T C_1(T) + \int_0^T \int_1^{r_0} a(y) f(s, y) \, dyds \right]$$

for all $t \in [0, T]$. Gronwall's lemma then gives

$$\xi(t, R, \varepsilon) \leq C_4(T, R) \left[\xi(0, R, \varepsilon) + \varepsilon^{(p_0-1)/p_0} + \varepsilon^{(p_0-1)/p_0} \int_0^T \int_1^{r_0} a(y) f(s, y) \, dyds \right] ,$$

as claimed. $\qquad\qquad\qquad\qquad\qquad\qquad\qquad\qquad\qquad\qquad\qquad\qquad\qquad\square$

We go on with the time equicontinuity.

Lemma 8.2.53. *Assume that the coagulation and fragmentation coefficients k, a and b satisfy (8.2.170), (8.2.171) and (8.2.172). Assume further that there is $r_0 > 1$ such that (8.2.173) holds true. Consider next $f^{in} \in L_1((0, r_0), (1 + x^{-2\sigma})dx)_+$ satisfying (8.2.174) and let $f \in C^1([0, \infty), L_1(0, r_0))$ be the corresponding solution to the C-F equation (8.0.1) given by Proposition 8.2.7. Given $T > 0$, there is $C_5(T) > 0$ depending only on σ, K_1 in (8.2.170), A_1 in (8.2.171), κ_1 in (8.2.172), C_0 in (8.2.175), and T such that, for $t_1 \in [0, T]$ and $t_2 \in [t_1, T]$,*

$$\int_0^R |f(t_2, x) - f(t_1, x)| \, dx \leq C_5(T)(t_2 - t_1) + (1 + \kappa_1) \int_{t_1}^{t_2} \int_1^{r_0} a(y) f(s, y) \, dyds .$$

Proof. Let $T > 0$ and $t \in [0, T]$. Owing to (8.2.170), (8.2.175) and Lemma 8.2.50,

$$\int_0^{r_0} |\mathcal{C} f(t, x)| \, dx \leq \frac{1}{2} \int_0^{r_0} \int_0^{r_0-y} k(x, y) f(t, x) f(t, y) \, dxdy$$

$$+ \int_0^{r_0} \int_0^{r_0} k(x, y) f(t, x) f(t, y) \, dydx$$

$$\leq \frac{3K_1}{2} \int_0^1 \int_0^1 (xy)^{-\sigma} f(t, x) f(t, y) \, dydx$$

$$+ 3K_1 \int_0^1 \int_1^{r_0} x^{-\sigma} y f(t, x) f(t, y) \, dydx$$

$$+ \frac{3K_1}{2} \int_1^{r_0} \int_1^{r_0} (x + y) f(t, x) f(t, y) \, dydx$$

$$\leq 3K_1 \left[C_1(T)^2 + C_0 C_1(T) + C_0^2 \right] .$$

For the fragmentation term, it follows from (8.2.171a), (8.2.172b) and Lemma 8.2.50 that

$$\int_0^{r_0} |\mathcal{F} f(t, x)| \, dx \leq \int_0^{r_0} a(y) f(t, y) \, dy + \int_0^{r_0} a(y) f(t, y) \int_0^y b(x, y) \, dxdy$$

$$\leq \int_0^{r_0} a(y) f(t,y) \left(1 + y^{2\sigma} \int_0^y x^{-2\sigma} b(x,y) \, \mathrm{d}x \right) \mathrm{d}y$$

$$\leq (1 + \kappa_1) \int_0^{r_0} a(y) f(t,y) \, \mathrm{d}y$$

$$\leq (1 + \kappa_1) \left(A_1 \int_0^1 f(t,y) \, \mathrm{d}y + \int_1^{r_0} a(y) f(t,y) \, \mathrm{d}y \right)$$

$$\leq (1 + \kappa_1) \left(A_1 C_1(T) + \int_1^{r_0} a(y) f(t,y) \, \mathrm{d}y \right) .$$

Consequently, by (8.0.1),

$$\int_0^R |\partial_t f(t,x)| \, \mathrm{d}x \leq C_6(T) + (1 + \kappa_1) \int_1^{r_0} a(y) f(t,y) \, \mathrm{d}y , \qquad t \in [0,T] ,$$

with $C_6(T) := 3K_1 \left[C_1(T)^2 + C_0 C_1(T) + C_0^2 \right] + (1 + \kappa_1) A_1 C_1(T)$. Lemma 8.2.53 then readily follows from the previous inequality after integration with respect to time over $[t_1, t_2]$ for $0 \leq t_1 \leq t_2 \leq T$. $\qquad \square$

Proof of Theorem 8.2.48. Let k, a and b be coagulation and fragmentation coefficients satisfying (8.2.170), (8.2.171) and (8.2.172), and consider $f^{in} \in X_{-2\sigma,+} \cap X_1$. We put

$$\varrho := M_1(f^{in}) \quad \text{and} \quad C_0 := M_{-2\sigma}(f^{in}) + \varrho .$$

Moreover, since $x \mapsto x \in L_1((0,\infty), f^{in}(x)\mathrm{d}x)$, the de la Vallée-Poussin theorem (Theorem 7.1.6) ensures the existence of $\psi \in \mathcal{C}_{VP,\infty}$ such that

$$C_0' := \int_0^\infty \psi(x) f^{in}(x) \, \mathrm{d}x < \infty .$$

Next, given $j \geq 1$, we define

$$k_j(x,y) := k_j(x,y) \mathbf{1}_{(0,j)}(x+y) \mathbf{1}_{(1/j,\infty)}(x) \mathbf{1}_{(1/j,\infty)}(y) , \qquad a_j(x) = a(x) \mathbf{1}_{(0,j)}(x)$$

and

$$f_j^{in}(x) = f^{in}(x) \mathbf{1}_{(0,j)}(x)$$

for $(x,y) \in (0,\infty)^2$. Clearly f_j^{in} is a nonnegative function in $L_1(0,j)$, and k_j and a_j satisfy (8.2.173), with $r_0 = j$, and are bounded according to (8.2.170) and (8.2.171). Then, by Proposition 8.2.7, the C-F equation (8.0.1) with coefficients (k_j, a_j, b) has a unique weak solution $f_j \in C^1([0,\infty), L_1(0,j))$ satisfying

$$\int_0^j x f_j(t,x) \, \mathrm{d}x = \int_0^j x f_j^{in}(x) \, \mathrm{d}x , \qquad t \geq 0 .$$

As before we extend f_j to $[0,\infty) \times (0,\infty)$ by setting $f_j(t,x) = 0$ for $t \geq 0$ and $x > j$.

Since $k_j \leq k$, $a_j \leq a$ and $f_j^{in} \leq f^{in}$, we infer from Lemmas 8.2.50–8.2.53 that we can argue as in Steps 1–3 of the proof of Theorem 8.2.23 to establish the existence of a nonnegative function $f \in C([0,\infty), X_{0,w} \cap X_{1,w})$ and a subsequence of $(f_j)_{j \geq 1}$ (not relabelled) such that

$$f_j \longrightarrow f \quad \text{in} \quad C([0,T], X_{0,w} \cap X_{1,w}) \tag{8.2.182}$$

for all $T > 0$ and

$$M_1(f(t)) = \varrho , \qquad t \geq 0 . \tag{8.2.183}$$

A further consequence of (8.2.182), Lemma 8.2.50 and Fatou's lemma is that, for all $T > 0$,

$$M_{-2\sigma}(f(t)) \leq C_1(T) , \qquad t \in [0, T] . \tag{8.2.184}$$

In order to show that f is a weak solution to the C-F equation (8.0.1) on $[0, \infty)$ with coefficients (k, a, b), we use the weak stability result (Theorem 8.2.11). To check the validity of (8.2.36) and (8.2.37), we need to control the behaviour of the sequence $(f_j)_{j \geq 1}$ for large sizes. This is done using (8.2.178), (8.2.179), Lemma 8.2.50, and the properties of ψ as in Step 4 of the proof of Theorem 8.2.23, with the help of Lemma 8.2.50 to handle the contribution of small sizes involved there. For the behaviour for small sizes, we infer from (8.2.170) and Lemma 8.2.50 that, for $T > 0$, $r > 1$ and $j \geq 1$,

$$\int_0^T \int_0^{1/r} \int_0^\infty k_j(x, y) f_j(t, x) f_j(t, y) \, dy dx dt$$

$$\leq K_1 \int_0^T \int_0^{1/r} \int_0^1 (xy)^{-\sigma} f_j(t, x) f_j(t, y) \, dy dx dt$$

$$+ K_1 \int_0^T \int_0^{1/r} \int_1^\infty x^{-\sigma} y f_j(t, x) f_j(t, y) \, dy dx dt$$

$$\leq K_1 r^{-\sigma} \int_0^T [M_{-2\sigma}(f_j(t)) M_{-\sigma}(f_j(t)) + M_{-2\sigma}(f_j(t)) M_1(f_j(t))] \, dt$$

$$\leq K_1 C_1(T)(C_1(T) + C_0) r^{-\sigma} ,$$

which guarantees that

$$\limsup_{r \to \infty} \left\{ \int_0^T \int_0^{1/r} \int_0^\infty k_j(x, y) f_j(t, x) f_j(t, y) \, dy dx dt \right\} = 0 ,$$

as required to apply Theorem 8.2.11. □

Combining the arguments from the proofs of Theorem 8.2.48 and Theorem 8.2.43, it is also possible to construct weak solutions to the C-F equation (8.0.1) on $[0, \infty)$ when the growth of k for large sizes exceeds the upper bound (8.2.170c). More precisely, one can prove the following result which somewhat extends [87].

Theorem 8.2.54. *Assume that the coagulation and fragmentation coefficients satisfy* (8.2.170a), (8.2.170b), (8.2.171a), (8.2.172a) *and* (8.2.172b), *as well as*

$$k(x, y) \leq K(1 + x)(1 + y) , \qquad (x, y) \in (1, \infty)^2 ,$$

and

$$\limsup_{x \to \infty \atop y > 1} \frac{k(x, y)}{xy} = \lim_{x \to \infty} \frac{a(x)}{x} = 0 .$$

Given $f^{in} \in X_{-2\sigma, +} \cap X_1$, *there exists at least one weak solution* f *to the C-F equation* (8.0.1) *on* $[0, \infty)$ *such that* $f \in L_\infty((0, T), X_{-2\sigma})$ *for all* $T > 0$ *and*

$$M_1(f(t)) \leq M_1(f^{in}) , \qquad t \geq 0 .$$

A final comment on Theorems 8.2.48 and 8.2.54 is that the range of their applicability for physically relevant fragmentation daughter distributions b is rather limited. In particular, when $\sigma > 1/2$ in (8.2.170), the daughter distribution function b given by $b(x, y) = b_\nu(x, y) = (\nu + 2) x^\nu y^{-\nu+1}$, $0 < x < y$, does not satisfy (8.2.172b) for $\nu \in (-2, 0]$, and this rules out realistic cases, see Lemma 2.2.3 in Volume I.

8.2.5   Uniqueness

We now turn to the uniqueness of weak solutions to the C-F equation. As already pointed out, one drawback of the construction of solutions to the C-F equation by compactness arguments is that uniqueness has to be established independently. This is in contrast to the semigroup approach, described in Section 8.1.2, where uniqueness and existence are obtained simultaneously.

We begin with the situation where no loss of matter (either due to gelation or shattering) is expected. Several results are available in that case [14, 74, 88, 104, 113, 120, 126, 127, 129, 160, 170, 223, 224, 253]. As mentioned in [170], the proof of the majority of these relies on the derivation of a differential inequality for the difference of two solutions in a weighted L_1-space, the choice of the weight being intimately related to the growth of the coagulation and fragmentation coefficients as well as to the decay of the solutions with respect to the size variable. Roughly speaking, given a suitably chosen weight ℓ, a natural functional setting guaranteeing uniqueness is introduced in [223, 224] and is the class of weak solutions f satisfying

$$f \in C([0,T], L_1((0,\infty), \ell(x)\mathrm{d}x)) \cap L_1((0,T) \times (0,\infty), \ell(x)^2\mathrm{d}x\mathrm{d}t) \ ,$$

when fragmentation is not taken into account. As indicated in the following statement, this can be easily extended to the full C-F equation (8.0.1).

Theorem 8.2.55. *Let $m_0 \in [0,1]$ and let k, a and b be coagulation and fragmentation coefficients such that b satisfies either (8.0.4) if $m_0 = 1$, or (8.0.4) and (8.2.7) if $m_0 \in [0,1)$. In addition, assume that there is a nonnegative and nondecreasing subadditive function ℓ defined on $(0,\infty)$, $\zeta \in [0,1]$, $K_0 \geq 0$, and $K_1 \geq 0$ such that*

$$\ell \in C^{0,m_0}([0,R]) \quad for\ each \quad R > 0 \quad and \quad \ell(0) = 0 \quad when \quad m_0 \in (0,1] \ , \tag{8.2.185}$$

$$k(x,y) \leq K_0 \ell(x)^\zeta \ell(y)^\zeta \ , \qquad (x,y) \in (0,\infty)^2 \ , \tag{8.2.186}$$

and

$$a(y) \int_0^y \left(\frac{\ell(x)}{x} - \frac{\ell(y)}{y} \right) x b(x,y) \ \mathrm{d}x \leq K_1 \ell(y) \ , \qquad y \in (0,\infty) \ . \tag{8.2.187}$$

Assume further that $\zeta = 1$ if $m_0 \in (0,1]$ and

$$\ell_* := \inf_{x>0} \{\ell(x)\} > 0 \quad if \quad m_0 = 0 \quad and \quad \zeta < 1 \ . \tag{8.2.188}$$

We set $\ell_* = 1$ when $\zeta = 1$.

Given $f^{in} \in X_{m_0,+} \cap X_1$ and $T > 0$, there is at most one weak solution f to the C-F equation (8.0.1) on $[0,T]$ possessing the following continuity and integrability properties: for $t \in (0,T)$,

$$f \in C([0,t], L_1((0,\infty), \ell(x)\mathrm{d}x)) \ , \tag{8.2.189a}$$

$$K_0 \int_0^t \int_0^\infty \ell(x)^{1+\zeta} f(s,x) \ \mathrm{d}x\mathrm{d}s < \infty \ , \tag{8.2.189b}$$

and

$$\lim_{R\to\infty} \int_0^t \int_R^\infty a(y) f(s,y) \int_0^R \ell(x) b(x,y) \ \mathrm{d}x\mathrm{d}y\mathrm{d}s = 0 \ . \tag{8.2.189c}$$

Let us emphasise here that, if $\ell(x)^{1+\zeta} \leq C(x^{m_0} + x)$, $x > 0$, for some $C > 0$, then any weak solution f to the C-F equation (8.0.1) satisfies the condition (8.2.189b), thanks to Definition 8.2.3. In contrast, if $\ell(x)^{1+\zeta}/x \to \infty$ as $x \to \infty$, then the condition (8.2.189b) requires somehow that, for all $t \in [0,T)$, $f(t)$ has a finite moment of sufficiently high order (and in any case of order higher than one). This property fails to be true for all times in general when gelation is expected to occur. The latter may indeed take place as (8.2.186) allows the coagulation coefficient k to grow sufficiently fast at infinity.

Proof of Theorem 8.2.55. Consider two initial conditions $f_i^{in} \in X_{m_0,+} \cap X_1$, $i = 1, 2$, and assume that, for each $i = 1, 2$, there is a weak solution f_i to the C-F equation (8.0.1) on $[0, T]$ with initial condition f_i^{in} satisfying the regularity properties (8.2.189). In particular,

$$I(t) := K_0 \int_0^t \int_0^\infty \ell(x)^{1+\varsigma} (f_1 + f_2)(s, x) \, \mathrm{d}x \mathrm{d}s \qquad (8.2.190)$$

is finite for all $t \in [0, T]$.

We next introduce

$$E(t, x) := (f_1 - f_2)(t, x) , \quad \sigma(t, x) := \mathrm{sign}(E(t, x)) ,$$
$$\ell_R(x) := \ell(x) \mathbf{1}_{(0,R)}(x) + \ell(R) \mathbf{1}_{[R,\infty)}(x)$$

for $(t, x) \in [0, T) \times (0, \infty)$ and $R > 1$. Fix $R > 1$. The definition of weak solutions (Definition 8.2.3) ensures that both $\mathcal{C} f_i$ and $\mathcal{F} f_i$ belong to $L_1((0, t) \times (0, R), x^{m_0} \mathrm{d}x \mathrm{d}s)$ for $i = 1, 2$ and $t \in (0, T)$. Therefore $\partial_t E \in L_1((0, t) \times (0, R), x^{m_0} \mathrm{d}x \mathrm{d}s)$ for all $t \in (0, T)$ and, since (8.2.185) guarantees that $\ell_R \in \Theta^{m_0}$, it follows from the C-F equation (8.0.1) that, for $t \in (0, T)$,

$$\frac{d}{dt} \int_0^\infty \ell_R(x) |E(t, x)| \, \mathrm{d}x = \int_0^\infty \ell_R(x) \sigma(t, x) \partial_t E(t, x) \, \mathrm{d}x$$
$$= \int_0^\infty (\mathcal{C} f_1 - \mathcal{C} f_2)(t, x) \ell_R(x) \sigma(t, x) \, \mathrm{d}x$$
$$+ \int_0^\infty (\mathcal{F} f_1 - \mathcal{F} f_2)(t, x) \ell_R(x) \sigma(t, x) \, \mathrm{d}x . \qquad (8.2.191)$$

For $t \in (0, T)$,

$$\int_0^\infty (\mathcal{C} f_1 - \mathcal{C} f_2)(t, x) \ell_R(x) \sigma(t, x) \, \mathrm{d}x$$
$$= \frac{1}{2} \int_0^\infty \int_0^\infty \ell_R(x + y) \sigma(t, x + y) k(x, y) (f_1 + f_2)(t, x) E(t, y) \, \mathrm{d}y \mathrm{d}x$$
$$- \frac{1}{2} \int_0^\infty \int_0^\infty (\ell_R(x) \sigma(t, x) + \ell_R(y) \sigma(t, y)) k(x, y) (f_1 + f_2)(t, x) E(t, y) \, \mathrm{d}y \mathrm{d}x$$
$$\leq \frac{1}{2} \int_0^\infty \int_0^\infty \ell_R(x + y) k(x, y) (f_1 + f_2)(t, x) |E(t, y)| \, \mathrm{d}y \mathrm{d}x$$
$$+ \frac{1}{2} \int_0^\infty \int_0^\infty (\ell_R(x) - \ell_R(y)) k(x, y) (f_1 + f_2)(t, x) |E(t, y)| \, \mathrm{d}y \mathrm{d}x .$$

Thanks to the subadditivity of ℓ_R, (8.2.186) and the property $\ell_R \leq \ell$, we deduce that

$$\int_0^\infty (\mathcal{C} f_1 - \mathcal{C} f_2)(t, x) \ell_R(x) \sigma(t, x) \, \mathrm{d}x$$
$$\leq \int_0^\infty \int_0^\infty \ell_R(x) k(x, y) (f_1 + f_2)(t, x) |E(t, y)| \, \mathrm{d}y \mathrm{d}x$$
$$\leq K_0 \int_0^\infty \int_0^\infty \ell(x)^{1+\varsigma} \ell(y)^\varsigma (f_1 + f_2)(t, x) |E(t, y)| \, \mathrm{d}y \mathrm{d}x .$$

Finally, either $\zeta = 1$ and $\ell(y)^\zeta = \ell(y)$, or $\zeta < 1$ and $\ell(y)^\zeta \le \ell_*^{\zeta-1}\ell(y)$ by (8.2.188), so that we end up with

$$\int_0^\infty (\mathcal{C}f_1 - \mathcal{C}f_2)(t,x)\ell_R(x)\sigma(t,x)\,\mathrm{d}x$$
$$\le K_0 \ell_*^{\zeta-1} \left(\int_0^\infty \ell(x)^{1+\zeta}(f_1 + f_2)(t,x)\,\mathrm{d}x \right) \int_0^\infty \ell(y)|E(t,y)|\,\mathrm{d}y \ . \tag{8.2.192}$$

Also, thanks to (8.0.4) and the linearity of the fragmentation operator $\mathscr{F}$,

$$\int_0^\infty (\mathscr{F}f_1 - \mathscr{F}f_2)(t,x)\ell_R(x)\sigma(t,x)\,\mathrm{d}x = \int_0^\infty \mathscr{F}E(t,x)\ell_R(x)\sigma(t,x)\,\mathrm{d}x$$
$$= \int_0^\infty a(y)E(t,y)\left(\int_0^y \ell_R(x)\sigma(t,x)b(x,y)\,\mathrm{d}x - \ell_R(y)\sigma(t,y) \right)\mathrm{d}y$$
$$\le \int_0^\infty a(y)|E(t,y)|\left(\int_0^y \ell_R(x)b(x,y)\,\mathrm{d}x - \frac{\ell_R(y)}{y}\int_0^y xb(x,y)\,\mathrm{d}x \right)\mathrm{d}y$$
$$\le \int_0^R a(y)|E(t,y)|\left[\int_0^y \left(\frac{\ell(x)}{x} - \frac{\ell(y)}{y} \right)xb(x,y)\,\mathrm{d}x \right]\mathrm{d}y$$
$$+ \int_R^\infty a(y)|E(t,y)|\int_0^R \ell(x)b(x,y)\,\mathrm{d}x\mathrm{d}y \ ,$$

hence, by (8.2.187),

$$\int_0^\infty (\mathscr{F}f_1 - \mathscr{F}f_2)(t,x)\ell_R(x)\sigma(t,x)\,\mathrm{d}x$$
$$\le K_1 \int_0^\infty \ell(y)|E(t,y)|\,\mathrm{d}y + \int_R^\infty a(y)(f_1 + f_2)(t,y)\int_0^R \ell(x)b(x,y)\,\mathrm{d}x\mathrm{d}y \ . \tag{8.2.193}$$

Combining (8.2.191), (8.2.192) and (8.2.193) gives, for $t \in [0,T]$,

$$\frac{d}{dt}\int_0^\infty \ell_R(x)|E(t,x)|\,\mathrm{d}x$$
$$\le \left[K_0\ell_*^{\zeta-1}\int_0^\infty \ell(x)^{1+\zeta}(f_1+f_2)(t,x)\,\mathrm{d}x + K_1 \right]\int_0^\infty \ell(y)|E(t,y)|\,\mathrm{d}y$$
$$+ \int_R^\infty a(y)(f_1+f_2)(t,y)\int_0^R \ell(x)b(x,y)\,\mathrm{d}x\mathrm{d}y \ ,$$

hence, after integration with respect to time,

$$\int_0^\infty \ell_R(x)|E(t,x)|\,\mathrm{d}x \le \int_0^\infty \ell_R(x)|E(0,x)|\,\mathrm{d}x$$
$$+ \int_0^t \left(K_0\ell_*^{\zeta-1}\int_0^\infty \ell(x)^{1+\zeta}(f_1+f_2)(s,x)\,\mathrm{d}x + K_1 \right)\int_0^\infty \ell(y)|E(s,y)|\,\mathrm{d}y\mathrm{d}s$$
$$+ \int_0^t \int_R^\infty a(y)(f_1+f_2)(s,y)\int_0^R \ell(x)b(x,y)\,\mathrm{d}x\mathrm{d}y\mathrm{d}s \ .$$

Owing to (8.2.189a) and (8.2.189c) we may pass to the limit as $R \to \infty$ in the above inequality and end up with

$$\int_0^\infty \ell(x)|E(t,x)|\,\mathrm{d}x \le \int_0^\infty \ell(x)|E(0,x)|\,\mathrm{d}x$$

$$+ \int_0^t \left(K_0 \ell_*^{\zeta-1} \int_0^\infty \ell(x)^{1+\zeta} (f_1 + f_2)(s, x) \, \mathrm{d}x + K_1 \right) \int_0^\infty \ell(y) |E(s, y)| \, \mathrm{d}y \mathrm{d}s .$$

We finally infer from (8.2.189a), (8.2.190) and Gronwall's lemma that

$$\int_0^\infty \ell(x) |(f_1 - f_2)(t, x)| \, \mathrm{d}x \le e^{\ell_*^{\zeta-1} I(T) + K_1 T} \int_0^\infty \ell(x) |(f_1^{in} - f_2^{in})(x)| \, \mathrm{d}x \qquad (8.2.194)$$

for $t \in [0, T)$, from which Theorem 8.2.55 readily follows. $\qquad\square$

Remark 8.2.56. *The proof of Theorem 8.2.55 and, in particular (8.2.194), actually provides the continuous dependence on the initial data in $L_1((0, \infty), \ell(x)\mathrm{d}x)$ for weak solutions possessing the additional regularity properties (8.2.189).*

A first consequence of Theorem 8.2.55 is the non-expansive property of the fragmentation equation in X_1 which is one of the building blocks of the semigroup approach developed in Volume I.

Proposition 8.2.57. *Assume that $k \equiv 0$ and let a and b be fragmentation coefficients satisfying (8.2.146b) and (8.2.146c). Consider two initial conditions $f_i^{in} \in X_{1,+}$, $i = 1, 2$, and assume that, for $i \in \{1, 2\}$, there is a mass-conserving weak solution f_i to the fragmentation equation on $[0, \infty)$ such that*

$$(t, x) \longmapsto x a(x) f_i(t, x) \in L_1((0, T) \times (0, \infty)) \qquad (8.2.195)$$

for all $T > 0$. Then

$$\|f_1(t) - f_2(t)\|_{[1]} \le \|f_1^{in} - f_2^{in}\|_{[1]} , \qquad t \ge 0 ,$$

recalling that $\| \cdot \|_{[1]}$ denotes the norm in X_1.

Proof. Setting $\ell(x) = x$, $x \in (0, \infty)$, the assumptions (8.2.185), (8.2.186), (8.2.187) and (8.2.188) are obviously satisfied with $m_0 = 1$, $K_0 = K_1 = 0$ and $\zeta = 1$, while (8.2.189) follows from (8.2.195) and the definition of weak solutions, see Definition 8.2.1 or Definition 8.2.3. Proposition 8.2.57 is then a straightforward consequence of (8.2.194) since $K_0 = 0$ implies that the function I defined in (8.2.190) vanishes identically. $\qquad\square$

Several uniqueness results available in the literature are proved along the lines of Theorem 8.2.55, though in a simpler setting. One of them is the following.

Corollary 8.2.58. *Let k, a and b be coagulation and fragmentation coefficients satisfying (8.2.146) as well as (8.2.147a) and (8.2.147b) with $m_0 = 0$. Assume further that there is $K > 0$ such that*

$$k(x, y) \le K\sqrt{1 + x}\sqrt{1 + y} , \qquad (x, y) \in (0, \infty)^2 , \qquad (8.2.196)$$

and

$$a(y) \int_0^y b(x, y)\sqrt{1 + x} \, \mathrm{d}x \le K\sqrt{1 + y} , \qquad y \in (0, \infty) . \qquad (8.2.197)$$

Given $f^{in} \in X_{0,1,+}$, there is a unique weak solution f to the C-F equation (8.0.1) on $[0, \infty)$.

With Corollary 8.2.58 we actually recover the uniqueness results established in [14, Theorem 4.1], [127, Theorem 3.1] and [253] for weak solutions with initial data in $X_{0,1,+}$.

Proof. First, Theorem 8.2.43 provides the existence of at least one weak solution to the C-F equation (8.0.1) on $[0, \infty)$ under the assumptions of Corollary 8.2.58. Next, the uniqueness statement in Corollary 8.2.58 is a straightforward consequence of Theorem 8.2.55 and of the definition of a weak solution (Definition 8.2.1). Indeed, setting $\ell(x) := \sqrt{1+x}$, $x \in (0, \infty)$, the function ℓ satisfies (8.2.185) with $m_0 = 0$, while the properties (8.2.186) with $\zeta = 1$ and (8.2.187) with $K_0 = K_1 = K$ readily follow from (8.2.196) and (8.2.197), respectively. The continuity property (8.2.189a) and the integrability condition required in (8.2.189b) are automatically satisfied for this particular choice of ℓ, according to Proposition 8.2.2 and the definition of a weak solution (Definition 8.2.1). Finally, we deduce from (8.2.197) that, for $t > 0$,

$$\int_0^t \int_R^\infty a(y) f(s,y) \int_0^R \ell(x) b(x,y) \, \mathrm{d}x \mathrm{d}y \mathrm{d}s \le K \int_0^t \int_R^\infty \sqrt{1+y} f(s,y) \, \mathrm{d}y \mathrm{d}s$$

$$\le \frac{K}{\sqrt{1+R}} \int_0^t \int_0^\infty (1+y) f(s,y) \, \mathrm{d}y \mathrm{d}s \; ,$$

and the right-hand side of the above inequality converges to zero by (8.2.189a), whence (8.2.189c). □

Also, from Theorem 8.2.55 with $\ell(x) = 1 + x$, $x \in (0, \infty)$, and $\zeta = 1$, one retrieves the uniqueness results from [104, Theorem 3], [139], [191], [192, Theorem 1], and [224, Theorem 2.1], which hold for initial data in $X_{0,m,+}$ for some suitably chosen $m \ge 2$.

Next, when the fragmentation is sufficiently strong, uniqueness results are also available, see [88, Theorem 6.1], [126] and [129], and the proofs also proceed along the lines of that of Theorem 8.2.55. For completeness, we state the corresponding result.

Corollary 8.2.59. *Consider fragmentation coefficients a and b satisfying (8.2.72c), (8.2.74), (8.2.75) with $m_0 = 0$, and (8.2.76) and assume that there is $\lambda \in [0, 2)$, $\gamma \in (\lambda/2, 1]$, $a_0 > 0$, and $K > 0$ such that*

$$k(x,y) \le K(1+x)^{\lambda/2}(1+y)^{\lambda/2} \; , \qquad (x,y) \in (0, \infty)^2 \; , \qquad (8.2.198)$$

and

$$a_0 y^\gamma \le a(y) \le K(1+y) \; , \qquad y \in (0, \infty) \; . \qquad (8.2.199)$$

Given $f^{in} \in X_{0,1,+}$, there is a unique mass-conserving weak solution f to the C-F equation (8.0.1) on $[0, \infty)$ satisfying

$$f \in C([0,t], X_{0,1}) \cap L_1((0,t), X_{(2+\lambda)/2}) \qquad (8.2.200)$$

for all $t > 0$.

Proof. That there is at least one mass-conserving weak solution to the C-F equation (8.0.1) on $[0, \infty)$ with the improved integrability property of f assumed in (8.2.200) is actually a consequence of the strong fragmentation assumption $\gamma > \lambda/2 \ge \lambda - 1$, see Theorem 8.2.38. It also satisfies the required continuity property due to Proposition 8.2.2 and the conservation of mass.

For the uniqueness issue, we set $\ell(x) := 1 + x$, $x \in (0, \infty)$, and $\zeta := \lambda/2 \in [0, 1]$, and check that we are in a position to apply Theorem 8.2.55 (with $m_0 = 0$). To begin with, ℓ

satisfies (8.2.185) with $m_0 = 0$, and (8.2.186) with $K_0 = K$ readily follows from (8.2.198). Next, (8.2.187) with $K_1 = \kappa_{0,1} K$ is deduced from (8.2.75) and (8.2.199), since

$$a(y) \int_0^y \left(\frac{\ell(x)}{x} - \frac{\ell(y)}{y} \right) xb(x,y) \, dx \leq a(y) \int_0^y \left(\frac{1}{x} - \frac{1}{y} \right) xb(x,y) \, dx$$

$$\leq a(y) \int_0^y b(x,y) \, dx \leq \kappa_{0,1} K(1+y) \ .$$

Also, (8.2.188) is satisfied with $\ell_* = 1$ according to the definition of ℓ. In addition, we deduce (8.2.189a) and (8.2.189b) from (8.2.200). It remains to check (8.2.189c). To this end, consider $t \in (0,T)$ and $R > 0$. It follows from (8.2.4d) with $\vartheta(x) = x \mathbf{1}_{(0,R)}(x)$, $x \in (0,\infty)$, that

$$\int_0^R x(f(t,x) - f^{in}(x)) \, dx = -I_1(t,R) - I_2(t,R) + I_3(t,R) \ , \tag{8.2.201}$$

where

$$I_1(t,R) := \frac{1}{2} \int_0^t \int_0^R \int_{R-x}^R (x+y)k(x,y)f(s,x)f(s,y) \, dy dx ds$$

$$I_2(t,R) := \int_0^t \int_0^R \int_R^\infty xk(x,y)f(s,x)f(s,y) \, dy dx ds \ ,$$

$$I_3(t,R) := \int_0^t \int_R^\infty a(y)f(s,y) \int_0^R xb(x,y) \, dx dy ds \ .$$

We infer from (8.2.198) that

$$I_2(t,R) \leq K \int_0^t \left(\int_0^R x(1+x)^{\lambda/2} f(s,x) \, dx \right) \left(\int_R^\infty (1+y)^{\lambda/2} f(s,y) \, dy \right) ds$$

$$\leq K \int_0^t \left[M_1(f(s)) + M_{(2+\lambda)/2}(f(s)) \right] \int_R^\infty (1+y)f(s,y) \, dy ds \ ,$$

and the right-hand side of the above inequality converges to zero as $R \to \infty$, thanks to (8.2.200) and Lebesgue's dominated convergence theorem. Thus,

$$\lim_{R \to \infty} I_2(t,R) = 0 \ . \tag{8.2.202}$$

Similarly, owing to the symmetry of k and (8.2.198),

$$I_1(t,R) = \int_0^t \int_0^R \int_{R-x}^R xk(x,y)f(s,x)f(s,y) \, dy dx ds$$

$$\leq \int_0^t \int_0^{R/2} \int_{R/2}^R xk(x,y)f(s,x)f(s,y) \, dy dx ds$$

$$+ \int_0^t \int_{R/2}^R \int_0^R xk(x,y)f(s,x)f(s,y) \, dy dx ds$$

$$\leq K \int_0^t \left[M_1(f(s)) + M_{(2+\lambda)/2}(f(s)) \right] \int_{R/2}^\infty (1+y)f(s,y) \, dy ds$$

$$+ K \int_0^t \left[M_0(f(s)) + M_1(f(s)) \right] \int_{R/2}^\infty \left(x + x^{(2+\lambda)/2} \right) f(s,y) \, dy ds \ .$$

We use again (8.2.200) and Lebesgue's dominated convergence theorem to deduce that the right-hand side of the above inequality converges to zero as $R \to \infty$ and thus conclude that

$$\lim_{R \to \infty} I_1(t, R) = 0 \ . \tag{8.2.203}$$

Since mass conservation guarantees that

$$\lim_{R \to \infty} \int_0^R x f(t, x) \, \mathrm{d}x = M_1(f^{in}) = \lim_{R \to \infty} \int_0^R x f^{in}(x) \, \mathrm{d}x \ ,$$

we may pass to the limit as $R \to \infty$ in (8.2.201) and infer from (8.2.202) and (8.2.203) that

$$\lim_{R \to \infty} \int_0^t \int_R^\infty a(y) f(s, y) \int_0^R x b(x, y) \, \mathrm{d}x \mathrm{d}y \mathrm{d}s = 0 \ . \tag{8.2.204}$$

Finally, since $a n_0 f \in L_1((0, t) \times (0, \infty))$ by Definition 8.2.1 and

$$a(y) f(s, y) \int_0^R b(x, y) \, \mathrm{d}x \le a(y) n_0(y) f(s, y) \ , \qquad (s, y) \in (0, t) \times (R, \infty) \ ,$$

Lebesgue's dominated convergence theorem ensures that

$$\lim_{R \to \infty} \int_0^t \int_R^\infty a(y) f(s, y) \int_0^R b(x, y) \, \mathrm{d}x \mathrm{d}y \mathrm{d}s = 0 \ . \tag{8.2.205}$$

Recalling that $\ell(x) = 1 + x$, $x \in (0, \infty)$, the property (8.2.189c) readily follows from (8.2.204) and (8.2.205). □

If the coagulation kernel vanishes in an appropriate way as $(x, y) \to 0$, then a variant of Corollary 8.2.59 without restriction on the growth of a is available and is adapted from [88, Theorem 6.1], where the discrete C-F equation is handled.

Corollary 8.2.60. *Consider fragmentation coefficients a and b satisfying (8.2.72c), (8.2.74), (8.2.75) with $m_0 = 0$, and (8.2.76), and assume that there is $\lambda \in [0, 2]$, $\gamma > \lambda/2$, $a_0 > 0$, and $K > 0$ such that*

$$k(x, y) \le K(xy)^{\lambda/2} \left(\min\{x, 1\} \min\{y, 1\} \right)^{(2-\lambda)/2} \ , \qquad (x, y) \in (0, \infty)^2 \ , \tag{8.2.206}$$

and

$$a_0 y^\gamma \le a(y) \ , \qquad y \in (0, \infty) \ . \tag{8.2.207}$$

Given $f^{in} \in X_{0,1,+}$, there is a unique mass-conserving weak solution f to the C-F equation (8.0.1) on $[0, \infty)$ satisfying (8.2.200).

Proof. The existence of at least one mass-conserving weak solution to the C-F equation (8.0.1) on $[0, \infty)$ satisfying (8.2.200) is again provided by Theorem 8.2.38. Next, setting $\ell(x) = x$, $x \in (0, \infty)$, and $\zeta = 1$, we deduce (8.2.186) from (8.2.206) while (8.2.187) is obviously satisfied with $K_1 = 0$, since its left-hand side is identically equal to zero. Checking that (8.2.200) and (8.2.206) imply (8.2.189) is done as in the proof of Corollary 8.2.59. □

Uniqueness results may also be obtained with the help of a weight function ℓ which is not subadditive, as in [35, 37], where $\ell(x) = 1 + x^m$, $x \in (0, \infty)$, is used for a suitable $m > 1$. Though such a function ℓ is no longer subadditive (since $m > 1$), a suitable control on the growth of $\ell(x + y) - \ell(y)$ is available and the following result can be proved along similar lines to Theorem 8.2.55.

Theorem 8.2.61. *Let $m > 1$ and set $\ell(x) := 1 + x^m$, $x \in (0, \infty)$. Assume that the fragmentation coefficients a and b satisfy (8.2.72c) and (8.2.187) with this choice of ℓ. Assume further that there is $\lambda \in [0, 1]$, and $K_0 > 0$ such that, for $(x, y) \in (0, \infty)^2$,*

$$k(x, y) \leq K_0(1 + x^\lambda + y^\lambda) \quad and \quad n_0(y) = \int_0^y b(x, y) \, \mathrm{d}x \leq K_0 . \tag{8.2.208}$$

For $f^{in} \in X_{0,1,+}$ and $T > 0$, there is at most one weak solution f to the C-F equation (8.0.1) on $[0, T)$ such that $f \in C([0, T), X_{0,m})$ and f possesses the integrability properties (8.2.189c) and

$$\int_0^t \int_0^\infty x^{m+\lambda} f(s, x) \, \mathrm{d}x \mathrm{d}s < \infty , \qquad t \in [0, T) . \tag{8.2.209}$$

Proof. Since the subadditivity property of ℓ is only used in the derivation of (8.2.192) in the proof of Theorem 8.2.55, we just indicate the modifications to be made. Since $m > 1$, we infer from the convexity of $x \mapsto x^m$ and Lemma 7.4.1 that

$$\ell(x + y) - \ell(y) \leq mx(x + y)^{m-1} \leq C(m) \left(x^m + xy^{m-1} \right)$$

for $(x, y) \in (0, \infty)^2$. Since

$$\ell_R(x + y) - \ell_R(y) \leq \max\{0, \ell(x + y) - \ell(x)\} , \qquad (x, y) \in (0, \infty)^2 ,$$

we may proceed as in the proof of Theorem 8.2.55 to derive the following upper bound for the coagulation term:

$$\int_0^\infty (\mathcal{C}f_1 - \mathcal{C}f_2)(t, x)\ell_R(x)\sigma(t, x) \, \mathrm{d}x$$

$$\leq (1 + C(m)) \int_0^\infty \int_0^\infty (1 + x^m + xy^{m-1})k(x, y)(f_1 + f_2)(t, x)|E(t, y)| \, \mathrm{d}y \mathrm{d}x .$$

Since

$$\begin{aligned}
(1 + x^m + xy^{m-1})k(x, y) &\leq K_0 \left[(1 + x^\lambda)(1 + x^m) + (1 + x^\lambda)xy^{m-1} \right] \\
&\quad + K_0 \left[(1 + x^m)y^\lambda + xy^{m-1+\lambda} \right] \\
&\leq 2K_0 \left[(1 + x^{m+\lambda})\ell(y) + (1 + x^{1+\lambda})\ell(y) \right] \\
&\quad + \quad K_0 \left[(1 + x^m)\ell(y) + (1 + x)\ell(y) \right] \\
&\leq 10K_0(1 + x^{m+\lambda})\ell(y)
\end{aligned}$$

by (8.2.208) and Young's inequality, we finally obtain

$$\int_0^\infty (\mathcal{C}f_1 - \mathcal{C}f_2)(t, x)\ell_R(x)\sigma(t, x) \, \mathrm{d}x$$

$$\leq 10K_0(1 + C(m)) \int_0^\infty \int_0^\infty (1 + x^{m+\lambda})\ell(y)(f_1 + f_2)(t, x)|E(t, y)| \, \mathrm{d}y \mathrm{d}x$$

instead of (8.2.192). The proof is then completed as for Theorem 8.2.55. □

Corollary 8.2.62. *Assume that*

$$k(x, y) = K \left(x^\alpha y^\beta + x^\beta y^\alpha \right) , \qquad a(y) = a_0 x^\gamma , \qquad (x, y) \in (0, \infty)^2 , \tag{8.2.210}$$

where $0 \le \alpha \le \beta \le 1 - \alpha$, $K > 0$, $\gamma > 0$ and $a_0 > 0$. *Assume further that the daughter distribution b satisfies* (8.2.72c) *and* (8.2.75) *with $m_0 = 0$ and $p_0 \in (1, 1 + \gamma] \cap (1, 2)$. Let $m > 1$ be such that $m \ge \gamma$. For $f^{in} \in X_{0,2m,+}$, there is a unique mass-conserving weak solution f to the C-F equation* (8.0.1) *on $[0, \infty)$ satisfying*

$$f \in C([0, t], X_{0,m}) \cap L_\infty((0, t), X_{2m}) , \qquad t > 0 . \tag{8.2.211}$$

Proof. Since $0 \le \alpha + \beta \le 1$, Young's inequality implies that

$$k(x, y) \le K \left(x^{\alpha + \beta} + y^{\alpha + \beta} \right) \le K(2 + x + y) , \qquad (x, y) \in (0, \infty)^2 , \tag{8.2.212}$$

and the existence of at least one mass-conserving weak solution f to the C-F equation (8.0.1) on $[0, \infty)$, which also belongs to $L_\infty((0, T), X_{2m})$ for all $T > 0$, follows from Theorem 8.2.23. Combining this additional property with Proposition 8.2.2 establishes the required time continuity of f.

Next, the growth condition (8.2.208) is satisfied according to (8.2.212) and so we now check that (8.2.187) holds true with $\ell(x) = 1 + x^m$, $x \in (0, \infty)$. In fact, since $m > 1$ and $m \ge \gamma$, it follows from (8.2.75) and Young's inequality that, for $y > 0$,

$$a(y) \int_0^y \left(\frac{1 + x^m}{x} - \frac{1 + y^m}{y} \right) x b(x, y) \, \mathrm{d}x \le a_0 y^\gamma \int_0^y b(x, y) \, \mathrm{d}x$$
$$\le a_0 \kappa_{0,1} y^\gamma \le a_0 \kappa_{0,1} \ell(y) ,$$

hence (8.2.187) with $K_1 = a_0 \kappa_{0,1}$. Also, for $t > 0$ and $R > 1$, we infer from (8.2.75) and (8.2.210) that

$$\int_0^t \int_R^\infty a(y) f(s, y) \int_0^R \ell(x) b(x, y) \, \mathrm{d}x \mathrm{d}y \mathrm{d}s \le 2 a_0 \int_0^t \int_R^\infty y^{m+\gamma} f(s, y) \int_0^R b(x, y) \, \mathrm{d}x \mathrm{d}y \mathrm{d}s$$
$$\le 2 a_0 \kappa_{0,1} \int_0^t \int_R^\infty y^{2m} f(s, y) \, \mathrm{d}y \mathrm{d}s ,$$

and the right-hand side of the above inequality converges to zero as $R \to \infty$ according to (8.2.211), thereby ensuring the validity of (8.2.189c). Finally, the integrability condition (8.2.209) is provided by (8.2.211) and an application of Theorem 8.2.61 completes the proof. □

Let us finally point out that uniqueness of weak solutions has also been established for coagulation kernels featuring a singularity for small sizes such as

$$k(x, y) = x^{-\alpha} y^\beta + x^\beta x^{-\alpha} , \qquad (x, y) \in (0, \infty)^2 ,$$

with $-1 \le \alpha < 0 \le \beta \le 1$ and $\alpha + \beta \in [0, 1)$, see [113, Theorem 2.9]. Uniqueness is proved with a weight function ℓ of the form $\ell(x) = c_1 x^{-\alpha} + c_2 x^m$, $x \in (0, \infty)$, where $m > 0$ is suitably chosen and $(c_1, c_2) \in [0, \infty)^2 \setminus \{(0, 0)\}$. While the proof of [113, Theorem 2.9] is similar to that of Theorem 8.2.55 when $m \le 1$, it borrows some steps of both Theorems 8.2.55 and 8.2.61 when $m > 1$.

A common feature of the previous uniqueness results is that they only deal with coagulation and fragmentation coefficients for which the gelation phenomenon is not expected to occur. Indeed, Corollaries 8.2.59, 8.2.60 and 8.2.62 only deal with mass-conserving solutions, while the growth condition on k in Corollary 8.2.58 prevents the occurrence of gelation. In fact, the uniqueness issue turns out to be far more difficult for solutions having a finite gelation time, the results being available so far only guaranteeing uniqueness prior to the gelation time. The only uniqueness which is also valid beyond the gelation time requires no fragmentation $a \equiv 0$ and the coagulation kernel k to be the multiplicative kernel $k_\times$. In that case, given a weak solution f to the C-F equation (8.0.1) on $[0, \infty)$ with an initial condition $f^{in} \in X_{1,+}$, the time evolution of $M_1(f(t))$ for all times $t \geq 0$ is shown to be determined uniquely by f^{in}, the proof relying on the Laplace transform [222, Proposition 2.6], [244, 264], see also [108, 148, 183, 282, 286]. This approach also applies to $k(x, y) = xy + A(x+y) + B$, $A \geq 0$, $B \geq 0$.

A different approach to uniqueness is developed in [74, 120]. For coagulation kernels that are homogeneous of order $\lambda \in (-\infty, 2]$ (or close to such a kernel) and in the absence of fragmentation, it is possible to control the evolution of

$$\int_0^\infty x^{\lambda-1} \left| \int_x^\infty (f_1 - f_2)(t, y) \, dy \right| \, dx \quad \text{for} \quad \lambda \in [0, 2] \, ,$$

or

$$\int_0^\infty x^{\lambda-1} \left| \int_0^x (f_1 - f_2)(t, y) \, dy \right| \, dx \quad \text{for} \quad \lambda < 0 \, ,$$

where f_1 and f_2 are two weak solutions to the C-F equation (8.0.1) on $[0, T]$ which additionally belong to $L_\infty((0, T), X_\lambda)$, and derive the uniqueness and continuous dependence results with respect to the above defined distance [120]. A similar result is available when fragmentation is included, provided that the overall fragmentation rate a is bounded and its derivative exists and decays appropriately as $x \to \infty$ [74]. Owing to the additional integrability required on f_1 and f_2, this approach also applies only prior to the gelation time.

Chapter 9

Gelation and Shattering

So far, our focus has been on the issues of existence and uniqueness of solutions to the C-F equation (8.0.1), and we have identified classes of coagulation and fragmentation coefficients for which mass-conserving solutions exist. In this chapter, we explore situations where an infringement of the physically expected property of mass conservation arises due to the existence of solutions that display mass loss. As discussed in Sections 1.1 and 2.3.1 of Volume I, a loss of matter may result either from a runaway growth driven by a rapidly growing coagulation kernel at large sizes (gelation) or from the appearance of dust due to the unboundedness of the overall fragmentation rate for small sizes (shattering). We devote the first section (Section 9.1) to gelation, and go on in the subsequent Section 9.2 to investigate the instantaneous occurrence of this phenomenon. Shattering is studied in the last section (Section 9.3).

9.1 Gelation

To begin with, let us recall the definition of the gelation time.

Definition 9.1.1. *Consider* $f^{in} \in X_{m_0,+} \cap X_1$ *for some* $m_0 \in [0,1]$ *and let* f *be a weak solution to the C-F equation (8.0.1) on* $[0,\infty)$ *with initial condition* f^{in} *in the sense of Definition 8.2.1 or Definition 8.2.3. The gelation time* T_{gel} *of* f *is defined by*

$$T_{gel} := \sup\left\{ t \geq 0 \; : \; M_1(f(t)) = M_1(f^{in}) \right\} . \tag{9.1.1}$$

Obviously, $T_{gel} = \infty$ for mass-conserving solutions to the C-F equation (8.0.1) on $[0,\infty)$ such as those constructed in Section 8.2.2. Now we are interested in the opposite situation, where the gelation time is finite.

We first study the occurrence of the gelation phenomenon for the following class of coagulation and fragmentation kernels: there is $\lambda \in (1,2]$, $\gamma \geq 0$, $K_1 > 0$, and $a_0 \geq 0$ such that

$$k(x,y) \geq K_1(xy)^{\lambda/2} , \qquad (x,y) \in (0,\infty)^2 , \tag{9.1.2}$$

and

$$a(x) \leq \begin{cases} a_0 x^{\lambda/2} , & x \in (0,1) , \\[2mm] a_0 x^\gamma , & x > 1 . \end{cases} \tag{9.1.3}$$

We further assume that there is a nonnegative function $h \in L_1((0,1), z\mathrm{d}z)$, and $m_0 \in [0,1)$ such that

$$h \geq 0 , \quad \int_0^1 z h(z) \, \mathrm{d}z = 1 , \quad \int_0^1 z^{m_0} h(z) \, \mathrm{d}z < \infty , \tag{9.1.4}$$

and

$$b(x,y) \le \frac{1}{y} h\left(\frac{x}{y}\right) , \qquad 0 < x < y . \tag{9.1.5}$$

For $m \ge m_0$ we set

$$\mathfrak{h}_m := 1 - \int_0^1 z^m h(z) \, \mathrm{d}z = \int_0^1 (z - z^m) h(z) \, \mathrm{d}z , \tag{9.1.6}$$

where the last equality follows from (9.1.4). Note that $\mathfrak{h}_1 = 0$ and $\mathfrak{h}_m(m-1) > 0$ for $m \in [m_0, \infty) \setminus \{1\}$.

We shall start our investigations with the case $\lambda = 2$, this being the first for which a rigorous proof of the occurrence of gelation was given, in [183] without fragmentation, and in [158] with fragmentation.

Proposition 9.1.2. *Assume that the coagulation and fragmentation coefficients k, a and b satisfy (9.1.2)–(9.1.5) with $\lambda = 2$ and $m_0 = 0$. Let $f^{in} \in X_{0,1,+}$ and let f be a weak solution to the C-F equation (8.0.1) on $[0, \infty)$ with initial condition f^{in}. If*

$$\gamma \in [0,1] \quad and \quad \varrho := M_1(f^{in}) > \frac{2a_0 |\mathfrak{h}_0|}{K_1} , \tag{9.1.7}$$

then $T_{gel} < \infty$, where $\mathfrak{h}_0 \in (-\infty, 0)$ is defined in (9.1.6).

It follows from Proposition 9.1.2 that, in the absence of fragmentation ($a_0 = 0$), gelation occurs for every non-zero initial condition. As expected, fragmentation slows down the runaway growth and the total mass ϱ needs to be sufficiently large for the gelation phenomenon to take place.

Proof. Let $t > 0$. On the one hand, it follows from (9.1.2) that

$$\int_0^\infty \mathcal{C}f(t,x) \, \mathrm{d}x = -\frac{1}{2} \int_0^\infty \int_0^\infty k(x,y) f(t,x) f(t,y) \, \mathrm{d}y \mathrm{d}x$$
$$\le -\frac{K_1}{2} M_1(f(t))^2 .$$

On the other hand, we infer from (9.1.3), (9.1.5), (9.1.6) and (9.1.7) that

$$\int_0^\infty \mathcal{F}f(t,x) \, \mathrm{d}x = \int_0^\infty a(y) f(t,y) \left[\int_0^y b(x,y) \, \mathrm{d}x - 1 \right] \mathrm{d}y$$
$$\le |\mathfrak{h}_0| \int_0^\infty a(y) f(t,y) \, \mathrm{d}y$$
$$\le a_0 |\mathfrak{h}_0| \left(\int_0^1 y f(t,y) \, \mathrm{d}y + \int_1^\infty y^\gamma f(t,y) \, \mathrm{d}y \right)$$
$$\le a_0 |\mathfrak{h}_0| M_1(f(t)) .$$

Therefore, for $t \in (0, T_{gel})$, we infer from (8.2.4d) (with $\vartheta \equiv 1$) that

$$M_0(f(t)) - M_0(f^{in}) = \int_0^t \int_0^\infty [\mathcal{C}f(s,x) + \mathcal{F}f(s,x)] \, \mathrm{d}x \mathrm{d}s$$
$$\le \int_0^t \left(a_0 |\mathfrak{h}_0| M_1(f(s)) - \frac{K_1}{2} M_1(f(s))^2 \right) \mathrm{d}s$$
$$\le \frac{\varrho t}{2} (2a_0 |\mathfrak{h}_0| - \varrho K_1) ,$$

hence

$$\varrho(\varrho K_1 - 2a_0|\mathfrak{h}_0|)t \le 2M_0(f^{in}) \ .$$

Owing to (9.1.7), we may let $t \to T_{gel}$ in the above inequality to conclude that

$$T_{gel} \le \frac{2M_0(f^{in})}{\varrho(\varrho K_1 - 2a_0|\mathfrak{h}_0|)} \ ,$$

and thus T_{gel} is finite as claimed. $\qquad\square$

We next turn to the case $\lambda \in (1,2)$ and readily see that the proof of Proposition 9.1.2 cannot be adapted in a straightforward way: indeed, a similar computation gives an upper bound for the coagulation term involving only $M_{\lambda/2}(f)$ which is a moment of order less than one. To handle this case, a more elaborate technique, encountered earlier in Lemma 8.2.14, was developed in [112].

Theorem 9.1.3. *Assume that the coagulation and fragmentation coefficients k, a and b satisfy (9.1.2)–(9.1.5) with $\lambda \in (1,2)$ and $m_0 = 0$. Let $f^{in} \in X_{0,1,+}$ and let f be a weak solution to the C-F equation (8.0.1) on $[0,\infty)$ with initial condition f^{in}. There is $\varrho_\star > 0$ depending only on K_1, λ, a_0, γ and h such that, if*

$$\gamma \in [0, \lambda - 1) \quad and \quad \varrho := M_1(f^{in}) > \varrho_\star \ , \tag{9.1.8}$$

then $T_{gel} < \infty$. Moreover, $\varrho_\star = 0$ if $a_0 = 0$ (i.e., no fragmentation occurs).

Proof. As $0 < 2 - \lambda < 1 - \gamma \le 1$, we may fix $m \in (2 - \lambda, 1 - \gamma] \cap (0,1)$. Let $t > 0$ and set $\vartheta(x) := \min\{x, x^m\}$, $x > 0$. Then $\chi_\vartheta(x,y) \le 0$ for all $(x,y) \in (0,\infty)^2$, so that

$$\int_0^\infty \mathcal{C}f(t,x)\vartheta(x) \ \mathrm{d}x$$
$$\le \frac{1}{2} \int_1^\infty \int_1^\infty [(x+y)^m - x^m - y^m] \, k(x,y)f(t,x)f(t,y) \ \mathrm{d}y\mathrm{d}x \ .$$

Since $1 < (m + \lambda)/2$, we infer from (9.1.2) and Lemma 8.2.14 (with $\mu = 1$) that

$$\int_0^\infty \mathcal{C}f(t,x)\vartheta(x) \ \mathrm{d}x \le -\frac{K_1}{C^2} \left(\int_1^\infty xf(t,x) \ \mathrm{d}x \right)^2 \ , \tag{9.1.9}$$

where C is a positive constant depending only on λ and m. Next, it follows from (9.1.4) and (9.1.5) that

$$\int_0^\infty \mathscr{F}f(t,x)\vartheta(x) \ \mathrm{d}x = \int_0^1 a(y)f(t,y) \left[\int_0^y xb(x,y) \ \mathrm{d}x - y \right] \ \mathrm{d}y$$
$$+ \int_1^\infty a(y)f(t,y) \left[\int_0^y \min\{x, x^m\}b(x,y) \ \mathrm{d}x - y^m \right] \ \mathrm{d}y$$
$$\le \int_1^\infty a(y)f(t,y) \left[\int_0^y x^m b(x,y) \ \mathrm{d}x - y^m \right] \ \mathrm{d}y$$
$$\le -\mathfrak{h}_m \int_1^\infty a(y)y^m f(t,y) \ \mathrm{d}y \ .$$

Since $\mathfrak{h}_m < 0$ due to $m < 1$, we use (9.1.3) to obtain

$$\int_0^\infty \mathscr{F}f(t,x)\vartheta(x) \ \mathrm{d}x \le a_0|\mathfrak{h}_m| \int_1^\infty y^{m+\gamma}f(t,y) \ \mathrm{d}y \ . \tag{9.1.10}$$

Combining (8.2.4d), (9.1.9) and (9.1.10) we are led to

$$\int_0^\infty \vartheta(x)[f(t,x) - f^{in}(x)]\,\mathrm{d}x \leq -\frac{K_1}{C^2}\int_0^t \left(\int_1^\infty xf(s,x)\,\mathrm{d}x\right)^2 \mathrm{d}s$$

$$+ a_0|\mathfrak{h}_m|\int_0^t\int_1^\infty y^{m+\gamma}f(s,y)\,\mathrm{d}y\mathrm{d}s\,,$$

hence

$$\frac{K_1}{C^2}\int_0^t\left(\int_1^\infty xf(s,x)\,\mathrm{d}x\right)^2\mathrm{d}s \leq \varrho + a_0|\mathfrak{h}_m|\int_0^t\int_1^\infty x^{m+\gamma}f(s,x)\,\mathrm{d}x\mathrm{d}s\,. \tag{9.1.11}$$

It remains to estimate the behaviour for small sizes. To this end, we argue as in the proof of Proposition 9.1.2. We first deduce from (9.1.2) that

$$\int_0^\infty \mathcal{C}f(t,x)\,\mathrm{d}x = -\frac{1}{2}\int_0^\infty\int_0^\infty k(x,y)f(t,x)f(t,y)\,\mathrm{d}y\mathrm{d}x$$

$$\leq -\frac{K_1}{2}M_{\lambda/2}^2(f(t))\,, \tag{9.1.12}$$

while (9.1.3), (9.1.4) and (9.1.5) ensure that

$$\int_0^\infty \mathcal{F}f(t,x)\,\mathrm{d}x \leq |\mathfrak{h}_0|\int_0^\infty a(y)f(t,y)\,\mathrm{d}y$$

$$\leq a_0|\mathfrak{h}_0|\left(\int_0^1 y^{\lambda/2}f(t,y)\,\mathrm{d}y + \int_1^\infty y^\gamma f(t,y)\,\mathrm{d}y\right)\,.$$

Since $\lambda > \gamma + 1 > \gamma + (\lambda/2)$ by (9.1.8), it follows that $\gamma < \lambda/2$. Thus, $y^\gamma \leq y^{\lambda/2}$ for $y > 1$ and we infer from the previous inequality and the Cauchy–Schwarz inequality that

$$\int_0^\infty \mathcal{F}f(t,x)\,\mathrm{d}x \leq a_0|\mathfrak{h}_0|M_{\lambda/2}(f(t)) \leq \frac{K_1}{4}M_{\lambda/2}^2(f(t)) + \frac{(a_0\mathfrak{h}_0)^2}{K_1}\,. \tag{9.1.13}$$

Therefore, for $t > 0$, we deduce from (8.2.4d) (with $\vartheta \equiv 1$), (9.1.12) and (9.1.13) that

$$M_0(f(t)) - M_0(f^{in}) = \int_0^t\int_0^\infty [\mathcal{C}f(s,x) + \mathcal{F}f(s,x)]\,\mathrm{d}x\mathrm{d}s$$

$$\leq \int_0^t\left[\frac{(a_0\mathfrak{h}_0)^2}{K_1} - \frac{K_1}{4}M_{\lambda/2}^2(f(s))\right]\mathrm{d}s$$

$$\leq \frac{(a_0\mathfrak{h}_0)^2}{K_1}t - \frac{K_1}{4}\int_0^t M_{\lambda/2}^2(f(s))\,\mathrm{d}s\,,$$

hence

$$\frac{K_1}{4}\int_0^t M_{\lambda/2}^2(f(s))\,\mathrm{d}s \leq M_0(f^{in}) + \frac{(a_0\mathfrak{h}_0)^2}{K_1}t\,. \tag{9.1.14}$$

Since $\lambda < 2$, we can now combine (9.1.11) and (9.1.14) to obtain, for $t > 0$,

$$\int_0^t M_1^2(f(s))\,\mathrm{d}s \leq 2\int_0^t\left(\int_0^1 xf(s,x)\,\mathrm{d}x\right)^2\mathrm{d}s + 2\int_0^t\left(\int_1^\infty xf(s,x)\,\mathrm{d}x\right)^2\mathrm{d}s$$

$$\leq 2\int_0^t\left(\int_0^1 x^{\lambda/2}f(s,x)\,\mathrm{d}x\right)^2\mathrm{d}s$$

$$+ \frac{2C^2}{K_1} \left[\varrho + a_0 |\mathfrak{h}_m| \int_0^t \int_1^\infty x^{m+\gamma} f(s,x) \, \mathrm{d}x \mathrm{d}s \right]$$

$$\leq \frac{8}{K_1} \left[M_0(f^{in}) + \frac{(a_0 \mathfrak{h}_0)^2}{K_1} t + \frac{\varrho C^2}{4} \right]$$

$$+ \frac{2C^2 a_0 |\mathfrak{h}_m|}{K_1} \int_0^t \int_1^\infty x^{m+\gamma} f(s,x) \, \mathrm{d}x \mathrm{d}s \, . \qquad (9.1.15)$$

Since $m+\gamma \leq 1$, we deduce from (9.1.15) and the elementary inequality $x^{m+\gamma} \leq x$ for $x > 1$ that, for $t \in [0, T_{gel})$,

$$\varrho^2 t \leq \frac{8}{K_1} M_0(f^{in}) + \frac{8(a_0 \mathfrak{h}_0)^2}{K_1^2} t + \frac{2\varrho C^2}{K_1} + \frac{2C^2 a_0 |\mathfrak{h}_m| \varrho}{K_1} t \, .$$

Equivalently, for $t \in [0, T_{gel})$,

$$\left(\varrho^2 - \frac{2C^2 a_0 |\mathfrak{h}_m|}{K_1} \varrho - \frac{8(a_0 \mathfrak{h}_0)^2}{K_1^2} \right) t \leq \frac{8}{K_1} M_0(f^{in}) + \frac{2\varrho C^2}{K_1} \, ,$$

from which it readily follows that $T_{gel} < \infty$ as soon as the factor of t in the left-hand side of the above inequality is positive; that is, if ϱ is large enough for $a_0 > 0$, or for all $\varrho > 0$ if $a_0 = 0$. $\qquad \square$

Remark 9.1.4. *In contrast to the case $\lambda = 2$, the choice $\gamma = \lambda - 1$ is excluded when $\lambda \in (1,2)$, so that the occurrence of gelation is an open problem when $\lambda \in (1,2)$ and $\gamma = \lambda - 1$. If it does occur, then the total mass of the initial condition has to be sufficiently large, as mass-conserving solutions are constructed in [155] when the total mass is small.*

We next turn to the case where (9.1.4) holds true with $m_0 \in (0,1)$, thus allowing the daughter distribution function b to be non-integrable, a case which has not been yet investigated in the literature. Clearly, we can no longer use $\vartheta \equiv 1$ as a test function and we have to proceed in a different way to estimate the contribution for small sizes. This, however, requires additional constraints on m_0 and λ.

Theorem 9.1.5. *Assume that the coagulation and fragmentation coefficients k, a and b satisfy (9.1.2)–(9.1.5) with $\lambda \in (1,2)$ and $m_0 \in (0, 2 - \lambda)$. Let $f^{in} \in X_{0,1,+}$ and let f be a weak solution to the C-F equation (8.0.1) on $[0,\infty)$ with initial condition f^{in}. There is $\varrho_\star > 0$ depending only on K_1, λ, a_0, γ and h such that, if*

$$\gamma \in [0, \lambda - 1] \quad and \quad \varrho := M_1(f^{in}) > \varrho_\star \, , \qquad (9.1.16)$$

then $T_{gel} < \infty$. Moreover, $\varrho_\star = 0$ if $a_0 = 0$ (i.e., no fragmentation occurs).

Proof. According to (9.1.16) and the property $m_0 < 2 - \lambda$, there exists $m \in (2 - \lambda, 1 - \gamma)$ such that $m > m_0$. We proceed as in the proof of the estimate (9.1.11) in Theorem 9.1.3 to establish that, for $t > 0$,

$$\frac{K_1}{C_m^2} \int_0^t \left(\int_1^\infty x f(s,x) \, \mathrm{d}x \right)^2 \mathrm{d}s \leq \varrho + a_0 |\mathfrak{h}_m| \int_0^t \int_1^\infty x^{m+\gamma} f(s,x) \, \mathrm{d}x \mathrm{d}s \, ,$$

where $C_m > 0$ depends only λ and m. Since $m + \gamma < 1$, we further obtain

$$\frac{K_1}{C_m^2} \int_0^t \left(\int_1^\infty x f(s,x) \, \mathrm{d}x \right)^2 \mathrm{d}s \leq \varrho + a_0 |\mathfrak{h}_m| \int_0^t \int_1^\infty x f(s,x) \, \mathrm{d}x \mathrm{d}s \, . \qquad (9.1.17)$$

Consider next

$$l \in [m_0, 2 - \lambda) \subset (0,1) , \qquad l \geq \frac{2 - \lambda}{2} , \tag{9.1.18}$$

and set $\vartheta(x) = \max\{x^l, 2^{l-1}x\}$, $x > 0$. Since $\chi_\vartheta(x,y) \leq 0$ for $(x,y) \in (0,\infty)^2$ and

$$(0,1) \times (0,1) \subset \{(x,y) \in (0,2)^2 \; : \; x + y < 2\} ,$$

we see that

$$\int_0^\infty \mathcal{C}f(t,x)\vartheta(x) \, \mathrm{d}x \leq \frac{1}{2} \int_0^1 \int_0^1 \left[(x+y)^l - x^l - y^l \right] k(x,y) f(t,x) f(t,y) \, \mathrm{d}y \mathrm{d}x .$$

Owing to (9.1.2) and the property $l < 2 - \lambda$, we infer from Lemma 8.2.12 (with $\mu = 1$) that there is $C_l > 0$ depending only on λ and l such that

$$\int_0^\infty \mathcal{C}f(t,x)\vartheta(x) \, \mathrm{d}x \leq -\frac{K_1}{C_l^2} \left(\int_0^1 x f(t,x) \, \mathrm{d}x \right)^2 . \tag{9.1.19}$$

Next, by (9.1.5),

$$\int_0^\infty \mathcal{F}f(t,x)\vartheta(x) \, \mathrm{d}x \leq \int_0^2 a(y)f(t,y) \left(\int_0^y \frac{x^l}{y} h\left(\frac{x}{y}\right) \, \mathrm{d}x - y^l \right) \, \mathrm{d}y$$

$$+ \int_2^\infty a(y)f(t,y) \left(\int_0^2 \frac{x^l}{y} h\left(\frac{x}{y}\right) \, \mathrm{d}x - 2^{l-1}y \right) \, \mathrm{d}y$$

$$+ 2^{l-1} \int_2^\infty a(y)f(t,y) \left(\int_2^y \frac{x}{y} h\left(\frac{x}{y}\right) \, \mathrm{d}x - y \right) \, \mathrm{d}y$$

$$\leq \int_0^2 y^l a(y)f(t,y) \left(\int_0^1 z^l h(z) \, \mathrm{d}z - 1 \right) \, \mathrm{d}y$$

$$+ \int_2^\infty a(y)f(t,y) \left(y^l \int_0^{2/y} z^l h(z) \, \mathrm{d}z - 2^{l-1}y \right) \, \mathrm{d}y$$

$$+ 2^{l-1} \int_2^\infty y a(y)f(t,y) \left(\int_{2/y}^1 z h(z) \, \mathrm{d}z - 1 \right) \, \mathrm{d}y .$$

Since $2^{l-1}y \geq y^l$ for $y \geq 2$, it follows from (9.1.4) and (9.1.6) that

$$\int_0^\infty \mathcal{F}f(t,x)\vartheta(x) \, \mathrm{d}x \leq |\mathfrak{h}_l| \int_0^2 y^l a(y)f(t,y) \, \mathrm{d}y$$

$$+ \int_2^\infty y^l a(y)f(t,y) \left(\int_0^{2/y} z^l h(z) \, \mathrm{d}z - 1 \right) \, \mathrm{d}y$$

$$\leq |\mathfrak{h}_l| \int_0^\infty y^l a(y)f(t,y) \, \mathrm{d}y .$$

Hence, on using (9.1.3),

$$\int_0^\infty \mathcal{F}f(t,x)\vartheta(x) \, \mathrm{d}x \leq a_0|\mathfrak{h}_l| \left[\int_0^1 x^{(2l+\lambda)/2} f(t,x) \, \mathrm{d}x + \int_1^\infty x^{l+\gamma} f(t,x) \, \mathrm{d}x \right] .$$

Since (9.1.18) implies that $2l + \lambda > 2$ and $l + \gamma < 1$, we deduce finally that

$$\int_0^\infty \mathcal{F}f(t,x)\vartheta(x) \, \mathrm{d}x \leq a_0|\mathfrak{h}_l| M_1(f(t)) . \tag{9.1.20}$$

Combining (8.2.4d), (9.1.19) and (9.1.20) gives, for $t > 0$,

$$\int_0^\infty \vartheta(x)[f(t,x) - f^{in}(x)] \, dx \leq -\frac{K_1}{C_l^2} \int_0^t \left(\int_0^1 xf(s,x) \, dx \right)^2 \, ds$$
$$+ a_0|\mathfrak{h}_l| \int_0^t M_1(f(s)) \, ds \ .$$

Hence

$$\frac{K_1}{C_l^2} \int_0^t \left(\int_0^1 xf(s,x) \, dx \right)^2 \, ds \leq M_l(f^{in}) + M_1(f^{in})$$
$$+ a_0|\mathfrak{h}_l| \int_0^t M_1(f(s)) \, ds \ . \tag{9.1.21}$$

It now follows from (9.1.17) and (9.1.21) that, for $t > 0$,

$$\int_0^t M_1(f(s))^2 \, ds \leq 2 \int_0^t \left(\int_0^1 xf(s,x) \, dx \right)^2 \, ds + 2 \int_0^t \left(\int_1^\infty xf(s,x) \, dx \right)^2 \, ds$$
$$\leq \frac{C_m^2 + C_l^2}{K_1} \varrho + \frac{C_l^2}{K_1} M_l(f^{in})$$
$$+ \frac{(C_m^2|\mathfrak{h}_m| + C_l^2|\mathfrak{h}_l|)a_0}{K_1} \int_0^t M_1(f(s)) \, ds \ .$$

Consequently, for $t \in (0, T_{gel})$,

$$\varrho \left(\varrho - \frac{(C_m^2|\mathfrak{h}_m| + C_l^2|\mathfrak{h}_l|)a_0}{K_1} \right) t \leq \frac{C_m^2 + C_l^2}{K_1} \varrho + \frac{C_l^2}{K_1} M_l(f^{in}) \ ,$$

and T_{gel} is finite provided ϱ is large enough. $\qquad\square$

Remark 9.1.6. *The requirement $m_0 < 2 - \lambda$ in Theorem 9.1.5 seems to be optimal. Indeed, it is shown in [173] that, when*

$$k(x,y) = xy \ , \quad a(x) = a_0 x \ , \quad b(x,y) = \frac{1}{x} \ , \quad 0 < x < y \ ,$$

there is a mass-conserving weak solution to the C-F equation (8.0.1) on $[0,\infty)$ for all initial conditions $f^{in} \in X_{0,1,+}$. The above coefficients satisfy (9.1.2)–(9.1.5) with $K_1 = 1$, $\lambda = 2$, $\gamma = 1$, $\mathfrak{h}(z) = 1/z$, $z \in (0,1)$ and any $m_0 \in (0,1)$.

So far we have focussed on the occurrence of gelation for coagulation kernels k satisfying (9.1.2) which, since $\lambda > 1$, implies in particular that

$$\lim_{x \to \infty} \frac{k(x,1)}{\sqrt{x}} = \infty \ . \tag{9.1.22}$$

It is then tempting to investigate whether the condition (9.1.22) is sufficient for the onset of gelation. The following result rules out the sufficiency of this condition.

Proposition 9.1.7. *[161, Proposition 33] Assume that $a \equiv 0$ and that*

$$k(x,y) = k_0(x)k_0(y) \ , \qquad (x,y) \in (0,\infty)^2 \ , \tag{9.1.23}$$

where $k_0 \in C([0, \infty)) \cap C^1((0, \infty))$ is a concave and positive function satisfying

$$\int_1^\infty \frac{dx}{k_0^2(x)} = \infty . \qquad (9.1.24)$$

Let $f^{in} \in X_{1,+} \cap X_2$. Then there exists a mass-conserving weak solution f to the coagulation equation on $[0, \infty)$ with initial condition f^{in}. Furthermore, $f \in L_\infty((0, T), X_2)$ for all $T > 0$.

Proposition 9.1.7 applies in particular to $k_0(x) = \sqrt{2 + x}[\ln(2 + x)]^\alpha$, $x \in (0, \infty)$, when $\alpha \in (0, 1/2]$, as the property (9.1.24) is clearly satisfied in this case. However, it leaves open the question of whether gelation takes place in finite time for such coagulation kernels if $f^{in} \in X_{1,+} \setminus X_2$.

The main idea behind the proof of Proposition 9.1.7 is the observation that, at least formally, if f is a weak solution to the coagulation equation with initial condition f^{in}, then its second moment $M_2(f)$ satisfies

$$\frac{d}{dt} M_2(f(t)) = \left(\int_0^\infty k_0(x) x f(t, x) \, dx \right)^2 , \qquad t > 0 .$$

Owing to the concavity of k_0 and Jensen's inequality, we obtain

$$\frac{d}{dt} M_2(f(t)) \leq M_1^2(f(t)) \left[k_0 \left(\frac{M_2(f(t))}{M_1(f(t))} \right) \right]^2 , \qquad t > 0 .$$

Then, as long as the total mass is conserved, that is, $M_1(f(t)) = M_1(f^{in})$, the condition (9.1.24) ensures that $M_2(f)$ is bounded locally in time, thereby preventing the occurrence of gelation. This formal proof can be made rigorous at the approximation level and we refer to [161, Proposition 33] for details.

When dealing with coagulation kernels with the product structure (9.1.23), the following result shows that it is possible to relax the power law growth required in (9.1.2).

Proposition 9.1.8. *[161, Proposition 36] Assume that $a \equiv 0$ and that k is given by (9.1.23), where $k_0 \in C([0, \infty)) \cap C^1((0, \infty))$ is a concave and positive function satisfying*

$$\int_1^\infty \frac{dx}{k_0(x)\sqrt{x}} < \infty , \qquad \lim_{x \to \infty} \frac{k_0(x)}{\sqrt{x}} = \lim_{x \to \infty} \frac{x}{k_0(x)} = \infty , \qquad (9.1.25)$$

and

$$k_0(x) \geq \delta x , \qquad x \in (0, 1) , \qquad (9.1.26)$$

for some $\delta > 0$. Let $f^{in} \in X_{0,1,+}$ and let f be a weak solution to the coagulation equation on $[0, \infty)$ with initial condition f^{in}. Then $T_{gel} < \infty$.

Proposition 9.1.8 applies in particular to $k_0(x) = \sqrt{\xi + x}[\ln(\xi + x)]^\alpha$, $x \in (0, \infty)$, when $\alpha > 1$ and $\xi > e^{2\sqrt{\alpha(\alpha - 1)}}$.

Instead of adapting the argument designed to prove Lemma 8.2.14 and then proceeding as in the proof of Theorem 9.1.3, we make use of the original approach developed in [112] which relies on the following result.

Proposition 9.1.9. *[112] Assume that $a \equiv 0$ and also that there is a nonnegative function $k_0 \in C([0, \infty))$ such that $k(x, y) \geq k_0(x) k_0(y)$ for $(x, y) \in (0, \infty)^2$. Let $f^{in} \in X_{1,+}$ and* *consider a weak solution f to the coagulation equation on $[0, \infty)$ with initial condition f^{in}. If $\zeta \in W^1_{\infty,loc}(0, \infty)$ is a nonnegative and nondecreasing function satisfying $\zeta(0) = 0$ and*

$$\mathcal{I}(\zeta) := \int_0^\infty \frac{\zeta'(z)}{\sqrt{z}} \, dz < \infty , \qquad (9.1.27)$$

then

$$\int_0^t \left(\int_0^\infty k_0(x)\zeta(x)f(s,x) \, dx \right)^2 ds \leq 2\mathcal{I}^2(\zeta)M_1(f^{in}) , \qquad t > 0 .$$

Proof. Let $t > 0$. Fix $z > 0$ and take $\vartheta(x) = \min\{x, z\}$, $x \in (0, \infty)$, in (8.2.4d). For this particular choice, $\chi_\vartheta \leq 0$ in $(0, \infty)^2$ and $\chi_\vartheta(x, y) = -z$ for $(x, y) \in (z, \infty)^2$. Since $\vartheta \leq$ id, $a \equiv 0$, and f is nonnegative, we obtain from (8.2.4d) that

$$\frac{z}{2} \int_0^t \int_z^\infty \int_z^\infty k(x, y)f(s, x)f(s, y) \, dydxds \leq M_1(f^{in}) .$$

Thanks to the lower bound on k, we end up with

$$\int_0^t \left(\int_z^\infty k_0(x)f(s, x) \, dx \right)^2 ds \leq \frac{2}{z} M_1(f^{in}) . \qquad (9.1.28)$$

We then infer from (9.1.28), the nonnegativity of ζ', Fubini's theorem and the Cauchy–Schwarz inequality that

$$\int_0^t \left(\int_0^\infty k_0(x)\zeta(x)f(s, x) \, dx \right)^2 ds \leq \int_0^t \left(\int_0^\infty \int_0^x k_0(x)\zeta'(z)f(s, x) \, dzdx \right)^2 ds$$

$$\leq \int_0^t \left(\int_0^\infty \zeta'(z) \int_z^\infty k_0(x)f(s, x) \, dxdz \right)^2 ds$$

$$\leq \mathcal{I}(\zeta) \int_0^t \int_0^\infty \zeta'(z)\sqrt{z} \left(\int_z^\infty k_0(x)f(s, x) \, dx \right)^2 dzds$$

$$\leq 2\mathcal{I}^2(\zeta)M_1(f^{in}) ,$$

as claimed. $\qquad \square$

Proposition 9.1.8 now follows as a direct consequence of Proposition 9.1.9.

Proof of Proposition 9.1.8. Let $t > 0$. Owing to (9.1.25) and the concavity and nonnegativity of k_0, we see that the function

$$\zeta(x) = \left(\frac{x}{k_0(x)} - \frac{1}{k_0(1)} \right)_+ , \qquad x \in (0, \infty)$$

satisfies the requirements of Proposition 9.1.9 and also that there is $x_0 > 1$ such that $\zeta(x) > x/2k_0(x)$ for $x \in (x_0, \infty)$. Consequently, by Proposition 9.1.9,

$$\frac{1}{4} \int_0^t \left(\int_{x_0}^\infty xf(s, x) \, dx \right)^2 ds \leq \int_0^t \left(\int_{x_0}^\infty k_0(x)\zeta(x)f(s, x) \, dx \right)^2 ds$$

$$\leq 2\mathcal{I}^2(\zeta)M_1(f^{in}) . \qquad (9.1.29)$$

We next take $\vartheta \equiv 1$ in (8.2.4d) and use (9.1.23) to obtain

$$\int_0^t \left(\int_0^\infty k_0(x) f(s,x) \, dx \right)^2 ds \leq 2 M_0(f^{in}) \ .$$

Since the positivity of k_0 and (9.1.26) ensure that $k_0(x) \geq \delta_0 x$ for $x \in (0, x_0)$ for some $\delta_0 \in (0, \delta)$, we further obtain

$$\int_0^t \left(\int_0^{x_0} x f(s,x) \, dx \right)^2 ds \leq \frac{2}{\delta_0^2} M_0(f^{in}) \ . \tag{9.1.30}$$

Combining (9.1.29) and (9.1.30), and noticing that both upper bounds do not depend on t, we see that $t \mapsto M_1(f(t))$ belongs to $L_1(0,\infty)$ and thus cannot be constant, and hence $T_{gel} < \infty$. $\square$

The outcome of Propositions 9.1.7 and 9.1.8 leaves a gap in the analysis of coagulation kernels of the form (9.1.23) for $k_0(x) = (2+x)[\ln(2+x)]^\alpha$, $x \in (0,\infty)$, for $\alpha \in (1/2, 1]$. The onset of gelation in such a case seems to be an open question.

9.2   Instantaneous Gelation

In the previous section (Proposition 9.1.2 and Theorem 9.1.3) we have shown that the gelation phenomenon takes place for any non-zero solution to the coagulation equation (without fragmentation) with a coagulation kernel satisfying

$$k(x,y) \geq K_1 (xy)^{\lambda/2} \ , \qquad (x,y) \in (0,\infty)^2 \ , \tag{9.2.1}$$

for some $\lambda \in (1,2]$ and $K_1 > 0$. More specifically, consider a nonnegative initial condition $f^{in} \in X_{0,1}$ and let f be a weak solution to the coagulation equation on $[0,\infty)$ with initial condition f^{in}. The results of Section 9.1 guarantee that the gelation time T_{gel} defined in Definition 9.1.1 is finite for coagulation kernels satisfying (9.2.1), and provide an upper bound on T_{gel} in terms of λ, K_1, $M_0(f^{in})$ and $\varrho = M_1(f^{in})$. However, no lower bound on T_{gel} (if any) is given and the purpose of this section is to investigate this issue. It turns out that a more precise assumption on k is needed: there is $0 \leq \alpha \leq \beta$, and $K_2 > 0$ such that $\lambda := \alpha + \beta > 1$ and

$$k(x,y) \geq K_2 \left(x^\alpha y^\beta + x^\beta y^\alpha \right) \ , \qquad (x,y) \in (0,\infty)^2 \ . \tag{9.2.2}$$

Then two different types of behaviour occur depending on the range of β in (9.2.2). We shall first prove that $T_{gel} = 0$ when $\beta > 1$; that is, the gelation phenomenon takes place instantaneously. On the contrary, when $\beta \leq 1$ (which implies that $\lambda \leq 2$), the gelation time can be either positive or zero, depending on the decay of the initial condition at infinity, see Proposition 9.2.7.

To show that instantaneous gelation takes place when $\beta > 1$, we follow the argument developed in [66, 260] for the discrete coagulation equation along with suitable modifications to handle the behaviour for small values of x.

Theorem 9.2.1. *Assume that k satisfies (9.2.2) for some $\beta > 1$, $a \equiv 0$, and that there is $\omega > 1$, and $K_3 > 0$ such that*

$$k(x,y) \leq K_3 (1+x)^\omega (1+y)^\omega \ , \qquad (x,y) \in (0,\infty)^2 \ . \tag{9.2.3}$$

Consider $f^{in} \in X_{0,1,+}$ and let f be a weak solution to the coagulation equation on $[0,\infty)$ with initial condition f^{in}. Then $T_{gel} = 0$.

The proof of Theorem 9.2.1 involves two steps. The starting point is to establish that, if $T_{gel} > 0$, then the lower bound (9.2.2) on k implies that any moment $M_m(f(t))$ is finite for $m \geq 1$ and $t \in [0, T_{gel})$, a feature which already places strong constraints on the initial condition. The second step is to show that $M_m(f)$ can only be finite on an interval $(0, t_m)$ which depends on m and that $t_m \to 0$ as $m \to \infty$, thereby contradicting the positivity of T_{gel}.

Lemma 9.2.2. *Assume that k satisfies (9.2.2) for some $\beta > 1$, $a \equiv 0$, and consider $f^{in} \in X_{0,1,+}$. Let f be a weak solution to the coagulation equation on $[0,\infty)$ with initial condition f^{in} such that $T_{gel} \in (0,\infty]$. Then, for all $m \geq 2$ and $t \in [0, T_{gel})$, $M_m(f(t))$ is finite. More precisely, for all $T \in (0, T_{gel})$,*

$$\sup_{t \in [0,T]} M_m(f(t)) < \infty .$$

Lemma 9.2.2 implies in particular that $M_m(f^{in}) < \infty$ for all $m \geq 2$, so that its first consequence is that $T_{gel} = 0$ as soon as f^{in} does not decay sufficiently fast as $x \to \infty$, thus establishing Theorem 9.2.1 in that case.

The proof of Lemma 9.2.2 relies on Dini's monotone convergence theorem [149, Proposition 8.24] which we first recall.

Theorem 9.2.3 (Dini's monotone convergence theorem). *Let X be a compact space and let $(g_j)_{j\geq 1}$ be a sequence of real-valued continuous functions on X. Assume that there is $g \in C(X)$ such that*

$$g_j(x) \leq g_{j+1}(x) , \qquad x \in X , \qquad j \geq 1 ,$$
$$\lim_{j\to\infty} g_j(x) = g(x) , \qquad x \in X .$$

Then $(g_j)_{j\geq 1}$ converges uniformly to g in X.

Proof of Lemma 9.2.2. Let $T \in (0, T_{gel})$. We first recall that (8.2.4d) with $\vartheta \equiv 1$ and $a \equiv 0$ implies that

$$M_0(f(t)) \leq M_0(f^{in}) , \qquad t \in [0,T] . \tag{9.2.4}$$

We next define

$$I_l(t) := \int_l^\infty x f(t,x)\, dx \quad \text{and} \quad J_l(t) := \int_0^l x f(t,x)\, dx$$

for $t \in [0,T]$ and $l \geq 1$. Since $T < T_{gel}$, the total mass conservation guarantees

$$I_l(t) + J_l(t) = \varrho := M_1(f^{in}) , \qquad t \in [0,T] , \tag{9.2.5}$$

as well as

$$\lim_{l\to\infty} J_l(t) = \varrho , \qquad t \in [0,T] .$$

In addition, $J_l \in C([0,T])$ and $J_l \leq J_{l+1}$ for all $l \geq 1$. Dini's monotone convergence theorem (Theorem 9.2.3) then ensures that $(J_l)_{l\geq 1}$ converges uniformly to ϱ in $[0,T]$. Therefore there is $l_0 \geq 1$ such that $\varrho - J_l(t) \leq \varrho/2$ for all $t \in [0,T]$ and $l \geq l_0$, hence

$$\frac{\varrho}{2} \leq J_l(t) , \qquad t \in [0,T] , \qquad l \geq l_0 . \tag{9.2.6}$$

Next let $l \geq l_0$. It follows from (8.2.4d) and (9.2.5) with $\vartheta(x) = x\mathbf{1}_{(0,l)}(x)$, $x \in (0, \infty)$, that, for $0 \leq t < \tau \leq T$,

$$I_l(t) - I_l(\tau) = J_l(\tau) - J_l(t) \leq -\int_t^\tau \int_0^l \int_l^\infty x k(x, y) f(s, x) f(s, y) \, \mathrm{d}y \mathrm{d}x \mathrm{d}s .$$

Owing to (9.2.2) with $\beta > 1$, we further obtain

$$I_l(t) - I_l(\tau) \leq -K_2 \int_t^\tau \left(\int_0^l x^{1+\alpha} f(s, x) \, \mathrm{d}x \right) \left(\int_l^\infty y^\beta f(s, y) \, \mathrm{d}y \right) \mathrm{d}s$$

$$\leq -K_2 l^{\beta-1} \int_t^\tau \left(\int_0^l x^{1+\alpha} f(s, x) \, \mathrm{d}x \right) I_l(s) \, \mathrm{d}s . \tag{9.2.7}$$

It next follows from Hölder's inequality, (9.2.4) and (9.2.6) that, for $s \in [0, T]$,

$$\frac{\varrho}{2} \leq J_l(s) \leq \left(\int_0^l x^{1+\alpha} f(s, x) \, \mathrm{d}x \right)^{1/(1+\alpha)} \left(\int_0^l f(s, x) \, \mathrm{d}x \right)^{\alpha/(1+\alpha)}$$

$$\leq \left(\int_0^l x^{1+\alpha} f(s, x) \, \mathrm{d}x \right)^{1/(1+\alpha)} M_0(f^{in})^{\alpha/(1+\alpha)} .$$

Combining the previous inequality and (9.2.7) gives

$$I_l(t) - I_l(\tau) \leq -K_2 \left(\frac{\varrho}{2} \right)^{1+\alpha} M_0(f^{in})^{-\alpha} l^{\beta-1} \int_t^\tau I_l(s) \, \mathrm{d}s ,$$

and therefore

$$I_l(t) + C_1 l^{\beta-1} \int_t^\tau I_l(s) \, \mathrm{d}s \leq I_l(\tau) , \qquad 0 \leq t < \tau \leq T , \tag{9.2.8}$$

where $C_1 := K_2(\varrho/2)^{1+\alpha} M_0(f^{in})^{-\alpha}$. Setting

$$Q_l(\tau) := I_l(t) + C_1 l^{\beta-1} \int_t^\tau I_l(s) \, \mathrm{d}s , \qquad \tau \in [t, T] ,$$

we deduce from (9.2.8) that

$$\frac{\mathrm{d}}{\mathrm{d}\tau} Q_l(\tau) = C_1 l^{\beta-1} I_l(\tau) \geq C_1 l^{\beta-1} Q_l(\tau) , \qquad \tau \in [t, T] .$$

Hence, on integrating with respect to τ over (t, T) and again making use of (9.2.8), it follows that

$$e^{-C_1 l^{\beta-1} t} I_l(t) = e^{-C_1 l^{\beta-1} t} Q_l(t) \leq e^{-C_1 l^{\beta-1} T} Q_l(T) \leq e^{-C_1 l^{\beta-1} T} I_l(T) .$$

Since $I_l(T) \leq M_1(f(T)) = \varrho$, we end up with

$$I_l(t) \leq e^{-C_1 l^{\beta-1}(T-t)} \varrho , \quad t \in [0, T] , \qquad l \geq l_0 . \tag{9.2.9}$$

Now, fix $m \geq 2$ and $t \in [0, T]$. For $l \geq L \geq l_0$,

$$\int_L^l x^m f(t, x) \, \mathrm{d}x \leq \sum_{j=L}^{l-1} (j+1)^{m-1} \int_j^{j+1} x f(t, x) \, \mathrm{d}x$$

$$= \sum_{j=L}^{l-1} (j+1)^{m-1} \left[I_j(t) - I_{j+1}(t) \right]$$

$$= \sum_{j=L}^{l-1} (j+1)^{m-1} I_j(t) - \sum_{j=L+1}^{l} j^{m-1} I_j(t)$$

$$\leq (L+1)^{m-1} I_L(t) + (m-1) \sum_{j=L+1}^{l-1} (j+1)^{m-2} I_j(t) \ .$$

Owing to (9.2.5) and (9.2.9) we further obtain

$$\int_L^l x^m f(t,x) \, \mathrm{d}x \leq (L+1)^{m-1} \varrho$$

$$+ (m-1)\varrho \sum_{j=L+1}^{l-1} \exp\left\{ (m-2)\ln(j+1) - C_1(T-t)j^{\beta-1} \right\} \ .$$

Let $T' \in (0,T)$. Since $\beta > 1$ there is $L(T') \geq l_0$ depending only on T, T', C_1, β and m such that $m\ln(j+1) - C_1(T-T')j^{\beta-1} \leq 0$ for all $j \geq L(T')$. Choosing $L = L(T')$ in the previous inequality gives, for $t \in [0, T']$,

$$\int_{L(T')}^l x^m f(t,x) \, \mathrm{d}x \leq (L(T')+1)^{m-1}\varrho + (m-1)\varrho \sum_{j=L(T')+1}^{l-1} \frac{1}{(j+1)^2}$$

$$\leq (L(T')+1)^{m-1}\varrho + \frac{\pi^2(m-1)\varrho}{4} \ .$$

Letting $l \to \infty$ in the above inequality and observing that

$$\int_0^{L(T')} x^m f(t,x) \, \mathrm{d}x \leq L(T')^{m-1}\varrho$$

allow us to conclude that $M_m(f)$ belongs to $L_\infty(0,T')$. Since T and T' were arbitrarily chosen in $[0, T_{gel})$ and $[0,T)$, respectively, the proof of Lemma 9.2.2 is complete. □

We next derive an equation for $M_m(f)$ for $m \geq 2$.

Lemma 9.2.4. *Assume that k satisfies (9.2.2) for some $\beta > 1$ as well as (9.2.3), $a \equiv 0$, and consider $f^{in} \in X_{0,1,+}$. Let f be a weak solution to the coagulation equation on $[0,\infty)$ with initial condition f^{in} such that $T_{gel} \in (0,\infty]$. Then, for all $m \geq 2$ and $0 \leq t \leq \tau < T_{gel}$,*

$$M_m(f(\tau)) - M_m(f(t))$$
$$= \frac{1}{2} \int_t^\tau \int_0^\infty \int_0^\infty \left[(x+y)^m - x^m - y^m \right] k(x,y) f(s,x) f(s,y) \, \mathrm{d}x\mathrm{d}y\mathrm{d}s \ .$$

Proof. Let $l \geq 1$, $\tau \in (0, T_{gel})$ and $t \in [0,\tau]$. We take $\vartheta(x) = x^m \mathbf{1}_{(0,l)}(x)$, $x \in (0,\infty)$, in (8.2.4d) to obtain

$$\int_0^l x^m (f(\tau,x) - f(t,x)) \, \mathrm{d}x = \int_t^\tau (I_{1,l}(s) + I_{2,l}(s) + I_{3,l}(s)) \, \mathrm{d}s \ , \tag{9.2.10}$$

where

$$I_{1,l}(s) = \frac{1}{2} \int_0^l \int_0^l \left[(x+y)^m - x^m - y^m \right] k(x,y) f(s,x) f(x,y) \, \mathrm{d}y\mathrm{d}x \ ,$$

$$I_{2,l}(s) = -\frac{1}{2} \int_0^l \int_{l-x}^l (x+y)^m k(x,y) f(s,x) f(s,y) \ \mathrm{d}y \mathrm{d}x \ ,$$

$$I_{3,l}(s) = -\int_0^l \int_l^\infty x^m k(x,y) f(s,x) f(s,y) \ \mathrm{d}y \mathrm{d}x \ .$$

Owing to the algebraic growth (9.2.3) of k, it follows from Lemma 7.4.1, Lemma 7.4.2, Lemma 9.2.2 and Lebesgue's dominated convergence theorem that

$$\lim_{l\to\infty} \int_t^\tau I_{1,l}(s) \ \mathrm{d}s$$

$$= \frac{1}{2} \int_t^\tau \int_0^\infty \int_0^\infty \left[(x+y)^m - x^m - y^m \right] k(x,y) f(s,x) f(x,y) \ \mathrm{d}y \mathrm{d}x \mathrm{d}s$$

and

$$\lim_{l\to\infty} \int_t^\tau \left(I_{2,l}(s) + I_{3,l}(s) \right) \ \mathrm{d}s = 0 \ .$$

We then pass to the limit as $l \to \infty$ in (9.2.10) and complete the proof of Lemma 9.2.4. $\quad\square$

We finally investigate in more detail the time evolution of $M_m(f)$ for $m \geq 2$.

Lemma 9.2.5. *Assume that k satisfies (9.2.2) for some $\beta > 1$ as well as (9.2.3), $a \equiv 0$, and consider $f^{in} \in X_{0,1,+}$. Let f be a weak solution to the coagulation equation on $[0,\infty)$ with initial condition f^{in} such that $T_{gel} \in (0,\infty]$. Then, for all $m \geq 2$ and $T \in (0, T_{gel})$,*

$$T \leq t + C_2 M_m(f(t))^{-(\beta-1)/(m-1)} \ , \qquad t \in [0,T] \ ,$$

where $C_2 > 0$ depends only on α, β, K_2, $\varrho = M_1(f^{in})$, and $M_0(f^{in})$.

Proof. Since

$$(x+y)^m - x^m - y^m = m(m-1) \int_0^x \int_0^y (x_* + y_*)^{m-2} \ \mathrm{d}y_* \mathrm{d}x_*$$

$$\geq \frac{m(m-1)}{2} \int_0^x \int_0^y \left(x_*^{m-2} + y_*^{m-2} \right) \ \mathrm{d}y_* \mathrm{d}x_*$$

$$\geq \frac{m}{2} \left(x^{m-1}y + xy^{m-1} \right)$$

for $(x,y) \in (0,\infty)^2$, we infer from (9.2.2), the symmetry of k, and Lemma 9.2.4 that, for $0 \leq t < \tau \leq T$,

$$M_m(f(\tau)) - M_m(f(t)) \geq \frac{m}{2} \int_t^\tau \int_0^\infty \int_0^\infty x^{m-1} y k(x,y) f(s,x) f(s,y) \ \mathrm{d}y \mathrm{d}x \mathrm{d}s$$

$$\geq \frac{mK_2}{2} \int_t^\tau M_{m-1+\beta}(f(s)) M_{1+\alpha}(f(s)) \ \mathrm{d}s \ . \qquad (9.2.11)$$

Since $m \in (1, m-1+\beta)$ and $1 \in (0, 1+\alpha]$, we infer from Hölder's inequality, (9.2.4) and the total mass conservation (9.2.5) that, for $s \in [t, \tau]$,

$$M_m(f(s)) \leq M_{m-1+\beta}(f(s))^{(m-1)/(m+\beta-2)} \varrho^{(\beta-1)/(m+\beta-2)}$$

and

$$\varrho \leq M_{1+\alpha}(f(s))^{1/(1+\alpha)} M_0(f^{in})^{\alpha/(1+\alpha)} \ .$$

Combining these lower bounds with (9.2.11) leads us to

$$M_m(f(\tau)) \geq M_m(f(t)) + \frac{mK_2}{2} \frac{\varrho^{1+\alpha} M_0(f^{in})^{-\alpha}}{\varrho^{(\beta-1)/(m-1)}} \int_t^\tau M_m(f(s))^{(m+\beta-2)/(m-1)} \, ds \;,$$

and, since

$$\varrho^{(\beta-1)/(m-1)} \leq (1+\varrho)^{(\beta-1)/(m-1)} \leq (1+\varrho)^{\beta-1} \;,$$

it follows that

$$M_m(f(\tau) \geq M_m(f(t)) + mC_3 \int_t^\tau M_m(f(s))^{(m+\beta-2)/(m-1)} \, ds \;, \qquad (9.2.12)$$

where

$$C_3 := \frac{K_2}{2} \frac{\varrho^{1+\alpha} M_0(f^{in})^{-\alpha}}{(1+\varrho)^{\beta-1}} \;.$$

Introducing

$$X_m(\tau) := M_m(f(t)) + mC_3 \int_t^\tau M_m(f(s))^{(m+\beta-2)/(m-1)} \, ds \;, \qquad \tau \in [t, T] \;,$$

we infer from (9.2.12) that $M_m(f(\tau)) \geq X_m(\tau)$ and

$$\frac{dX_m}{d\tau}(\tau) = mC_3 M_m(f(\tau))^{(m+\beta-2)/(m-1)} \geq mC_3 X_m(\tau)^{(m+\beta-2)/(m-1)}$$

for $\tau \in [t, T]$. After integration with respect to τ over (t, T), we obtain

$$0 \leq X_m(T)^{-(\beta-1)/(m-1)} \leq X_m(t)^{-(\beta-1)/(m-1)} - \frac{m(\beta-1)C_3}{m-1}(T-t) \;.$$

Consequently,

$$T - t \leq \frac{m-1}{m(\beta-1)C_3} X_m(t)^{-(\beta-1)/(m-1)} \;,$$

from which Lemma 9.2.5 readily follows with $1/C_2 := (\beta-1)C_3$. $\qquad\square$

Proof of Theorem 9.2.1. Assume for contradiction that $T_{gel} \in (0, \infty]$, and then consider $T \in (0, T_{gel})$ and $m \geq 2$. On the one hand, by Lemma 9.2.5,

$$T \leq t + C_2 M_m(f(t))^{-(\beta-1)/(m-1)} \;, \qquad t \in [0, T] \;. \qquad (9.2.13)$$

On the other hand, for all $l \geq 1$ and $t \in (0, T)$,

$$M_m(f(t))^{1/m} \geq \left(\int_l^\infty x^m f(t, x) \, dx \right)^{1/m} \geq l \left(\int_l^\infty f(t, x) \, dx \right)^{1/m} \;. \qquad (9.2.14)$$

Let $t \in [0, T]$. Owing to (9.2.2), (9.2.3) and Lemma 9.2.2, we infer from Proposition 8.2.20 that

$$\int_l^\infty f(t, x) \, dx > 0 \;,$$

so that we may let $m \to \infty$ in (9.2.14) to obtain

$$\liminf_{m \to \infty} M_m(f(t))^{1/m} \geq l \;.$$

The above inequality being valid for all $l \geq 1$, we conclude that

$$\lim_{m \to \infty} M_m(f(t))^{1/m} = \infty \;.$$

We then let $m \to \infty$ in (9.2.13) and end up with $T \leq t$ for all $t \in (0, T)$, which is a contradiction. Therefore, $T_{gel} = 0$. $\qquad\square$

As noticed in [66] for the discrete coagulation equation, a consequence of Theorem 9.2.1 is the non-existence of solutions to the coagulation equation for some specific choices of coagulation kernel.

Corollary 9.2.6. *Assume that $a \equiv 0$ and there is $\lambda > 1$ such that*

$$k(x,y) = x^\lambda + y^\lambda , \qquad (x,y) \in (0,\infty)^2 . \tag{9.2.15}$$

Let $T > 0$ and consider $f^{in} \in X_{0,1,+}$, $f^{in} \not\equiv 0$. There is no weak solution to the coagulation equation on $[0,T)$ in the sense of Definition 8.2.1.

Corollary 9.2.6 shows that there is a striking difference between "sum" coagulation kernels such as (9.2.15) and "product" coagulation kernels such as $k(x,y) = (xy)^{\lambda/2}$, $(x,y) \in (0,\infty)^2$, when their growth is superlinear for large sizes. Indeed, no solution to the coagulation equation exists for the former when $\lambda > 1$, even locally in time. In contrast, according to Theorem 8.2.47, weak solutions to the coagulation equation on $[0,\infty)$ exist for the latter whatever the value of $\lambda > 2$.

The proof of Corollary 9.2.6 relies on the crucial observation made in [14, Section 3] that the specific form (9.2.15) of the coagulation kernel implies that, in this particular case, any weak solution has to be a mass-conserving solution to the coagulation equation, thereby contradicting Theorem 9.2.1.

Proof. Let f be a weak solution to the coagulation equation on $[0,T)$ in the sense of Definition 8.2.1. Since $f^{in} \not\equiv 0$ and $M_0(f) \in C([0,T))$, there is $T_0 \in [0,T)$, and $\varepsilon_0 > 0$ such that

$$M_0(f(t)) \geq \varepsilon_0 , \qquad t \in [0,T_0] . \tag{9.2.16}$$

Combining (8.2.4b), (9.2.15) and (9.2.16) ensures that

$$\varepsilon_0 \int_0^{T_0} M_\lambda(f(t)) \, \mathrm{d}t \leq \int_0^{T_0} M_\lambda(f(t)) M_0(f(t)) \, \mathrm{d}t$$

$$\leq \frac{1}{2} \int_0^{T_0} \int_0^\infty \int_0^\infty k(x,y) f(t,x) f(t,y) \, \mathrm{d}y \mathrm{d}x \mathrm{d}t < \infty ,$$

and thus

$$M_\lambda(f) \in L_1(0,T_0) . \tag{9.2.17}$$

We next define

$$J(R) := \int_0^{T_0} \int_R^\infty \int_0^\infty (x^\lambda y + xy^\lambda) f(t,x) f(t,y) \, \mathrm{d}y \mathrm{d}x \mathrm{d}t , \qquad R \in (0,\infty) ,$$

and claim that

$$\lim_{R\to\infty} J(R) = 0 . \tag{9.2.18}$$

Indeed, according to (8.2.4a), $M_1(f) \in L_\infty(0,T_0)$ and

$$J(R) \leq \sup_{s \in [0,T_0]} \{M_1(f(s))\} \int_0^{T_0} \int_R^\infty x^\lambda f(t,x) \, \mathrm{d}x \mathrm{d}t$$

$$+ \int_0^{T_0} M_\lambda(f(t)) \int_R^\infty y f(t,y) \, \mathrm{d}y \mathrm{d}t .$$

On the one hand, the first term on the right-hand side of the previous inequality converges to zero as $R \to \infty$ by (9.2.17). On the other hand, a further use of (8.2.4a) and (9.2.17) along with Lebesgue's dominated convergence theorem implies that the second term on the right-hand side of the previous inequality converges also to zero as $R \to \infty$, thereby completing the proof of the claim (9.2.18).

Now, let $t \in [0, T_0]$, $R > 0$, and take $\vartheta(x) = \min\{x, R\}$, $x \in (0, \infty)$, in the weak formulation (8.2.4d) to obtain

$$
\int_0^\infty \min\{x, R\}[f(t, x) - f^{in}(x)] \, dx
$$

$$
= -\frac{1}{2} \int_0^t \int_0^R \int_{R-x}^R (x + y - R)k(x, y)f(s, x)f(s, y) \, dydxds
$$

$$
- \int_0^t \int_0^R \int_R^\infty xk(x, y)f(s, x)f(s, y) \, dydxds
$$

$$
- \frac{R}{2} \int_0^t \int_R^\infty \int_R^\infty k(x, y)f(s, x)f(s, y) \, dydxds \, .
$$

Owing to (9.2.15),

$$
I_1(t, R) := \frac{1}{2} \int_0^t \int_0^R \int_{R-x}^R (x + y - R)k(x, y)f(s, x)f(s, y) \, dydxds
$$

$$
\leq \frac{1}{2} \int_0^t \int_0^R \int_{R-x}^R (x^\lambda y + xy^\lambda)f(s, x)f(s, y) \, dydxds
$$

$$
\leq \int_0^t \int_0^R \int_{R-x}^R x^\lambda y f(s, x)f(s, y) \, dydxds
$$

$$
\leq \int_0^t \int_0^{R/2} \int_{R/2}^R x^\lambda y f(s, x)f(s, y) \, dydxds
$$

$$
+ \int_0^t \int_{R/2}^R \int_0^R x^\lambda y f(s, x)f(s, y) \, dydxds
$$

$$
\leq J(R/2) \, .
$$

Also,

$$
I_2(t, R) := \int_0^t \int_0^R \int_R^\infty xk(x, y)f(s, x)f(s, y) \, dydxds
$$

$$
\leq \int_0^t \int_0^R \int_R^\infty (x^{\lambda+1} + xy^\lambda)f(s, x)f(s, y) \, dydxds
$$

$$
\leq J(R) \, ,
$$

and

$$
I_3(t, R) := \frac{R}{2} \int_0^t \int_R^\infty \int_R^\infty k(x, y)f(s, x)f(s, y) \, dydxds \leq \frac{J(R)}{2} \, .
$$

Collecting the above estimates we conclude that

$$
\left| \int_0^\infty \min\{x, R\}[f(t, x) - f^{in}(x)] \, dx \right| \leq J(R/2) + 2J(R) \, .
$$

Since both $f(t)$ and f^{in} belong to X_1, we may pass to the limit as $R \to \infty$ in the previous inequality and deduce from (9.2.18) that

$$
M_1(f(t)) = M_1(f^{in}) \, , \qquad t \in [0, T_0] \, . \tag{9.2.19}
$$

Consequently, f is a mass-conserving solution to the coagulation equation on $[0, T_0)$. However, the coagulation kernel k given by (9.2.15) satisfies (9.2.2) (with $\beta = \lambda$ and $\alpha = 0$) and (9.2.3) (with $\omega = \lambda$), and Theorem 9.2.1 implies that the gelation time for f is equal to zero, clearly contradicting (9.2.19). $\qquad\square$

We next identify a class of coagulation kernels for which a positive lower bound on the gelation time is available, so that instantaneous gelation is excluded.

Proposition 9.2.7. *Assume that $a \equiv 0$ and there is $0 \leq \alpha \leq \beta \leq 1$, and $K_4 > 0$ such that $\lambda := \alpha + \beta \in (1, 2]$ and*

$$k(x, y) \leq K_4 \left(x^\alpha y^\beta + x^\beta y^\alpha \right) , \qquad (x, y) \in (0, \infty)^2 . \tag{9.2.20}$$

Consider $f^{in} \in X_{0,1,+}$ such that

$$M_{2\beta}(f^{in}) < \infty . \tag{9.2.21}$$

There is $C_4 > 0$ depending only on α, β and K_4 such that the coagulation equation has a unique mass-conserving weak solution f on $[0, T_0)$ with initial condition f^{in} satisfying $M_{2\beta}(f) \in L_\infty(0, T)$ for all $T \in (0, T_0)$, where

$$T_0 := C_4 M_1(f^{in})^{-(\beta-\alpha)/(2\beta-1)} M_{2\beta}(f^{in})^{-(\lambda-1)/(2\beta-1)} .$$

In particular, $T_{gel} \geq T_0$.

It is unlikely that Proposition 9.2.7 extends to arbitrary initial conditions $f^{in} \in X_{0,1,+}$. Indeed, as we shall see below in Proposition 9.2.8, it can be shown that $T_{gel} = 1/M_2(f^{in})$ when $k(x, y) = xy$, so that $T_{gel} = 0$ if $f^{in} \notin X_2$ [183, 190, 198, 224, 243]. For $\lambda \in (1, 2)$, a natural conjecture in this direction is that $T_{gel} > 0$ if and only if $M_\lambda(f^{in}) < \infty$. It is supported by the results of [262] where a result similar to Proposition 9.2.7 is shown for the coagulation kernel $K(x, y) = (xy)^{\lambda-1}(x + y)^{2-\lambda}$, $(x, y) \in (0, \infty)^2$, when $\lambda \in (1, 2]$ and the moment $M_{2\alpha}(f)$ is replaced by $M_\lambda(f)$. This kernel clearly satisfies (9.2.20) with $\alpha = \lambda - 1$ and $\beta = 1$.

Proof. For $j \geq 1$ and $(x, y) \in (0, \infty)^2$ we define $k_j(x, y) := k(x, y)\mathbf{1}_{(0,j)}(x + y)$ and also $f_j^{in}(x) := f^{in}(x)\mathbf{1}_{(0,j)}(x)$. Let f_j be the solution to the coagulation equation with coagulation kernel k_j and initial condition f_j^{in}, given by Proposition 8.2.7. We recall that

$$M_1(f_j(t)) = M_1(f_j^{in}) \leq \varrho := M_1(f^{in}) , \qquad t \geq 0 , \ j \geq 1 . \tag{9.2.22}$$

Since $2\beta \in (1, 2]$, it follows from (8.2.13), (9.2.20), the positivity (7.4.2) of $\chi_{2\beta}$, and Lemma 7.4.5 (with $m = 2\beta$) that, for $j \geq 1$ and $t > 0$,

$$\frac{\mathrm{d}}{\mathrm{d}t} M_{2\beta}(f_j(t)) \leq \frac{K_4(4^\beta - 2)}{2} \int_0^j \int_0^{j-x} (xy)^\beta \left(x^\alpha y^\beta + x^\beta y^\alpha \right) f_j(t, x) f_j(t, y) \, \mathrm{d}y \mathrm{d}x$$

$$\leq 4^\beta K_4 \int_0^j \int_0^{j-x} x^{2\beta} y^\lambda f_j(t, x) f_j(t, y) \, \mathrm{d}y \mathrm{d}x$$

$$\leq 4^\beta K_4 M_{2\beta}(f_j(t)) M_\lambda(f_j(t)) .$$

Since $\lambda \in (1, 2\beta]$, we infer from Hölder's inequality and (9.2.22) that

$$M_\lambda(f_j(t)) \leq M_{2\beta}(f_j(t))^{(\lambda-1)/(2\beta-1)} M_1(f_j(t))^{(2\beta-\lambda)/(2\beta-1)}$$

$$\leq \varrho^{(\beta-\alpha)/(2\beta-1)} M_{2\beta}(f_j(t))^{(\lambda-1)/(2\beta-1)} .$$

Combining the previous two inequalities we end up with

$$\frac{\mathrm{d}}{\mathrm{d}t} M_{2\beta}(f_j(t)) \leq 4^\beta K_4 \varrho^{(\beta-\alpha)/(2\beta-1)} M_{2\beta}(f_j(t))^{(\lambda+2\beta-2)/(2\beta-1)} .$$

Since $\lambda + 2\beta - 2 > 2\beta - 1$, integrating the above differential inequality and using (9.2.21) leads to

$$M_{2\beta}(f_j(t))^{-(\lambda-1)/(2\beta-1)} \geq M_{2\beta}(f_j^{in})^{-(\lambda-1)/(2\beta-1)} - C_4 \varrho^{(\beta-\alpha)/(2\beta-1)} t$$
$$\geq M_{2\beta}(f^{in})^{-(\lambda-1)/(2\beta-1)} - C_4 \varrho^{(\beta-\alpha)/(2\beta-1)} t$$

for $t \geq 0$, with $C_4 = 4^\beta K_4 (\lambda - 1)/(2\beta - 1)$. Therefore, introducing

$$T_0 := \frac{1}{C_4} M_{2\beta}(f^{in})^{-(\lambda-1)/(2\beta-1)} \varrho^{-(\beta-\alpha)/(2\beta-1)} ,$$

we see that for each $T \in (0, T_0)$ there is a positive constant $C(T)$, depending only on α, β, K_4, f^{in} and T, such that

$$\sup_{t \in [0,T], \, j \geq 1} M_{2\beta}(f_j(t)) \leq C(T) . \qquad (9.2.23)$$

Since $2\beta \geq \lambda > 1$, we next argue as in the proof of Theorem 8.2.23 to conclude that there is a mass-conserving weak solution f to the coagulation equation on $[0, T_0)$ such that $M_{2\beta}(f) \in L_\infty(0, T)$ for all $T \in (0, T_0)$, the latter being a consequence of (9.2.23). The uniqueness of such a solution follows from Theorem 8.2.55 (with $\ell(x) = (1 + x)^\beta$ and $\zeta = 1$). $\qquad \square$

We conclude this section with the computation of the gelation time for the multiplicative kernel $k_\times$. On the one hand, it shows that the finiteness of a moment of the initial condition of order higher than one is necessary to exclude instantaneous gelation, such as (9.2.21) in Proposition 9.2.7. On the other hand, it shows that instantaneous gelation and gelation after a finite time may occur simultaneously for a given kernel and depends on the integrability properties of the initial condition for large sizes.

Proposition 9.2.8. *Let $k = k_\times$, $a \equiv 0$, and consider $f^{in} \in X_{0,1,+}$, $f^{in} \not\equiv 0$. If f is a weak solution to the coagulation equation on $[0, \infty)$ with initial condition f^{in} such that $M_1(f(t)) \leq \varrho := M_1(f^{in})$ for all $t > 0$, then $T_{gel} = 1/M_2(f^{in})$ when $M_2(f^{in}) < \infty$ and $T_{gel} = 0$ otherwise.*

Proof. Let f be a weak solution to the coagulation equation with multiplicative coagulation kernel $k = k_\times$ (and $a \equiv 0$). In this case we can use the Laplace transform to reduce the nonlocal integro-differential equation (8.0.1) to a partial differential equation. Specifically, we set

$$\mathscr{L} f(t, \xi) := \int_0^\infty x e^{-x\xi} f(t, x) \, \mathrm{d}x , \qquad (t, \xi) \in [0, \infty)^2 .$$

Since $x \mapsto x e^{-x\xi}$ belongs to $L_\infty(0, \infty)$ for $\xi > 0$, we infer from (8.2.4d) with $\vartheta(x) = x e^{-x\xi}$ that

$$\partial_t \mathscr{L} f(t, \xi) = \int_0^\infty \int_0^\infty x e^{-x\xi} \left(e^{-y\xi} - 1 \right) xy f(t, x) f(t, y) \, \mathrm{d}y \mathrm{d}x$$
$$= \left(\int_0^\infty x^2 e^{-x\xi} f(t, x) \, \mathrm{d}x \right) \left(\int_0^\infty y e^{-y\xi} f(t, y) \, \mathrm{d}y - M_1(f(t)) \right) .$$

Since

$$\partial_\xi \mathscr{L} f(t,\xi) = -\int_0^\infty x^2 e^{-x\xi} f(t,x) \, \mathrm{d}x \, , \qquad (t,\xi) \in (0,\infty)^2 \, ,$$

we conclude that $\mathscr{L} f$ is a strong solution to the non-autonomous scalar conservation law

$$\partial_t \mathscr{L} f(t,\xi) = -\left[\mathscr{L} f(t,\xi) - M_1(f(t)) \right] \partial_\xi \mathscr{L} f(t,\xi) \, , \qquad (t,\xi) \in (0,\infty)^2 \, . \qquad (9.2.24)$$

We shall now use the associated characteristic equation to study the behaviour of $\mathscr{L} f(t,\xi)$ as $\xi \to 0$ for $t \geq 0$. Specifically, for $\xi_0 \in (0,\infty)$, let $\Xi(\cdot;\xi_0)$ be the unique positive solution to the characteristic equation

$$\begin{aligned}
\frac{d\Xi}{dt}(t;\xi_0) &= \mathscr{L} f(t,\Xi(t;\xi_0)) - M_1(f(t)) \, , \qquad t \in (0, T(\xi_0)) \, , \\
\Xi(0;\xi_0) &= \xi_0 \, ,
\end{aligned} \qquad (9.2.25)$$

which is well defined on a maximal time interval $[0, T(\xi_0))$, $T(\xi_0) > 0$, with the alternative

$$T(\xi_0) = \infty \qquad \text{or} \qquad T(\xi_0) < \infty \quad \text{and} \quad \lim_{t \to T(\xi_0)} \Xi(t;\xi_0) = 0 \, . \qquad (9.2.26)$$

Indeed, blowup in finite time of Ξ is excluded, since $\Xi(t;\xi_0) \leq \xi_0$ for all $t \in [0, T(\xi_0))$ by (9.2.25), the latter guaranteeing that $d\Xi(t;\xi_0)/dt \leq 0$ for $t \in [0, T(\xi_0))$. Owing to (9.2.24), $t \mapsto \mathscr{L} f(t,\Xi(t;\xi_0))$ is constant on $[0, T(\xi_0))$; that is,

$$\mathscr{L} f(t,\Xi(t;\xi_0)) = \mathscr{L} f^{in}(\xi_0) \, , \qquad t \in [0, T(\xi_0)) \, . \qquad (9.2.27)$$

Furthermore, if $\xi_0 > 0$ is such that $T(\xi_0) < \infty$, then (9.2.26) and (9.2.27) imply that

$$\mathscr{L} f(T(\xi_0), 0) = \mathscr{L} f^{in}(\xi_0) \, . \qquad (9.2.28)$$

Combining (9.2.25) and (9.2.27) leads us, after integration with respect to time, to

$$\Xi(t;\xi_0) = \xi_0 + \mathscr{L} f^{in}(\xi_0) t - \int_0^t M_1(f(s)) \, \mathrm{d}s \, , \qquad t \in [0, T(\xi_0)) \, . \qquad (9.2.29)$$

Since $M_1(f(t)) \leq \varrho := M_1(f^{in})$ for all $t > 0$, we deduce from (9.2.29) that

$$T(\xi_0) \geq T_*(\xi_0) := \frac{\xi_0}{\varrho - \mathscr{L} f^{in}(\xi_0)} \, , \qquad \xi_0 > 0 \, . \qquad (9.2.30)$$

Introducing

$$T_2 := \frac{1}{M_2(f^{in})} \quad \text{if } M_2(f^{in}) < \infty \qquad \text{and} \qquad T_2 := 0 \text{ otherwise,}$$

we notice that

$$\frac{dT_*}{d\xi_0}(\xi_0) = \frac{1}{(\varrho - \mathscr{L} f^{in}(\xi_0))^2} \int_0^\infty \left(e^{x\xi_0} - 1 - x\xi_0 \right) x e^{-x\xi_0} f^{in}(x) \, \mathrm{d}x > 0$$

and

$$\lim_{\xi_0 \to 0} T_*(\xi_0) = \inf_{\xi_0 > 0} T_*(\xi_0) = T_2 \, , \qquad \lim_{\xi_0 \to \infty} T_*(\xi_0) = \infty \, .$$

Consequently, T_* is an increasing C^1-diffeomorphism from $(0,\infty)$ onto (T_2, ∞). We denote its inverse by T_*^{-1}.

Step 1. Assume for contradiction that $T_{gel} > T_2$ and then fix $t \in (T_2, T_{gel})$. It follows that $M_1(f(s)) = \varrho = M_1(f^{in})$ for all $s \in [0, t]$. Moreover, (9.2.30) ensures that we have $t = T_*(T_*^{-1}(t)) \leq T(T_*^{-1}(t))$. Consequently, owing to (9.2.29),

$$\lim_{\xi_0 \to T_*^{-1}(t)} \Xi(t; \xi_0) = T_*^{-1}(t) - [\varrho - \mathcal{L}f^{in}(T_*^{-1}(t))]\, t = 0 \;,$$

from which we deduce that $T(T_*^{-1}(t)) = t$. Now, thanks to (9.2.28),

$$\varrho = \mathcal{L}f(t, 0) = \mathcal{L}f(t, \Xi(t; T_*^{-1}(t))) = \mathcal{L}f^{in}(T_*^{-1}(t)) \;,$$

which contradicts the strict positivity of $T_*^{-1}(t)$. We have thus established that $T_{gel} \leq T_2$. In particular, $T_{gel} = 0$ when $T_2 = 0$; that is, when $M_2(f^{in}) = \infty$.

Step 2. First, if $T_2 = 0$, then $T_{gel} \geq T_2$. Assume next that $T_2 > 0$. Setting

$$H(t, \xi_0) := \xi_0 + \mathcal{L}f^{in}(\xi_0)t - \int_0^t M_1(f(s))\, ds \;, \qquad (t, \xi_0) \in [0, \infty)^2 \;,$$

we note that

$$H(t, 0) = \varrho t - \int_0^t M_1(f(s))\, ds \geq 0 \;,$$

and, recalling (9.2.29),

$$\Xi(t, \xi_0) = H(t, \xi_0) \;, \qquad (t, \xi_0) \in [0, T(\xi_0)) \;. \tag{9.2.31}$$

Consider now $t \in (0, T_2)$. Then

$$\partial_{\xi_0} H(t, \xi_0) = 1 - t \int_0^\infty x^2 e^{-x\xi_0} f^{in}(x)\, dx > 1 - \frac{t}{T_2} > 0 \;, \qquad \xi_0 > 0 \;.$$

Therefore, $H(t, \xi_0) > 0$ for $\xi_0 > 0$ and we infer from (9.2.26) and (9.2.31) that $T(\xi_0) > t$ for all $\xi_0 > 0$. Consequently, using again (9.2.31),

$$\Xi(t; \xi_0) = \xi_0 + \mathcal{L}f^{in}(\xi_0)t - \int_0^t M_1(f(s))\, ds$$

for all $\xi_0 > 0$. Letting $\xi_0 \to 0$ in the above identity and observing that $\Xi(t; \xi_0) \to 0$ as $\xi_0 \to 0$, we end up with

$$\varrho t = \int_0^t M_1(f(s))\, ds \;.$$

The previous identity being valid for all $t \in (0, T_2)$, we conclude that $M_1(f(t)) = \varrho$ for $t \in [0, T_2)$. Consequently, $T_{gel} \geq T_2$, which completes the proof. $\qquad \square$

We summarise the outcome of the previous analysis for the typical "sum" and "product" coagulation kernels. For the sum kernels $k(x, y) = x^\lambda + y^\lambda$, $(x, y) \in (0, \infty)^2$, it follows from Theorem 8.2.23 and Corollary 9.2.6 that:

sum kernel	$k(x, y) = x^\lambda + y^\lambda$
$\lambda \in [0, 1]$	global existence of mass-conserving solutions ($T_{gel} = \infty$)
$\lambda > 1$	non-existence, even locally

For the product kernels $k(x,y) = (xy)^{\lambda/2}$, $(x,y) \in (0,\infty)^2$, we infer from Theorem 8.2.23, Theorem 8.2.47, Proposition 9.1.2, Theorem 9.1.3 and Theorem 9.2.1 that:

product kernel	$k(x,y) = (xy)^{\lambda/2}$
$\lambda \in [0,1]$	global existence of mass-conserving solutions $(T_{gel} = \infty)$
$\lambda \in (1,2]$	global existence of weak solutions, $T_{gel} < \infty$
$\lambda > 2$	global existence of weak solutions, $T_{gel} = 0$

9.3   Shattering

As already mentioned, fragmentation may also lead to a loss of matter in finite time due to the formation of dust; that is, particles of size zero. In an analogous manner to the onset of gelation for the coagulation equation, this phenomenon, usually referred to as disintegration or shattering, occurs only when the overall rate of fragmentation increases without bound, but this time as the size decreases to zero, for instance when $a(x) = x^\gamma$, $x \in (0,\infty)$, for some $\gamma \in (-\infty, 0)$ [77, 116, 188]. The fact that shattering can arise in fragmentation models is particularly evident in explicit solutions computed in [188] for some specific fragmentation coefficients, with refined criteria for its occurrence, together with qualitative information on the resulting loss of matter, subsequently derived in [116, 132, 133, 134, 135, 145] by a probabilistic approach. It transpires that shattering is equivalent to the dishonesty of the semigroup associated with the fragmentation equation. This equivalence was first observed in [16] for fragmentation models with power law coefficients, and this led to the subsequent formulation of necessary and sufficient criteria for shattering in [18], see Section 5.1.3 in Volume I for an updated presentation of the results. These results are stronger than the probabilistic ones in that they do not require monotonicity of the fragmentation rate a close to $x = 0$ but, on the other hand, they are applicable for a narrower class of fragmentation kernels b. The advantage of using semigroup methods becomes more visible when handling fragmentation processes with growth or decay, where only limited probabilistic results are available, [132], but a fairly comprehensive treatment is given in [11] for the decay-fragmentation case and in [19] in the growth-fragmentation one; see also [26, Chapter 9] and an updated account in Section 5.2 in Volume I.

In this section, we provide an alternative proof of the occurrence of shattering which is in the same vein as those presented above for gelation, as it relies on moment estimates combined with a technique that was developed in [112] to prove that gelation takes place for the coagulation equation. Unfortunately, it is not as powerful as the probabilistic arguments used in [116, 132, 133] or the semigroup approach detailed in Section 5.2.4, particularly since it does not show that shattering occurs instantaneously. More precisely, we assume that the overall fragmentation rate a is differentiable in $(0,\infty)$, and that there is $m \geq 1$, $(\varepsilon_0, \varepsilon_\infty) \in (0,1)^2$, $y_0 > 0$, and $\omega_0 \in [0,\infty)$ such that

$$y \mapsto y^m a(y) \text{ is nondecreasing in } (0, 2y_0) , \tag{9.3.1a}$$

$$y^m a(y) \leq (1 - \varepsilon_0)(2y_0)^m a(2y_0) , \qquad y \in (0, y_0) , \tag{9.3.1b}$$

$$\varepsilon_\infty \leq y^m a(y) \leq \varepsilon_\infty^{-1} y , \qquad y \in (y_0, \infty) , \tag{9.3.1c}$$

and

$$g := \int_0^{2y_0} \frac{\mathrm{d}x}{xa(x)} < \infty , \qquad \lim_{x \to 0} \frac{1}{a(x)} = \omega_0 . \tag{9.3.1d}$$

Example 9.3.1. The particular choice $a(x) = x^\gamma$, $x \in (0, \infty)$, satisfies (9.3.1) when $\gamma < 0$ with $m = 1 - \gamma$, $y_0 = 1$, $\varepsilon_0 = \varepsilon_\infty = 1/2$, and $\omega_0 = 0$. $\diamond$

As for the fragmentation daughter distribution b, we assume that there is $\delta_2 > 0$ such that

$$\int_0^y xb(x, y) \, \mathrm{d}x = y \quad \text{and} \quad N_2(y) = \int_0^y x(y - x)b(x, y) \, \mathrm{d}x \geq \delta_2 y^2 \tag{9.3.2}$$

for $y \in (0, \infty)$.

Theorem 9.3.2. *Let a and b be fragmentation coefficients satisfying (9.3.1) and (9.3.2) and consider $f^{in} \in X_{1,+} \cap X_{m+1}$, $f^{in} \not\equiv 0$. Let $0 \leq f \in L_\infty((0, \infty), X_1 \cap X_{m+1})$ be a weak solution to the fragmentation equation on $[0, \infty)$ in the following sense: for all $t \in [0, \infty)$ and $\vartheta \in L_\infty(0, \infty)$ with compact support in $(0, \infty)$, there holds*

$$\int_0^\infty [f(t, x) - f^{in}(x)]\vartheta(x) \, \mathrm{d}x = -\int_0^t \int_0^\infty a(y)N_\vartheta(y)f(s, y) \, \mathrm{d}y\mathrm{d}s . \tag{9.3.3}$$

Then $T_{sh} < \infty$, where the shattering time T_{sh} is defined by

$$T_{sh} := \inf \left\{ t \geq 0 \; : \; M_1(f(t)) < M_1(f^{in}) \right\} .$$

According to the explicit solutions computed in [188], one expects that $T_{sh} = 0$ for a large class of fragmentation coefficients and initial data. This, indeed, is confirmed in Corollary 5.2.25.

Proof. Let us first observe that (9.3.2) implies that, for $y \in (0, \infty)$,

$$N_{m+1}(y) = \int_0^y (y^m - x^m) \, xb(x, y) \, \mathrm{d}x \geq \int_0^y y^{m-1}(y - x)xb(x, y) \, \mathrm{d}x$$

$$= y^{m-1} N_2(y) \geq \delta_2 y^{m+1} . \tag{9.3.4}$$

Step 1: Estimate for small sizes. Let $t > 0$ and $r \in (0, \infty)$. Since the solution f belongs to $L_\infty((0, \infty), X_1 \cap X_{m+1})$, it follows from (9.3.1b) and (9.3.1c), and an approximation argument, that we can take $\vartheta(x) = \min\{x^{m+1}, r^m x\}$, $x \in (0, \infty)$, in (9.3.3). For $y \in (0, r)$,

$$N_\vartheta(y) = y^{m+1} - \int_0^y x^{m+1}b(x, y) \, \mathrm{d}x = N_{m+1}(y) \geq \delta_2 y^{m+1}$$

by (9.3.4) while, for $y \geq r$, (9.3.2) ensures that

$$N_\vartheta(y) = r^m y - \int_0^r x^{m+1}b(x, y) \, \mathrm{d}x - r^m \int_r^y xb(x, y) \, \mathrm{d}x$$

$$= \int_0^r \left(r^m x - x^{m+1} \right) b(x, y) \, \mathrm{d}x \geq 0 .$$

We then infer from (9.3.3) that

$$\delta_2 \int_0^t \int_0^r y^{m+1} a(y) f(s,y) \, dy ds \leq \int_0^t \int_0^r a(y) N_\vartheta(y) f(s,y) \, dy ds$$

$$\leq \int_0^t \int_0^\infty a(y) N_\vartheta(y) f(s,y) \, dy ds$$

$$= \int_0^\infty [f^{in}(x) - f(t,x)] \vartheta(x) \, dx$$

$$\leq r^m M_1(f^{in}) \, ,$$

hence

$$\int_0^t \int_0^r y^{m+1} a(y) f(s,y) \, dy ds \leq \frac{M_1(f^{in})}{\delta_2} r^m \, . \tag{9.3.5}$$

At this stage, we proceed as in the proof of [112, Theorem 2.2] and multiply (9.3.5) by $-\zeta'(r)$ before integrating with respect to r over $(0, 2y_0)$ for some suitably chosen function ζ, see Propositions 9.1.8 and 9.1.9. The crucial point is to find a function ζ which is convenient for our purpose. Here an appropriate choice turns out to be

$$\zeta(r) = \frac{1}{r^m a(r)} - \frac{1}{(2y_0)^m a(2y_0)} \, , \quad r \in (0, 2y_0) \, , \quad \zeta(r) = 0 \, , \quad r \geq 2y_0 \, .$$

Owing to (9.3.1a) and the differentiability of a, ζ' is nonpositive in $(0, 2y_0)$ and we deduce from (9.3.5) that

$$-\int_0^{2y_0} \int_0^t \int_0^r \zeta'(r) y^{m+1} a(y) f(s,y) \, dy ds dr \leq -\frac{M_1(f^{in})}{\delta_2} \int_0^{2y_0} r^m \zeta'(r) \, dr \, . \tag{9.3.6}$$

On the one hand, Fubini's theorem, (9.3.1a) and (9.3.1b) imply that

$$\int_0^{2y_0} \int_0^t \int_0^r (-\zeta'(r)) y^{m+1} a(y) f(s,y) \, dy ds dr$$

$$= \int_0^t \int_0^{2y_0} y^{m+1} a(y) f(s,y) \int_y^{2y_0} (-\zeta'(r)) \, dr dy ds$$

$$= \int_0^t \int_0^{2y_0} y^{m+1} a(y) [\zeta(y) - \zeta(2y_0)] f(s,y) \, dy ds$$

$$\geq \int_0^t \int_0^{y_0} y^{m+1} a(y) \zeta(y) f(s,y) \, dy ds$$

$$\geq \varepsilon_0 \int_0^t \int_0^{y_0} y f(s,y) \, dy ds \, .$$

On the other hand, by (9.3.1d),

$$-\int_0^{2y_0} r^m \zeta'(r) \, dr = m \int_0^{2y_0} r^{m-1} \zeta(r) \, dr + \lim_{r \to 0} r^m \zeta(r)$$

$$\leq m\mathcal{I} + \omega_0 \, .$$

Consequently,

$$\int_0^t \int_0^{y_0} y f(s,y) \, dy ds \leq \frac{M_1(f^{in})}{\delta_2 \varepsilon_0} (\omega_0 + m\mathcal{I}) \, . \tag{9.3.7}$$

Step 2: Estimate for large sizes. Let $t > 0$. On using once more the fact that $f \in L_\infty((0, \infty), X_1 \cap X_{m+1})$, it follows from (9.3.1b) and (9.3.1c), and an approximation argument, that we may take $\vartheta(x) = x^{m+1}$, $x \in (0, \infty)$, in (9.3.3) and deduce from (9.3.4) that

$$\delta_2 \int_0^t \int_0^\infty y^{m+1} a(y) f(s, y) \, \mathrm{d}y \mathrm{d}s \leq \int_0^t \int_0^\infty a(y) N_{m+1}(y) f(s, y) \, \mathrm{d}y \mathrm{d}s$$
$$= M_{m+1}(f^{in}) - M_{m+1}(f(t)) \leq M_{m+1}(f^{in}) \ .$$

Combining the previous inequality with (9.3.1c) leads us to

$$\int_0^t \int_{y_0}^\infty y f(s, y) \, \mathrm{d}y \mathrm{d}s \leq \frac{1}{\varepsilon_\infty} \int_0^t \int_{y_0}^\infty y^{m+1} a(y) f(s, y) \, \mathrm{d}y \mathrm{d}s$$
$$\leq \frac{M_{m+1}(f^{in})}{\delta_2 \varepsilon_\infty} \ . \tag{9.3.8}$$

Step 3: Finiteness of the shattering time. Summing (9.3.7) and (9.3.8), we end up with

$$\int_0^t M_1(f(s)) \, \mathrm{d}s \leq \frac{M_1(f^{in})}{\delta_2 \varepsilon_0} (\omega_0 + m\mathcal{I}) + \frac{M_{m+1}(f^{in})}{\delta_2 \varepsilon_\infty} \tag{9.3.9}$$

for all $t > 0$. Either $T_{sh} = 0$ and it is then clearly finite, or $T_{sh} > 0$ and, for any $t \in (0, T_{sh})$, it follows from (9.3.9) that

$$t M_1(f^{in}) \leq \frac{M_1(f^{in})}{\delta_2 \varepsilon_0} (\omega_0 + m\mathcal{I}) + \frac{M_{m+1}(f^{in})}{\delta_2 \varepsilon_\infty} \ .$$

Letting $t \to T_{sh}$ in the previous inequality implies that T_{sh} is finite and completes the proof. $\qquad\square$

Besides Example 9.3.1, Theorem 9.3.2 also applies to $a(x) = 1 + |\ln x|^q$, $q > 1$. Indeed, it satisfies (9.3.1) with $m = 1$. In fact, according to [116, 132], when the overall fragmentation rate a is monotone decreasing in a neighbourhood of zero, shattering occurs if and only if there is $y_* > 0$ such that

$$\int_0^{y_*} \frac{\mathrm{d}x}{x a(x)} < \infty \ . \tag{9.3.10}$$

Though the assumptions (9.3.1) required to apply Theorem 9.3.2 include (9.3.10), additional assumptions are needed but are likely to be of a technical nature and could possibly be relaxed, as one can see in Theorem 5.1.42, at least for the case of separable fragmentation kernels, such as (2.2.60).

Chapter 10

Long-Term Behaviour

A central issue in the use of C-F models is their potential to provide predicted outcomes through the qualitative information that can be retrieved from their analysis. For instance, at first glance, the outcome of Smoluchowski's coagulation equation is the decay to zero of the total number of particles, along with the conservation of matter for all times, for moderately growing coagulation coefficients, or the occurrence of gelation at a finite time T_{gel} for coagulation coefficients increasing sufficiently rapidly with the particle sizes. Such information is, however, too crude and does not allow specific features among the different choices of coagulation coefficients to be singled out. More precise investigations are then needed and lead, in particular, to the so-called dynamical scaling hypothesis, which has already been discussed in Section 2.3.4. This hypothesis states that for large times, or near the gelation time, details of the initial distribution of particles no longer matter and the dynamics only retains the mean size $\sigma(t)$ at time t of the distribution of particles in the following sense: there is a scaling exponent $\tau \in \mathbb{R}$, and a scaling profile φ such that "generic" solutions (in a sense to be made precise) to the coagulation equation satisfy

$$f(t,x) \sim \frac{1}{\sigma(t)^\tau} \varphi\left(\frac{x}{\sigma(t)}\right) \quad \text{as} \quad t \to \infty \quad \text{or} \quad t \to T_{gel} . \tag{10.0.1}$$

Thus, in the long-term limit, or at the gelation time, the behaviour of a suitably rescaled version of an (almost) arbitrary solution to the coagulation equation is expected to be accurately described by the mean size $\sigma(t)$ at time t and a single function depending on a single variable, with any influence of the initial state having faded away. Of course, the first requirement when checking the validity of (10.0.1) is to determine the exponent τ, the mean size σ and the scaling profile φ, and this turns out to be a rather complex task, even in the few cases for which explicit solutions are available. We point out here that, as a result of coagulation events, the mean size σ is expected to increase without bound with time, that is,

$$\lim_{t \to \infty} \sigma(t) = \infty \quad \text{or} \quad \lim_{t \to T_{gel}} \sigma(t) = \infty . \tag{10.0.2}$$

The first step towards (10.0.1) is then to figure out whether there are solutions of the form $(t,x) \mapsto \sigma(t)^{-\tau}\varphi(x/\sigma(t))$ to the coagulation equation. This clearly requires the coagulation coefficients to be homogeneous; that is, there is $\lambda \in \mathbb{R}$ such that

$$k(\xi x, \xi y) = \xi^\lambda k(x,y) , \qquad (x,y,\xi) \in (0,\infty)^3 . \tag{10.0.3}$$

Typical examples of homogeneous coagulation coefficients include Smoluchowski's original kernel

$$k(x,y) = \left(x^{1/3} + y^{1/3}\right)\left(x^{-1/3} + y^{-1/3}\right),$$

see [247, 248], as well as the constant, additive and multiplicative kernels, together with their generalisation $k(x,y) = x^\alpha y^\beta + x^\beta y^\alpha$. Inserting the self-similarity ansatz

$$\frac{1}{\sigma(t)^\tau} \varphi\left(\frac{x}{\sigma(t)}\right) \tag{10.0.4}$$

into the coagulation equation, we obtain

$$\sigma(t)^{\tau-\lambda-2}\frac{d\sigma}{dt}(t)\left(-\tau\varphi(y)-y\frac{d\varphi}{dy}(y)\right)=\frac{1}{2}\int_0^y k(y-y_*,y_*)\varphi(y-y_*)\varphi(y_*)\,dy_*$$
$$-\int_0^\infty k(y,y_*)\varphi(y)\varphi(y_*)\,dy_*$$

for $t>0$ and $y\in(0,\infty)$. Separating variables and recalling (10.0.2) we obtain the existence of a positive constant w such that σ solves the ordinary differential equation

$$\sigma(t)^{\tau-\lambda-2}\frac{d\sigma}{dt}(t)=w>0\,, \tag{10.0.5}$$

while φ satisfies the integro-differential equation

$$w\left(-\tau\varphi(y)-y\frac{d\varphi}{dy}(y)\right)=\frac{1}{2}\int_0^y k(y-y_*,y_*)\varphi(y-y_*)\varphi(y_*)\,dy_*$$
$$-\int_0^\infty k(y,y_*)\varphi(y)\varphi(y_*)\,dy_*\,,\qquad y\in(0,\infty)\,. \tag{10.0.6}$$

First, solving (10.0.5) subject to (10.0.2) of course depends on the yet undetermined value of τ and leads to three different cases:

$$\tau>\lambda+1\ \text{ with }\ \sigma(t)=(1+w(\tau-\lambda-1)t)^{1/(\tau-\lambda-1)}\,,\qquad t\geq0\,, \tag{10.0.7a}$$
$$\tau=\lambda+1\ \text{ with }\ \sigma(t)=e^{wt}\,,\qquad t\geq0\,, \tag{10.0.7b}$$

or

$$\tau<\lambda+1\ \text{ with }\ \sigma(t)=[1-w(\lambda+1-\tau)t]^{1/(\tau-\lambda-1)} \tag{10.0.7c}$$

for $t\in[0,1/w(\lambda+1-\tau))$, where σ is normalised to satisfy $\sigma(0)=1$. At this point, the parameter τ is still largely undetermined but (10.0.7) already sheds some light on the possible values it can take. Indeed, on the one hand, divergence of the mean size σ at a finite time, which is expected at the onset of gelation, only happens when $\tau<\lambda+1$, thus excluding *a priori* scaling parameters ranging in $[\lambda+1,\infty)$ in the gelling regime. On the other hand, an elementary computation shows that the only value of τ which leaves invariant the total mass is $\tau=2$ and this specific choice of the exponent τ will be of particular importance in the forthcoming analysis. In fact, $\tau=2$ is formally a necessary condition for a non-zero solution φ to (10.0.6) with finite mass to exist. Indeed, multiplying (10.0.6) by y and then integrating over $(0,\infty)$ formally gives $(2-\tau)M_1(\varphi)=0$ and thus $\tau=2$ if $\varphi\in X_1$. Also, still according to (10.0.7), the mean size σ is defined for all positive times and $\tau=2$ provided $\lambda\leq1$. This is perfectly consistent with the growth conditions required on the coagulation kernel k in Section 8.2.2.4 to guarantee the existence of mass-conserving solutions to the coagulation equation.

As for equation (10.0.6) having the scaling profile φ as solution, it is a nonlinear and nonlocal integro-differential equation and the value of φ at an arbitrary $y\in(0,\infty)$ depends on the values of φ both on (y,∞) and $(0,y)$. It is thus unlikely that (10.0.6) can be written as an initial-value problem. In addition, $\varphi\equiv0$ is always a solution to (10.0.6) but this is obviously useless for our purpose. These two features are among the mathematical difficulties encountered in the search for self-similar solutions to the coagulation equation, which will be discussed in Section 10.2. Furthermore, let us finally point out that, thanks to the homogeneity (10.0.3) of k, equation (10.0.6) satisfies the following invariance property:

If φ is a solution to (10.0.6) with parameter w and $(\kappa,l)\in(0,\infty)^2$, then $\varphi_{\kappa,l}:y\mapsto\kappa\varphi(ly)$ is a solution to (10.0.6) with parameter $w\kappa/l^{\lambda+1}$ and $M_1(\varphi_{\kappa,l})=\kappa l^{-2}M_1(\varphi)$. $\qquad(10.0.8)$

This property will be used later to prescribe the value of w, or that of the total mass of φ (if finite), by an appropriate choice of the parameters κ and l. It is worth pointing out here the special feature of the exponent $\lambda = 1$, for which the rescaling of both w and the total mass are the same, so that only one of these two parameters can be given *a priori* an arbitrary value.

A similar issue can be raised for the fragmentation equation and the quest for self-similar solutions also requires some homogeneity assumptions on the fragmentation coefficients. More specifically, we assume that there is $\gamma \in \mathbb{R}$, and a nonnegative function $h \in L_1((0,1), z\mathrm{d}z)$ such that

$$a(x) = x^\gamma \quad \text{and} \quad b(x,y) = \frac{1}{y}h\left(\frac{x}{y}\right) , \qquad 0 < x < y , \tag{10.0.9}$$

with

$$\int_0^1 zh(z)\mathrm{d}z = 1 . \tag{10.0.10}$$

Recall that (10.0.9) and (10.0.10) imply

$$y = \int_0^y xb(x,y) \, \mathrm{d}x , \qquad y \in (0,\infty) ,$$

and thus guarantee that there is no loss of matter during fragmentation of a single particle. Clearly $h(z) = h_\nu(z) := (\nu + 2)z^\nu$ satisfies (10.0.10) for $\nu > -2$. We now introduce the yet undetermined mean size $\sigma(t)$ at time t of the particle distribution, which is now expected to decay to zero as $t \to \infty$,

$$\lim_{t\to\infty} \sigma(t) = 0 . \tag{10.0.11}$$

Inserting the self-similarity ansatz,

$$\frac{1}{\sigma(t)^\tau}\varphi\left(\frac{x}{\sigma(t)}\right) , \tag{10.0.12}$$

in the fragmentation equation gives

$$-\frac{1}{\sigma(t)^{\gamma+1}}\frac{d\sigma}{dt}(t)\left(\tau\varphi(y) + y\frac{d\varphi}{dy}(y)\right) = -a(y)\varphi(y) + \int_y^\infty a(y_*)b(y,y_*)\varphi(y_*) \, \mathrm{d}y_* .$$

As before, we separate variables and take into account (10.0.11) to obtain the existence of a positive constant w such that

$$\sigma(t)^{-\gamma-1}\frac{d\sigma}{dt}(t) = -w , \qquad t \geq 0 , \tag{10.0.13}$$

and

$$w\left(\tau\varphi(y) + y\frac{d\varphi}{dy}(y)\right) = -a(y)\varphi(y) + \int_y^\infty a(y_*)b(y,y_*)\varphi(y_*) \, \mathrm{d}y_* \tag{10.0.14}$$

for $y \in (0,\infty)$. Owing to the linearity of the fragmentation equation, there is no dependence of the mean size σ on τ and we deduce from (10.0.13) that we have the following two cases:

$$\gamma > 0 \quad \text{and} \quad \sigma(t) = (1 + \gamma wt)^{-1/\gamma} , \qquad t \geq 0 , \tag{10.0.15a}$$

$$\gamma = 0 \quad \text{and} \quad \sigma(t) = e^{-wt} , \qquad t \geq 0 . \tag{10.0.15b}$$

The case $\gamma < 0$ is excluded since (10.0.11) and (10.0.13) are then not compatible. Observe that mass-conserving self-similar solutions correspond to the choice $\tau = 2$. As in the case of the profile equation (10.0.6) stemming from the coagulation equation, the profile equation (10.0.14) is unlikely to be written as an initial-value problem and also $\varphi \equiv 0$ is always a solution, so that the mathematical analysis of (10.0.14) shares similar difficulties with that of (10.0.6). However the linearity of the fragmentation equation allows one to use a broader variety of mathematical tools, and, as a result, more complete results are available in this case, as we shall see in Section 10.1 below.

When both coagulation and fragmentation are turned on, we recall that the former increases the mean size of the particles, while the latter decreases it. We might thus expect a balance between these two competing mechanisms resulting, in particular, in the existence of stationary solutions. This feature, however, is unlikely to show up for arbitrary non-zero coagulation and fragmentation coefficients as either coagulation or fragmentation may dominate the dynamics. For instance, as shown in Section 9.1, gelation occurs in the presence of fragmentation, thus excluding the existence of stationary solutions for large values of the total mass. Nevertheless, there is a special class of coagulation and fragmentation coefficients satisfying the detailed balance condition (6.1.20), for which the existence of a continuum of stationary solutions is built into the model, and their properties (including stability) are the subject of Sections 10.3.1 and 10.3.2, see also Section 2.3.3 in Volume I. We also discuss the existence of stationary solutions and self-similar solutions to the C-F equation in Sections 10.3.3 and 10.3.4, respectively.

10.1   Continuous Fragmentation

10.1.1   Self-Similar Profiles

The purpose of this section is to construct mass-conserving self-similar solutions to the fragmentation equation

$$\partial_t f(t,x) = -a(x)f(t,x) + \int_x^\infty a(y)b(x,y)f(t,y) \, \mathrm{d}y , \qquad (t,x) \in (0,\infty)^2 ; \qquad (10.1.1)$$

that is, solutions of the form

$$\frac{1}{\sigma(t)^2} \varphi \left(\frac{x}{\sigma(t)} \right) , \qquad (t,x) \in (0,\infty)^2 , \qquad (10.1.2)$$

which corresponds to the choice $\tau = 2$ in (10.0.12). According to the discussion at the beginning of this chapter, we assume that the fragmentation rate a and the daughter distribution function b possess the following homogeneity properties: there is $\gamma > 0$, and a nonnegative function $h \in L_1((0,1), z\mathrm{d}z)$ such that

$$a(x) = x^\gamma \ \text{ and } \ b(x,y) = \frac{1}{y} h \left(\frac{x}{y} \right) , \qquad 0 < x < y , \qquad (10.1.3a)$$

with

$$\int_0^1 z h(z) \mathrm{d}z = 1 . \qquad (10.1.3b)$$

Note that we restrict the forthcoming analysis to $\gamma > 0$, which is included in the admissible range (10.0.15). In the borderline case $\gamma = 0$ the solutions behave differently and we refer to [101, 188, 284] for a thorough discussion of this case. Choosing $w = 1/\gamma > 0$, it follows from (10.0.15a) that

$$\sigma(t) = (1+t)^{-1/\gamma}, \qquad t > 0, \tag{10.1.4}$$

while the profile φ solves (10.0.14) with $\tau = 2$ and $w = 1/\gamma$:

$$2\varphi(y) + y\frac{d\varphi}{dy}(y) = -\gamma a(y)\varphi(y) + \gamma \int_y^\infty a(y_*)b(y, y_*)\varphi(y_*)\, dy_* \tag{10.1.5}$$

for $y \in (0, \infty)$. We further require that φ is a nonnegative function in $(0, \infty)$ and has a finite total mass which may be normalised to one due to the linearity of (10.1.5). We thus look for a (weak) solution φ to (10.1.5) in the following sense.

Definition 10.1.1. *Given a and b satisfying (10.1.3), a self-similar profile to the fragmentation equation (10.1.1) is a function $\varphi \in X_1 \cap X_{\gamma+1}$ satisfying*

$$\varphi \geq 0 \quad a.e. \ in \ (0, \infty) \quad and \quad M_1(\varphi) = 1, \tag{10.1.6}$$

and

$$\int_0^\infty \left(y\frac{d\vartheta}{dy}(y) - \vartheta(y) \right) \varphi(y)\, dy = \gamma \int_0^\infty a(y)N_\vartheta(y)\varphi(y)\, dy \tag{10.1.7}$$

for all $\vartheta \in \Theta^1 = \{\vartheta \in C^{0,1}([0, \infty)) : \vartheta(0) = 0\}$, recalling that N_ϑ is defined in (8.2.3).

The right-hand side of (10.1.7) is well defined since $\varphi \in X_{\gamma+1}$ and

$$a(y)N_\vartheta(y) \leq y^\gamma \left[y + \int_0^y y_*b(y_*, y)\, dy_* \right] \left\| \frac{d\vartheta}{dy} \right\|_\infty \leq 2y^{\gamma+1} \left\| \frac{d\vartheta}{dy} \right\|_\infty,$$

according to (10.1.3). Also, note that (10.1.7) is a weak formulation of (10.1.5).

An alternative equation for self-similar profiles is obtained from (10.1.5) after multiplication by y and integration.

Proposition 10.1.2. *Given a and b satisfying (10.1.3), let φ be a self-similar profile to the fragmentation equation in the sense of Definition 10.1.1. Then*

$$y^2\varphi(y) = \gamma \int_y^\infty y_*^{1+\gamma}\varphi(y_*) \int_0^{y/y_*} zh(z)\, dz dy_*, \qquad y \in (0, \infty). \tag{10.1.8}$$

Moreover, $\varphi \in C((0, \infty))$ and $d\varphi/dy \in X_2$.

Proof. Let $\zeta \in C_0^\infty((0, \infty))$ and take $\vartheta(y) = y\zeta(y)$, $y \in (0, \infty)$, in (10.1.7). Since

$$N_\vartheta(y_1) = \int_0^{y_1} [\zeta(y_1) - \zeta(y_*)]y_*b(y_*, y_1)\, dy_* = \int_0^{y_1} \int_{y_*}^{y_1} y_*b(y_*, y_1)\frac{d\zeta}{dy}(y)\, dy dy_*$$

$$= \int_0^{y_1} \frac{d\zeta}{dy}(y) \int_0^y y_*b(y_*, y_1)\, dy_* dy$$

for $y_1 \in (0, \infty)$ by (10.1.3), it follows from (10.1.7) that

$$\int_0^\infty y^2\varphi(y)\frac{d\zeta}{dy}(y)\, dy = \gamma \int_0^\infty a(y_1)\varphi(y_1) \int_0^{y_1} \frac{d\zeta}{dy}(y) \int_0^y y_*b(y_*, y_1)\, dy_* dy dy_1$$

$$= \gamma \int_0^\infty \frac{d\zeta}{dy}(y) \int_y^\infty a(y_1)\varphi(y_1) \int_0^y y_*b(y_*, y_1)\, dy_* dy_1 dy.$$

The above identity being valid for all $\zeta \in C_0^\infty((0,\infty))$, there is a constant $C \in \mathbb{R}$ such that

$$C = y^2\varphi(y) - \gamma \int_y^\infty a(y_1)\varphi(y_1) \int_0^y y_* b(y_*, y_1) \, dy_* dy_1 \,, \qquad y \in (0,\infty) \,. \qquad (10.1.9)$$

Owing to (10.1.3) and the property $\varphi \in X_{\gamma+1}$, we deduce from (10.1.9) that

$$\left| y^2\varphi(y) - C \right| \le \gamma \int_y^\infty y_1^{\gamma+1}\varphi(y_1) \, dy_1 \xrightarrow[y\to\infty]{} 0 \,,$$

which is only compatible with the property $\varphi \in X_1$ for $C = 0$. Now, using the self-similar form (10.1.3a) of b in (10.1.9) gives (10.1.8). The claimed continuity of φ is then a straightforward consequence of (10.1.8) and the integrability properties of h and φ.

Finally, we infer from (10.1.3) and (10.1.8) that

$$\frac{d}{dy}\left(y^2\varphi(y)\right) = -\gamma y^{\gamma+1}\varphi(y) + \gamma y \int_y^\infty y_*^\gamma\varphi(y_*)b(y,y_*) \, dy_*$$

for a.e. $y \in (0,\infty)$. Since $\varphi \in X_{\gamma+1}$, it follows from (10.1.3) and Fubini–Tonelli's theorem that

$$\int_0^\infty y \int_y^\infty y_*^\gamma\varphi(y_*)b(y,y_*) \, dy_* dy = \int_0^\infty y_*^{\gamma+1}\varphi(y_*) \, dy_* < \infty \,.$$

Consequently, $y \mapsto d\left(y^2\varphi(y)\right)/dy \in X_0$ with

$$\int_0^\infty \left| \frac{d}{dy}\left(y^2\varphi(y)\right) \right| \, dy \le 2\gamma M_{\gamma+1}(\varphi) \,.$$

Since

$$y^2\frac{d\varphi}{dy}(y) = \frac{d}{dy}\left(y^2\varphi(y)\right) - 2y\varphi(y)$$

and $\varphi \in X_1$, we conclude that

$$\left\| \frac{d\varphi}{dy} \right\|_{[2]} = \int_0^\infty y^2 \left| \frac{d\varphi}{dy}(y) \right| \, dy \le 2\gamma M_{\gamma+1}(\varphi) + 2M_1(\varphi) < \infty \,,$$

which completes the proof.   □

Equation (10.1.8) seems simpler than (10.1.5) but it mostly proves useful in analysing the behaviour of $\varphi(y)$ as $y \to 0$ and $y \to \infty$ [12, 77, 258]. Nevertheless, part of the forthcoming existence result for self-similar profiles of the fragmentation equation relies on (10.1.8). Before investigating the existence issue, let us point out an interesting consequence of (10.1.8), namely, that it implies that self-similar profiles of the fragmentation equation in the sense of Definition 10.1.1 do not exist for arbitrary nonnegative functions $h \in L_1((0,1), z dz)$ satisfying (10.1.3b). This phenomenon is uncovered in [43] by a probabilistic approach and we provide here an alternative proof relying on deterministic arguments. It is also worth pointing out here that condition (10.1.10) below appears in a different context, but in a similar way, in Section 5.2.7 in Volume I.

Proposition 10.1.3. *Given a and b satisfying (10.1.3), let φ be a self-similar profile to the fragmentation equation in the sense of Definition 10.1.1. Then*

$$\ell_h := \int_0^1 zh(z)|\ln z| \, dz < \infty \qquad (10.1.10)$$

and

$$M_{\gamma+1}(\varphi) = \frac{1}{\gamma\ell_h} \,. \qquad (10.1.11)$$

Proof. Consider $y_0 \in (0, \infty)$. We divide both sides of (10.1.8) by $y \in (y_0, \infty)$ and integrate the resulting identity with respect to y over (y_0, ∞) to obtain

$$\int_{y_0}^{\infty} y\varphi(y) \, \mathrm{d}y = \gamma \int_{y_0}^{\infty} \frac{1}{y} \int_{y}^{\infty} y_*^{1+\gamma} \varphi(y_*) \int_{0}^{y/y_*} zh(z) \, \mathrm{d}z \mathrm{d}y_* \mathrm{d}y \ .$$

It then follows from Fubini–Tonelli's theorem that

$$\int_{y_0}^{\infty} y\varphi(y) \, \mathrm{d}y = \gamma \int_{0}^{1} zh(z) \int_{y_0}^{\infty} y_*^{1+\gamma} \varphi(y_*) \int_{zy_*}^{y_*} \mathbf{1}_{(y_0, \infty)}(y) \frac{\mathrm{d}y}{y} \, \mathrm{d}y_* \mathrm{d}z$$

$$= \gamma \int_{0}^{1} zh(z) \int_{y_0}^{y_0/z} y_*^{1+\gamma} \varphi(y_*) \ln\left(\frac{y_*}{y_0}\right) \, \mathrm{d}y_* \mathrm{d}z$$

$$+ \gamma \int_{0}^{1} zh(z) |\ln z| \int_{y_0/z}^{\infty} y_*^{1+\gamma} \varphi(y_*) \, \mathrm{d}y_* \mathrm{d}z \ . \tag{10.1.12}$$

Owing to (10.1.6), the left-hand side of (10.1.12) is bounded from above by 1, hence, since both h and φ are nonnegative,

$$1 \geq \gamma \int_{0}^{1} zh(z) |\ln z| \int_{y_0/z}^{\infty} y_*^{1+\gamma} \varphi(y_*) \, \mathrm{d}y_* \mathrm{d}z \ .$$

Since $\varphi \in X_{1+\gamma}$ with $M_{\gamma+1}(\varphi) > 0$, we may pass to the limit as $y_0 \to 0$ in the previous inequality and deduce from Fatou's lemma that $1 \geq \gamma \ell_h M_{\gamma+1}(\varphi)$. Consequently, ℓ_h is finite and we may then pass to the limit as $y_0 \to 0$ in (10.1.12) to derive (10.1.11). $\qquad \square$

According to Proposition 10.1.3, stronger integrability of h is needed for a self-similar profile to exist and this condition turns out to be sufficient as well [43, 116].

Theorem 10.1.4. *Assume that the overall fragmentation rate a and the daughter distribution function b satisfy (10.1.3), with $\gamma > 0$, and (10.1.10). Then there exists a self-similar profile φ to the fragmentation equation in the sense of Definition 10.1.1 satisfying*

$$\varphi \in \bigcap_{m \geq 1} X_m \ .$$

This profile is unique if $h > 0$ in $(0, 1)$.

Let us recall here that explicit or closed-form formulas are available for φ for suitable choices of the function h [116, 188, 258, 284, 285], see also Section 2.3.2.1 in Volume I.

The route towards Theorem 10.1.4 is not straightforward and requires three steps. We start with an existence result which is established in [113, Theorem 3.1] and requires more on h than the mere finiteness (10.1.10) of ℓ_h, see (10.1.13) below, but this additional assumption implies better integrability for the profile.

Theorem 10.1.5. *Assume that the overall fragmentation rate a and the daughter distribution function b satisfy (10.1.3) with $\gamma > 0$. Assume further that there is $m_0 \in [0, 1)$ such that*

$$h \in L_1((0, 1), z^{m_0} \mathrm{d}z) \ . \tag{10.1.13}$$

Then there exists a self-similar profile φ to the fragmentation equation in the sense of Definition 10.1.1 satisfying

$$\varphi \in L_\infty((0, \infty), y^{m_0+1} \mathrm{d}y) \cap \bigcap_{m \geq m_0} X_m \ .$$

Moreover, φ is unique if $h > 0$ in $(0, 1)$.

To construct such a solution it turns out to be convenient to use the dynamical systems approach alluded to in Section 7.3.2. This approach requires us to first identify an evolution problem having φ as a stationary solution. To this end, we use a rather classical idea, the so-called transformation to self-similar variables, or scaling variables. More precisely, we introduce the self-similar variables

$$s := \frac{1}{\gamma} \ln (1 + t) , \qquad y := x(1 + t)^{1/\gamma} , \tag{10.1.14}$$

and the rescaled function

$$g(s, y) := e^{-2s} f \left(e^{\gamma s} - 1, y e^{-s} \right) , \qquad (s, y) \in [0, \infty) \times (0, \infty) . \tag{10.1.15}$$

Equivalently,

$$f(t, x) = (1 + t)^{2/\gamma} g \left(\frac{1}{\gamma} \ln (1 + t), x(1 + t)^{1/\gamma} \right) \tag{10.1.16}$$

for $(t, x) \in [0, \infty) \times (0, \infty)$. We infer from (10.1.1) and (10.1.16) that g solves

$$\partial_s g(s, y) + y \partial_y g(s, y) + 2g(s, y)$$
$$= -\gamma a(y) g(s, y) + \gamma \int_y^\infty a(y_*) b(y, y_*) g(s, y_*) \, \mathrm{d}y_* \tag{10.1.17}$$

for $(s, y) \in (0, \infty)^2$. According to (10.1.5), self-similar profiles φ are indeed stationary solutions to (10.1.17). We supplement (10.1.17) with an initial condition

$$g(0, y) = g^{in}(y) , \qquad y \in (0, \infty) . \tag{10.1.18}$$

Besides the well-posedness of (10.1.17)–(10.1.18), the other building block of the dynamical systems approach to the existence of steady states is the construction of a subset of the positive cone $X_{1,+}$ of X_1, which is invariant under the flow associated with (10.1.17)–(10.1.18) and satisfies the convexity and compactness properties required to apply the results of Section 7.3.2. The compactness issue actually requires an additional assumption on the function h in (10.1.3a): in addition to (10.1.3b) and (10.1.13), we also assume that there is $m_1 \in [1, 2)$ such that

$$1 + m_0 - \gamma \le m_1 \quad \text{and} \quad H_1 := \int_0^1 z^{m_1} \left| \frac{dh}{dz}(z) \right| \, \mathrm{d}z < \infty . \tag{10.1.19}$$

Thanks to this additional assumption, a W_1^1-estimate is available for solutions to (10.1.17) and allows us to construct an invariant set which is compact in X_1. We then prove Theorem 10.1.5 by a dynamical systems approach under this additional assumption. In a second step, the differentiability assumption (10.1.19) is removed by an approximation procedure and a compactness argument completes the proof of Theorem 10.1.5. For further use, we define

$$\mathfrak{h}_m := \int_0^1 (z - z^m) h(z) \, \mathrm{d}z = 1 - \int_0^1 z^m h(z) \, \mathrm{d}z , \qquad m \ge m_0 , \tag{10.1.20}$$

and note that $\mathfrak{h}_m > 0$ if and only if $m > 1$.

Owing to the connection (10.1.15) and (10.1.16) between the solutions to the fragmentation equation (10.1.1) and to (10.1.17)–(10.1.18), the notion of weak solutions to the latter is readily deduced from that to the former, given in Definitions 8.2.1 and 8.2.3.

Definition 10.1.6. *Assume that the fragmentation coefficients a and b satisfy (10.1.3) with $\gamma > 0$ and consider $m \in [m_0, 1)$, $g^{in} \in X_{1,+} \cap X_m$, and $T \in (0, \infty]$. A mass-conserving weak solution to (10.1.17)–(10.1.18) on $[0, T)$ is a nonnegative function $g \in C([0, T), X_{m,w} \cap X_{1,w})$ such that $g(0) = g^{in}$,*

$$(s, y) \longmapsto y^{\gamma+m} g(s, y) \in L_1((0, t) \times (0, \infty)) , \qquad M_1(g(t)) = M_1(g^{in}) ,$$

and

$$\int_0^\infty \vartheta(x)[g(t, y) - g^{in}(y)] \, dy = \int_0^t \int_0^\infty \left(y \frac{d\vartheta}{dy}(y) - \vartheta(y) - y^\gamma N_\vartheta(y) \right) g(s, y) \, dyds$$

for all $\vartheta \in \Theta^m \cap C^{0,1}(0, \infty)$ and $t \in (0, T)$.

We may now rely on the analysis carried out in Chapter 8 to obtain the well-posedness of (10.1.17)–(10.1.18) in the positive cone of $X_{m_*} \cap X_1$ for all $m_* \in (m_0, 1)$.

Proposition 10.1.7. *Assume that the fragmentation coefficients a and b satisfy (10.1.3) with $\gamma > 0$, and that h satisfies (10.1.13) and (10.1.19). Let $m_* \in (m_0, 1)$. Given $g^{in} \in X_{1,+} \cap X_{m_*}$ satisfying $M_1(g^{in}) = 1$, there is a unique mass-conserving solution g to (10.1.17)–(10.1.18). In particular,*

$$M_1(g(s)) = M_1(g^{in}) = 1 , \qquad s \geq 0 . \tag{10.1.21}$$

Furthermore, $\{g^{in} \mapsto g(s)\}_{s \geq 0}$ is a dynamical system in the positive cone of $X_{m_*} \cap X_1$ equipped with the strong topology of X_1.

Proof. We first observe that (10.1.19) implies that, for $z \in (0, 1)$,

$$h(z) = h(1) - \int_z^1 \frac{dh}{dz_*}(z_*) \, dz_* \leq h(1) + z^{-m_1} \int_z^1 z_*^{m_1} \left| \frac{dh}{dz_*}(z_*) \right| \, dz_*$$
$$\leq [h(1) + H_1] z^{-m_1} .$$

Next, fix $m_* \in (m_0, 1)$ and define $p_* := \min\{(m_1 + m_* - m_0)/m_1, 1 + \gamma\} > 1$. Then, for $y \in (0, \infty)$ and $p \in [1, p_*]$,

$$\int_0^y x^{m_*} b(x, y)^p \, dx = y^{m_*+1-p} \int_0^1 z^{m_*} h(z)^p \, dz$$
$$\leq [h(1) + H_1]^{p-1} y^{m_*+1-p} \int_0^1 z^{m_*-m_1(p-1)} h(z) \, dz$$
$$\leq [h(1) + H_1]^{p-1} y^{m_*+1-p} \int_0^1 z^{m_0} h(z) \, dz .$$

Consequently, a and b satisfy the assumptions (8.2.74) and (8.2.75) with (m_*, p_*) instead of (m_0, p_0), as well as (8.2.76) with $\delta_2 = \mathfrak{h}_2 > 0$, where $\mathfrak{h}_2$ is defined in (10.1.20). Owing to the connection (10.1.15) between solutions to (10.1.17)–(10.1.18) and solutions to the fragmentation equation (10.1.1), the existence of a mass-conserving solution g to (10.1.17)–(10.1.18) is then a straightforward consequence of Theorem 8.2.30 (as $m_* > m_0 \geq 0$). Uniqueness and the dynamical system features are guaranteed by Proposition 8.2.57. $\quad\square$

We now derive several estimates satisfied by mass-conserving solutions to (10.1.17)–(10.1.18) when the initial condition satisfies

$$g^{in} \in X_{1,+} \cap X_{m_0} \quad \text{and} \quad M_1(g^{in}) = 1 . \tag{10.1.22}$$

The existence and uniqueness of a mass-conserving solution g to (10.1.17)–(10.1.18) in this case are guaranteed by Proposition 10.1.7 as $X_1 \cap X_{m_0} \subset X_1 \cap X_{m_*}$ for any $m_* \in (m_0, 1)$.

We first study the time evolution of moments.

Lemma 10.1.8. *Assume that a, b and h satisfy (10.1.3) with $\gamma > 0$ and (10.1.13), and consider an initial condition g^{in} satisfying (10.1.22). The corresponding solution g to (10.1.17)–(10.1.18) satisfies the following integrability properties:*

(a) If $m > 1$ and $M_m(g^{in}) < \infty$, then

$$M_m(g(s)) \leq \max\{\mu_{1,m}, M_m(g^{in})\} , \qquad s \geq 0 , \tag{10.1.23}$$

where $\mu_{1,m}$ depends only on γ, m and $\mathfrak{h}_m$, the latter being defined in (10.1.20).

(b) If $m \in [m_0, 1)$ and $M_{1+\gamma}(g^{in}) < \infty$, then

$$M_m(g(s)) \leq \max\left\{ M_m(g^{in}), \mu_{1,m} \sup_{\sigma \geq 0}\{M_{1+\gamma}(g(\sigma))\} \right\} \tag{10.1.24}$$

for $s \geq 0$, where $\mu_{1,m}$ depends only on γ, m and $\mathfrak{h}_m$.

Proof. Let $m \geq m_0$. We take $\vartheta(y) = y^m$ in the weak formulation of (10.1.17) in Definition 10.1.6, and use (10.1.3) and (10.1.20) to obtain

$$\frac{d}{ds} M_m(g(s)) = (m-1)M_m(g(s)) - \gamma \mathfrak{h}_m M_{m+\gamma}(g(s)) , \qquad s > 0 . \tag{10.1.25}$$

Let us first consider the case $m > 1$. Then $\mathfrak{h}_m > 0$ and, since

$$M_m(g(s)) \leq M_1(g(s))^{\gamma/(\gamma+m-1)} M_{m+\gamma}(g(s))^{(m-1)/(\gamma+m-1)} , \qquad s \geq 0 ,$$

by Hölder's inequality, we deduce from (10.1.21), (10.1.25) and Young's inequality that, for $s > 0$,

$$\frac{d}{ds} M_m(g(s)) \leq \frac{(m-1)\gamma \mathfrak{h}_m}{\gamma+m-1} M_m(g(s))^{(\gamma+m-1)/(m-1)} + C(\gamma, m, \mathfrak{h}_m)$$
$$\qquad - \gamma \mathfrak{h}_m M_m(g(s))^{(\gamma+m-1)/(m-1)}$$
$$\leq C(\gamma, m, \mathfrak{h}_m) - \frac{\gamma^2 \mathfrak{h}_m}{\gamma+m-1} M_m(g(s))^{(\gamma+m-1)/(m-1)} ,$$

from which (10.1.23) readily follows.

Let us now consider $m \in [m_0, 1)$. Then $\mathfrak{h}_m < 0$ and we deduce from Hölder's inequality that

$$M_{\gamma+m}(g(s)) \leq M_m(g(s))^{(1-m)/(\gamma+1-m)} M_{1+\gamma}(g(s))^{\gamma/(\gamma+1-m)} , \qquad s \geq 0 ,$$

since $m + \gamma \in (m, \gamma+1)$. We then infer from (10.1.25) and Young's inequality that, for $s > 0$,

$$\frac{d}{ds} M_m(g(s)) + (1-m)M_m(g(s))$$
$$\leq \gamma|\mathfrak{h}_m|M_m(g(s))^{(1-m)/(\gamma+1-m)} M_{1+\gamma}(g(s))^{\gamma/(\gamma+1-m)}$$
$$\leq \frac{1-m}{2} M_m(g(s)) + C(\gamma, m, \mathfrak{h}_m)M_{1+\gamma}(g(s)) .$$

Since $M_{1+\gamma}(g)$ is bounded by (10.1.23), we obtain

$$\frac{d}{ds}M_m(g(s)) + \frac{1-m}{2}M_m(g(s)) \leq C(\gamma, m, \mathfrak{h}_m) \sup_{\sigma \geq 0}\{M_{1+\gamma}(g(\sigma))\}$$

for $s > 0$, which gives (10.1.24) after integration. □

We next turn to the uniform integrability issue and use the differentiability of h to derive a weighted L_1-estimate on $\partial_y g$.

Lemma 10.1.9. *Assume that a, b and h satisfy (10.1.3), (10.1.13) and (10.1.19). Consider an initial condition g^{in} satisfying (10.1.22), and such that*

$$g^{in} \in X_{m_1+\gamma-1} \cap W_1^1((0,\infty), y^{m_1}dy) ,$$

where the parameter $m_1 \in [1,2)$ is introduced in (10.1.19). Then the corresponding solution g to (10.1.17)–(10.1.18) belongs to $W_1^1((0,\infty), y^{m_1}dy)$ for all times and

$$M_{m_1}(|\partial_y g(s)|) \leq \max\left\{ M_{m_1}\left(\left|\frac{dg^{in}}{dy}\right|\right), \mu_2 \sup_{\sigma \geq 0}\{M_{m_1+\gamma-1}(g(\sigma))\}\right\} \qquad (10.1.26)$$

for all $s \geq 0$, where μ_2 only depends on γ and h.

Proof. It follows from (10.1.17) that $G := \partial_y g$ solves

$$\partial_s G(s,y) = -y\partial_y G(s,y) - (3 + \gamma a(y))G(s,y) - \gamma g(s,y)\frac{da}{dy}(y)$$
$$- \gamma a(y)b(y,y)g(s,y) + \gamma \int_y^\infty a(y_*)\partial_y b(y,y_*)g(s,y_*) \, dy_*$$

for $(s,y) \in (0,\infty)^2$. Then, since $a(y) = y^\gamma$ by (10.1.3a) and $\partial_y|G| = \text{sign}(G)\partial_y G$, we use an integration by parts to obtain

$$\frac{d}{ds}M_{m_1}(|G(s)|) = \int_0^\infty y^{m_1}\text{sign}(G(s,y))\partial_s G(s,y) \, dy$$
$$\leq -(2-m_1)M_{m_1}(|G(s)|) - \gamma M_{\gamma+m_1}(|G(s)|)$$
$$+ \gamma(\gamma + h(1))M_{\gamma+m_1-1}(g(s))$$
$$+ \gamma \int_0^\infty y_*^{\gamma-2}g(s,y_*) \int_0^{y_*} y^{m_1}\left|\frac{dh}{dz}\left(\frac{y}{y_*}\right)\right| \, dy dy_*$$
$$\leq -(2-m_1)M_{m_1}(|G(s)|) + \gamma(\gamma + h(1) + H_1)M_{\gamma+m_1-1}(g(s)) .$$

Since $\gamma + m_1 - 1 \geq m_0$ according to (10.1.19), the moment $M_{\gamma+m_1-1}(g(s))$ is finite for all $s \geq 0$ and uniformly bounded by Lemma 10.1.8 (b), and we end up with

$$\frac{d}{ds}M_{m_1}(|G(s)|) + (2-m_1)M_{m_1}(|G(s)|) \leq C(\gamma, h) \sup_{\sigma \geq 0}\{M_{\gamma+m_1-1}(g(\sigma))\} , \qquad s > 0 .$$

Integrating the above differential inequality gives (10.1.26), thanks to the positivity of $2-m_1$. Finally, as $m_1 \geq 1$, the boundedness of g in X_{m_1} is a consequence of Lemma 10.1.8 (a). □

We are now in a position to apply Theorem 7.3.6 to prove Theorem 10.1.5, when h satisfies the additional assumption (10.1.19).

Proposition 10.1.10. *Assume that a, b and h satisfy (10.1.3), (10.1.13) and (10.1.19). Then there is a self-similar profile φ to the fragmentation equation (10.1.1), in the sense of Definition 10.1.1, satisfying*

$$\varphi \in L_\infty((0,\infty), y^{m_0+1}\mathrm{d}y) \cap \bigcap_{m \geq m_0} X_m .$$

Furthermore,

$$M_{m_0}(\varphi) \leq \bar{\mu}_{1,m_0}\bar{\mu}_{1,1+\gamma} \quad and \quad M_m(\varphi) \leq \bar{\mu}_{1,m} , \qquad m > 1 , \tag{10.1.27}$$

and

$$y^{m_0+1}\varphi(y) \leq \gamma M_{\gamma+m_0}(\varphi)\left(\int_0^1 z^{m_0}h(z)\,\mathrm{d}z\right) , \qquad y \in (0,\infty) , \tag{10.1.28}$$

where $\bar{\mu}_{1,m} := \max\{\mu_{1,m}, \Gamma(m+1)\}$ for $m \geq m_0$, the constant $\mu_{1,m}$ being defined in Lemma 10.1.8 and depending only on γ, m and $\mathfrak{h}_m$.

Proof. Let Z be the subset of $X_{1,+}$ defined as follows: $\zeta \in Z$ if

$$\zeta \in W_1^1((0,\infty), y^{m_1}\mathrm{d}y) \cap \bigcap_{m \geq m_0} X_m , \quad \zeta \geq 0 \text{ a.e. in } (0,\infty) ,$$

$$M_1(\zeta) = 1 , \qquad M_{m_0}(\zeta) \leq \bar{\mu}_{1,m_0}\bar{\mu}_{1,1+\gamma} ,$$

$$M_m(\zeta) \leq \bar{\mu}_{1,m} , \qquad m > 1 , \qquad M_{m_1}\left(\left|\frac{d\zeta}{dy}\right|\right) \leq \mu_3 ,$$

with

$$\mu_3 := \mu_2 \bar{\mu}_{1,m_0}^{(2-m_1)/(\gamma+1-m_0)} \bar{\mu}_{1,1+\gamma}^{(\gamma+m_1-m_0-1)/(\gamma+1-m_0)} + \Gamma(m_1+1) .$$

Clearly, Z is a closed convex subset of $X_{1,+}$ which contains $x \mapsto e^{-x}$, and is compact in X_1 by Proposition 7.2.2, since $m_0 < 1 \leq m_1 < 2$.

Consider now $g^{in} \in Z$ and let g be the corresponding solution to (10.1.17)–(10.1.18). We readily infer from (10.1.21) and Lemma 10.1.8 (a) that

$$M_1(g(s)) = 1 \quad and \quad M_m(g(s)) \leq \bar{\mu}_{1,m} , \qquad m > 1 , \tag{10.1.29}$$

for all $s \geq 0$. We next deduce from Lemma 10.1.8 (b) (with $m = m_0$) and (10.1.29) that

$$M_{m_0}(g(s)) \leq \bar{\mu}_{1,m_0}\bar{\mu}_{1,1+\gamma} , \qquad s \geq 0 , \tag{10.1.30}$$

while Lemma 10.1.9, (10.1.29), (10.1.30), and the inequality

$$M_{m_1+\gamma-1}(g(s)) \leq M_{m_0}(g(s))^{(2-m_1)/(\gamma+1-m_0)} M_{\gamma+1}(g(s))^{(\gamma+m_1-1-m_0)/(\gamma+1-m_0)}$$

ensure that $M_{m_1}(|\partial_y g(s)|) \leq \mu_3$ for all $s \geq 0$. We have thus shown that, if $g^{in} \in Z$, then $g(s) \in Z$ for all $s \geq 0$. Consequently, Z is a compact and convex subset of $X_{1,+}$, which is positively invariant for the dynamical system associated with (10.1.17)–(10.1.18), and the existence of a stationary solution $\varphi \in Z$ to (10.1.17) follows from Theorem 7.3.6. Moreover, since $\varphi \in Z$, it satisfies (10.1.27).

To prove (10.1.28), we recall that, according to Proposition 10.1.2, φ satisfies (10.1.8); that is,

$$y^2\varphi(y) = \gamma \int_y^\infty y_*^{\gamma+1}\varphi(y_*)\int_0^{y/y_*} zh(z)\,\mathrm{d}z\mathrm{d}y_* , \qquad y \in (0,\infty) .$$

Since

$$\int_0^{y/y_*} zh(z)\,\mathrm{d}z \le \left(\frac{y}{y_*}\right)^{1-m_0} \int_0^1 z^{m_0}h(z)\,\mathrm{d}z \ ,$$

by (10.1.13), we conclude that φ satisfies

$$y^2\varphi(y) \le \gamma y^{1-m_0} M_{\gamma+m_0}(\varphi)\left(\int_0^1 z^{m_0}h(z)\,\mathrm{d}z\right) \ ,$$

from which (10.1.28) follows. $\qquad\qquad\qquad\qquad\qquad\qquad\qquad\qquad\qquad\qquad\qquad$ $\square$

We now remove the differentiability condition (10.1.19) on h and prove Theorem 10.1.5. To this end, we construct a sequence $(h_\varepsilon)_{\varepsilon\in(0,1)}$ of approximations of h which satisfy (10.1.3b) and (10.1.13), uniformly with respect to $\varepsilon \in (0,1)$, as well as (10.1.19) but with a strong dependence on ε. According to Proposition 10.1.10, there is a self-similar profile φ_ε to the fragmentation equation (10.1.1) with fragmentation coefficients $(a, b_\varepsilon, h_\varepsilon)$ for all $\varepsilon \in (0,1)$. Moreover, according to (10.1.27) and (10.1.28), φ_ε is bounded in weighted L_1- and L_∞-spaces, which guarantees the compactness necessary to pass to the limit as $\varepsilon \to 0$.

Proof of Theorem 10.1.5. Consider a nonnegative function $h \in L_1((0,1), z\mathrm{d}z)$ satisfying (10.1.3b) and (10.1.13). We first construct suitable approximations of h. To this end, let $\zeta \in C_0^\infty(\mathbb{R})$ be an even and nonnegative function satisfying

$$\int_\mathbb{R} \zeta(r)\,\mathrm{d}r = 1 \quad \text{and} \quad \operatorname{supp}\zeta \subset (-1,1) \ .$$

For $\varepsilon \in (0,1)$, $r \in \mathbb{R}$ and $z \in (0,1)$, we define $\zeta_\varepsilon(r) = \varepsilon^{-2}\zeta(r\varepsilon^{-2})$,

$$\nu_\varepsilon := \int_0^1 z \int_\varepsilon^1 \zeta_\varepsilon(z - z_*)h(z_*)\mathrm{d}z_*\mathrm{d}z \ ,$$

$$h_\varepsilon(z) := \frac{1}{\nu_\varepsilon}\int_\varepsilon^1 \zeta_\varepsilon(z - z_*)h(z_*)\mathrm{d}z_* \ .$$

Then

$$\lim_{\varepsilon\to 0}\nu_\varepsilon = 1 \ , \qquad \lim_{\varepsilon\to 0}\int_0^1 z^m|h(z) - h_\varepsilon(z)|\,\mathrm{d}z = 0 \ , \quad m \ge m_0 \ , \qquad (10.1.31)$$

and $dh_\varepsilon/dz \in L_1((0,1), z^{m_1}\mathrm{d}z)$, where $m_1 := 1+m_0 \in [1,2)$. Since $m_1 \ge m_0+1 \ge m_0+1-\gamma$, we conclude that a and b_ε, defined by $b_\varepsilon(y,y_*) := h_\varepsilon(y/y_*)/y_*$, $0 < y < y_*$, satisfy (10.1.3a), (10.1.3b), (10.1.13) and (10.1.19), for each $\varepsilon \in (0,1)$. It then follows from Proposition 10.1.10 that, for each $\varepsilon \in (0,1)$, there is a solution φ_ε to (10.1.5), with b_ε instead of b, which satisfies (10.1.6). Furthermore, it follows from (10.1.31) that, for all $m \ge m_0$, the family $(\mathfrak{h}_{m,\varepsilon})_{\varepsilon\in(0,1)}$, defined by

$$\mathfrak{h}_{m,\varepsilon} := \int_0^1 (z - z^m)h_\varepsilon(z)\,\mathrm{d}z \ ,$$

is bounded from above and below by positive constants depending only on m and h. Combining this observation with (10.1.27) and (10.1.28), and recalling that the constants in (10.1.27) only depend on γ, m, and $\mathfrak{h}_{m,\varepsilon}$, we conclude that the family $\mathcal{E} := \{\varphi_\varepsilon \ : \ \varepsilon \in (0,1)\}$ is bounded in X_m for all $m \ge m_0$, and in $L_\infty((0,\infty), y^{m_0+1}\mathrm{d}y)$. Owing to these bounds, $\mathcal{E}$ satisfies the requirements of the Dunford–Pettis theorem (Theorem 7.1.3) in X_1 and there is a sequence $(\varepsilon_j)_{j\ge 1}$, and $\varphi \in X_1$ such that

$$\lim_{j\to\infty}\varepsilon_j = 0 \quad \text{and} \quad \varphi_{\varepsilon_j} \rightharpoonup \varphi \quad \text{in} \quad X_1 \text{ as } j \to \infty \ . \qquad (10.1.32)$$

We deduce from the properties of $\mathcal{E}$ and the convergence (10.1.32) that φ lies in $X_{1,+}$ with $M_1(\varphi) = 1$ and

$$\varphi \in L_\infty((0,\infty), y^{m_0+1}\mathrm{d}y) \cap \bigcap_{m \geq m_0} X_m \ .$$

Moreover, for any $m > m_0$,

$$\varphi_{\varepsilon_j} \rightharpoonup \varphi \quad \text{in} \quad X_m \ . \tag{10.1.33}$$

It remains to check that φ is a weak solution to (10.1.5). To this end, consider $\vartheta \in \Theta^1$ and recall that

$$\int_0^\infty \varphi_{\varepsilon_j}(y) \left[\vartheta(y) - y\frac{d\vartheta}{dy}(y) + \gamma y^\gamma \vartheta(y) \right] \ \mathrm{d}y$$

$$= \gamma \int_0^\infty y_*^{\gamma-1} \varphi_{\varepsilon_j}(y_*) \int_0^{y_*} \vartheta(y) h_{\varepsilon_j}\left(\frac{y}{y_*}\right) \ \mathrm{d}y\mathrm{d}y_*$$

$$= \gamma \int_0^\infty y_*^\gamma \varphi_{\varepsilon_j}(y_*) \int_0^1 \vartheta(y_* z) h_{\varepsilon_j}(z) \ \mathrm{d}z\mathrm{d}y_* \ . \tag{10.1.34}$$

Introducing

$$\xi_j(y_*) := \frac{1}{y_*} \int_0^1 \vartheta(y_* z) h_{\varepsilon_j}(z) \ \mathrm{d}z \ , \quad y_* \in (0,\infty) \ ,$$

we see that (10.1.3b) (for h_{ε_j}), (10.1.31) and the properties of ϑ imply that, for $y_* \in (0,\infty)$,

$$|\xi_j(y_*)| \leq \left\| \frac{d\vartheta}{dy} \right\|_\infty \quad \text{and} \quad \lim_{j \to \infty} \xi_j(y_*) = \frac{1}{y_*} \int_0^1 \vartheta(y_* z) h(z) \ \mathrm{d}z \ .$$

Recalling (10.1.33) (with $m = \gamma + 1$), we may apply Proposition 7.1.12 to conclude that

$$\lim_{j \to \infty} \int_0^\infty y_*^{\gamma+1} \varphi_{\varepsilon_j}(y_*) \zeta_j(y_*) \ \mathrm{d}y_* = \int_0^\infty y_*^\gamma \varphi(y_*) \int_0^1 \vartheta(y_* z) h(z) \ \mathrm{d}z\mathrm{d}y_* \ .$$

Since passing to the limit in the left-hand side of (10.1.34) is easy thanks to (10.1.33), we have thus shown that φ is indeed a self-similar profile to (10.1.1) in the sense of Definition 10.1.1 and satisfies the properties listed in Theorem 10.1.5. $\qquad\square$

We are now in a position to complete the proof of Theorem 10.1.4, which is performed by a compactness method similar to that of Theorem 10.1.5. The main difficulty to overcome is the behaviour for small sizes which is no longer controlled by an estimate in X_{m_0} for some $m_0 \in [0,1)$.

Proof of Theorem 10.1.4. Consider a nonnegative function $h \in L_1((0,1), z\mathrm{d}z)$ satisfying (10.1.3b) and (10.1.10).

Step 1: Approximation. For $j \geq 1$, define

$$h_j(z) := \frac{1}{\nu_j} h(z) \mathbf{1}_{(1/j,1)}(z) \ , \quad z \in (0,1) \ , \tag{10.1.35a}$$

and

$$\nu_j := \int_{1/j}^1 z h(z) \ \mathrm{d}z \ . \tag{10.1.35b}$$

Then $h_j \in L_1((0,1), z\mathrm{d}z)$ satisfies (10.1.3b) and it readily follows from (10.1.12), (10.1.35) and the properties of h that

$$\lim_{j\to\infty} \nu_j = 1 , \qquad \lim_{j\to\infty} \int_0^1 zh_j(z)|\ln z| \ \mathrm{d}z = \ell_h , \tag{10.1.36}$$

and that there is $j_0 \geq 1$ such that

$$\nu_j \geq \frac{1}{2} , \qquad 2\ell_h \geq \ell_{h_j} := \int_0^1 zh_j(z)|\ln z| \ \mathrm{d}z \geq \frac{\ell_h}{2} , \qquad j \geq j_0 , \tag{10.1.37}$$

the parameter ℓ_h being defined in (10.1.10). In addition,

$$\int_0^1 h_j(z) \ \mathrm{d}z \leq j \int_{1/j}^1 zh_j(z) \ \mathrm{d}z \leq j ,$$

so that $h_j \in L_1(0,1)$ and thus satisfies (10.1.13) with $m_0 = 0$. Thanks to Theorem 10.1.5, there is a self-similar profile $\varphi_j \in X_1$ to the fragmentation equation (10.1.1) with fragmentation coefficients a and b_j, the latter being defined by $b_j(y, y_*) := h_j(y/y_*)/y_*$ for $0 < y < y_*$. Moreover, by Theorem 10.1.5,

$$\varphi_j \in L_\infty((0,\infty), y\mathrm{d}y) \cap \bigcap_{m\geq 0} X_m . \tag{10.1.38}$$

Our aim now is to derive estimates on $(\varphi_j)_{j\geq 1}$ that will guarantee the weak compactness of this sequence in X_1. We first deduce from (10.1.6), (10.1.11) and (10.1.37) that

$$M_1(\varphi_j) = 1 , \qquad M_{\gamma+1}(\varphi_j) = \frac{1}{\gamma\ell_{h_j}} \leq \frac{2}{\gamma\ell_h} , \qquad j \geq j_0 . \tag{10.1.39}$$

Step 2: Control of higher-order moments. Let $m > 0$ and $j \geq j_0$. We multiply (10.1.8) by y^{m-1} and find, after integration over $(0,\infty)$ and using Fubini–Tonelli's theorem,

$$M_{m+1}(\varphi_j) = \gamma \int_0^\infty y^{m-1} \int_y^\infty y_*^{1+\gamma} \varphi_j(y_*) \int_0^{y/y_*} zh_j(z) \ \mathrm{d}z\mathrm{d}y_*\mathrm{d}y$$

$$= \gamma \int_0^1 zh_j(z) \int_0^\infty y_*^{1+\gamma} \varphi_j(y_*) \int_{zy_*}^{y_*} y^{m-1} \ \mathrm{d}y\mathrm{d}y_*\mathrm{d}z$$

$$= \frac{\gamma}{m} M_{m+\gamma+1}(\varphi_j) \int_0^1 z(1-z^m)h_j(z) \ \mathrm{d}z .$$

Since

$$1 - z^m \geq -m\ln(z)e^{m\ln z} \geq m|\ln z|/2 , \qquad z \in [2^{-1/m}, 1) ,$$

by the convexity of $x \mapsto e^x$ and $1 - z^m \geq 1/2$ for $z \in (0, 2^{-1/m})$, we further obtain

$$M_{m+1}(\varphi_j) \geq \frac{\gamma}{m} M_{m+\gamma+1}(\varphi_j) \int_0^1 z\zeta_m(z)h_j(z) \ \mathrm{d}z$$

with

$$\zeta_m(z) = \frac{1}{2}\mathbf{1}_{(0,2^{-1/m})}(z) + \frac{m}{2}|\ln z|\mathbf{1}_{[2^{-1/m},1)}(z) , \qquad z \in (0,1) .$$

Since $\zeta_m(z) \leq m(1 + |\ln z|)/2$ for $z \in (0,1)$, we infer from (10.1.3b), (10.1.10), (10.1.35a) and Lebesgue's dominated convergence theorem that

$$\lim_{j\to\infty} \int_0^1 z\zeta_m(z)h_j(z) \ \mathrm{d}z = 2\varepsilon_m := \int_0^1 z\zeta_m(z)h(z) \ \mathrm{d}z > 0 .$$

Therefore, there is $j_m \geq j_0$ large enough such that

$$\int_0^1 z\zeta_m(z)h_j(z)\,\mathrm{d}z \geq \varepsilon_m \quad \text{for} \quad j \geq j_m \ ,$$

and thus

$$M_{m+1}(\varphi_j) \geq \frac{\gamma\varepsilon_m}{m} M_{m+\gamma+1}(\varphi_j) \ , \qquad j \geq j_m \ . \tag{10.1.40}$$

In particular, since $\gamma > 0$, it follows by induction from (10.1.39) and (10.1.40) that, for $i \geq 2$ and $j \geq J_i := \max\{j_{l\gamma} \ : \ l \in \{1,\ldots,i-1\}\}$,

$$M_{i\gamma+1}(\varphi_j) \leq M_{\gamma+1}(\varphi_j) \prod_{l=1}^{i-1} \left(\frac{l}{\varepsilon_{l\gamma}}\right) \leq \frac{2}{\gamma\ell_h} \prod_{l=1}^{i-1} \left(\frac{l}{\varepsilon_{l\gamma}}\right) \ . \tag{10.1.41}$$

Step 3: Behaviour for small sizes. It remains to control the behaviour for small sizes so as to ensure that no mass escapes as $y \to 0$. Fix $r \in (0,1)$ and consider $j \geq j_0$. By (10.1.8) and Fubini–Tonelli's theorem,

$$\int_0^r y\varphi_j(y)\,\mathrm{d}y = \gamma \int_0^1 zh_j(z) \int_0^\infty y_*^{1+\gamma}\varphi_j(y_*) \int_{zy_*}^{y_*} \frac{\mathbf{1}_{(0,r)}(y)}{y}\,\mathrm{d}y\mathrm{d}y_*\mathrm{d}z$$

$$= \gamma \int_0^1 zh_j(z)|\ln z| \int_0^r y_*^{1+\gamma}\varphi_j(y_*)\mathrm{d}y_*\mathrm{d}z$$

$$+ \gamma \int_0^1 zh_j(z) \int_r^{r/z} y_*^{1+\gamma}\varphi_j(y_*)\ln\left(\frac{r}{zy_*}\right)\mathrm{d}y_*\mathrm{d}z \ .$$

On the one hand, it follows from (10.1.37) and (10.1.39) that

$$\gamma \int_0^1 zh_j(z)|\ln z| \int_0^r y_*^{1+\gamma}\varphi_j(y_*)\mathrm{d}y_*\mathrm{d}z \leq 2\gamma\ell_h r^\gamma \ . \tag{10.1.42}$$

On the other hand,

$$\gamma \int_0^1 zh_j(z) \int_r^{r/z} y_*^{1+\gamma}\varphi_j(y_*)\ln\left(\frac{r}{zy_*}\right)\mathrm{d}y_*\mathrm{d}z = \gamma\left[I_1(r,j) + I_2(r,j)\right] \ , \tag{10.1.43}$$

where

$$I_1(r,j) := \int_0^{\sqrt{r}} zh_j(z) \int_r^{r/z} y_*^{1+\gamma}\varphi_j(y_*)\ln\left(\frac{r}{zy_*}\right)\mathrm{d}y_*\mathrm{d}z \ ,$$

$$I_2(r,j) := \int_{\sqrt{r}}^1 zh_j(z) \int_r^{r/z} y_*^{1+\gamma}\varphi_j(y_*)\ln\left(\frac{r}{zy_*}\right)\mathrm{d}y_*\mathrm{d}z \ .$$

Since $r/(zy_*) \leq 1/z$ for $y_* \in (r,r/z)$, we infer from (10.1.35a), (10.1.37) and (10.1.39) that

$$I_1(r,j) \leq \int_0^{\sqrt{r}} zh_j(z) \int_r^{r/z} y_*^{1+\gamma}\varphi_j(y_*)|\ln z|\mathrm{d}y_*\mathrm{d}z$$

$$\leq \frac{1}{\nu_j} M_{\gamma+1}(\varphi_j) \int_0^{\sqrt{r}} zh(z)|\ln z|\,\mathrm{d}z$$

$$\leq \frac{4}{\gamma\ell_B} \int_0^{\sqrt{r}} zh(z)|\ln z|\,\mathrm{d}z \ . \tag{10.1.44}$$

Next, since $zy_* > zr > r^{3/2}$ for $y_* \in (r, r/z)$ and $z \in (\sqrt{r}, 1)$, it follows from (10.1.3b), (10.1.35a), (10.1.37) and (10.1.39) that

$$I_2(r,j) \leq \int_{\sqrt{r}}^1 z h_j(z) \int_r^{r/z} y_* \left(\frac{r}{z}\right)^\gamma \varphi_j(y_*) \ln\left(\frac{1}{\sqrt{r}}\right) \mathrm{d}y_* \mathrm{d}z$$

$$\leq \frac{r^{\gamma/2}|\ln r|}{2} M_1(\varphi_j) \int_{\sqrt{r}}^1 z h_j(z) \, \mathrm{d}z$$

$$\leq r^{\gamma/2}|\ln r| \, . \tag{10.1.45}$$

Collecting (10.1.42), (10.1.43), (10.1.44) and (10.1.45), we end up with

$$\int_0^r y\varphi_j(y) \, \mathrm{d}y \leq \omega(r) := 2\gamma \ell_h r^\gamma + r^{\gamma/2}|\ln r| + \frac{4}{\gamma \ell_h} \int_0^{\sqrt{r}} z h(z)|\ln z| \, \mathrm{d}z \, . \tag{10.1.46}$$

Owing to the integrability property (10.1.10) of h and the positivity of γ,

$$\lim_{r \to 0} \omega(r) = 0 \, . \tag{10.1.47}$$

Step 4: Compactness and convergence. Let us now establish that $\mathcal{E} := \{\varphi_j : j \geq j_0\}$ is relatively sequentially weakly compact in X_1. To this end, observe that a further consequence of (10.1.8), (10.1.35a) and (10.1.39) is that, for $j \geq j_0$,

$$y^2 \varphi_j(y) \leq \gamma M_{\gamma+1}(\varphi_j) \leq \frac{2}{\ell_h} \, , \qquad y \in (0, \infty) \, . \tag{10.1.48}$$

Consequently, given a measurable subset E of $(0, \infty)$, $r \in (0, 1)$ and $R > 1$, we infer from (10.1.39), (10.1.46) and (10.1.48) that, for $j \geq j_0$,

$$\int_E y\varphi_j(y) \, \mathrm{d}y \leq \int_0^r y\varphi_j(y) \, \mathrm{d}y + \int_r^R \mathbf{1}_E(y) y\varphi_j(y) \, \mathrm{d}y + \int_R^\infty y\varphi_j(y) \, \mathrm{d}y$$

$$\leq \omega(r) + \frac{2}{\ell_h r}|E| + \frac{2}{\gamma \ell_h R^\gamma} \, .$$

Hence,

$$\eta\{\mathcal{E}; X_1\} \leq \omega(r) + \frac{2}{\gamma \ell_h R^\gamma} \, ,$$

the modulus of uniform integrability $\eta\{\mathcal{E}; X_1\}$ of $\mathcal{E}$ in X_1 being defined in Definition 7.1.2. The above inequality being valid for all $r \in (0,1)$ and $R > 1$, we may let $r \to 0$ and $R \to \infty$ and use (10.1.47) to conclude that

$$\eta\{\mathcal{E}; X_1\} = 0 \, .$$

Furthermore, arguing as above gives

$$\lim_{R \to \infty} \sup_{j \geq j_0} \left\{ \int_R^\infty y\varphi_j(y) \, \mathrm{d}y \right\} = 0 \, ,$$

and we have thus proved the relative sequential weak compactness of $\mathcal{E}$ in X_1, see Theorem 7.1.3. Therefore, there is $\varphi \in X_1$, and a subsequence of $(\varphi_j)_{j \geq j_0}$ (not relabelled) such that

$$\varphi_j \rightharpoonup \varphi \quad \text{in} \quad X_1 \, . \tag{10.1.49}$$

A first consequence of (10.1.49) is that φ is nonnegative a.e. in $(0, \infty)$ and also satisfies $M_1(\varphi) = 1$. Furthermore, it follows from (10.1.41) and a truncation argument that $\varphi \in X_{i\gamma+1}$ and $\varphi_j \rightharpoonup \varphi$ in $X_{i\gamma+1}$ for all $i \geq 2$. We then proceed as in the proof of Theorem 10.1.5 to show that φ is a self-similar profile to (10.1.1) in the sense of Definition 10.1.1 and complete the proof. □

We next exploit the contractivity properties of the fragmentation equation to prove that (10.1.5) has a unique solution satisfying (10.1.6).

Proposition 10.1.11. *Consider a and b satisfying (10.1.3) and assume further that $h > 0$ in $(0, 1)$. Then there is at most one self-similar profile to the fragmentation equation (10.1.1) in the sense of Definition 10.1.1.*

Proof. Let φ_i, $i = 1, 2$, be two self-similar profiles to (10.1.1) in the sense of Definition 10.1.1. Introducing

$$f_i(t, x) := (1 + t)^{2/\gamma} \varphi_i \left(x(1 + t)^\gamma \right) , \qquad (t, x) \in [0, \infty) \times (0, \infty) , \ i = 1, 2 ,$$

we infer from Definition 10.1.1 that

$$\frac{d}{dt} \int_0^\infty f_i(t, x) \vartheta(x) \, dx = - \int_0^\infty a(y) N_\vartheta(y) f_i(s, y) \, dy \tag{10.1.50}$$

for all $\vartheta \in \Theta^1$ and

$$M_1(f_i(t)) = 1 \ \text{ and } \ M_{\gamma+1}(f_i(t)) = \frac{M_{\gamma+1}(\varphi_i)}{1 + t} , \qquad t \geq 0 , \ i = 1, 2 . \tag{10.1.51}$$

In particular, since $a(x) = x^\gamma$ for $x \in (0, \infty)$ by (10.1.3a), we infer from (10.1.51) and Fubini's theorem that, for $i = 1, 2$ and $t > 0$,

$$x \mapsto -xa(x) f_i(t, x) \ \text{ and } \ x \mapsto x \int_x^\infty a(y) b(x, y) f_i(t, y) \, dy$$

belong to X_0. It then follows from (10.1.50) that $\partial_t f_i(t)$ belongs to X_1 for all $t > 0$ and

$$x \partial_t f_i(t, x) = -xa(x) f_i(t, x) + x \int_x^\infty a(y) b(x, y) f(t, y) \, dy \quad \text{for a.e. } x \in (0, \infty) . \tag{10.1.52}$$

We now introduce $E = f_1 - f_2$ and $\Sigma = \text{sign}(E)$. Since $E \in W_\infty^1((0, \infty), X_1)$ by (10.1.51) and (10.1.52), we infer from (10.1.3), (10.1.52) and Fubini's theorem that, for $t > 0$,

$$\frac{d}{dt} \int_0^\infty x|E(t, x)| \, dx = \int_0^\infty x\Sigma(t, x) \partial_t E(t, x) \, dx$$

$$= - \int_0^\infty xa(x)|E(t, x)| \, dx$$

$$+ \int_0^\infty a(y) E(t, y) \int_0^y x\Sigma(t, x) b(x, y) \, dxdy$$

$$= - \int_0^\infty a(y)|E(t, y)| \int_0^y (1 - \Sigma(t, x)\Sigma(t, y)) xb(x, y) \, dxdy . \tag{10.1.53}$$

Since

$$\int_0^\infty x|E(t, x)| \, dx = (1 + t)^{2/\gamma} \int_0^\infty x \left| (\varphi_1 - \varphi_2) \left(x(1 + t)^{1/\gamma} \right) \right| \, dx$$

$$= \int_0^\infty y |(\varphi_1 - \varphi_2)(y)| \, \mathrm{d}y \, ,$$

the left-hand side of (10.1.53) vanishes and we conclude that

$$\int_0^\infty a(y)|E(t,y)| \int_0^y (1 - \Sigma(t,x)\Sigma(t,y))xb(x,y) \, \mathrm{d}x\mathrm{d}y = 0$$

for all $t > 0$. Recalling the continuity of φ_i, $i = 1, 2$, established in Proposition 10.1.2, we may pass to the limit as $t \to 0$ in the previous identity and conclude that

$$\int_0^\infty a(y)|(\varphi_1 - \varphi_2)(y)| \int_0^y (1 - \Sigma_0(x)\Sigma_0(y))xb(x,y) \, \mathrm{d}x\mathrm{d}y = 0$$

with $\Sigma_0 := \Sigma(0, \cdot) = \mathrm{sign}(\varphi_1 - \varphi_2)$. Clearly $\Sigma_0(x)\Sigma_0(y) \leq 1$ for all $0 < x < y$ and we end up with

$$|(\varphi_1 - \varphi_2)(y)| \int_0^y (1 - \Sigma_0(x)\Sigma_0(y))xb(x,y) \, \mathrm{d}x = 0 \, , \qquad y \in (0,\infty) \setminus \mathcal{Z} \, , \qquad (10.1.54)$$

where $\mathcal{Z}$ is a measurable subset of $(0,\infty)$ with $|\mathcal{Z}| = 0$.

Now, define

$$\mathcal{P} := \{y \in (0,\infty) \setminus \mathcal{Z} \; : \; \varphi_1(y) > \varphi_2(y)\} \, ,$$
$$\mathcal{N} := \{y \in (0,\infty) \setminus \mathcal{Z} \; : \; \varphi_1(y) < \varphi_2(y)\} \, .$$

For $y \in \mathcal{P}$, $\Sigma_0(y) = 1$ and we infer from (10.1.54) and the positivity of h that $\Sigma_0(x)\Sigma_0(y) = 1$ for almost every $x \in (0, y)$. Consequently, $|(0, y) \cap \mathcal{P}^c| = 0$ for $y \in \mathcal{P}$ and a similar argument shows that $|(0, y) \cap \mathcal{N}^c| = 0$ for $y \in \mathcal{N}$. Consequently, either $|\mathcal{P}| = 0$ or $|\mathcal{N}| = 0$, so that

$$\int_0^\infty x|(\varphi_1 - \varphi_2)(x)| \, \mathrm{d}x = \int_0^\infty x(\varphi_1 - \varphi_2)(x) \, \mathrm{d}x = 0 \, ,$$

by (10.1.6), and thus $\varphi_1 = \varphi_2$. $\qquad\qquad\square$

Finally we supplement Theorem 10.1.5 with lower bounds on the profile φ, recalling that Theorem 10.1.5 already provides an upper bound.

Proposition 10.1.12. *Assume that the overall fragmentation rate a and the daughter distribution function b satisfy (10.1.3) with $\gamma > 0$ and (10.1.13). Let $\varphi \in X_1$ be a self-similar profile to the fragmentation equation (10.1.1) and assume further that $h > 0$ a.e. in $(0,1)$. Then $\varphi > 0$ in $(0,\infty)$ and, for all $y \in (0,\infty)$,*

$$y_*^2 e^{y_*^\gamma} \varphi(y_*) \geq y^2 e^{y^\gamma} \varphi(y) \, , \qquad y_* \in (y, \infty) \, .$$

In Proposition 10.1.12, the positivity of h is assumed for simplicity. It is actually still valid if h is only positive on a subinterval of $(0, 1)$ [202].

Proof. Since $\partial_y \varphi \in X_2$ by Proposition 10.1.2, we can deduce from (10.1.8) that, for a.e. $y \in (0,\infty)$,

$$\partial_y \left(y^2 e^{y^\gamma} \varphi(y)\right) = \gamma \int_y^\infty a(y_*)b(y, y_*)\varphi(y_*) \, \mathrm{d}y_* \geq 0 \, . \qquad (10.1.55)$$

Now, since $\varphi \not\equiv 0$ by (10.1.6), there exists $y_0 \in (0, \infty)$ such that $\varphi(y_0) > 0$. On the one hand, for $y \in (y_0, \infty)$, it readily follows from (10.1.55) that

$$y^2 e^{y^\gamma} \varphi(y) \geq y_0^2 e^{y_0^\gamma} \varphi(y_0) > 0 . \qquad (10.1.56)$$

On the other hand, assume for contradiction that there is $y_1 \in (0, y_0)$ such that $\varphi(y_1) = 0$ and $\varphi(y) > 0$ for $y \in (y_1, y_0]$. We then deduce from (10.1.8) that

$$\int_{y_1}^{\infty} a(y_*) \varphi(y_*) \int_0^{y_1/y_*} z h(z) \, \mathrm{d}z \mathrm{d}y_* = 0 .$$

Owing to the positivity of a and h, this identity readily implies that $\varphi(y) = 0$ for $y \in (y_1, \infty)$ and contradicts the definition of y_1. Consequently, φ is positive in $(0, y_0)$ and thus in $(0, \infty)$ thanks to (10.1.56). Integrating (10.1.55) gives the last statement of Proposition 10.1.12. $\square$

Theorems 10.1.4 and 10.1.5 provide rough information on the behaviour of the profile φ as $y \to 0$ and $y \to \infty$. More precise results in this direction are available when $h \in L_1(0, 1)$ which we report now, following [12], see also [47] for a stochastic approach and [77, 258] for asymptotic expansions derived formally. These results turn out to be sensitive to the behaviour of h as $z \to 0$ and $z \to 1$.

Proposition 10.1.13. *[12] Assume that the overall fragmentation rate a and the daughter distribution function b satisfy (10.1.3) with $\gamma > 0$ and (10.1.13) with $m_0 = 0$. Assume further that there is $\nu > -1$, $\mu \geq 0$, $\delta > 0$, $h_0 > 0$, and $h_1 > 0$ such that*

$$h(z) \sim h_0 z^\nu \quad \text{as} \ z \to 0 ,$$
$$h(z) = h_1(1 - z)^\mu + O\left(((1 - z)^{\mu + \delta}\right) \quad \text{as} \ z \to 1 .$$

Let φ be a self-similar profile to the fragmentation equation in the sense of Definition 10.1.1. Then there is $C_0 > 0$, and $C_\infty > 0$ such that

$$\varphi(y) \sim C_\infty y^{p-2} e^{-y^\gamma} \quad \text{as} \ y \to \infty \quad \text{and} \quad \varphi(y) \sim C_0 y^\nu \quad \text{as} \ y \to 0 ,$$

with $p = h_1$ if $\mu = 0$ and $p = 0$ if $\mu > 0$.

10.1.2   Convergence and Decay Rates

We next turn to the stability of the self-similar solution to the fragmentation equation (10.1.1) constructed in Section 10.1.1. Several authors have investigated this issue, using either a stochastic approach [43, 116] or a deterministic one [202]. The proof given below relies on arguments from the theory of dynamical systems and is taken from [113, 202].

We assume throughout this section that the overall fragmentation rate a and the daughter distribution function b satisfy (10.1.3), with $\gamma > 0$, and (10.1.10), as well as $h > 0$ a.e. in $(0, 1)$. According to Theorem 10.1.4 and Proposition 10.1.11, there is a unique self-similar profile φ to the fragmentation equation in the sense of Definition 10.1.1 and thus a unique self-similar solution φ_s given by $\varphi_s(t, x) := \sigma(t)^{-2} \varphi(x/\sigma(t))$ for $(t, x) \in [0, \infty) \times (0, \infty)$, where $\sigma(t) = (1 + t)^{-1/\gamma}$, see (10.1.4). We also assume that there is $m_0 \in [0, 1)$, and $p_0 \in (1, 1 + \gamma)$ such that

$$h \in L_{p_0}((0, 1), z^{m_0} \mathrm{d}z) . \qquad (10.1.57)$$

Consider next an initial condition f^{in} satisfying

$$f^{in} \in X_{m_0, +} \cap X_{1 + \gamma} \quad \text{with} \ M_1(f^{in}) = 1 . \qquad (10.1.58)$$

By Theorem 8.2.23 ($m_0 = 0$) or Theorem 8.2.30 ($m_0 \in (0,1)$) and Proposition 8.2.57, there is a unique mass-conserving weak solution f to the fragmentation equation (10.1.1).

Our aim now is to study the behaviour of $f(t) - \varphi_s(t)$ as $t \to \infty$ and, in particular, to show that it converges to zero in a suitable topology. Instead of estimating the difference between the two time-dependent functions $f(t)$ and $\varphi_s(t)$, it turns out to be more convenient to exploit the specific, and simpler, dependence of φ_s on time and size and perform the same transformation as in Section 10.1.1. We therefore introduce the self-similar variables

$$s := \frac{1}{\gamma} \ln(1+t), \qquad y := x(1+t)^{1/\gamma}, \tag{10.1.59}$$

and the rescaled function

$$g(s,y) := e^{-2s} f\left(e^{\gamma s} - 1, ye^{-s}\right), \qquad (s,y) \in [0,\infty) \times (0,\infty). \tag{10.1.60}$$

Equivalently,

$$f(t,x) = (1+t)^{2/\gamma} g\left(\frac{1}{\gamma} \ln(1+t), x(1+t)^{1/\gamma}\right) \tag{10.1.61}$$

for $(t,x) \in [0,\infty) \times (0,\infty)$. An interesting feature of this classical technique is the identity

$$\|f(t) - \varphi_s(t)\|_{[1]} = \|g(\ln(1+t)/\gamma) - \varphi\|_{[1]}, \qquad t \geq 0, \tag{10.1.62}$$

where $\|\cdot\|_{[1]}$ denotes the norm in X_1, which clearly transforms the convergence to zero of $\|f(t) - \varphi_s(t)\|_{[1]}$ as $t \to \infty$, to the convergence in X_1 of $g(s)$ to the time-independent function φ as $s \to \infty$. Recalling (10.1.17) and (10.1.18), g solves

$$\partial_s g(s,y) + y\partial_y g(s,y) + 2g(s,y)$$
$$= -\gamma a(y)g(s,y) + \gamma \int_y^\infty a(y_*)b(y,y_*)g(s,y_*)\, \mathrm{d}y_* \tag{10.1.63a}$$

for $(s,y) \in (0,\infty)^2$, with initial condition

$$g(0,y) = f^{in}(y), \qquad y \in (0,\infty), \tag{10.1.63b}$$

in the sense of Definition 10.1.6. Furthermore, due to (10.1.61), the fact that f is a mass-conserving solution to (10.1.1) implies

$$M_1(g(s)) = M_1(f(e^{\gamma s} - 1)) = M_1(f^{in}) = 1, \qquad s \geq 0. \tag{10.1.64}$$

From the analysis carried out in Section 10.1.1, the self-similar profile φ is a stationary solution to (10.1.63a) in a weak sense, see (10.1.5) and (10.1.7), and we devote the remainder of this section to the analysis of the behaviour of $g(s) - \varphi$ as $s \to \infty$. The first step is to show that this difference indeed converges to zero as $s \to \infty$ in X_1, thereby establishing the stability of φ with respect to that topology. This will be the content of Section 10.1.2.1. Once the stability is proved, the next step is to study the rate of convergence to zero of $g(s) - \varphi$ (if any). We summarise the results available so far in the literature in Section 10.1.2.2.

To begin with, let us state two useful properties of g. On the one hand, it readily follows from (10.1.62) and Proposition 8.2.57 that

$$s \longmapsto \|g(s) - \varphi\|_{[1]} \quad \text{is non-increasing.} \tag{10.1.65}$$

On the other hand, since $f^{in} \in X_{1+\gamma}$ and h satisfies (10.1.57), we deduce from Lemma 10.1.8 that there is $\mu_{1+\gamma} > 0$, depending only on γ, h and $M_{1+\gamma}(f^{in})$, such that

$$M_{1+\gamma}(g(s)) \leq \mu_{1+\gamma}, \qquad s \geq 0. \tag{10.1.66}$$

10.1.2.1   Convergence

The aim of this section is the following convergence result [202, Theorem 3.2]. We recall that, since a and b satisfy (10.1.3) and (10.1.10), and $h > 0$ in $(0, 1)$, there is a unique self-similar profile φ to (10.1.1), see Theorem 10.1.4 and Proposition 10.1.11.

Theorem 10.1.14. *There holds*

$$\lim_{t \to \infty} \| f(t) - \varphi_s(t) \|_{[1]} = \lim_{s \to \infty} \| g(s) - \varphi \|_{[1]} = 0 \ .$$

The starting point of the proof of Theorem 10.1.14 is the observation that an infinite number of Lyapunov functionals are available for (10.1.63) [202].

Proposition 10.1.15. *Let $\Phi \in C^1(\mathbb{R})$ be a nonnegative convex function satisfying $\Phi(1) = 0$ and assume that*

$$\mathcal{J} := \int_0^\infty y\varphi(y) \Phi \left(\frac{f^{in}(y)}{\varphi(y)} \right) \ \mathrm{d}y < \infty \ .$$

Then, for $s \geq 0$,

$$\mathcal{H}_\Phi[g(s), \varphi] + \gamma \int_0^s \mathcal{D}_\Phi[g(s_*), \varphi] \ \mathrm{d}s_* = \mathcal{H}_\Phi[f^{in}, \varphi] \ ,$$

where

$$\mathcal{H}_\Phi[\xi, \varphi] := \int_0^\infty y\varphi(y) \Phi \left(\frac{\xi(y)}{\varphi(y)} \right) \ \mathrm{d}y \ ,$$

$$\mathcal{D}_\Phi[\xi, \varphi] := \int_0^\infty a(y_*)\varphi(y_*) \int_0^{y_*} yb(y, y_*) \Psi \left(\frac{\xi(y)}{\varphi(y)}, \frac{\xi(y_*)}{\varphi(y_*)} \right) \mathrm{d}y\mathrm{d}y_* \ ,$$

and

$$\Psi(X, Y) := \Phi(Y) - \Phi(X) - \Phi'(X)(Y - X) \geq 0 \ , \qquad (X, Y) \in \mathbb{R}^2 \ ,$$

the nonnegativity of Ψ being a straightforward consequence of the convexity of Φ.

Proof. We first recall that the assumed positivity of h in $(0, 1)$ guarantees that $\varphi > 0$ in $(0, \infty)$ by Proposition 10.1.12. Then $G(s) := g(s)/\varphi$ for $s > 0$, and $G^{in} := f^{in}/\varphi$ are both well defined. To convey the main idea of the proof but avoid too many technical steps, the proof we provide now is mostly formal and requires classical approximation and truncation arguments, the latter in the spirit of those used in Chapter 8, to be fully justified.

It follows from (10.1.63a) that

$$\partial_s \left(y\varphi(y)\Phi(G(s, y)) \right) = -\partial_y \left(y^2\varphi(y)\Phi(G(s, y)) \right) - [G\Phi'(G) - \Phi(G)](s, y)\frac{d}{dy}\left(y^2\varphi(y) \right)$$

$$- \gamma y a(y)\Phi'(G(s, y))g(s, y)$$

$$+ \gamma y \Phi'(G(s, y)) \int_y^\infty a(y_*)b(y, y_*)g(s, y_*) \ \mathrm{d}y_* \ .$$

Since

$$\frac{d}{dy}\left(y^2\varphi(y) \right) = -\gamma y a(y)\varphi(y) + \gamma y \int_y^\infty a(y_*)b(y, y_*)\varphi(y_*) \ \mathrm{d}y_*$$

for $y \in (0, \infty)$, we further obtain

$$\partial_s \left(y\varphi(y)\Phi(G(s, y)) \right) + \partial_y \left(y^2\varphi(y)\Phi(G(s, y)) \right)$$

$$= -\gamma y a(y)\varphi(y)\Phi(G(s, y)) + \gamma y \Phi(G(s, y)) \int_y^\infty a(y_*)b(y, y_*)\varphi(y_*) \ \mathrm{d}y_*$$

$$- \gamma y \Phi'(G(s,y))G(s,y) \int_y^\infty a(y_*)b(y,y_*)\varphi(y_*)\ dy_*$$

$$+ \gamma y \Phi'(G(s,y)) \int_y^\infty a(y_*)b(y,y_*)g(s,y_*)\ dy_*\ .$$

Integrating the above identity with respect to $y \in (0,\infty)$, and assuming that $G(s)$ decays sufficiently rapidly to zero for large sizes, the contribution of the second term on the left-hand side vanishes and we may use Fubini's theorem to find

$$\frac{d}{ds} \int_0^\infty y\varphi(y)\Phi(G(s,y))\ dy$$

$$= -\gamma \int_0^\infty y_* a(y_*)\varphi(y_*)\Phi(G(s,y_*))\ dy_*$$

$$+ \gamma \int_0^\infty a(y_*)\varphi(y_*) \int_0^{y_*} yb(y,y_*)\Phi(G(s,y))\ dy dy_*$$

$$- \gamma \int_0^\infty a(y_*)\varphi(y_*) \int_0^{y_*} yb(y,y_*)\Phi'(G(s,y))\left[G(s,y) - G(s,y_*)\right]\ dy dy_*\ .$$

Proposition 10.1.15 now follows from the above identity together with (10.1.3). $\qquad\square$

We next prove the convergence of $g(s)$ to φ as $s \to \infty$ for a particular class of initial data, for which Proposition 10.1.15 can be used with $\Phi(r) = (r-1)^2/2$, $r \in \mathbb{R}$.

Proposition 10.1.16. *In addition to (10.1.58), assume that the initial condition f^{in} satisfies*

$$f^{in}/\varphi \in L_2((0,\infty), y\varphi(y)dy)\ . \tag{10.1.67}$$

Then

$$\lim_{t \to \infty} \|f(t) - \varphi_s(t)\|_{[1]} = \lim_{s \to \infty} \|g(s) - \varphi\|_{[1]} = 0\ . \tag{10.1.68}$$

Proof. The proof follows similar lines to that of the LaSalle's invariance principle (Theorem 7.3.5), and begins with the compactness of the set

$$\mathcal{G} := \{g(s) - \varphi\ :\ s \geq 0\}$$

with respect to the size variable.

Step 1: Sequential weak compactness in X_1. Proposition 10.1.15 with $\Phi(r) = (r-1)^2/2$, $r \in \mathbb{R}$, and the nonnegativity of $\mathcal{D}_2 := \mathcal{D}_\Phi$ enable us to infer that, for $s \geq 0$,

$$\int_0^\infty y\varphi(y)\left(\frac{g(s,y) - \varphi(y)}{\varphi(y)}\right)^2\ dy \leq C_1 := \int_0^\infty y\varphi(y)\left(\frac{f^{in}(y) - \varphi(y)}{\varphi(y)}\right)^2\ dy\ , \tag{10.1.69}$$

where C_1 is finite due to the assumed integrability (10.1.67) of f^{in} and $\varphi \in X_1$. A first consequence of (10.1.69) and the positivity of φ in $(0,\infty)$, established in Proposition 10.1.12, is that, for any $r > 1$, there is $C_2(r) > 0$ depending only on C_1, φ, and r such that

$$\int_{1/r}^r |g(s,y) - \varphi(y)|^2\ dy \leq C_2(r)\ , \qquad s \geq 0\ . \tag{10.1.70}$$

It also follows from (10.1.69) that if E is a measurable subset of $(0,\infty)$ and $\zeta > 0$, then, for $s \geq 0$,

$$\int_E y|g(s,y) - \varphi(y)| \, dy = \int_{E \cap \{|g(s) - \varphi| \leq \zeta\varphi\}} y|g(s,y) - \varphi(y)| \, dy$$

$$+ \int_{E \cap \{|g(s) - \varphi| > \zeta\varphi\}} y|g(s,y) - \varphi(y)| \, dy$$

$$\leq \zeta \int_E y\varphi(y) \, dy + \int_E y\frac{|g(s,y) - \varphi(y)|^2}{\zeta\varphi(y)} \, dy$$

$$\leq \zeta \int_E y\varphi(y) \, dy + \frac{C_1}{\zeta} \, . \tag{10.1.71}$$

Therefore, since $\varphi \in X_1$,

$$\eta\{\mathcal{G}; X_1\} \leq \frac{C_1}{\zeta}$$

for all $\zeta > 0$, and hence

$$\eta\{\mathcal{G}; X_1\} = 0 \, . \tag{10.1.72}$$

In addition, we infer from (10.1.71), with $E = (r, \infty)$ and $r > 1$, that

$$\sup_{s \geq 0} \left\{ \int_r^\infty y|g(s,y) - \varphi(y)| \, dy \right\} \leq \zeta \int_r^\infty y\varphi(y) \, dy + \frac{C_1}{\zeta} \, .$$

Using once more that $\varphi \in X_1$, we first let $r \to \infty$ and then $\zeta \to \infty$ to conclude that

$$\lim_{r \to \infty} \sup_{s \geq 0} \left\{ \int_r^\infty y|g(s,y) - \varphi(y)| \, dy \right\} = 0 \, . \tag{10.1.73}$$

As a consequence of (10.1.72), (10.1.73) and the Dunford–Pettis theorem (Theorem 7.1.3), we deduce that

$$\mathcal{G} \text{ is relatively sequentially weakly compact in } X_1 \, . \tag{10.1.74}$$

Step 2: Time compactness. Consider $\vartheta \in W^1_\infty(0, \infty)$. It follows from (10.1.3), (10.1.57) and (10.1.63a) that, for $s \geq 0$,

$$\left| \int_0^\infty y\vartheta(y)\partial_s g(s,y) \, dy \right| \leq \left| \int_0^\infty y^2[g(s,y) - \varphi(y)]\frac{d\vartheta}{dy}(y) \, dy \right|$$

$$+ \gamma \int_0^\infty a(y_*)|g(s,y_*) - \varphi(y_*)||N_\vartheta(y_*)| \, dy_*$$

$$\leq \left\| \frac{d\vartheta}{dy} \right\|_\infty M_2(|g(s) - \varphi|)$$

$$+ \gamma \left(1 + \int_0^1 h(z) \, dz \right) \|\vartheta\|_\infty M_{1+\gamma}(|g(s) - \varphi|) \, .$$

Since

$$M_2(|g(s) - \varphi|) \leq \sqrt{M_3(\varphi)} \left(\int_0^\infty y\frac{|g(s,y) - \varphi(y)|^2}{\varphi(y)} \right)^{1/2} \leq \sqrt{C_1 M_3(\varphi)}$$

by (10.1.69) and the Cauchy–Schwarz inequality, and

$$M_{1+\gamma}(|g(s) - \varphi|) \leq M_{1+\gamma}(g(s)) + M_{1+\gamma}(\varphi) \leq \mu_{1+\gamma} + M_{1+\gamma}(\varphi)$$

by (10.1.66), we end up with

$$\left| \int_0^\infty y\vartheta(y)\partial_s g(s,y) \, dy \right| \le C_3 \|\vartheta\|_{W^1_\infty} \,, \qquad s \ge 0 \,. \tag{10.1.75}$$

Consider next $\vartheta \in L_\infty(0,\infty)$. There is a sequence of functions $(\vartheta_j)_{j\ge1}$ in $W^1_\infty(0,\infty)$ such that

$$\sup_{j\ge1} \|\vartheta_j\|_\infty \le \|\vartheta\|_\infty \quad \text{and} \quad \vartheta_j \to \vartheta \text{ a.e. in } (0,\infty) \,. \tag{10.1.76}$$

Let $\delta \in (0,1)$ and $r > 1$. According to Egorov's theorem (Theorem 7.1.11), we infer from (10.1.76) that there is $E_{\delta,r} \subset (0,r)$ such that $|E_{\delta,r}| \le \delta$ and

$$\lim_{j\to\infty} \sup_{y\in(0,r)\setminus E_{\delta,r}} \{|\vartheta_j(y) - \vartheta(y)|\} = 0 \,. \tag{10.1.77}$$

We now infer from (10.1.75) and (10.1.76) that, for $s_0 \ge 0$, $s > 0$, $r > 0$, $\delta \in (0,1)$ and $j \ge 1$,

$$\left| \int_0^\infty y\vartheta(y)[g(s+s_0,y) - g(s_0,y)] \, dy \right|$$

$$\le \left| \int_0^\infty y\vartheta_j(y)[g(s+s_0,y) - g(s_0,y)] \, dy \right|$$

$$+ \int_0^\infty y[g(s+s_0,y) + g(s_0,y)]|\vartheta_j(y) - \vartheta(y)| \, dy$$

$$\le \left| \int_{s_0}^{s+s_0} \int_0^\infty y\vartheta_j(y)\partial_{s_*} g(s_*,y) \, dy ds_* \right|$$

$$+ 2\|\vartheta\|_\infty \int_r^\infty y[g(s+s_0,y) + g(s_0,y)] \, dy$$

$$+ 2\|\vartheta\|_\infty \int_{E_{\delta,r}} y[g(s+s_0,y) + g(s_0,y)] \, dy$$

$$+ \int_{(0,r)\setminus E_{\delta,r}} y[g(s+s_0,y) + g(s_0,y)]|\vartheta_j(y) - \vartheta(y)| \, dy$$

$$\le C_3 s\|\vartheta_j\|_{W^1_\infty} + 4\|\vartheta\|_\infty \sup_{s_*\ge0} \left\{ \int_r^\infty yg(s_*,y) \, dy \right\} + 4\|\vartheta\|_\infty \eta\{\mathcal{G}, \delta; X_1\}$$

$$+ [M_1(g(s+s_0)) + M_1(g(s_0))] \sup_{y\in(0,r)\setminus E_{\delta,r}} \{|\vartheta_j(y) - \vartheta(y)|\} \,.$$

Since $M_1(g(s+s_0)) = M_1(g(s_0)) = 1$ by (10.1.64), we may let $s \to 0$ in the previous inequality and find that, for $r > 0$, $\delta \in (0,1)$ and $j \ge 1$,

$$\limsup_{s\to0} \sup_{s_0\ge0} \left\{ \left| \int_0^\infty y\vartheta(y)[g(s+s_0,y) - g(s_0,y)] \, dy \right| \right\}$$

$$\le 4\|\vartheta\|_\infty \sup_{s_*\ge0} \left\{ \int_r^\infty yg(s_*,y) \, dy \right\} + 4\|\vartheta\|_\infty \eta\{\mathcal{G}, \delta; X_1\}$$

$$+ 2 \sup_{y\in(0,r)\setminus E_{\delta,r}} \{|\vartheta_j(y) - \vartheta(y)|\} \,.$$

We now let $j \to \infty$ and deduce from (10.1.77) that for $r > 0$ and $\delta \in (0,1)$,

$$\limsup_{s\to0} \sup_{s_0\ge0} \left\{ \left| \int_0^\infty y\vartheta(y)[g(s+s_0,y) - g(s_0,y)] \, dy \right| \right\}$$

$$\leq 4\|\vartheta\|_\infty \sup_{s_* \geq 0} \left\{ \int_r^\infty y g(s_*,y) \, dy \right\} + 4\|\vartheta\|_\infty \eta\{\mathcal{G}, \delta; X_1\} \;. \tag{10.1.78}$$

Now,

$$\lim_{\delta \to 0} \eta\{\mathcal{G}, \delta; X_1\} = \eta\{\mathcal{G}; X_1\} = 0$$

by (10.1.72), while (10.1.73) and $\varphi \in X_1$ imply that

$$\limsup_{r \to \infty} \sup_{s_* \geq 0} \left\{ \int_r^\infty y g(s_*,y) \, dy \right\} \leq \lim_{r \to \infty} \sup_{s_* \geq 0} \left\{ \int_r^\infty y |g(s_*,y) - \varphi(y)| \, dy \right\}$$

$$+ \lim_{r \to \infty} \int_r^\infty y\varphi(y) \, dy = 0 \;.$$

Therefore, the right-hand side of (10.1.78) converges to zero as $\delta \to 0$ and $r \to \infty$, hence

$$\lim_{s \to 0} \sup_{s_0 \geq 0} \left\{ \left| \int_0^\infty y\vartheta(y)[g(s + s_0, y) - g(s_0, y)] \, dy \right| \right\} = 0 \;. \tag{10.1.79}$$

Step 3: Convergence. Consider now a sequence $(s_j)_{j \geq 1}$ of positive real numbers in $(1, \infty)$ such that $s_j \to \infty$ as $j \to \infty$ and define

$$g_j(s,y) := g(s_j + s, y) \;, \qquad (s,y) \in (-1,1) \times (0,\infty) \;.$$

According to (10.1.74), $(g_j(s))_{j \geq 1}$ is relatively sequentially weakly compact in X_1 for all $s \in (-1,1)$, while (10.1.79) implies that, for all $s_0 \in (-1,1)$ and $\vartheta \in L_\infty(0,\infty)$,

$$\lim_{s \to 0} \sup_{j \geq 1} \left\{ \left| \int_0^\infty y\vartheta(y)[g_j(s + s_0, y) - g_j(s_0, y)] \, dy \right| \right\}$$

$$= \lim_{s \to 0} \sup_{j \geq 1} \left\{ \left| \int_0^\infty y\vartheta(y)[g(s + s_j + s_0, y) - g(s_j + s_0, y)] \, dy \right| \right\} = 0 \;.$$

Consequently, we can apply the variant of Arzelà–Ascoli's theorem recalled in Theorem 7.1.16 to conclude that the sequence $(g_j)_{j \geq 1}$ is compact in $C([-1,1], X_{1,w})$. Therefore, there is $g_\infty \in C([-1,1], X_{1,w})$, and a subsequence of $(g_j)_{j \geq 1}$ (not relabelled) such that

$$g_j \longrightarrow g_\infty \quad \text{in } C([-1,1], X_{1,w}) \;. \tag{10.1.80}$$

In particular, recalling (10.1.64),

$$M_1(g_\infty(s)) = 1 \;, \qquad s \in [-1,1] \;. \tag{10.1.81}$$

Furthermore, after possibly extracting a further subsequence, we infer from (10.1.70) that, for all $r \in (1, \infty)$,

$$g_j \rightharpoonup g_\infty \quad \text{in } L_2((-1,1) \times (1/r, r)) \;. \tag{10.1.82}$$

Step 4: Identification of g_∞. We now exploit information from the dissipation term $\mathscr{D}_2(g, \varphi)$, recalling that $\mathscr{D}_2 = \mathscr{D}_\Phi$ for $\Phi(r) = (r-1)^2/2$, $r \in \mathbb{R}$. By Proposition 10.1.15 and (10.1.69),

$$\mathscr{D}_2[g(s), \varphi] = \frac{1}{2} \int_0^\infty a(y_*)\varphi(y_*) \int_0^{y_*} y b(y, y_*) \left[\frac{g(s,y)}{\varphi(y)} - \frac{g(s,y_*)}{\varphi(y_*)} \right]^2 dy\,dy_* \tag{10.1.83}$$

for $s \geq 0$, and

$$\int_0^\infty \mathscr{D}_2[g(s), \varphi] \, ds \leq \frac{C_1}{\gamma} \;.$$

On the one hand, it readily follows from the previous integrability property and the definition of g_j that

$$\lim_{j\to\infty} \int_{-1}^{1} \mathcal{D}_2[g_j(s), \varphi]\, ds = \lim_{j\to\infty} \int_{-1+s_j}^{1+s_j} \mathcal{D}_2[g(s), \varphi]\, ds = 0 \ . \tag{10.1.84}$$

On the other hand, setting

$$H_j(s, y, y_*) := \frac{g_j(s, y)}{\varphi(y)} - \frac{g_j(s, y_*)}{\varphi(y_*)} \ , \qquad H_\infty(s, y, y_*) := \frac{g_\infty(s, y)}{\varphi(y)} - \frac{g_\infty(s, y_*)}{\varphi(y_*)}$$

for $(s, y, y_*) \in (-1, 1) \times (0, \infty)^2$ and $j \geq 1$, and

$$\tau(r, z) := z \min\{r, h(z)\} \ , \qquad (r, z) \in (1, \infty) \times (0, 1) \ , \tag{10.1.85}$$

we infer from (10.1.82) and the positivity of φ, established in Proposition 10.1.12, that

$$H_j \rightharpoonup H_\infty \quad \text{in} \quad L_2((-1, 1) \times (1/r, r)^2)$$

for all $r > 1$. Therefore, for all $r > 1$,

$$0 \leq \int_{-1}^{1} \int_{1/r}^{r} a(y_*)\varphi(y_*) \int_{1/r}^{y_*} \tau\left(r, \frac{y}{y_*}\right) H_\infty^2(s, y, y_*)\, dy\, dy_*\, ds$$

$$\leq \liminf_{j\to\infty} \int_{-1}^{1} \int_{1/r}^{r} a(y_*)\varphi(y_*) \int_{1/r}^{y_*} \tau\left(r, \frac{y}{y_*}\right) H_j^2(s, y, y_*)\, dy\, dy_*\, ds$$

$$\leq 2 \liminf_{j\to\infty} \int_{-1}^{1} \mathcal{D}_2[g_j(s), \varphi]\, ds \ .$$

Combining (10.1.84) with the above inequality gives

$$a(y_*)\varphi(y_*) \int_{1/r}^{y_*} \tau\left(r, \frac{y}{y_*}\right) H_\infty^2(s, y, y_*)\, dy = 0$$

for almost every $(s, y_*) \in (-1, 1) \times (1/r, r)$. This property being valid for all $r > 1$, we end up with

$$a(y_*)\varphi(y_*)yb(y, y_*)H_\infty^2(s, y, y_*)\mathbf{1}_{(0,y_*)}(y) = 0 \quad \text{a.e. in} \quad (-1, 1) \times (0, \infty)^2 \ .$$

The positivity of a, b and φ, along with the symmetry of H_∞ with respect to the (rescaled) size variables, further implies that

$$H_\infty(s, y, y_*) = 0 \quad \text{a.e. in} \quad (-1, 1) \times (0, \infty)^2 \ .$$

Recalling the time continuity of g_∞, we finally obtain

$$\frac{g_\infty(s, y)}{\varphi(y)} = \frac{g_\infty(s, y_*)}{\varphi(y_*)} \quad \text{a.e. in} \quad (0, \infty)^2$$

for all $s \in [-1, 1]$. Consequently, for each $s \in [-1, 1]$, there is a constant $c(s)$ such that $g_\infty(s, y) = c(s)\varphi(y)$ for a.e. $y \in (0, \infty)$. But, since $M_1(g_\infty(s)) = 1 = M_1(\varphi)$ by (10.1.81), we readily conclude that $c(s) = 1$ for all $s \in [-1, 1]$, so that $g_\infty(s) = \varphi$ for all $s \in [-1, 1]$. Recalling (10.1.80), we have shown that $g(s_j) \rightharpoonup \varphi$ in X_1 as $j \to \infty$ and that zero is the only cluster point as $s \to \infty$ of $\mathcal{G}$ for the sequential weak topology of X_1. Owing to the relative sequential weak compactness (10.1.74) of $\mathcal{G}$ in X_1, we conclude that

$$g(s) \rightharpoonup \varphi \quad \text{in } X_1 \text{ as } s \to \infty . \tag{10.1.86}$$

Step 5: Strong convergence. We are left with strengthening the weak convergence (10.1.86). To this end we exploit further the dissipation functional $\mathcal{D}_2$ given by (10.1.83). Indeed, recalling (10.1.3a) and the definition (10.1.85) of the truncated version τ of $z \mapsto zh(z)$, we infer from (10.1.83) and the inequality

$$(\zeta - \zeta_*)^2 = (\zeta - 1)^2 + (\zeta_* - 1)^2 - 2(\zeta - 1)(\zeta_* - 1)$$
$$\geq (\zeta_* - 1)^2 - 2(\zeta - 1)(\zeta_* - 1) , \qquad (\zeta, \zeta_*) \in \mathbb{R}^2 ,$$

that, for $j \geq 1$ and $r > 1$,

$$2 \int_{-1}^{1} \mathcal{D}_2[g_j(s), \varphi] \, \mathrm{d}s$$

$$\geq \int_{-1}^{1} \int_{1/r}^{r} \int_{1/r^2}^{y_*} a(y_*)\varphi(y_*)\tau\left(r, \frac{y}{y_*}\right) \left[\frac{g_j(s, y)}{\varphi(y)} - \frac{g_j(s, y_*)}{\varphi(y_*)}\right]^2 \mathrm{d}y\mathrm{d}y_*\mathrm{d}s$$

$$\geq \frac{1}{r^\gamma} \int_{-1}^{1} \int_{1/r}^{r} \int_{1/r^2}^{y_*} \varphi(y_*)\tau\left(r, \frac{y}{y_*}\right) \left[\frac{g_j(s, y)}{\varphi(y)} - \frac{g_j(s, y_*)}{\varphi(y_*)}\right]^2 \mathrm{d}y\mathrm{d}y_*\mathrm{d}s$$

$$\geq \frac{1}{r^\gamma} \int_{-1}^{1} \int_{1/r}^{r} \int_{1/r^2}^{y_*} \varphi(y_*)\tau\left(r, \frac{y}{y_*}\right) \left[\frac{g_j(s, y_*)}{\varphi(y_*)} - 1\right]^2 \mathrm{d}y\mathrm{d}y_*\mathrm{d}s - \frac{2S_j(r)}{r^\gamma} ,$$

where

$$S_j(r) := \int_{-1}^{1} \int_{1/r}^{r} \int_{1/r^2}^{y_*} \frac{1}{\varphi(y)}\tau\left(r, \frac{y}{y_*}\right) [g_j(s, y) - \varphi(y)] [g_j(s, y_*) - \varphi(y_*)] \, \mathrm{d}y\mathrm{d}y_*\mathrm{d}s .$$

A further change of variables gives

$$2r^\gamma \int_{-1}^{1} \mathcal{D}_2[g_j(s), \varphi] \, \mathrm{d}s$$

$$\geq \int_{-1}^{1} \int_{1/r}^{r} \frac{y_*}{\varphi(y_*)} [g_j(s, y_*) - \varphi(y_*)]^2 \int_{1/(y_*r^2)}^{1} \tau(r, z) \, \mathrm{d}z\mathrm{d}y_*\mathrm{d}s - 2S_j(r)$$

$$\geq \int_{-1}^{1} \int_{1/r}^{r} \frac{y_*}{\varphi(y_*)} [g_j(s, y_*) - \varphi(y_*)]^2 \int_{1/r}^{1} \tau(r, z) \, \mathrm{d}z\mathrm{d}y_*\mathrm{d}s - 2S_j(r) .$$

Thanks to the positivity of φ, it follows from the weak convergences (10.1.82) and (10.1.86) that

$$\lim_{j \to \infty} S_j(r) = 0 .$$

Together with (10.1.84) and the previous inequality, this implies

$$\lim_{j \to \infty} \int_{-1}^{1} \int_{1/r}^{r} \frac{y_*}{\varphi(y_*)} [g_j(s, y_*) - \varphi(y_*)]^2 \int_{1/r}^{1} \tau(r, z) \, \mathrm{d}z\mathrm{d}y_*\mathrm{d}s = 0 \tag{10.1.87}$$

for all $r > 1$. Now, since

$$\lim_{r \to \infty} \int_{1/r}^{1} \tau(r, z) \, \mathrm{d}z = \int_{0}^{1} zh(z) \, \mathrm{d}z = 1$$

by (10.1.3b), there exists $r_1 > 1$ such that

$$\int_{1/r}^1 \tau(r, z) \, \mathrm{d}z \geq \frac{1}{2} , \qquad r \geq r_1 .$$

We then deduce from (10.1.87) that, for $r \geq r_1$,

$$\lim_{j \to \infty} \int_{-1}^1 \int_{1/r}^r \frac{y_*}{\varphi(y_*)} [g_j(s, y_*) - \varphi(y_*)]^2 \, \mathrm{d}y_* \mathrm{d}s = 0 ,$$

hence, using again the positivity of φ,

$$g_j \longrightarrow \varphi \quad \text{in} \quad L_2((-1, 1) \times (1/r, r)) \tag{10.1.88}$$

for all $r \geq r_1$. To finish the proof, we use the contraction property (10.1.65) and the Cauchy–Schwarz inequality to obtain, for $r \geq r_1$,

$$
\begin{aligned}
\|g(s_j) - \varphi\|_{[1]} &\leq \int_{-1}^0 \|g_j(s) - \varphi\|_{[1]} \, \mathrm{d}s \\
&\leq \int_{-1}^0 \int_0^{1/r} y |g_j(s, y) - \varphi(y)| \, \mathrm{d}y \mathrm{d}s \\
&\quad + r \left(\int_{-1}^0 \int_{1/r}^r |g_j(s, y) - \varphi(y)|^2 \, \mathrm{d}y \mathrm{d}s \right)^{1/2} \\
&\quad + \int_{-1}^0 \int_r^\infty y |g_j(s, y) - \varphi(y)| \, \mathrm{d}y \mathrm{d}s \\
&\leq \eta\{\mathcal{G}, 1/r; X_1\} + \sup_{s \geq 0} \left\{ \int_r^\infty y |g(s, y) - \varphi(y)| \, \mathrm{d}y \right\} \\
&\quad + r \left(\int_{-1}^0 \int_{1/r}^r |g_j(s, y) - \varphi(y)|^2 \, \mathrm{d}y \mathrm{d}s \right)^{1/2} .
\end{aligned}
$$

Owing to (10.1.88), we may pass to the limit as $j \to \infty$ in the above inequality to obtain

$$\limsup_{j \to \infty} \|g(s_j) - \varphi\|_{[1]} \leq \eta\{\mathcal{G}, 1/r; X_1\} + \sup_{s \geq 0} \left\{ \int_r^\infty y |g(s, y) - \varphi(y)| \, \mathrm{d}y \right\} .$$

We then use (10.1.72) and (10.1.73) to let $r \to \infty$ and conclude that $(g(s_j))_{j \geq 1}$ converges to φ in X_1. This convergence implies (10.1.68) by (10.1.62) and the contraction property (10.1.65). □

We are now in a position to complete the proof of Theorem 10.1.14 which follows from the contraction property stated in Proposition 8.2.57 and Proposition 10.1.16 by a density argument.

Proof of Theorem 10.1.14. Consider $f^{in} \in X_{m_0,+} \cap X_{1+\gamma}$ and a sequence $(f_l^{in})_{l \geq 1}$ in $X_{m_0,+} \cap X_{1+\gamma}$ such that

$$M_1(f_l^{in}) = 1 , \qquad \frac{f_l^{in}}{\varphi} \in L_2((0, \infty), y\varphi(y)\mathrm{d}y) , \qquad l \geq 1 , \tag{10.1.89}$$

and
$$\lim_{l\to\infty}\|f_l^{in}-f^{in}\|_{[1]}=0\;. \tag{10.1.90}$$

For $l\geq 1$, let f_l be the weak solution to the fragmentation equation with initial condition f_l^{in} given by Theorem 8.2.23 ($m_0=0$), or Theorem 8.2.30 ($m_0\in(0,1)$) and Proposition 8.2.57. Then, for $l\geq 1$ and $t\geq 0$, it follows from Proposition 8.2.57 that

$$\|f(t)-\varphi_s(t)\|_{[1]}\leq\|f(t)-f_l(t)\|_{[1]}+\|f_l(t)-\varphi_s(t)\|_{[1]}$$
$$\leq\|f^{in}-f_l^{in}\|_{[1]}+\|f_l(t)-\varphi_s(t)\|_{[1]}\;.$$

Owing to (10.1.89), we are in a position to apply Proposition 10.1.16 to pass to the limit as $t\to\infty$ and obtain

$$\limsup_{t\to\infty}\|f(t)-\varphi_s(t)\|_{[1]}\leq\|f^{in}-f_l^{in}\|_{[1]}\;.$$

The previous inequality being valid for all $l\geq 1$, we let $l\to\infty$ to complete the proof, taking (10.1.90) into account. $\qquad\square$

Remark 10.1.17. *Proposition 10.1.16 (and hence also Theorem 10.1.14) is valid under weaker positivity assumptions on h, see [202, Theorem 3.2]. We also refer to [43, 116] for a different proof, based on a stochastic approach.*

10.1.2.2   Decay Rates

Having elucidated the question of convergence to self-similarity for the fragmentation equation (10.1.1) in the previous section, the next issue is to figure out whether a rate of convergence can be identified. It possibly depends not only on the overall fragmentation rate a and on the daughter distribution function b, but also on specific features of the initial condition f^{in}. Here again, it turns out to be more convenient to work with equation (10.1.63) in self-similar variables. The aim is then to find a Banach space Y and positive real numbers $\zeta>0$ and $C\geq 1$ such that the solution g to (10.1.63) satisfies

$$\|g(s)-\varphi\|_Y\leq Ce^{-\zeta s}\|f^{in}-\varphi\|_Y\;,\qquad s\in(0,\infty)\;, \tag{10.1.91}$$

under suitable conditions on f^{in}, including $f^{in}\in X_1$ with $M_1(f^{in})=1$. As far as we know, the most general result is obtained in [204] by a semigroup approach. More precisely, consider $m_0\in[0,1)$, $m>1$, and $h\in C([0,1])\cap W_1^1(0,1)$. Then there is $\zeta_*>0$ such that (10.1.91) holds true with $Y=X_{m_0}\cap X_m$ for $\zeta\in(0,\zeta_*)$ and $C=C(\zeta)$, the initial condition being in Y.

A different approach is used in [12, 56, 57] where the validity of (10.1.91) is shown in the Hilbert space $Y=L_2((0,\infty),y\varphi^{-1}(y)dy)$ (with $\varphi^{-1}=1/\varphi$), when $\gamma\in(0,2)$ in (10.1.3a) and both h and $1/h$ belong to $L_\infty(0,1)$. In that case, $C=1$ and $\zeta=\zeta_2$ is given by

$$2\zeta_2:=\inf_{\xi\in Y}\left\{\frac{\mathcal{D}_2[\xi,\varphi]}{\mathcal{H}_2[\xi,\varphi]}\right\}\;,$$

where $\mathcal{H}_2=\mathcal{H}_\Phi$ and $\mathcal{D}_2=\mathcal{D}_\Phi$ are as defined in Proposition 10.1.15 with, once again, $\Phi(r)=(r-1)^2/2$, $r\in\mathbb{R}$. The cornerstone of the proof is the positivity of ζ_2. Once the decay in $L_2((0,\infty),y\varphi^{-1}(y)dy)$ is established, further decay rates may be derived and, under the same assumptions, (10.1.91) is also valid in $Y=L_2((0,\infty),(y+y^m)dy)$ for any $\zeta\in(0,\zeta_2)$ and with $C=C(\zeta)$ as soon as $m\geq 3$ is large enough. In the same vein, in the particular case $h\equiv 2$, it is shown in [122] that (10.1.91) holds true in $Y=L_2((0,\infty),ydy)$ with $\zeta=C=1$ for $\gamma\geq 2$ in (10.1.3a).

10.2 Coagulation

As already pointed out, in the absence of fragmentation ($a \equiv 0$), the driving mechanism of the dynamics of the coagulation equation

$$\partial_t f(t,x) = \frac{1}{2} \int_0^x k(x_*, x - x_*) f(t, x_*) f(t, x - x_*) \, \mathrm{d}x_*$$
$$- \int_x^\infty k(x, x_*) f(t,x) f(t,x_*) \, \mathrm{d}x_* , \qquad (t,x) \in (0,\infty)^2 , \qquad (10.2.1)$$

is merging of particles, which leads not only to an increase of the mean particle size but also to a decrease of the total number of particles as time goes by. A sufficient condition guaranteeing such a decay requires some positivity on the coagulation kernel and is given in the next result.

Proposition 10.2.1. *Assume that the coagulation kernel k is a nonnegative and symmetric measurable function on $(0,\infty)^2$ possessing the positivity property: for any $r_2 > r_1 > 0$, there is $\delta(r_1, r_2) > 0$ such that*

$$k(x,y) \geq \delta(r_1, r_2) , \qquad (x,y) \in (r_1, r_2)^2 . \qquad (10.2.2)$$

Consider $f^{in} \in X_{0,1,+}$ and let f be a weak solution to the coagulation equation (10.2.1) on $[0,\infty)$ with initial condition f^{in} satisfying

$$M_1(f(t)) \leq M_1(f^{in}) , \qquad t \geq 0 . \qquad (10.2.3)$$

Then

$$\lim_{t\to\infty} M_0(f(t)) = 0 .$$

Proof. For $t \geq 0$ and $R \in (0,\infty)$, we set

$$I(t,R) := \int_0^R f(t,x) \, \mathrm{d}x .$$

Given $t_2 \geq t_1 \geq 0$, the boundedness of $\mathbf{1}_{(0,R)}$ allows one to take $\vartheta = \mathbf{1}_{(0,R)}$ in (8.2.4d) and obtain

$$I(t_2, R) - I(t_1, R) \leq -\frac{1}{2} \int_{t_1}^{t_2} \int_0^R \int_0^\infty k(x,y) f(s,x) f(s,y) \, \mathrm{d}y\mathrm{d}x\mathrm{d}s . \qquad (10.2.4)$$

A first consequence of (10.2.4) is that $t \mapsto I(t,R)$ is a non-increasing function of time, and it is obviously bounded from below by zero. Therefore, there is $I_\infty(R) \geq 0$ such that

$$\lim_{t\to\infty} I(t,R) = I_\infty(R) , \qquad R \in (0,\infty) . \qquad (10.2.5)$$

We next infer from (10.2.2) and (10.2.4) that, for $R \in (0,\infty)$, $r \in (0,R)$ and $t > 0$,

$$\int_0^t [I(s,R) - I(s,r)]^2 \, \mathrm{d}s = \int_0^t \left(\int_r^R f(s,x) \, \mathrm{d}x \right)^2 \mathrm{d}s$$
$$\leq \frac{1}{\delta(r,R)} \int_0^t \int_r^R \int_r^R k(x,y) f(s,x) f(s,y) \, \mathrm{d}y\mathrm{d}x\mathrm{d}s$$

$$\leq \frac{2}{\delta(r,R)} I(0,R) \ ,$$

hence

$$\int_0^\infty [I(s,R) - I(s,r)]^2 \ \mathrm{d}s \leq \frac{2}{\delta(r,R)} M_0(f^{in}) \ . \tag{10.2.6}$$

Since $I(t,R) - I(t,r)$ converges to $I_\infty(R) - I_\infty(r)$ as $t \to \infty$ by (10.2.5), the time integrability property (10.2.6) implies that $I_\infty(R) = I_\infty(r)$. Consequently, recalling the time monotonicity of $t \mapsto I(t,r)$, we have shown that

$$0 \leq I_\infty(R) = I_\infty(r) \leq I(0,r) = \int_0^r f^{in}(x) \ \mathrm{d}x \ , \qquad r \in (0,R) \ .$$

Since $f^{in} \in X_0$, we may let $r \to 0$ in the previous inequality and conclude that $I_\infty(R) = 0$ for all $R \in (0,\infty)$. Thus (10.2.5) actually means

$$\lim_{t\to\infty} I(t,R) = 0 \ , \qquad R \in (0,\infty) \ . \tag{10.2.7}$$

Finally, owing to (10.2.3),

$$M_0(f(t)) \leq I(t,r) + \frac{1}{r} \int_r^\infty x f(t,x) \ \mathrm{d}x \leq I(t,r) + \frac{M_1(f^{in})}{r} \ , \qquad r \in (0,\infty) \ .$$

Letting first $t \to \infty$, using (10.2.7) and then letting $r \to \infty$ in the above inequality completes the proof. □

The just established decay to zero of the total number of particles for large times does not provide much information on the time evolution of solutions to the coagulation equation (10.2.1), and looking at finer scales seems necessary. In this connection, the dynamical scaling hypothesis (10.0.1)

$$f(t,x) \sim \frac{1}{\sigma(t)^\tau} \varphi\left(\frac{x}{\sigma(t)}\right) \qquad \text{as} \quad t \to \infty \quad \text{or} \quad t \to T_{gel} \ ,$$

might give some clues about this issue when the coagulation kernel is assumed to be homogeneous; that is, there is $\lambda \in \mathbb{R}$ such that

$$k(\xi x, \xi y) = \xi^\lambda k(x,y) \ , \qquad (\xi, x, y) \in (0,\infty)^3 \ . \tag{10.2.8}$$

This, however, first requires the identification of the involved scales, namely, the mean size $\sigma(t)$ at time t and the scaling exponent $\tau \in \mathbb{R}$, as well as that of the separating constant w and the scaling profile φ, see (10.0.5) and (10.0.6). Since the mean size σ is fully determined by the exponent τ and the separating constant w according to (10.0.7), let us now focus on the parameter τ and the scaling profile φ, the separating constant w playing an important role only for $\lambda = 1$, see (10.0.8). Due to the gelation phenomenon, there is a threshold value of the parameter λ as gelation is only expected to take place for $\lambda > 1$. Therefore, (10.0.7c) is, in principle, excluded for $\lambda \leq 1$. Since $\lambda = 1$ appears to be the borderline case, we split the discussion according to whether $\lambda \in (-\infty, 1)$, $\lambda = 1$, or $\lambda > 1$.

• $\lambda \in (-\infty, 1)$, $\tau = 2$. When $\lambda \in (-\infty, 1)$, no gelation is expected to occur and the starting point is to look for mass-conserving self-similar solutions, which corresponds to $\tau = 2$ and $\varphi \in X_1$. As $2 \geq \lambda + 1$ in this case, the mean size σ is given by (10.0.7a) and, after a suitable rescaling based on (10.0.8), we look for a self-similar profile φ satisfying

$$\varphi \in X_{1,+} \ , \qquad M_1(\varphi) = 1 \ , \tag{10.2.9}$$

and solving in a suitable sense

$$x\frac{d\varphi}{dx}(x) + 2\varphi(x) + C\varphi(x) = 0 , \qquad x \in (0, \infty) . \tag{10.2.10}$$

Owing to the complexity of the nonlinear and nonlocal integro-partial differential equation (10.2.10), the existence of self-similar profiles satisfying (10.2.9) and (10.2.10) is far from being straightforward and, for a long time, a positive answer was only known for the constant coagulation kernel $k \equiv 2$. Indeed, for $k \equiv 2$, there is an explicit solution to (10.2.10) satisfying (10.2.9) given by $\varphi(x) = e^{-x}$, $x \in (0, \infty)$, derived in [240], and it is the only one, see [198]. Numerical simulations are also performed in [121] for Smoluchowski's coagulation kernel $k(x, y) = (x^{1/3} + y^{1/3})(x^{-1/3} + y^{-1/3})$ and provide evidence of the existence of a self-similar profile φ in this case as well. Also, formal asymptotics reported in [121] indicate that φ decays exponentially fast as $x \to \infty$ and vanishes very rapidly as $x \to 0$, namely,

$$\varphi(x) \sim C_0 x^{-1.06} e^{-C_1 x^{-1/3}} \quad \text{as} \quad x \to 0 ,$$

the positive constants C_0 and C_1 being computed numerically. It is actually possible to use formal asymptotic expansions to derive a rather complete classification of the behaviours as $x \to \infty$ and $x \to 0$ of solutions to (10.2.10) satisfying (10.2.9) (if any!). Such a study is successfully undertaken in [263], thereby providing further insight into the complexity of the analysis of (10.2.10). To this end, an alternative formulation of (10.2.10) is derived in [263], which is somewhat related to the mass conservation and is in the same spirit as that reported in Proposition 10.1.2 for the fragmentation equation. It is best seen by multiplying (10.2.10) by $x\vartheta(x)$, where ϑ is an arbitrary smooth test function, and integrating with respect to x over $(0, \infty)$. Formally, this leads to

$$\int_0^\infty \vartheta(x)\partial_x \left(x^2\varphi(x)\right) \, \mathrm{d}x = -\int_0^\infty \int_0^\infty xk(x, y) \left[\vartheta(x + y) - \vartheta(x)\right]\varphi(x)\varphi(y) \, \mathrm{d}y\mathrm{d}x \tag{10.2.11}$$

$$= -\int_0^\infty \int_0^\infty \int_x^{x+y} xk(x, y)\varphi(x)\varphi(y)\frac{d\vartheta}{dz}(z) \, \mathrm{d}z\mathrm{d}y\mathrm{d}x .$$

Fubini's theorem then gives

$$\int_0^\infty z^2\varphi(z)\frac{d\vartheta}{dz}(z) \, \mathrm{d}z = \int_0^\infty \frac{d\vartheta}{dz}(z)\int_0^z \int_{z-x}^\infty xk(x, y)\varphi(x)\varphi(y) \, \mathrm{d}y\mathrm{d}x\mathrm{d}z .$$

Hence, using the properties of the derivative in the sense of distributions and assuming that $z^2\varphi(z) \to 0$ as $z \to \infty$, a property which complies with (10.2.9), to eliminate the arbitrary constant, we obtain

$$z^2\varphi(z) = \int_0^z \int_{z-x}^\infty xk(x, y)\varphi(x)\varphi(y) \, \mathrm{d}y\mathrm{d}x , \qquad z \in (0, \infty) . \tag{10.2.12}$$

As for the fragmentation equation, the alternative formula (10.2.12) does not seem to be helpful in the construction of self-similar profiles. It nevertheless proves useful in identifying, by formal asymptotics, the behaviour of φ for large and small sizes and we now summarise the outcome of [263], see also the survey article [180]. The analysis performed in [263] shows that, besides the homogeneity degree λ, the behaviour of the coagulation kernel $k(x, y)$ as $y \to 0$ or $y \to \infty$ for fixed $x \in (0, \infty)$ also plays an important role. This leads to the following definition.

Definition 10.2.2. *[263] Let k be a nonnegative and symmetric function on $(0, \infty)^2$, which is homogeneous with degree $\lambda \in \mathbb{R}$, see (10.2.8). Define*

$$l(z) := k(1, z) , \qquad z \in (0, \infty) , \tag{10.2.13a}$$

and assume that there is $\alpha \in [\lambda - 1, \infty)$, and $l_0 > 0$ such that

$$l(z) \sim l_0 z^\alpha \quad as \quad z \to 0 . \tag{10.2.13b}$$

The coagulation kernel k is said to belong to class I, class II, or class III if the corresponding function l defined in (10.2.13a) satisfies (10.2.13b) with $\alpha > 0$, $\alpha = 0$, or $\alpha < 0$, respectively.

A connection between the behaviour of l as $z \to 0$ and as $z \to \infty$ follows from the symmetry and homogeneity (10.2.8) of k, which guarantee that

$$l(z) = z^\lambda l(1/z) , \qquad z \in (0, \infty) . \tag{10.2.14}$$

Together with (10.2.13b), the identity (10.2.14) implies that

$$l(z) \sim l_0 z^{\lambda - \alpha} \quad as \quad z \to \infty . \tag{10.2.15}$$

Since it is physically relevant to assume that l grows at most linearly at infinity, the requirement $\alpha \geq \lambda - 1$ complies with this constraint.

To illustrate Definition 10.2.2, the product coagulation kernel $(xy)^{\lambda/2}$ belongs to class I for $\lambda > 0$, and to class III for $\lambda < 0$ (with $\alpha = \lambda/2$ in both cases), while the sum kernel $k(x, y) = x^\lambda + y^\lambda$ belongs to class II for $\lambda > 0$, and to class III for $\lambda < 0$.

Concerning the large size behaviour, self-similar profiles φ are expected to decay exponentially fast as $x \to \infty$ [263]. More precisely, there is $A_\infty > 0$, and $\delta > 0$ such that

$$\varphi(x) \sim A_\infty x^{-\lambda} e^{-\delta x} \quad as \quad x \to \infty , \tag{10.2.16}$$

except when $\alpha = \lambda - 1$ for which the situation is more complicated according to [261, Section 5]. Though exponential decay is not excluded in [261, Section 5] when $\alpha = \lambda - 1$, it is recently shown in [45] that, for $\rho > 0$ small enough, there are self-similar profiles $\varphi \in X_1$ decaying algebraically for large sizes, namely, $\varphi(x) \sim C(\rho) x^{-2-\rho}$ as $x \to \infty$. It is thus likely that there is coexistence of mass-conserving self-similar profiles with algebraic and exponential decay for large sizes when $\alpha = \lambda - 1$, a typical kernel belonging to this class being $k(x, y) = x^{\lambda-1} y + x y^{\lambda-1}$, $(x, y) \in (0, \infty)^2$. A similar feature is already known for the additive coagulation kernel k_+, see [42, 198] and the paragraph on the case $\lambda = 1$ below.

The small size behaviour is sensitive to the sign of α, according to formal asymptotic expansions computed in [263]: there is $A_0 > 0$ such that, as $x \to 0$,

$$\varphi(x) \sim A_0 \frac{M_{\lambda-\alpha}(\varphi)}{x^2} \exp\left(\frac{M_{\lambda-\alpha}(\varphi)}{\alpha} x^\alpha \right) \qquad if \quad \alpha < 0 , \tag{10.2.17a}$$

$$\varphi(x) \sim A_0 x^{-2+M_\lambda(\varphi)} \qquad if \quad \alpha = 0 , \tag{10.2.17b}$$

$$\varphi(x) \sim A_0 x^{-1-\lambda} \qquad if \quad \alpha > 0 . \tag{10.2.17c}$$

The constant A_0 is *a priori* not known in general, except when $\alpha > 0$. In that case, it is given by

$$\frac{1}{A_0} := \int_0^1 \int_{1-x}^\infty x k(x, y) (xy)^{-1-\lambda} \, dy dx . \tag{10.2.17d}$$

The outcome of the previous formal computations is that mass-conserving self-similar profiles are expected to feature a singularity as $z \to 0$ when $\alpha \geq 0$, and to vanish rapidly as $z \to 0$ when $\alpha < 0$. The former is actually a difficulty that has to be overcome in the existence proof. An additional problem stems from the fact that, when $\alpha > 0$, the function $x \mapsto A_0 x^{-1-\lambda}$ is an explicit solution to (10.2.12), but with infinite total mass, so that it does not comply with (10.2.9). Nevertheless, existence of self-similar profiles satisfying (10.2.9) and (10.2.10) is proved in [113, 118] for a large class of coagulation kernels; a detailed proof is given in Section 10.2.4. Whether these solutions are regular and satisfy (10.2.16) and (10.2.17) is investigated in [63, 85, 119, 193, 215, 219]. We shall come back to this issue in Section 10.2.5. Uniqueness of self-similar profiles is still a widely open problem, except for small perturbations of the constant kernel [214, 220], for product kernels $(xy)^{\lambda/2}$ with $\lambda < 0$ [163], and for sum kernels [63], the latter requiring a prior knowledge of $M_\lambda(\varphi)$. We finally mention that the stability of self-similar profiles is also a widely open question, except for the constant kernel [90, 151, 198], with numerical evidence of convergence to self-similarity provided in [115, 121, 152, 178].

• $\lambda \in (-\infty, 1)$, $\tau \neq 2$. The existence of self-similar solutions with a scaling exponent $\tau \neq 2$ was only considered recently, after the pioneering works [42, 198] dealing with the additive coagulation kernel k_+ and the constant kernel $k \equiv 2$. While the former is discussed later, as it has homogeneity of degree one, the outcome for the latter is that there are also self-similar solutions with scaling exponent $\tau \in (1, 2)$, their profiles having infinite mass. More precisely, given $\theta \in (0, 1)$, there is a self-similar solution to the coagulation equation with $k \equiv 2$ of the form

$$(t, x) \longmapsto (1 + t)^{-(1+\theta)/\theta} \varphi_\theta(x(1 + t)^{-1/\theta}) ,$$

the self-similar profile $\varphi_\theta = dF_\theta/dx$ being the derivative of the Mittag–Leffler distribution

$$F_\theta(x) := \sum_{j=1}^{\infty} \frac{(-1)^{j+1} x^{j\theta}}{\Gamma(1 + j\theta)} , \qquad x \in (0, \infty) .$$

A thorough study of these solutions and their basins of attraction is performed in [198] and a rough account of the results obtained therein is provided in Section 10.2.1. The discovery of these odd self-similar solutions, which were unexpected from the physical point of view, gave impetus to further research in that direction and a series of papers by Niethammer and Velázquez shows that the existence of self-similar solutions with a scaling exponent $\tau \neq 2$ is not a peculiarity of the constant kernel but rather a generic feature [213, 216, 217, 218]. More precisely, when $\tau \neq 2$, the self-similar profile φ is a nonnegative function solving, after a suitable rescaling based on (10.0.8),

$$x \frac{d\varphi}{dx}(x) + \tau\varphi(x) + C\varphi(x) = 0 , \qquad x \in (0, \infty) , \qquad (10.2.18)$$

instead of (10.2.10). As in the case $\tau = 2$, an alternative formulation of (10.2.18) can be derived and is given by

$$z^2\varphi(z) + (\tau - 2) \int_0^z x\varphi(x) \, \mathrm{d}x = \int_0^z \int_{z-x}^{\infty} xk(x, y)\varphi(x)\varphi(y) \, \mathrm{d}y\mathrm{d}x \qquad (10.2.19)$$

for $z \in (0, \infty)$.

A first consequence of (10.2.18) is that $\varphi \notin X_1$. Indeed, multiplying (10.2.18) by x, and then integrating over $(0, \infty)$, gives, at least formally, $(\tau - 2)M_1(\varphi) = 0$, thereby precluding the existence of any non-zero solutions to (10.2.18) in X_1. Since the divergence of the first moment of φ is due to a slow decay as $z \to \infty$, we are led to postulate that φ features a slow algebraic decay as $z \to \infty$. In fact, based on formal asymptotic expansions and already available results, it is expected that

$$\varphi(z) \sim A_\infty z^{-\tau} \quad \text{as} \quad z \to \infty , \qquad (10.2.20)$$

for some $A_\infty > 0$. Owing to the non-integrability of $z \mapsto z\varphi(z)$, we infer from (10.2.20) that necessarily $\tau < 2$, while (10.0.7) implies that $\tau \geq 1 + \lambda$. It is worth mentioning at this point that the large size behaviour (10.2.20) is solely prescribed by the linear terms in the left-hand side of (10.2.19), the contribution of the coagulation term being actually negligible for large sizes. Concerning the small size behaviour, it is expected to be still predicted by (10.2.17), though with different constants, possibly depending on τ.

The expected slow decay (10.2.20) of the self-similar profiles solving (10.2.18) is an additional difficulty when trying to construct them. Still, existence results are available for some coagulation kernels satisfying (10.2.13b) with $\alpha \in [0, 1)$ for $\tau \in (1 + \lambda, 2)$ [217, 218], and when $\alpha < 0$ for $\tau \in (1 + (\lambda - \alpha)_+, 2)$ [213], and the constructed solutions satisfy (10.2.20). In fact, when $\alpha \in (-1, 1)$, it is shown in [257] that any self-similar profile solving (10.2.18) in a suitable weak sense for some $\tau \in (1 + \max\{\lambda, \lambda - \alpha\}, 2)$ decays as prescribed by (10.2.20). We shall provide a more detailed account of these results in Section 10.2.6. Let us finally mention that the physical relevance of these self-similar solutions seems unclear due to their infinite mass. It is nevertheless of true mathematical interest to elucidate their role in the dynamics of the coagulation equation.

• $\lambda = 1$. Here again, no gelation is expected to occur and the starting point is to set $\tau = 2$ and investigate the existence of mass-conserving self-similar solutions. Then, according to (10.0.7b) and (10.0.8), the mean size σ strongly depends on the separation constant w and it cannot be assigned a particular value if the total mass is prescribed. We are thus looking for a positive real number w such that the equation

$$w \left(x \frac{d\varphi}{dx}(x) + 2\varphi(x) \right) + \mathcal{C}\varphi(x) = 0 , \qquad x \in (0, \infty) , \qquad (10.2.21)$$

has a (weak) solution φ satisfying

$$\varphi \in X_{1,+} , \qquad M_1(\varphi) = 1 . \qquad (10.2.22)$$

The only coagulation kernel k with homogeneity degree equal to one to be completely understood is the additive coagulation kernel k_+. As shown in [42, 198], there is a one-parameter family of self-similar profiles $(G_\theta)_{\theta \in (0,1/2]}$ such that

$$(t, x) \longmapsto \kappa e^{-2t/\theta} G_\theta(\sqrt{\kappa} x e^{-t/\theta}) , \qquad (t, x) \in \mathbb{R} \times (0, \infty) , \qquad (10.2.23)$$

is a mass-conserving self-similar solution to the coagulation equation, where $\kappa > 0$ is arbitrary. For $\theta \in (0, 1/2]$, the function G_θ is given by

$$G_\theta(x) := \frac{1}{\pi x^2} \sum_{j=1}^{\infty} \frac{(-1)^{j-1}}{j!} x^{j\theta} \Gamma(1 + j - j\theta) \sin(\pi j \theta) , \qquad x \in (0, \infty) ,$$

and satisfies (10.2.21) with $w = 1/\theta$, as well as (10.2.22). In particular,

$$G_{1/2}(x) = \frac{1}{\sqrt{4\pi}} x^{-3/2} e^{-x/4} , \qquad x \in (0, \infty) ,$$

which is already derived in [3, 263] (and also implicitly in [130]) and decays exponentially fast as $x \to \infty$. In contrast, for $\theta \in (0, 1/2)$, $G_\theta(x)$ behaves as $x^{-(2-\theta)/(1-\theta)}$ as $x \to \infty$ and $M_2(G_\theta) = \infty$. The stability of these self-similar solutions is studied in [198, 251] and a short account of the relevant results is given in Section 10.2.2.

For general coagulation kernels with homogeneity of degree one, it has already been observed in [263, Section 6] that the situation differs markedly depending on whether k belongs to class II (as the additive coagulation kernel k_+) or to class I (such as $k(x, y) = \sqrt{xy}$,

$(x,y) \in (0,\infty)^2$). More precisely, the small size behaviour for a mass-conserving self-similar profile φ when k lies in class I is expected to be $\varphi(x) \sim A_0 x^{-2}$ as $x \to 0$ (compare with (10.2.17c)) which is clearly not compatible with $\varphi \in X_1$. The scaling ansatz $\sigma(t)^{-2}\varphi(x/\sigma(t))$ is then not appropriate and alternative choices are derived in [263, Section 6] and [180, Section 3.5.6] (with a mean size $\sigma(t)$ of the form $e^{C\sqrt{t}}$ instead of e^{wt}). Therefore, equation (10.2.21) is likely to have no solution satisfying (10.2.22) at all but solutions to (10.2.21) with infinite mass are expected to exist and play a role in the description of the large time dynamics. This is supported by the analysis performed in [141, 221], where formal asymptotics and numerical simulations are used to identify the long-term behaviour.

In contrast, the situation for coagulation kernels with homogeneity of degree one in class II is expected to be similar to that of the additive coagulation kernel k_+: though a complete result is not yet available, it is shown in [46] that, for $w > w^* > 1$ large enough, there is at least one solution φ_w to (10.2.21), (10.2.22) such that

$$\varphi_w(x) \sim A_0 x^{-(2w-1)/w} \text{ as } x \to 0 \text{ and } \varphi_w(x) \sim A_\infty x^{-(2w-1)/(w-1)} \text{ as } x \to \infty .$$

Furthermore, no solution to (10.2.21), (10.2.22) exists for $w \in (1, w_*)$ for some $w_* > 1$.

• $\lambda \in (1,2]$. In this case, gelation is expected to take place, but the restriction $\lambda \leq 2$ prevents it from occurring instantaneously, at least for some initial data in $X_{1,+}$, see Proposition 9.2.7. The mean size σ is given by (10.0.7c) for some $\tau < \lambda + 1$ and blows up as time approaches the gelation time $T_{gel} > 0$. The first difficulty met here is the identification of the scaling parameter τ, as it can no longer be prescribed by the conservation of matter for initial data in $X_{1,+}$ as in the previously discussed cases. Computing τ turns out to be a challenging issue which has not received a satisfactory answer yet. In fact, it seems to be the major open problem related to the dynamics of coagulation, though there is again a particular case for which a complete description is available. Specifically, consider the multiplicative coagulation kernel $k_\times$, and an initial condition $f^{in} \in X_{1,+} \cap X_2$. Denoting the corresponding solution to the coagulation equation (10.2.1) with multiplicative coagulation kernel $k_\times$ by $\bar{f}$, we recall that $T_{gel} = 1/M_2(f^{in}) < \infty$ by Proposition 9.2.8, so that

$$M_1(\bar{f}(t)) = \varrho := M_1(f^{in}) , \qquad t \in [0, T_{gel}).$$

Introducing

$$f(s,x) := \varrho T_{gel} e^{-\varrho s} x \bar{f}(T_{gel}(1 - e^{-\varrho s}), x) , \qquad (s,x) \in [0,\infty) \times (0,\infty) , \qquad (10.2.24a)$$

or equivalently,

$$x\bar{f}(t,x) = \frac{1}{\varrho(T_{gel} - t)} f\left(\frac{\ln(T_{gel}) - \ln(T_{gel} - t)}{\varrho}, x\right) \qquad (10.2.24b)$$

for $(t,x) \in [0, T_{gel}) \times (0,\infty)$, a straightforward computation reveals a piece of good fortune, namely, that f is a mass-conserving solution on $[0,\infty)$ to the coagulation equation (10.2.1) with additive coagulation kernel k_+ and satisfies $M_1(f(s)) = \varrho$ for all $s \geq 0$ [94, 102, 283]. A first consequence of (10.2.24) is that self-similar solutions to the coagulation equation with multiplicative coagulation kernel $k_\times$ are given by

$$(t,x) \longmapsto \frac{\kappa^{3/2}}{\varrho T}\left(1 - \frac{t}{T}\right)_+^{-1+3(w/\varrho)} \varphi_{\times,w}\left(\sqrt{\kappa}x\left(1 - \frac{t}{T}\right)_+^{w/\varrho}\right) ,$$

where $T > 0$, $\varrho > 0$, and $\kappa > 0$ are arbitrary, $w \in [2,\infty)$, and $\varphi_{\times,w}(x) := G_{1/w}(x)/x$, $x \in (0,\infty)$, the functions $G_{1/w}$ being defined in (10.2.23). In particular,

$$\varphi_{\times,2}(x) = \frac{G_{1/2}(x)}{x} = \frac{1}{\sqrt{4\pi}} x^{-5/2} e^{-x/4} , \qquad x \in (0,\infty) ,$$

which is already derived in [263] and is the only self-similar profile (up to scaling) that decays exponentially fast for large sizes, the others corresponding to $w > 2$ and having an infinite third moment. The transformation (10.2.24) is also useful when investigating the stability of self-similar solutions for the coagulation equation with multiplicative coagulation kernel $k_\times$ [198, 251]. Results in that direction are of course deduced from those available for the coagulation equation with additive coagulation kernel k_+.

For self-similar solutions having a profile with an exponential decay for large sizes, the separating constant is $w = 2$, so that the scaling exponent τ in (10.0.7c) is given by $\tau = 5/2 = (\lambda + 3)/2$, the homogeneity of the multiplicative coagulation kernel $k_\times$ being $\lambda = 2$. Formal arguments are given in the physical literature to put forward the choice $\tau = (\lambda + 3)/2$ in the general case as the scaling exponent for which self-similar solutions with an exponentially decaying profile should exist [263]. However, no further convincing argument has been found in recent years to support this conjecture and the determination of the parameter τ for general values of λ is a challenging problem. As pointed out in a series of papers by Leyvraz [180, 181, 182], the sought-for scaling exponent τ should rather be determined as the one for which the nonlinear integral equation (10.0.6) has a nonnegative solution which decays exponentially fast for large sizes. This conjecture is supported by numerical simulations performed in [178, 181, 182] and it forms a basis of the strategy followed in [50] to obtain an existence result for the coagulation kernel $(xy)^{\lambda/2}$, $(x, y) \in (0, \infty)^2$, when $\lambda = 2 - 2\varepsilon$ and ε is small enough. The smallness of ε is required due to the approach used, which is a perturbative one, its starting point being the equation solved by the Bernstein transform of $x \mapsto x\varphi_{\times,2}(x)$.

10.2.1   The Constant Coagulation Kernel $k(x, y) = 2$

A detailed analysis of the dynamics of the coagulation equation (10.2.1) with constant coagulation kernel $k \equiv 2$,

$$\partial_t f(t, x) = \int_0^x f(t, x - y) f(t, y) \, dy - 2 f(t, x) \int_0^\infty f(t, y) \, dy \qquad (10.2.25a)$$

for $(t, x) \in (0, \infty)^2$, with initial condition

$$f(0, x) = f^{in}(x) , \qquad x \in (0, \infty) , \qquad (10.2.25b)$$

is available [198]. This is mainly due to the well known fact that the Laplace transform of any weak solution to (10.2.25) solves an ordinary differential equation which has an explicit solution [196, 240, 241]. In particular, it is well known that (10.2.25) has explicit mass-conserving self-similar solutions, given by

$$(t, x) \longmapsto \frac{1}{\varrho t^2} e^{-x/(\varrho t)} , \qquad (t, x) \in (0, \infty)^2 , \qquad (10.2.26)$$

the total mass ϱ being an arbitrary positive real number. The stability of these self-similar solutions is studied in [151], where each is shown to attract any solution to (10.2.25) with initial data having the same mass $M_1(f^{in}) = \varrho$ and decaying exponentially fast as $x \to \infty$, see also [90, 151] for a related analysis devoted to the discrete coagulation equation with constant kernel. Improved convergence results are subsequently derived in [64, 171, 198, 199, 251]. A surprising feature is uncovered in [198]: there exist other self-similar solutions scaling differently and, in particular, having infinite first moment. Their basin of attraction is also identified in [198] and a complete picture of the dynamics is provided in [200].

Let us now give a rough account of the analysis performed in [198]. We begin with the well-posedness of (10.2.25) in the larger space X_0.

Proposition 10.2.3. *Let $f^{in} \in X_{0,+}$. Then there exists a unique nonnegative function $f \in C^1([0, \infty), X_0)$ solving (10.2.25). Furthermore,*

$$M_0(f(t)) = \frac{M_0(f^{in})}{1 + M_0(f^{in})t} \, , \qquad t \geq 0 \, , \tag{10.2.27}$$

and the Laplace transform $\mathcal{L}f$ and the Bernstein function Λf associated with f, given by

$$\mathcal{L}f(t, \xi) := \int_0^\infty e^{-x\xi} f(t, x) \, dx \, , \qquad \Lambda f(t, \xi) := \int_0^\infty \left(1 - e^{-x\xi}\right) f(t, x) \, dx \tag{10.2.28}$$

for $(t, \xi) \in [0, \infty)^2$, satisfy

$$\mathcal{L}f(t, \xi) = M_0(f(t)) - \Lambda f(t, \xi) \quad and \quad \Lambda f(t, \xi) = \frac{\Lambda f^{in}(\xi)}{1 + \Lambda f^{in}(\xi)t} \, . \tag{10.2.29}$$

Proof. The well-posedness of (10.2.25) in X_0 is proved as in Proposition 8.2.7, after noticing that the coagulation operator $\mathcal{C}$ is locally Lipschitz continuous in X_0 thanks to the boundedness of k. Next, integrating (10.2.25a) over $(0, \infty)$ gives

$$\frac{d}{dt} M_0(f(t)) = -M_0(f(t))^2 \, , \qquad t > 0 \, ,$$

from which (10.2.27) readily follows. Finally, we multiply (10.2.25a) by $1 - e^{-x\xi}$, where $\xi \geq 0$, and integrate over $(0, \infty)$ with respect to x to obtain $\partial_t \Lambda f = -(\Lambda f)^2$. Solving this differential equation completes the proof of (10.2.29). $\qquad\square$

We now look for self-similar solutions to (10.2.25a) of the form (10.0.4); that is,

$$(t, x) \longmapsto \frac{1}{\sigma(t)^\tau} \varphi\left(\frac{x}{\sigma(t)}\right) \, , \qquad (t, x) \in (0, \infty)^2 \, , \tag{10.2.30}$$

the exponent τ, the mean size σ and the profile $\varphi \in X_0$ being yet to be determined. Thanks to (10.2.29), it is possible to identify all such nonnegative integrable solutions to (10.2.25a) [198, Section 4].

Proposition 10.2.4. *[198] Consider a self-similar solution to (10.2.25a) of the form (10.2.30) and assume that the profile φ belongs to $X_{0,+}$ and that $\sigma(0) = 1$. Then, there is $\theta \in (0, 1]$, $\zeta_0 > 0$, and $\zeta_* > 0$ such that $\tau = 1 + \theta \in (1, 2]$ and*

$$\sigma(t) = (1 + \zeta_0 t)^{1/\theta} \, , \qquad \varphi(x) = \zeta_0 \zeta_* \frac{dF_\theta}{dx}(x\zeta_*) \, , \qquad (t, x) \in [0, \infty) \times (0, \infty) \, ,$$

where F_θ is the Mittag–Leffler distribution

$$F_\theta(x) := \sum_{j=1}^\infty \frac{(-1)^{j+1} x^{j\theta}}{\Gamma(1 + j\theta)} \, , \qquad x \in (0, \infty) \, .$$

When, in Proposition 10.2.4, we let $\theta = 1$, then it is clear that $F_1(x) = 1 - e^{-x}$ and $dF_1(x)/dx = e^{-x}$, $x \in (0, \infty)$, so that we recover the self-similar solutions given by (10.2.26).

Proof. Set
$$\Phi(t,x) := \frac{1}{\sigma(t)^\tau} \varphi\left(\frac{x}{\sigma(t)}\right) , \qquad (t,x) \in [0,\infty) \times (0,\infty) .$$

For $t > 0$ and $\xi \geq 0$, we observe that
$$\Lambda\Phi(t,\xi) = \sigma(t)^{1-\tau} \Lambda\varphi(\xi\sigma(t))$$

and, since $\sigma(0) = 1$, we deduce from (10.2.29) that
$$\sigma(t)^{1-\tau} \Lambda\varphi(\xi\sigma(t)) = \frac{\Lambda\varphi(\xi)}{1 + t\Lambda\varphi(\xi)} . \qquad (10.2.31)$$

Since $\varphi \in X_0$ and $\Lambda\varphi(\xi) \to M_0(\varphi)$ as $\xi \to \infty$, we may pass to the limit as $\xi \to \infty$ in (10.2.31) and obtain
$$\sigma(t)^{\tau-1} = 1 + \zeta_0 t, \qquad t \geq 0 , \qquad (10.2.32)$$

with $\zeta_0 := M_0(\varphi)$. Observe that (10.2.32) complies with (10.0.7) only for $\tau > 1$.

We next take $t \geq 0$ and $\xi = 1$ in (10.2.31) and use (10.2.32) to find
$$\Lambda\varphi(\sigma(t)) = \sigma(t)^{\tau-1} \frac{\zeta_0 \Lambda\varphi(1)}{\zeta_0 + \Lambda\varphi(1)\left(\sigma(t)^{\tau-1} - 1\right)} ,$$

hence, since σ maps $[0,\infty)$ onto $[1,\infty)$ and is given by (10.2.32),
$$\Lambda\varphi(\xi) = \frac{\zeta_0 \zeta_1 \xi^{\tau-1}}{\zeta_0 + \zeta_1\left(\xi^{\tau-1} - 1\right)} , \qquad \xi \geq 1 ,$$

with $\zeta_1 := \Lambda\varphi(1) < M_0(\varphi) = \zeta_0$. Similarly, consider $\xi \in (0,1]$ and let $t_\xi > 0$ be such that $\xi = 1/\sigma(t_\xi)$. We then infer from (10.2.31) (with $t = t_\xi$) that
$$\Lambda\varphi(\xi) = \frac{\zeta_0 \zeta_1 \xi^{\tau-1}}{\zeta_0 + \zeta_1\left(\xi^{\tau-1} - 1\right)} , \qquad \xi \in (0,1] .$$

Since $\Lambda\varphi = M_0(\varphi) - \mathcal{L}\varphi = \zeta_0 - \mathcal{L}\varphi$, we end up with the following formula for the Laplace transform of φ:
$$\mathcal{L}\varphi(\xi) = \frac{\zeta_0}{1 + \zeta_2 \xi^{\tau-1}} , \qquad \xi > 0 , \qquad (10.2.33)$$

with $\zeta_2 := \zeta_1/(\zeta_0 - \zeta_1) > 0$.

On the one hand, for $\xi > 0$,
$$\frac{d\mathcal{L}\varphi}{d\xi}(\xi) = -\zeta_0\zeta_2(\tau - 1)\frac{\xi^{\tau-2}}{\left(1 + \zeta_2\xi^{\tau-1}\right)^2} ,$$
$$\frac{d^2\mathcal{L}\varphi}{d\xi^2}(\xi) = \zeta_0\zeta_2(\tau - 1)\frac{\xi^{\tau-3}}{\left(1 + \zeta_2\xi^{\tau-1}\right)^2}\left[-(\tau - 2) + \tau\zeta_2\xi^{\tau-1}\right] .$$

If $\tau \leq 1$, then $d\mathcal{L}\varphi(\xi)/d\xi \geq 0$, while $d^2\mathcal{L}\varphi(\xi)/d\xi^2 < 0$ in a right neighbourhood of zero when $\tau > 2$. Consequently, $\mathcal{L}\varphi$ is not completely monotone, which contradicts the fact that $\mathcal{L}\varphi$ is the Laplace transform of an integrable function [114, XIII.4, Theorem 1]. These values of τ are thus excluded.

On the other hand, if $\tau \in (1,2]$, then $\xi \mapsto 1/(1 + \xi^{\tau-1})$ is the Laplace transform of the derivative $dF_{\tau-1}/dx$ of the Mittag–Leffler distribution $F_{\tau-1}$. Scaling properties of the Laplace transform then imply that
$$\varphi(x) = \zeta_0\zeta_2^{-1/(\tau-1)}\frac{dF_{\tau-1}}{dx}\left(x\zeta_2^{-1/(\tau-1)}\right) , \qquad x > 0 .$$

Setting $\theta := \tau - 1 \in (0,1]$ and $\zeta_* := \zeta_2^{-1/\theta}$ completes the proof.   □

Let us finally elucidate the role played by these special solutions in the dynamics [198, Theorem 5.1].

Theorem 10.2.5. *[198] Consider $f^{in} \in X_{0,+}$ and denote the corresponding solution to (10.2.25) by f. Assume further that there is $\theta \in (0, 1]$, and $C_\infty > 0$ such that*

$$Q(x) := \int_0^x y f^{in}(y) \, \mathrm{d}y \sim C_\infty x^{1-\theta} \quad \text{as } x \to \infty . \tag{10.2.34}$$

Then

$$\lim_{t \to \infty} \int_0^{xt^{1/\theta}} f(t, y) \, \mathrm{d}y = \int_0^x \varphi(y) \, \mathrm{d}y , \qquad x \in (0, \infty) ,$$

where

$$\varphi(x) = \zeta_2^{-1/\theta} \frac{dF_\theta}{dx} \left(x \zeta_2^{-1/\theta} \right) , \qquad x \in (0, \infty) ,$$

and $\zeta_2 := C_\infty \Gamma(2 - \theta)/\theta$.

In particular, if $f^{in} \in X_{0,1,+}$, then the condition (10.2.34) is obviously satisfied with $\theta = 1$ and $C_\infty = M_1(f^{in})$, and Theorem 10.2.5 provides the expected convergence (in a weak sense) to the mass-conserving self-similar solution (10.2.26) with $\varrho = M_1(f^{in})$.

Proof. We first observe that, for $\xi > 0$,

$$\partial_\xi (\Lambda f)(0, \xi) = \xi \int_0^\infty Q(x) e^{-x\xi} \, \mathrm{d}x = \xi^{\theta-1} \int_0^\infty \left(\frac{y}{\xi} \right)^{\theta-1} Q\left(\frac{y}{\xi} \right) y^{1-\theta} e^{-y} \, \mathrm{d}y .$$

Now, (10.2.34) ensures that there is $R > 0$ such that

$$\left(\frac{y}{\xi} \right)^{\theta-1} Q\left(\frac{y}{\xi} \right) \le 2C_\infty , \qquad y > \xi R ,$$

while

$$\left(\frac{y}{\xi} \right)^{\theta-1} Q\left(\frac{y}{\xi} \right) \le \left(\frac{y}{\xi} \right)^\theta M_0(f^{in}) \le R^\theta M_0(f^{in}) , \qquad y \in [0, \xi R] .$$

Since $y \mapsto y^{1-\theta} e^{-y}$ belongs to X_0, we may use Lebesgue's dominated convergence theorem to conclude that

$$\partial_\xi (\Lambda f)(0, \xi) \sim \theta \zeta_2 \xi^{\theta-1} \quad \text{as } \xi \to 0 .$$

Hence, using $\Lambda f(0, \xi) \to 0$ as $\xi \to 0$,

$$\Lambda f(0, \xi) \sim \zeta_2 \xi^\theta \quad \text{as } \xi \to 0 . \tag{10.2.35}$$

It readily follows from (10.2.35) that, for all $\xi > 0$,

$$\lim_{t \to \infty} t \Lambda f\left(0, \frac{\xi}{t^{1/\theta}} \right) = \zeta_2 \xi^\theta .$$

Recalling (10.2.29) we see at once that

$$\lim_{t \to \infty} t \Lambda f\left(t, \frac{\xi}{t^{1/\theta}} \right) = \frac{\zeta_2 \xi^\theta}{1 + \zeta_2 \xi^\theta} ,$$

hence, thanks to (10.2.27) and (10.2.29),

$$\lim_{t\to\infty} t\mathcal{L}f\left(t, \frac{\xi}{t^{1/\theta}}\right) = \frac{1}{1 + \zeta_2 \xi^\theta} = \mathcal{L}\varphi(\xi)$$

for all $\xi \in (0, \infty)$. According to [114, XIII.1, Theorem 2a], we can now conclude that $y \mapsto t^{1/\theta} f(t, yt^{1/\theta})/M_0(f(t))$ converges to $\varphi/M_0(\varphi)$ as $t \to \infty$ in the weak topology of measures. Combining this convergence with the absolute continuity of φ with respect to the Lebesgue measure on $(0, \infty)$ completes the proof.   □

The results obtained in [198] actually go way beyond Theorem 10.2.5 as they provide a complete characterisation of the basin of attraction of each self-similar solution described in Proposition 10.2.4. In particular, the mass-conserving self-similar solution (10.2.26) (with profile dF_1/dx) is also an attractor for solutions emanating from initial data having infinite mass, but for this, the function Q defined in (10.2.34) should slowly diverge as $x \to \infty$, see [198] for details. Convergence rates are derived in [251].

As a final remark, let us mention that the Laplace transform is not used in [64, 171]. Instead, convergence of mass-conserving solutions to the mass-conserving self-similar solution (10.2.26) is proved in [171] by a dynamical systems approach and is based on the construction of an appropriate Lyapunov functional. In [64], the convergence rates are derived by showing that the linearised coagulation operator in self-similar variables has a spectral gap in well-chosen L_2-spaces with exponential weights.

10.2.2   The Additive Coagulation Kernel $k_+(x, y) = x + y$

In this section, we carry out an analysis similar to that of the previous section but now for the additive coagulation kernel k_+, in which case the coagulation equation can be written as

$$\partial_t f(t, x) = \frac{x}{2} \int_0^x f(t, x - y) f(t, y) \, dy - [x M_0(f(t)) + M_1(f(t))] f(t, x) \qquad (10.2.36a)$$

for $(t, x) \in (0, \infty)^2$, with initial condition

$$f(0, x) = f^{in}(x), \qquad x \in (0, \infty). \qquad (10.2.36b)$$

The well-posedness of (10.2.36), and the long-term behaviour of its solutions, can also be studied with the help of the Laplace transform as noticed in [130, 241] and thoroughly used in [42, 198]. The analysis is, however, more involved than for the constant kernel, due to the fact that the equation solved by the Laplace transform of the solution to (10.2.36) is a scalar conservation law instead of an ordinary differential equation.

Proposition 10.2.6. *Consider $f^{in} \in X_{0,1,+}$ with $\varrho := M_1(f^{in}) > 0$ and let f be a mass-conserving weak solution on $[0, \infty)$ to (10.2.36). The Laplace transform $\mathcal{L}f$ and the Bernstein function Λf associated with f*

$$\mathcal{L}f(t, \xi) := \int_0^\infty e^{-x\xi} f(t, x) \, dx, \qquad \Lambda f(t, \xi) := \int_0^\infty (1 - e^{-x\xi}) f(t, x) \, dx,$$

defined for $(t, \xi) \in [0, \infty)^2$, satisfy

$$\mathcal{L}f(t, \xi) = M_0(f(t)) - \Lambda f(t, \xi), \qquad M_0(f(t)) = M_0(f^{in})e^{-\varrho t}, \qquad (10.2.37)$$

and Λf solves the Burgers equation with linear damping

$$\partial_t \Lambda f - \Lambda f \partial_\xi (\Lambda f) = -\varrho \Lambda f \quad in \quad (0,\infty)^2 , \tag{10.2.38a}$$

$$\Lambda f(0) = \Lambda f^{in} \quad in \quad (0,\infty) . \tag{10.2.38b}$$

Thanks to Theorem 8.2.23, there exists at least one mass-conserving weak solution on $[0,\infty)$ to (10.2.36) with an initial condition in $X_{0,1,+}$, and it is unique by Theorem 8.2.55 (with $\ell = 1 + \mathrm{id}$ and $\zeta = 1$) and Theorem 8.2.23 under the additional assumption $f^{in} \in X_2$. In fact, this result can be improved and the well-posedness of (10.2.36) in $X_{1,+}$ is shown in [198, Theorem 2.8]. The proof relies on the solvability of (10.2.38) and requires a different notion of weak solution, since the constructed solution need not belong to X_0 for positive times. In such a case, its Laplace transform $\mathscr{L}f$ is not well defined but its Bernstein function Λf is, thanks to the elementary inequality

$$1 - e^{-x\xi} \le x\xi , \qquad (x,\xi) \in [0,\infty)^2 . \tag{10.2.39}$$

Proof of Proposition 10.2.6. On the one hand, the formula for $M_0(f)$ readily follows from the conservation of mass and the weak formulation (8.2.4d) of (10.2.36) with $\vartheta \equiv 1$. On the other hand, given $\xi \in (0,\infty)$, the function $x \mapsto \vartheta(x) = 1 - e^{-x\xi}$, $x \in (0,\infty)$, satisfies

$$\chi_\vartheta(x,y) = -(1 - e^{-x\xi})(1 - e^{-y\xi}) , \qquad (x,y) \in (0,\infty)^2 .$$

Consequently, for $(s,\xi) \in (0,\infty)^2$,

$$\frac{1}{2} \int_0^\infty \int_0^\infty \chi_\vartheta(x,y) k_+(x,y) f(s,x) f(s,y) \, dy dx$$
$$= -\int_0^\infty \int_0^\infty x(1 - e^{-x\xi})(1 - e^{-y\xi}) f(s,x) f(s,y) \, dy dx$$
$$= -\Lambda f(s,\xi) \left[M_1(f(s)) - \partial_\xi \Lambda f(s,\xi) \right] .$$

Since f is mass-conserving, we readily deduce (10.2.38) from (8.2.4d) and the previous identity. $\square$

Being a scalar conservation law which can be studied by the method of characteristics, equation (10.2.38) paves the way not only towards a representation formula for f but also towards a complete classification of self-similar solutions. We first focus on the latter and begin by pointing out that, the natural functional setting for (10.2.36) being X_1 as previously mentioned, we look for self-similar solutions with a profile $\varphi \in X_1$ and thus for mass-conserving self-similar solutions according to the discussion in the preamble of Chapter 10. In addition, as the homogeneity degree of k_+ is equal to one, we infer from (10.0.7) that mass-conserving self-similar solutions to (10.2.36a) are to be sought in the form

$$(t,x) \longmapsto e^{-2wt} \varphi(x e^{-wt}) , \tag{10.2.40}$$

the parameter $w > 0$ and the (nonnegative) profile $\varphi \in X_1$ to be determined.

Proposition 10.2.7. *[42, 198] Fix $\varrho > 0$ and consider a self-similar solution to (10.2.36a) of the form (10.2.40) with $w > 0$ and a profile $\varphi \in X_{1,+}$, satisfying $M_1(\varphi) = \varrho$. Then there is $\theta \in (0, 1/2]$, and $\zeta > 0$ such that*

$$w = \frac{\varrho}{\theta} \in [2\varrho, \infty) , \qquad \varphi(x) = \frac{\zeta^2}{\varrho} G_\theta \left(\frac{\zeta x}{\varrho} \right) , \qquad x \in (0,\infty) ,$$

where

$$G_\theta(x) := \frac{1}{\pi x^2} \sum_{j=1}^{\infty} \frac{(-1)^{j-1}}{j!} x^{j\theta} \Gamma(1+j-j\theta) \sin(\pi j\theta) , \qquad x \in (0,\infty) .$$

In particular,

$$G_{1/2}(x) = \frac{e^{-x/4}}{\sqrt{4\pi} x^{3/2}} , \qquad x \in (0,\infty) .$$

It is worth pointing out that only $G_{1/2}$ decays exponentially fast for large sizes and has a finite second moment. When $\theta \in (0, 1/2)$, one has

$$G_\theta(x) \sim \frac{1}{(1-\theta)|\Gamma(\theta/(\theta-1))|} x^{-(2-\theta)/(1-\theta)} \quad \text{as } x \to \infty ,$$

and $G_\vartheta \notin X_2$ [198, Eq. (6.16)].

Proof of Proposition 10.2.7. Let $\Phi(t,x) := e^{-2wt}\varphi(xe^{-wt})$ for $(t,x) \in [0,\infty) \times (0,\infty)$, and note that

$$\Lambda\Phi(t,\xi) = e^{-wt}\Lambda\varphi(\xi e^{wt}) , \qquad (t,\xi) \in [0,\infty)^2 . \tag{10.2.41}$$

Also, since $\varphi \in X_1$ and $M_1(\varphi) = \varrho$, it follows from (10.2.39) that

$$\Lambda\varphi(\xi) < \varrho\xi , \qquad 0 < \frac{d\Lambda\varphi}{d\xi}(\xi) < \varrho , \qquad \xi \in (0,\infty) , \tag{10.2.42}$$

$$\Lambda\varphi(0) = 0 , \qquad \frac{d\Lambda\varphi}{d\xi}(0) = \varrho , \qquad \lim_{\xi\to\infty} \frac{d\Lambda\varphi}{d\xi}(\xi) = 0 . \tag{10.2.43}$$

Inserting the formula (10.2.41) in (10.2.38a) shows that

$$(\varrho - w)\Lambda\varphi(\xi) + (w\xi - \Lambda\varphi(\xi))\frac{d\Lambda\varphi}{d\xi}(\xi) = 0 , \qquad \xi \in [0,\infty) . \tag{10.2.44}$$

First, it follows from (10.2.42) and (10.2.44) that

$$w \neq \varrho . \tag{10.2.45}$$

Indeed, assume for contradiction that $w = \varrho$. We then infer from (10.2.42) and (10.2.44) that $\Lambda\varphi(\xi) = \varrho\xi$ for all $\xi \in [0,\infty)$, which contradicts (10.2.42).

Next, owing to the monotonicity (10.2.42) of $\Lambda\varphi$ and (10.2.43), there is $z_\infty \in (0,\infty]$ such that $\Lambda\varphi$ is a homeomorphism from $[0,\infty)$ onto $[0, z_\infty)$ and a C^1-diffeomorphism from $(0,\infty)$ onto $(0, z_\infty)$. Denoting its inverse function by ψ and taking $\xi = \psi(z)$ in (10.2.44) for some $z \in (0, z_\infty)$ we see that ψ solves

$$(\varrho - w)z\psi'(z) + w\psi(z) - z = 0 , \qquad z \in (0, z_\infty) ,$$

with initial value $\psi(0) = 0$. Since $w \neq \varrho$ by (10.2.45), integration of the ordinary differential equation for ψ shows that there is a constant $c \in \mathbb{R}$ such that

$$\varrho\psi(z) = z + cz^{w/(w-\varrho)} , \qquad z \in (0, z_\infty) . \tag{10.2.46}$$

Recalling that ψ is increasing, $\psi(0) = 0$, and $\varrho\psi \neq \text{id}$, by (10.2.42) and (10.2.43), we deduce from (10.2.45) and (10.2.46) that $w > \varrho$ and $c > 0$. It also follows from (10.2.46), and the definition of z_∞, that $z_\infty = \infty$. Thus,

$$\lim_{\xi\to\infty} \Lambda\varphi(\xi) = \infty , \qquad \varphi \notin X_0 , \tag{10.2.47}$$

and we infer from (10.2.46) that

$$\varrho\xi = \Lambda\varphi(\xi) + c(\Lambda\varphi(\xi))^{w/(w-\varrho)} , \qquad \xi \in [0,\infty) . \tag{10.2.48}$$

Furthermore, differentiating twice (10.2.48) gives

$$\left[1 + \frac{cw}{w-\varrho}(\Lambda\varphi)^{\varrho/(w-\varrho)}\right]\frac{d^2\Lambda\varphi}{d\xi^2} = -\frac{cw\varrho}{(w-\varrho)^2}(\Lambda\varphi)^{(2\varrho-w)/(w-\varrho)}\left[\frac{d\Lambda\varphi}{d\xi}\right]^2 .$$

If $w \in (\varrho, 2\varrho)$, then it follows from (10.2.43) that

$$\lim_{\xi\to 0}\left[1 + \frac{cw}{w-\varrho}(\Lambda\varphi(\xi))^{\varrho/(w-\varrho)}\right] = 1$$

and

$$\lim_{\xi\to 0}(\Lambda\varphi(\xi))^{(2\varrho-w)/(w-\varrho)}\left[\frac{d\Lambda\varphi}{d\xi}(\xi)\right]^2 = 0 ,$$

so that

$$\lim_{\xi\to 0}\frac{d^2\Lambda\varphi}{d\xi^2}(\xi) = 0 . \tag{10.2.49}$$

However, since $d\Lambda\varphi/d\xi$ is completely monotone as it is the Laplace transform of the integrable function $x \mapsto x\varphi(x)$, the function $d^2\Lambda\varphi/d\xi^2$ is non-positive and non-decreasing, which fails to comply with (10.2.49). Therefore,

$$w \geq 2\varrho \quad \text{and} \quad \theta := \frac{\varrho}{w} \in \left(0,\frac{1}{2}\right] . \tag{10.2.50}$$

To proceed further, we set $U(\xi) := \Lambda\varphi(\xi)/\xi$ for $\xi \in (0,\infty)$ and $U(0) = \varrho$ and deduce from (10.2.42) and (10.2.48) that

$$U(\xi) < \varrho \quad \text{and} \quad U(\xi) - c^{\theta-1}\xi^{-\theta}\omega(U(\xi)) = 0 , \qquad \xi \in (0,\infty) ,$$

where $\omega(r) := (\varrho-r)^{1-\theta}$ for $r \in [0,\varrho]$. Using Lagrange's inversion formula [1, Equation 3.6.6] leads to the closed-form representation

$$U(\xi) = \sum_{j=1}^{\infty}\frac{u_j}{j!}\left(c^{\theta-1}\xi^{-\theta}\right)^j \quad \text{with} \quad u_j := \left(\frac{d^{j-1}}{dr^{j-1}}\omega^j\right)(0) , \qquad j \geq 1 .$$

Fortunately, u_j is explicitly computable for all $j \geq 1$ with

$$u_1 = \varrho^{1-\theta} \quad \text{and} \quad u_j = (-1)^{j-1}\varrho^{1-j\theta}\prod_{i=2}^{j}(i-j\theta) , \qquad j \geq 2 .$$

Therefore, for $\xi \in (0,\infty)$,

$$\Lambda\varphi(\xi) = c^{\theta-1}(\varrho\xi)^{1-\theta} + \sum_{j=2}^{\infty}\frac{(-1)^{j-1}c^{j(\theta-1)}(\varrho\xi)^{1-j\theta}}{j!}\prod_{i=2}^{j}(i-j\theta)$$

and, after differentiation with respect to ξ,

$$\frac{d\Lambda\varphi}{d\xi}(\xi) = \sum_{j=1}^{\infty}(-1)^{j-1}\frac{c^{j(\theta-1)}\varrho^{1-j\theta}\xi^{-j\theta}}{j!}\prod_{i=1}^{j}(i-j\theta) .$$

Since

$$\prod_{i=1}^{j}(i - j\theta) = \frac{\Gamma(1 + j - j\theta)}{\Gamma(1 - j\theta)} = \Gamma(1 + j - j\theta)\Gamma(1 + j\theta)\frac{\sin(j\pi\theta)}{j\pi\theta}$$

$$= \frac{1}{\pi}\Gamma(1 + j - j\theta)\Gamma(j\theta)\sin(j\pi\theta)$$

by the reflection formula for Γ, see [1, Formula 6.1.17], we end up with

$$\frac{d\Lambda\varphi}{d\xi}(\xi) = \frac{\varrho}{\pi}\sum_{j=1}^{\infty}\frac{(-1)^{j-1}}{j!}c^{j(\theta-1)}(\varrho\xi)^{-j\theta}\Gamma(1 + j - j\theta)\Gamma(j\theta)\sin(j\pi\theta) .$$

Since $d\Lambda\varphi/d\xi$ is the Laplace transform of $x \mapsto x\varphi(x)$ and $\xi \mapsto \Gamma(j\theta)\xi^{-j\theta}$ is the Laplace transform of $x \mapsto x^{j\theta-1}$ for all $j \geq 1$, the previous identity and a term-by-term identification show that, for $x \in (0, \infty)$,

$$x\varphi(x) = \frac{\varrho}{\pi}\sum_{j=1}^{\infty}\frac{(-1)^{j-1}}{j!}c^{j(\theta-1)}\varrho^{-j\theta}x^{j\theta-1}\Gamma(1 + j - j\theta)\sin(j\pi\theta) ,$$

hence

$$\varphi(x) = \frac{\zeta^2}{\varrho}G_\theta\left(\frac{\zeta x}{\varrho}\right) , \qquad x \in (0, \infty) ,$$

as claimed, with $\zeta := c^{(\theta-1)/\theta}$.   □

Another consequence of (10.2.38) is a representation formula for Λf which is derived by the method of characteristics.

Proposition 10.2.8. *Consider $f^{in} \in X_{0,1,+}$ with $\varrho := M_1(f^{in}) > 0$ and let f be a mass-conserving weak solution on $[0, \infty)$ to (10.2.36). The Bernstein function Λf of f, defined in Proposition 10.2.6, is given by*

$$\Lambda f(t, \xi) = e^{-\varrho t}\Lambda f^{in}(\Sigma(t, \xi)) , \qquad (t, \xi) \in [0, \infty)^2 , \tag{10.2.51}$$

where $\Sigma(t, \xi)$ is the unique solution in $[0, \infty)$ to

$$\varrho\xi = \varrho\Sigma(t, \xi) - (1 - e^{-\varrho t})\Lambda f^{in}(\Sigma(t, \xi)) . \tag{10.2.52}$$

Proof. The characteristics associated with (10.2.38) are given by

$$\frac{dS}{dt}(t, \xi_0) = -\Lambda f(t, S(t, \xi_0)) , \qquad t > 0 , \qquad S(0, \xi_0) = \xi_0 , \tag{10.2.53}$$

where $\xi_0 \in [0, \infty)$. We infer from (10.2.38a) and (10.2.53) that, for $t > 0$,

$$\frac{d}{dt}\Lambda f(t, S(t, \xi_0)) = -\varrho\Lambda f(t, S(t, \xi_0)) ,$$

hence, by (10.2.38b),

$$\Lambda f(t, S(t, \xi_0)) = e^{-\varrho t}\Lambda f^{in}(\xi_0) , \qquad t \in [0, \infty) . \tag{10.2.54}$$

Combining (10.2.53) and (10.2.54) leads us to

$$\varrho S(t, \xi_0) = \varrho\xi_0 - (1 - e^{-\varrho t})\Lambda f^{in}(\xi_0) , \qquad (t, \xi_0) \in [0, \infty)^2 . \tag{10.2.55}$$

Also, since $f^{in} \in X_1$ with $M_1(f^{in}) = \varrho$, it follows from (10.2.39) that

$$\Lambda f^{in}(0) = 0 < \Lambda f^{in}(\xi) < \varrho\xi \quad \text{and} \quad 0 < \frac{d\Lambda f^{in}}{d\xi}(\xi) < \varrho = \frac{d\Lambda f^{in}}{d\xi}(0) \qquad (10.2.56)$$

for $\xi \in (0, \infty)$. By (10.2.55) and (10.2.56) we have

$$e^{-\varrho t}\xi_0 \leq S(t, \xi_0) \leq \xi_0 , \qquad \xi_0 \in [0, \infty) ,$$

$$\varrho \partial_{\xi_0} S(t, \xi_0) = \varrho - (1 - e^{-\varrho t})\frac{d\Lambda f^{in}}{d\xi}(\xi_0) > \varrho e^{-\varrho t} , \qquad \xi_0 \in (0, \infty) ,$$

for each $t \in [0, \infty)$, from which we readily deduce that $\xi_0 \mapsto S(t, \xi_0)$ is a homeomorphism from $[0, \infty)$ onto $[0, \infty)$ and a C^1-diffeomorphism from $(0, \infty)$ onto $(0, \infty)$. Denoting its inverse function by $\Sigma(t, \cdot)$; that is,

$$\Sigma(t, S(t, \xi_0)) = \xi_0 , \qquad (t, \xi_0) \in [0, \infty)^2 , \qquad (10.2.57)$$

the identities (10.2.51) and (10.2.52) are straightforward consequences of (10.2.54), (10.2.55) and (10.2.57). □

Thanks to Proposition 10.2.8, we are now in a position to investigate the role in the dynamics played by the self-similar solutions described in Proposition 10.2.7. In this direction, we report the following result which is included in [198, Theorem 7.1].

Theorem 10.2.9. *[198] Consider $f^{in} \in X_{0,1,+}$ with $\varrho := M_1(f^{in}) > 0$ and let f be a mass-conserving weak solution on $[0, \infty)$ to (10.2.36). Assume further that there is $\mu \in (0, 1]$, and $C_\infty > 0$ such that*

$$Q(x) := \int_0^x y^2 f^{in}(y) \, dy \sim C_\infty x^{1-\mu} \quad \text{as} \quad x \to \infty . \qquad (10.2.58)$$

Then

$$\lim_{t \to \infty} \int_0^{xe^{wt}} yf(t, y) \, dy = \int_0^x y\varphi(y) \, dy , \qquad x \in (0, \infty) ,$$

where $w := (1 + \mu)\varrho/\mu$ and

$$\varphi(x) := \frac{\zeta^2}{\varrho}G_{\mu/(1+\mu)}\left(\frac{\zeta x}{\varrho}\right) , \qquad x \in (0, \infty) , \qquad \zeta := \left(\frac{C_\infty\Gamma(2-\mu)}{\mu(\mu+1)\varrho^{1+\mu}}\right)^{-1/\mu} ,$$

the function $G_{\mu/(1+\mu)}$ being defined in Proposition 10.2.7.

As for the constant coagulation kernel, a complete characterisation of the basin of attraction of each self-similar profile is provided in [198, Theorem 7.1]. In particular, $G_{1/2}$ and its dilations not only attract solutions to (10.2.36) with initial condition $f^{in} \in X_0 \cap X_2$ which obviously satisfy (10.2.58) with $\mu = 1$, but also solutions to (10.2.36) with initial condition $f^{in} \in X_{0,1}$ for which Q slowly diverges to infinity as $x \to \infty$.

To prove Theorem 10.2.9, we shall find $w \geq 2\varrho$ such that, as $t \to \infty$, the rescaled Bernstein function $\xi \mapsto e^{wt}\Lambda f(t, \xi e^{-wt})$ of f converges pointwise to the Bernstein function $\Lambda\varphi$ of one of the self-similar profiles described in Proposition 10.2.7. It turns out to be more convenient to study the large time limit (if any) of $\xi \mapsto \partial_\xi \Lambda f(t, \xi e^{-wt})$, for which we establish the following lemma.

Lemma 10.2.10. *Consider $f^{in} \in X_{0,1,+}$ with $\varrho := M_1(f^{in}) > 0$ and let f be a mass-conserving weak solution on $[0, \infty)$ to (10.2.36). Assume further that f^{in} satisfies (10.2.58). Then, for all $\xi \in (0, \infty)$,*

$$\lim_{t \to \infty} \partial_\xi \Lambda f(t, \xi e^{-wt}) = \frac{1}{(\psi' \circ \psi_\star)(\xi)} \ , \tag{10.2.59}$$

where

$$\psi(\xi) := \frac{\xi + c\xi^{1+\mu}}{\varrho} \ , \qquad \xi \in [0, \infty) \ , \tag{10.2.60}$$

$$w := \frac{1+\mu}{\mu}\varrho \ , \qquad c := \frac{C_\infty \Gamma(2 - \mu)}{\mu(\mu+1)\varrho^{1+\mu}} \ , \tag{10.2.61}$$

and $\psi_\star$ denotes the inverse function of ψ; that is, $\psi \circ \psi_\star = \mathrm{id}$.

Proof. Introducing $u := \partial_\xi \Lambda f$ and $u^{in} := d\Lambda f^{in}/d\xi$, we infer from (10.2.58) and the properties of f^{in} that

$$\varrho - u^{in}(\xi) \sim \frac{C_\infty \Gamma(2-\mu)}{\mu}\xi^\mu \quad \text{as} \quad \xi \to 0 \ , \tag{10.2.62}$$

and, since $\Lambda f^{in}(\xi) \to 0$ as $\xi \to 0$,

$$\varrho\xi - \Lambda f^{in}(\xi) \sim \frac{C_\infty \Gamma(2-\mu)}{\mu(\mu+1)}\xi^{1+\mu} \quad \text{as} \quad \xi \to 0 \ . \tag{10.2.63}$$

It also follows from (10.2.51) and (10.2.52) that, for $(t, \xi) \in [0, \infty)^2$,

$$u(t, \xi) = \frac{\varrho e^{-\varrho t} u^{in}(\Sigma(t, \xi))}{\varrho - (1 - e^{-\varrho t}) u^{in}(\Sigma(t, \xi))} \ ,$$

the function Σ being defined in Proposition 10.2.8. Thus,

$$\frac{\varrho}{u(t, \xi)} = 1 + \frac{e^{\varrho t}\left[\varrho - u^{in}(\Sigma(t, \xi))\right]}{u^{in}(\Sigma(t, \xi))} \ , \qquad (t, \xi) \in (0, \infty)^2 \ . \tag{10.2.64}$$

Introducing

$$\Xi(t, \xi) := e^{wt} S\left(t, \frac{\xi e^{(\varrho-w)t}}{\varrho}\right) \ , \qquad (t, \xi) \in (0, \infty)^2 \ , \tag{10.2.65}$$

where S is given by (10.2.55), we note that

$$\Sigma(t, \Xi(t, \xi)e^{-wt}) = \frac{\xi e^{(\varrho-w)t}}{\varrho} \ , \qquad (t, \xi) \in (0, \infty)^2 \ , \tag{10.2.66}$$

by (10.2.57). We now fix $\xi > 0$ and deduce from (10.2.55) and (10.2.65) that

$$\varrho\Xi(t, \xi) = e^{wt}\left[\xi e^{(\varrho-w)t} - \Lambda f^{in}\left(\frac{\xi e^{(\varrho-w)t}}{\varrho}\right)\right] + e^{(w-\varrho)t}\Lambda f^{in}\left(\frac{\xi e^{(\varrho-w)t}}{\varrho}\right) \ . \tag{10.2.67}$$

On the one hand, the property $w \geq 2\varrho > \varrho$ (see (10.2.61)) and (10.2.63) imply that, as $t \to \infty$,

$$e^{wt}\left[\xi e^{(\varrho-w)t} - \Lambda f^{in}\left(\frac{\xi e^{(\varrho-w)t}}{\varrho}\right)\right] \sim ce^{[w+(1+\mu)(\varrho-w)]t}\xi^{1+\mu} \ ,$$

the constant c being defined in (10.2.61), and the choice (10.2.61) of w guarantees that the right-hand side of the previous formula does not depend on time. Consequently,

$$\lim_{t\to\infty} e^{wt}\left[\xi e^{(\varrho-w)t} - \Lambda f^{in}\left(\frac{\xi e^{(\varrho-w)t}}{\varrho}\right)\right] = c\xi^{1+\mu} . \qquad (10.2.68)$$

On the other hand, since $\Lambda f^{in}(0) = 0$ and $d\Lambda f^{in}(0)/d\xi = \varrho$, there holds

$$\lim_{t\to\infty} e^{(w-\varrho)t}\Lambda f^{in}\left(\frac{\xi e^{(\varrho-w)t}}{\varrho}\right) = \xi . \qquad (10.2.69)$$

Combining (10.2.67), (10.2.68) and (10.2.69) gives

$$\lim_{t\to\infty} \Xi(t,\xi) = \frac{\xi + c\xi^{1+\mu}}{\varrho} = \psi(\xi) . \qquad (10.2.70)$$

We next deduce from (10.2.64) (with $\Xi(t,\xi)e^{-wt}$ instead of ξ) and (10.2.66) that

$$\frac{\varrho}{u(t,\Xi(t,\xi)e^{-wt})} = 1 + \frac{e^{\varrho t}\left[\varrho - u^{in}(\xi e^{(\varrho-w)t}/\varrho)\right]}{u^{in}(\xi e^{(\varrho-w)t}/\varrho)}$$

and from (10.2.61) and (10.2.62) that

$$\lim_{t\to\infty} e^{\varrho t}\left[\varrho - u^{in}(\xi e^{(\varrho-w)t}/\varrho)\right] = \frac{C_\infty\Gamma(2-\mu)}{\mu\varrho^\mu}\xi^\mu .$$

Therefore

$$\lim_{t\to\infty} \frac{\varrho}{u(t,\Xi(t,\xi)e^{-wt})} = 1 + \frac{C_\infty\Gamma(2-\mu)}{\mu\varrho^{1+\mu}}\xi^\mu = \varrho\psi'(\xi) ,$$

hence

$$\lim_{t\to\infty} u(t,\Xi(t,\xi)e^{-wt}) = \frac{1}{\psi'(\xi)} . \qquad (10.2.71)$$

We now combine (10.2.70) and (10.2.71) to identify the limit of $u(t,\psi(\xi)e^{-wt})$ as $t\to\infty$. To this end, we consider $\delta \in (0,\xi)$ and infer from (10.2.70) and the monotonicity of ψ that there is $t_\delta(\xi) > 0$ such that, for $t \geq t_\delta(\xi)$,

$$\Xi(t,\xi-\delta) - \psi(\xi-\delta) \leq \psi(\xi) - \psi(\xi-\delta) ,$$
$$\psi(\xi) - \psi(\xi+\delta) \leq \Xi(t,\xi+\delta) - \psi(\xi+\delta) ,$$

and hence

$$\Xi(t,\xi-\delta) \leq \psi(\xi) \leq \Xi(t,\xi+\delta) .$$

Owing to the monotonicity of $\xi_0 \mapsto u(t,\xi_0)$ for each $t \in [0,\infty)$, we further obtain that

$$u(t,\Xi(t,\xi+\delta)e^{-wt}) \leq u(t,\psi(\xi)e^{-wt}) \leq u(t,\Xi(t,\xi-\delta)e^{-wt}) , \qquad t \geq t_\delta(\xi) ,$$

hence, by (10.2.71),

$$\frac{1}{\psi'(\xi+\delta)} \leq \liminf_{t\to\infty} u(t,\psi(\xi)e^{-wt}) \leq \limsup_{t\to\infty} u(t,\psi(\xi)e^{-wt}) \leq \frac{1}{\psi'(\xi-\delta)} .$$

Letting $\delta \to 0$ in the previous inequalities gives

$$\lim_{t\to\infty} u(t,\psi(\xi)e^{-wt}) = \frac{1}{\psi'(\xi)} .$$

Since the previous property holds true for all $\xi > 0$ and ψ is clearly a homeomorphism from $(0,\infty)$ onto $(0,\infty)$, Lemma 10.2.10 readily follows after substituting $\psi(\xi)$ for ξ. $\qquad\square$

Theorem 10.2.9 is now a simple consequence of Lemma 10.2.10 and the classification of self-similar solutions performed in Proposition 10.2.7.

Proof of Theorem 10.2.9. Keeping the notation introduced in Lemma 10.2.10, it remains to connect ψ to the Bernstein function of one of the self-similar profiles described in Proposition 10.2.7. To this end, we note that (10.2.60) and (10.2.61) imply that $\psi_\star$ solves

$$\varrho\xi = \psi_\star(\xi) + c\psi_\star^{w/(w-\varrho)}(\xi) , \qquad \xi \in [0, \infty) ;$$

that is, it solves (10.2.48). According to Proposition 10.2.7 and its proof, we conclude that $\psi_\star = \Lambda\varphi$ with

$$\varphi(x) := \frac{\zeta^2}{\varrho}G_\theta\left(\frac{\zeta x}{\varrho}\right) , \qquad x \in (0, \infty) ,$$

where $\zeta := c^{(\theta-1)/\theta}$ and $\theta = \varrho/w = \mu/(1+\mu) \in (0, 1/2]$. Recalling (10.2.59), we have shown that

$$\lim_{t\to\infty} \partial_\xi \Lambda f(t, \xi e^{-wt}) = \frac{d\Lambda\varphi}{d\xi}(\xi) , \qquad \xi \in (0, \infty) . \tag{10.2.72}$$

Since $\xi \mapsto \partial_\xi \Lambda f(t, \xi e^{-wt})$ is the Laplace transform of $x \mapsto e^{2wt}xf(t, xe^{wt})$, we infer from [114, XIII.1, Theorem 2a], the conservation of mass, and the convergence (10.2.72) that $x \mapsto e^{2wt}xf(t, xe^{wt})$ converges to $x \mapsto x\varphi(x)$ as $t \to \infty$ in the weak topology of measures. Combining this convergence with the absolute continuity of φ with respect to the Lebesgue measure on $(0, \infty)$ completes the proof. $\qquad\square$

10.2.3   The Diagonal Coagulation Kernel $k(x, y) = r(x)\delta_{x-y}$

Before dealing with a general class of coagulation kernels, let us stop for a moment to discuss the very peculiar case where pairwise coagulation involves only particles of the same size; that is, the coagulation kernel k is given by

$$k(x, y) = r(x)\delta_{x-y} , \qquad (x, y) \in (0, \infty)^2 , \tag{10.2.73}$$

where r is a nonnegative function and δ is the Dirac mass. Despite the fact that the coagulation kernel (10.2.73) is unlikely to have any physical relevance, it is somewhat an extreme case of a coagulation kernel for which the rate of coalescence is the highest when the sizes of the incoming particles are comparable, a more realistic example being the product kernel $k(x, y) = r(x)r(y)$, $(x, y) \in (0, \infty)^2$. As far as we know, the diagonal coagulation kernel (10.2.73) was introduced in [179] to study the onset of gelation for the discrete coagulation equations, see also [52, 53] for further results in that direction. It was further studied later on in [181], from a completely different perspective though. More precisely, assuming additionally that k is homogeneous of order $\lambda < 1$, that is,

$$r(x) = x^{1+\lambda} , \qquad x \in (0, \infty) , \tag{10.2.74}$$

in (10.2.73)[1], the aim of [181] is, in particular, to construct mass-conserving self-similar solutions of the form $\varphi_s(t, x) = \sigma(t)^{-2}\varphi(x\sigma(t)^{-1})$ with the hope that a simpler analysis is possible. Indeed, when k is defined via (10.2.73) and (10.2.74), the coagulation equation is reduced to

$$\partial_t f(t, x) = \frac{1}{4}\left(\frac{x}{2}\right)^{1+\lambda}\left[f\left(t, \frac{x}{2}\right)\right]^2 - x^{1+\lambda}[f(t, x)]^2 , \qquad (t, x) \in (0, \infty)^2 , \tag{10.2.75}$$

---

[1]The exponent is $\lambda + 1$ as the Dirac mass is homogeneous of order -1.

which is a nonlinear delay differential equation (instead of a nonlinear nonlocal equation). The existence of mass-conserving self-similar solutions to (10.2.75) is established in [181] when $\lambda < 1$, the uniqueness up to scaling being shown in [172].

Proposition 10.2.11. *Let $\lambda < 1$ and let $\omega > 0$ be the unique positive solution to*

$$\frac{1+\omega}{2} = \frac{1 - 2^{(\lambda-1)(1+\omega)}}{1 - 2^{\lambda-1}} .$$

There is a mass-conserving self-similar solution

$$\varphi_s : (t, x) \longmapsto \frac{1}{(1 + (1-\lambda)t)^{2/(1-\lambda)}} \varphi \left(\frac{x}{(1 + (1-\lambda)t)^{1/(1-\lambda)}} \right) \tag{10.2.76}$$

to (10.2.75) such that

- $\varphi(x) = x^{-1-\lambda} \left(\dfrac{1-\lambda}{1 - 2^{\lambda-1}} - A_0 x^\omega + o(x^\omega) \right)$ *as* $x \to 0$,

- $x \mapsto x^{1+\lambda} \varphi(x)$ *is decreasing on* $[0, \infty)$,

- $\varphi(x) \le A_\infty x^{-1-\lambda} e^{-ax}$, $\qquad x \in (0, \infty)$,

for some $A_0 > 0$, $A_\infty > 0$, and $a > 0$. In addition, φ_s is unique up to a scaling; that is, if $\tilde{\varphi}_s$ is a mass-conserving self-similar solution to (10.2.75) of the form (10.2.76) with profile $\tilde{\varphi}$, then there is $\zeta > 0$ such that $\varphi(x) = \zeta^{1+\lambda} \tilde{\varphi}(\zeta x)$ for $x \in (0, \infty)$.

The existence proof takes advantage of the fact that φ solves a differential equation with delay, and relies on a shooting method.

The existence of other self-similar solutions is obtained in [216].

Proposition 10.2.12. *Let $\lambda < 1$ and $\beta > 1/(1-\lambda)$. There is a self-similar solution*

$$(t, x) \longmapsto \frac{1}{(1 + t/\beta)^{1+\beta(1+\lambda)}} \varphi \left(\frac{x}{(1 + t/\beta)^\beta} \right)$$

to (10.2.75) such that

- $\varphi(x) = x^{-1-\lambda} \left(\dfrac{1}{\beta(1 - 2^{\lambda-1})} - A_0 x^\omega + o(x^\omega) \right)$ *as* $x \to 0$,

- $\varphi(x) \le A_\infty x^{-(1+\beta(1+\lambda))/\beta}$, $\qquad x \in (0, \infty)$,

for some $A_0 > 0$ and $A_\infty > 0$, where ω is the unique positive solution to

$$\frac{1 + \beta(1-\lambda)\omega}{2} = \frac{1 - 2^{(\lambda-1)(1+\omega)}}{1 - 2^{\lambda-1}} .$$

While the behaviour for small sizes is of the same order for both self-similar profiles, the difference between the profiles described in Proposition 10.2.11 and in Proposition 10.2.12 is the slow decay for large sizes of those constructed in the latter, which prevents them from being in X_1.

It is worth emphasizing at this point that Propositions 10.2.11 and 10.2.12 were the first results providing the existence of a variety of self-similar solutions to the coagulation equation with a coagulation kernel which does not belong to the so-called solvable ones.

Once the existence of these special solutions is established, the next step is obviously to consider their role in the large time dynamics, if any. In particular, investigating the stability of mass-conserving self-similar solutions might shed some light on the dynamical scaling hypothesis (10.0.1) as $t \to \infty$. The dynamics of (10.2.75) turns out to be far more complex and is described in the following result [172].

Theorem 10.2.13. *[172] Let $\lambda < 1$ and consider $f^{in} \in X_{1,+}$ satisfying the additional properties*

$$x \mapsto 2^{(1+\lambda)x} f^{in}(2^x) \in L_\infty(0, \infty) , \qquad (10.2.77a)$$

$$F_0 \in L_1(0, 1) , \qquad (10.2.77b)$$

where

$$F_0(\theta) := (1 - \lambda) \ln 2 \sum_{j \in \mathbb{Z}} 4^{j+\theta} f^{in}(2^{j+\theta}) , \qquad \theta \in (0, 1) . \qquad (10.2.77c)$$

Denoting the weak solution to (10.2.75) with initial condition f^{in} by f, there exists a function $\nu : [0, \infty) \times (0, \infty) \to [0, \infty)$, depending only on f^{in}, such that

$$\lim_{t \to \infty} \int_0^\infty x \left| f(t, x) - \frac{\nu(t, x)^{(1+\lambda)/(1-\lambda)}}{t^{2/(1-\lambda)}} \varphi \left(\frac{x\nu(t, x)^{1/(1-\lambda)}}{t^{1/(1-\lambda)}} \right) \right| \, dx = 0 ,$$

where φ is the self-similar profile in X_1 satisfying $M_1(\varphi) = 1/(1 - \lambda)$, constructed in Proposition 10.2.11. More precisely,

$$\frac{1}{\nu(t, x)} := (F_0 \circ \Xi) \left(\frac{\ln t}{(1 - \lambda) \ln 2} , \frac{\ln x}{\ln 2} - \frac{\ln t}{(1 - \lambda) \ln 2} \right)$$

for $(t, x) \in [0, \infty) \times (0, \infty)$, where $\Xi(s, y) := y + s - \lfloor y + s \rfloor$ for $(s, y) \in \mathbb{R}^2$.

Since the function ν depends in general on both t and x, a consequence of Theorem 10.2.13 is that the dynamical scaling hypothesis (10.0.1) is only valid for initial conditions f^{in} for which F_0 is a constant. This is obviously a non-generic feature which is, in addition, highly unstable with respect to perturbations. For instance, given $\alpha > 1$, the initial condition f^{in} given by $f^{in}(x) = x^{-1-\alpha} \mathbf{1}_{[1,\infty)}(x)$, $x \in (0, \infty)$, belongs to X_1 and satisfies (10.2.77) with

$$F_0(\theta) = (1 - \lambda) \ln 2 \, \frac{2^{(1-\alpha)\theta}}{1 - 2^{1-\alpha}} , \qquad \theta \in [0, 1] .$$

In this particular case, the long-term behaviour is not self-similar. However, owing to the 1-periodicity of Ξ with respect to its two variables, one has

$$\nu(2^{\lambda-1}t, x) = \nu(t, x) , \qquad (t, x) \in [0, \infty) \times (0, \infty) ,$$

so that, though time-dependent, the dynamics is in the end cyclic with respect to time.

To conclude this section, it is worth mentioning that there are initial conditions for which F_0 is a constant. A somewhat artificial example is the following: let $(\zeta_j)_{j \in \mathbb{Z}}$ be a sequence of nonnegative real numbers in $\ell_1(\mathbb{Z})$ such that $(2^{(1+\lambda)j} \zeta_j)_{j \geq 1}$ is bounded and set

$$f^{in}(x) := \frac{1}{x^2} \sum_{j \in \mathbb{Z}} \zeta_j \mathbf{1}_{(2^j, 2^{j+1})}(x) , \qquad x \in (0, \infty) .$$

Then $F_0(\theta) = \sum_{j \in \mathbb{Z}} \zeta_j$ for all $\theta \in (0, 1)$. It is however clear that this nice property is destroyed by suitable slight perturbations of f^{in}.

10.2.4 Mass-Conserving Self-Similar Profiles

The purpose of this section is to construct mass-conserving self-similar profiles to the co-agulation equation (10.2.1) for homogeneous coagulation kernels of order $\lambda < 1$. Specifically, we consider a coagulation kernel $k \in C((0,\infty)^2)$ satisfying

$$k(x,y) = k(y,x) \geq 0 , \qquad (x,y) \in (0,\infty)^2 , \tag{10.2.78}$$

and assume that there is $\lambda \in (-\infty, 1)$ such that

$$k(\xi x, \xi y) = \xi^\lambda k(x,y) , \qquad (\xi, x, y) \in (0,\infty)^3 . \tag{10.2.79}$$

We then look for a nonnegative function $\varphi \in X_1$ satisfying $M_1(\varphi) = 1$ and solving

$$x\frac{d\varphi}{dx}(x) + 2\varphi(x) + C\varphi(x) = 0 , \qquad x \in (0,\infty) , \tag{10.2.80}$$

in a suitable sense, which we make precise now. Recall that since $\lambda < 1$, we may always set $w = 1$ in (10.0.6) and $M_1(\varphi) = 1$ according to (10.0.8).

Definition 10.2.14. *A mass-conserving self-similar profile to the coagulation equation is a function φ such that*

$$\varphi \in X_{1,+} , \qquad M_1(\varphi) = 1 , \tag{10.2.81}$$

and

$$(x,y) \longmapsto xyk(x,y)\varphi(x)\varphi(y) \in L_1((0,\infty)^2) , \tag{10.2.82}$$

and which satisfies (10.2.80) *in the following weak sense:*

$$\int_0^\infty x^2\varphi(x)\frac{d\vartheta}{dx}(x) \, dx = \int_0^\infty \int_0^\infty xk(x,y)[\vartheta(x+y) - \vartheta(x)]\varphi(x)\varphi(y) \, dydx \tag{10.2.83}$$

for all $\vartheta \in W_\infty^1(0,\infty)$ with compact support in $[0,\infty)$.

Remark 10.2.15. *Owing to the integrability* (10.2.81) *of φ, the weak formulation* (10.2.83) *is actually satisfied by all $\vartheta \in W_\infty^1(0,\infty)$ such that $x \mapsto x(d\vartheta/dx)(x) \in L_\infty(0,\infty)$.*

A first consequence of Definition 10.2.14 is the validity of (10.2.12).

Proposition 10.2.16. *Let φ be a mass-conserving self-similar profile to the coagulation equation in the sense of Definition 10.2.14. Then $\varphi \in X_2$ and*

$$z^2\varphi(z) = \int_0^z \int_{z-x}^\infty xk(x,y)\varphi(y)\varphi(x) \, dydx \quad for \ a.e. \ z \in (0,\infty) . $$

Proof. Introducing

$$\mathscr{R}(z) := \int_0^z \int_{z-x}^\infty xk(x,y)\varphi(y)\varphi(x) \, dydx , \qquad z \in (0,\infty) , $$

it follows from the nonnegativity of k and φ, and Tonelli's theorem, that

$$\int_0^\infty \mathscr{R}(z) \, dz = \int_0^\infty \int_0^\infty xk(x,y)\varphi(y)\varphi(x) \int_x^{x+y} dz \, dydx$$

$$= \int_0^\infty \int_0^\infty xyk(x,y)\varphi(y)\varphi(x) \, dydx ,$$

and the right-hand side of the above identity is finite, according to (10.2.82). We next infer from (10.2.83) and Fubini's theorem that, for $\vartheta \in W^1_\infty(0, \infty)$ with compact support,

$$\int_0^\infty z^2 \varphi(z) \frac{d\vartheta}{dz}(z) \, dz = \int_0^\infty \int_0^\infty x k(x, y) \varphi(y) \varphi(x) \int_x^{x+y} \frac{d\vartheta}{dz}(z) dz dy dx$$
$$= \int_0^\infty \mathscr{R}(z) \frac{d\vartheta}{dz}(z) \, dz \ .$$

The previous identity being valid for all compactly supported $\vartheta \in W^1_\infty(0, \infty)$ and both $z \mapsto z^2 \varphi(z)$ and $\mathscr{R}$ being locally integrable on $(0, \infty)$, we deduce that there is a constant $C \in \mathbb{R}$ such that

$$z^2 \varphi(z) = \mathscr{R}(z) + C \quad \text{for a.e. } z \in (0, \infty) \ . \tag{10.2.84}$$

If $C > 0$, then it follows from (10.2.84) and the nonnegativity of $\mathscr{R}$ that $z\varphi(z) \geq C/z$ for all $z \in (0, \infty)$, hence contradicting $\varphi \in X_1$. If $C < 0$, then (10.2.84) and the nonnegativity of φ imply that $\mathscr{R}(z) \geq |C| = -C$, which contradicts the just established integrability of $\mathscr{R}$. Therefore, $C = 0$ and we readily deduce from (10.2.84) and the integrability of $\mathscr{R}$ that $\varphi \in X_2$, as claimed. $\qquad\square$

As far as we know, there is no direct proof of the existence of a solution to (10.2.80) satisfying (10.2.81), and the proofs that are available require two steps: first, (10.2.80) is replaced by an approximation to which one can apply the dynamical systems approach, alluded to in Section 7.3.2, and thereby construct a stationary solution to this approximation. The derivation of suitable estimates is next carried out and a solution to (10.2.80) satisfying (10.2.81) is finally obtained by a compactness argument. Two different approximations have been designed: in [113], the drift term $y(d\varphi/dy)(y) + 2\varphi(y)$ is approximated by a fragmentation operator, while the analysis of [118] relies on a size discrete scheme which we describe below. We collect the existence results from [113, 118] in the next theorem, focusing on the following five classes of coagulation kernels.

(k1) Smoluchowski's coagulation kernel and its variants: there is $\alpha \in [0, 1)$, and $\beta \in (0, \infty)$ such that

$$k_1(x, y) := (x^\alpha + y^\alpha)\left(x^{-\beta} + y^{-\beta}\right) \ , \qquad (x, y) \in (0, \infty)^2 \ . \tag{10.2.85a}$$

Then $\lambda = \alpha - \beta \in (-\infty, 1)$ and the original derivation of Smoluchowski corresponds to $\alpha = \beta = 1/3$ [247, 248].

(k2) Singular product kernel: $\lambda < 0$ and

$$k_2(x, y) := (xy)^{\lambda/2} \ , \qquad (x, y) \in (0, \infty)^2 \ . \tag{10.2.85b}$$

(k3) Singular sum kernel: $\lambda < 0$ and

$$k_3(x, y) := (x + y)^\lambda \ , \qquad (x, y) \in (0, \infty)^2 \ . \tag{10.2.85c}$$

(k4) Sum kernel and its variants: there is $\alpha \in (0, \infty)$, and $\beta \in (0, 1/\alpha)$ such that

$$k_4(x, y) := (x^\alpha + y^\alpha)^\beta \ , \qquad (x, y) \in (0, \infty)^2 \ . \tag{10.2.85d}$$

Then $\lambda = \alpha\beta \in (0, 1)$, the classical sum kernel corresponding to $\beta = 1$.

(k5) Product kernel and its variants: there is $\alpha \in (0, 1/2)$, and $\beta \in [\alpha, 1 - \alpha)$ such that

$$k_5(x, y) := x^\alpha y^\beta + x^\beta y^\alpha , \qquad (x, y) \in (0, \infty)^2 . \qquad (10.2.85e)$$

Then $\lambda = \alpha + \beta \in (0, 1)$, the classical product kernel corresponding to $\alpha = \beta$.

Theorem 10.2.17. *Assume that there is $i \in \{1, \cdots, 5\}$ such that the coagulation kernel k is given by $k = k_i$. Then there exists at least one mass-conserving self-similar profile φ to the coagulation equation (10.2.1) which satisfies the following additional properties.*

(a) *If $k \in \{k_1, k_2\}$, then $\varphi \in L_\infty(0, \infty)$ and $\varphi \in X_m$ for all $m \in \mathbb{R}$.*

(b) *If $k = k_3$, then $\varphi \in L_\infty(0, \infty)$ and $\varphi \in X_m$ for all $m \geq \lambda$.*

(c) *If $k = k_4$, then $\varphi \in L_p((0, \infty), x^{\lambda - 1 + 2p}\mathrm{d}x)$ for all $p \in (1, \lambda)$ and $\varphi \in X_m$ for all $m \geq \lambda$.*

(d) *If $k = k_5$, then $\varphi \in L_p((0, \infty), x^{\lambda - 1 + 2p}\mathrm{d}x)$ for all $p \in (1, \lambda)$ and $\varphi \in X_m$ for all $m > \lambda$.*

Instead of working with φ, it turns out to be more convenient to work with the function $Q(x) := x\varphi(x)$, $x \in (0, \infty)$, which satisfies

$$\int_0^\infty xQ(x)\frac{\mathrm{d}\vartheta}{\mathrm{d}x}\,\mathrm{d}x = \int_0^\infty \int_0^\infty \frac{k(x, y)}{y}\,[\vartheta(x + y) - \vartheta(x)]\,Q(x)Q(y)\,\mathrm{d}y\mathrm{d}x \qquad (10.2.86)$$

for all $\vartheta \in W_\infty^1(0, \infty)$ with compact support in $[0, \infty)$, according to (10.2.83). In particular, the discrete approximation introduced below corresponds to (10.2.86).

10.2.4.1 A Discrete Approximation Scheme

Let $I \geq 1$ be a positive integer and consider two families of nonnegative real numbers $(v_i)_{1 \leq i \leq I+1}$ and $(k_{i,j})_{1 \leq i,j \leq I}$ such that

$$v_1 = v_{I+1} = 0 , \qquad k_{i,j} = k_{j,i} , \qquad 1 \leq i, j \leq I . \qquad (10.2.87)$$

We next define the function $H = (H_i)_{1 \leq i \leq I} \in C(\mathbb{R}^I, \mathbb{R}^I)$ by

$$H_i(q) := v_{i+1}q_{i+1} - v_iq_i + \sum_{j=1}^{i-1} \frac{k_{j,i-j}}{j}q_{i-j}q_j - \sum_{j=1}^{I-i} \frac{k_{i,j}}{j}q_iq_j \qquad (10.2.88)$$

for $q = (q_i)_{1 \leq i \leq I} \in \mathbb{R}^I$ and $1 \leq i \leq I$. In (10.2.88), the sums where the upper limit is strictly lower than the lower limit are to be taken to be equal to zero. Specifically, this occurs for $i = 1$ and $i = I$ and, taking (10.2.87) into account, $H_1(q)$ and $H_I(q)$ are given by

$$H_1(q) = v_2q_2 - \sum_{j=1}^{I-1} \frac{k_{1,j}}{j}q_1q_j ,$$

$$H_I(q) = -v_Iq_I + \sum_{j=1}^{I_1} \frac{k_{j,I-j}}{j}q_{I-j}q_j .$$

Let us first prove the existence of a zero of H in the positive cone of $\mathbb{R}^I$ with the help of the dynamical systems approach, outlined in Section 7.3.2.

Proposition 10.2.18. *Let $I \geq 1$ and consider two families of nonnegative real numbers $(v_i)_{1 \leq i \leq I+1}$ and $(k_{i,j})_{1 \leq i,j \leq I}$ satisfying (10.2.87). There exists $q \in [0, \infty)^I$ such that*

$$H(q) = 0 \quad and \quad \sum_{i=1}^{I} q_i = 1 .$$

Proof. Clearly H is a locally Lipschitz continuous function in $\mathbb{R}^I$ and we deduce from (10.2.87) that

$$\sum_{i=1}^{I} H_i(q) = 0 , \qquad q \in \mathbb{R}^I . \tag{10.2.89}$$

First, the Cauchy–Lipschitz theorem ensures that, for each $q \in \mathbb{R}^I$, there is a unique solution $\Psi(\cdot, q) = (\Psi_i(\cdot, q))_{1 \leq i \leq I} \in C^1([0, T(q)), \mathbb{R}^I)$ to the system of ordinary differential equations

$$\frac{d}{dt}\Psi(t, q) = H(\Psi(t, q)) , \qquad \Psi(0, q) = q ,$$

which is defined on a maximal time interval $[0, T(q))$ and satisfies the following alternative: either $T(q) = \infty$ or

$$T(q) < \infty \quad and \quad \lim_{t \to T(q)} \sum_{i=1}^{I} |\Psi_i(t, q)| = \infty ,$$

see [4, Theorem (7.6)] for instance. Moreover, it readily follows from (10.2.89) that

$$\sum_{i=1}^{I} \Psi_i(t, q) = \sum_{i=1}^{I} q_i , \qquad t \in [0, T(q)) . \tag{10.2.90}$$

Next, for $q \in [0, \infty)^I$ such that $q_i = 0$ for some $1 \leq i \leq I$, the definition (10.2.88) of H implies that $H_i(q) \geq 0$, so that H is quasi-positive. Consequently, if $q \in [0, \infty)^I$, then $\Psi(t, q) \in [0, \infty)^I$ for all $t \in [0, T(q))$, see [4, Theorem (16.5)]. Combining this property with (10.2.90) implies that, for $q \in [0, \infty)^I$ and $t \in [0, T(q))$,

$$0 \leq \sum_{i=1}^{I} |\Psi_i(t, q)| = \sum_{i=1}^{I} \Psi_i(t, q) = \sum_{i=1}^{I} q_i < \infty ,$$

and guarantees that $T(q) = \infty$ according to the above alternative. We have thus shown that $\Psi : [0, \infty) \times [0, \infty)^I \to [0, \infty)^I$ is a dynamical system. Furthermore, introducing the non-empty convex and compact subset Z of $\mathbb{R}^I$ defined by

$$Z := \left\{ q \in [0, \infty)^I \ : \ \sum_{i=1}^{I} q_i = 1 \right\} ,$$

the previous analysis and (10.2.90) ensure that Z is a positively invariant set for the dynamical system Ψ. We then infer from Theorem 7.3.6 that there is $q \in Z$ such that $\Psi(t, q) = q$ for all $t \geq 0$. In other words, $H(q) = 0$ and the proof is complete since $q \in Z$. $\square$

We now specify the approximation to (10.2.86) which is suitable for carrying out the proof of Theorem 10.2.17. Let $J \geq 1$ be a positive integer and put $I = J^2$, $v_{I+1}^J = 0$,

$$v_i^J := (i-1)/J , \qquad k_{i,j}^J := k(i/J, j/J) , \qquad 1 \leq i, j \leq I .$$

By (10.2.78), the families $(v_i^J)_{1 \le i \le I+1}$ and $(k_{i,j}^J)_{1 \le i,j \le I}$ satisfy (10.2.87) and we infer from Proposition 10.2.18 that there exists $q^J = (q_i^J)_{1 \le i \le I} \in [0, \infty)^I$ such that

$$\sum_{j=1}^{i-1} \frac{1}{j} k((i-j)/J, j/J) q_{i-j}^J q_j^J - \sum_{j=1}^{I-i} \frac{1}{j} k(i/J, j/J) q_i^J q_j^J$$

$$= -\frac{1}{J} \left[i \mathbf{1}_{[1,I-1]}(i) q_{i+1}^J - (i-1) q_i^J \right] \tag{10.2.91}$$

for $1 \le i \le I$ and

$$\sum_{i=1}^{I} q_i^J = 1 . \tag{10.2.92}$$

We then define the probability measure $Q^J(\mathrm{d}x)$ on $(0, \infty)$ by

$$Q^J(\mathrm{d}x) := \sum_{i=1}^{I} q_i^J \delta_{i/J}(\mathrm{d}x) , \tag{10.2.93}$$

and first show that, as expected, $Q^J(\mathrm{d}x)$ solves an approximation of (10.2.86). In the following, for a function $\vartheta \in C((0, \infty))$, we use the notation

$$\langle Q^J(\mathrm{d}x), \vartheta(x) \rangle = \sum_{i=1}^{I} q_i^J \vartheta(i/J) .$$

Lemma 10.2.19. *Let* $\vartheta \in C((0, \infty))$. *Then*

$$\left\langle Q^J(\mathrm{d}x) Q^J(\mathrm{d}y), \frac{k^J(x,y)}{y} [\vartheta(x+y) - \vartheta(x)] \right\rangle$$

$$= \left\langle Q^J(\mathrm{d}x), (Jx-1)[\vartheta(x) - \vartheta(x-1/J)] \right\rangle \tag{10.2.94}$$

with $k^J(x,y) := k(x,y) \mathbf{1}_{[0,J]}(x+y)$, $(x,y) \in (0, \infty)^2$.

Proof. For each $i \in \{1, \ldots, I\}$, we multiply (10.2.91) by $J\vartheta(i/J)$ and sum the resulting identities to obtain

$$0 = \sum_{j=1}^{I-1} \sum_{i=j+1}^{I} q_j^J q_{i-j}^J \frac{k((i-j)/J, j/J)}{(i-j)/J} \vartheta(i/J) - \sum_{i=1}^{I} \sum_{j=1}^{I-i} q_i^J q_j^J \frac{k(i/J, j/J)}{j/J} \vartheta(i/J)$$

$$+ \sum_{j=1}^{I} (j-1) q_j^J \left[\vartheta((j-1)/J) - \vartheta(j/J) \right]$$

$$= \sum_{j=1}^{I-1} \sum_{i=1}^{I-j} q_j^J q_i^J \frac{k(i/J, j/J)}{i/J} \vartheta((i+j)/J) - \sum_{i=1}^{I} \sum_{j=1}^{I-i} q_i^J q_j^J \frac{k(i/J, j/J)}{j/J} \vartheta(i/J)$$

$$+ \left\langle Q^J(\mathrm{d}x), (Jx-1)[\vartheta(x-1/J) - \vartheta(x)] \right\rangle$$

$$= \left\langle Q^J(\mathrm{d}x) Q^J(\mathrm{d}y), \frac{k^J(x,y)}{y} [\vartheta(x+y) - \vartheta(x)] \right\rangle$$

$$+ \left\langle Q^J(\mathrm{d}x), (Jx-1)[\vartheta(x-1/J) - \vartheta(x)] \right\rangle ,$$

and the proof is complete. $\qquad \square$

10.2.4.2   Moment Estimates

We now investigate the compactness properties of the sequence $(Q^J(\mathrm{d}x))_{J \geq 1}$ of probability measures constructed in the previous section, and begin with moment estimates. As a first step, we recall that

$$\langle Q^J(\mathrm{d}x), 1 \rangle = 1 , \qquad J \geq 1 , \tag{10.2.95}$$

by (10.2.92) and (10.2.93). While no specific assumption on the coagulation kernel k was needed in Section 10.2.4.1 besides (10.2.78) and (10.2.79), the moment estimates which we derive now depend on the growth of k for large and small sizes. We begin with an estimate of the moment of order $\lambda - 1$ for $Q^J(\mathrm{d}x)$ (corresponding to $M_\lambda(\varphi)$) which plays a central role in the forthcoming analysis when $k \in \{k_1, k_2, k_3, k_4\}$. We note that such an estimate is not true for k_5, according to (10.2.17c).

Lemma 10.2.20. *Assume that $k \in \{k_1, k_2, k_3, k_4\}$. Then there is $\mu_{1,\lambda-1} > 0$, depending only on k, such that*

$$\langle Q^J(\mathrm{d}x), x^{\lambda-1} \rangle \leq \mu_{1,\lambda-1} , \qquad J \geq 1 .$$

In addition, given $m \in (\lambda - 1, 0)$, there is $\mu_{1,m} > 0$, depending only on k and m, such that

$$\langle Q^J(\mathrm{d}x), x^m \rangle \leq \mu_{1,m} , \qquad J \geq 1 .$$

Proof. A common property shared by $\{k_1, k_2, k_3, k_4\}$ is that there is $C_1 > 0$, depending only on k, such that

$$k(x,y) \geq C_1 (\max\{x,y\})^\lambda , \qquad (x,y) \in (0,\infty)^2 . \tag{10.2.96}$$

Now, take $\vartheta(x) = x^{\lambda-1}$, $x \in (0,\infty)$, in (10.2.94). Since $\lambda < 1$, it follows from (10.2.96) that, for $J \geq 1$ and $y \geq x$,

$$\frac{k^J(x,y)}{y} \left[x^{\lambda-1} - (x+y)^{\lambda-1} \right] \geq C_1 y^{\lambda-1} (1 - 2^{\lambda-1}) x^{\lambda-1} \mathbf{1}_{[0,J]}(x+y) .$$

Therefore, for $J \geq 1$,

$$
\begin{aligned}
L_J &:= \left\langle Q^J(\mathrm{d}x) Q^J(\mathrm{d}y), \frac{k^J(x,y)}{y} [x^{\lambda-1} - (x+y)^{\lambda-1}] \right\rangle \\
&\geq \left\langle Q^J(\mathrm{d}x) Q^J(\mathrm{d}y), \frac{k^J(x,y)}{y} \mathbf{1}_{[0,\infty)}(y-x) [x^{\lambda-1} - (x+y)^{\lambda-1}] \right\rangle \\
&\geq (1 - 2^{\lambda-1}) C_1 \left\langle Q^J(\mathrm{d}x) Q^J(\mathrm{d}y), (xy)^{\lambda-1} \mathbf{1}_{[0,\infty)}(y-x) \mathbf{1}_{[0,J]}(x+y) \right\rangle \\
&= (1 - 2^{\lambda-1}) C_1 \left\langle Q^J(\mathrm{d}x) Q^J(\mathrm{d}y), (xy)^{\lambda-1} \mathbf{1}_{[0,\infty)}(y-x) \right\rangle \\
&\quad - (1 - 2^{\lambda-1}) C_1 \left\langle Q^J(\mathrm{d}x) Q^J(\mathrm{d}y), (xy)^{\lambda-1} \mathbf{1}_{[0,\infty)}(y-x) \mathbf{1}_{(J,\infty)}(x+y) \right\rangle \\
&\geq (1 - 2^{\lambda-1}) \frac{C_1}{2} \left\langle Q^J(\mathrm{d}x), x^{\lambda-1} \right\rangle^2 \\
&\quad - (1 - 2^{\lambda-1}) C_1 \left\langle Q^J(\mathrm{d}x) Q^J(\mathrm{d}y), (xy)^{\lambda-1} \mathbf{1}_{(J/2,\infty)}(y) \right\rangle .
\end{aligned}
$$

Thanks to (10.2.95),

$$
\begin{aligned}
\left\langle Q^J(\mathrm{d}x) Q^J(\mathrm{d}y), (xy)^{\lambda-1} \mathbf{1}_{(J/2,\infty)}(y) \right\rangle &\leq \left(\frac{2}{J} \right)^{1-\lambda} \left\langle Q^J(\mathrm{d}x) Q^J(\mathrm{d}y), x^{\lambda-1} \right\rangle \\
&= 2^{1-\lambda} J^{\lambda-1} \left\langle Q^J(\mathrm{d}x), x^{\lambda-1} \right\rangle ,
\end{aligned}
$$

and we conclude that

$$L_J \geq (1 - 2^{\lambda-1})\frac{C_1}{2}\left\langle Q^J(\mathrm{d}x), x^{\lambda-1}\right\rangle^2 - (2^{1-\lambda} - 1)C_1 J^{\lambda-1}\left\langle Q^J(\mathrm{d}x), x^{\lambda-1}\right\rangle .$$

Next, for $J \geq 1$ and $x \geq 2/J$,

$$(Jx - 1)\left[(x - 1/J)^{\lambda-1} - x^{\lambda-1}\right] = (1 - \lambda)(Jx - 1)\int_{x-1/J}^{x} y^{\lambda-2}\,\mathrm{d}y$$

$$\leq (1 - \lambda)(x - 1/J)^{\lambda-1} \leq (1 - \lambda)\left(\frac{x}{2}\right)^{\lambda-1}$$

$$\leq (1 - \lambda)2^{1-\lambda}x^{\lambda-1} ,$$

so that

$$R_J := \left\langle Q^J(\mathrm{d}x), (Jx - 1)\left[(x - 1/J)^{\lambda-1} - x^{\lambda-1}\right]\right\rangle$$
$$\leq (1 - \lambda)2^{1-\lambda}\left\langle Q^J(\mathrm{d}x), x^{\lambda-1}\right\rangle .$$

Since $L_J = R_J$ by (10.2.94), we end up with

$$(1 - 2^{\lambda-1})\frac{C_1}{2}\left\langle Q^J(\mathrm{d}x), x^{\lambda-1}\right\rangle^2$$
$$\leq \left[(2^{1-\lambda} - 1)C_1 J^{\lambda-1} + (1 - \lambda)2^{1-\lambda}\right]\left\langle Q^J(\mathrm{d}x), x^{\lambda-1}\right\rangle ,$$

from which the first statement of Lemma 10.2.20 follows since $J \geq 1$ and $\lambda < 1$.

Consider next $m \in (\lambda - 1, 0)$. By Hölder's inequality and (10.2.95),

$$\left\langle Q^J(\mathrm{d}x), x^m\right\rangle \leq \left\langle Q^J(\mathrm{d}x), x^{\lambda-1}\right\rangle^{|m|/(1-\lambda)}\left\langle Q^J(\mathrm{d}x), 1\right\rangle^{(m+1-\lambda)(1-\lambda)}$$
$$\leq \mu_{1,\lambda-1}^{|m|/(1-\lambda)} ,$$

and the proof of Lemma 10.2.20 is complete. □

Since $\lambda < 1$, an immediate consequence of Lemma 10.2.20 is that, for $r > 0$,

$$\left\langle Q^J(\mathrm{d}x), \mathbf{1}_{[0,r]}(x)\right\rangle \leq r^{1-\lambda}\left\langle Q^J(\mathrm{d}x), x^{\lambda-1}\mathbf{1}_{[0,r]}(x)\right\rangle \leq \mu_{1,\lambda-1}r^{1-\lambda} ,$$

so that the sequence $(Q^J(\mathrm{d}x))_{J\geq 1}$ does not concentrate mass at $x = 0$; that is,

$$\lim_{r\to 0}\sup_{J\geq 1}\left\langle Q^J(\mathrm{d}x), \mathbf{1}_{[0,r]}(x)\right\rangle = 0 .$$

As already mentioned, Lemma 10.2.20 cannot be true for k_5 due to (10.2.17c), but the following weaker result can be proved.

Lemma 10.2.21. *Assume that $k = k_5$ and consider $m \in (\lambda - 1, 0)$. There is $\mu_{1,m} > 0$, depending only on k and m, such that*

$$\left\langle Q^J(\mathrm{d}x), x^m\right\rangle \leq \mu_{1,m} , \qquad J \geq 2 .$$

Proof. Let $J \geq 2$. Since $m < 0$, we argue as in the proof of Lemma 10.2.20 to show that, for $x \geq 2/J$,

$$(Jx - 1)\left[(x - 1/J)^m - x^m\right] \leq |m|2^{-m}x^m.$$

Consequently,

$$R_J := \left\langle Q^J(\mathrm{d}x), (Jx - 1)\left[(x - 1/J)^m - x^m\right]\right\rangle \leq |m|2^{-m}\left\langle Q^J(\mathrm{d}x), x^m\right\rangle . \qquad (10.2.97)$$

On the other hand,

$$k(x,y) \geq (xy)^{\lambda/2} \quad \text{and} \quad x^m - (x+y)^m \geq |m|y(x+y)^{m-1}, \qquad (x,y) \in (0,\infty)^2 .$$

Consequently,

$$L_J := \left\langle Q^J(dx)Q^J(dy), \frac{k^J(x,y)}{y}[x^m - (x+y)^m] \right\rangle$$

$$\geq |m| \left\langle Q^J(dx)Q^J(dy), \frac{(xy)^{\lambda/2}}{(x+y)^{1-m}} \mathbf{1}_{[0,J]}(x+y) \right\rangle$$

$$\geq |m| \left\langle Q^J(dx)Q^J(dy), \frac{(xy)^{\lambda/2}}{(x+y)^{1-m}} \mathbf{1}_{(0,1]}(x) \mathbf{1}_{(0,1]}(y) \right\rangle ,$$

having used that $(0,1]^2 \subset \{(x,y) \in (0,\infty)^2 : x+y \leq J\}$ since $J \geq 2$. At this point we argue as in the proof of Lemma 8.2.12 and define $\theta := 2/(1-\lambda+m) > 0$ and $x_i := i^{-\theta}$, $i \geq 1$. Then

$$L_J \geq |m| \sum_{i=1}^{\infty} \left\langle Q^J(dx)Q^J(dy), \frac{(xy)^{\lambda/2}}{(x+y)^{1-m}} \mathbf{1}_{(x_{i+1},x_i]}(x) \mathbf{1}_{(x_{i+1},x_i]}(y) \right\rangle$$

$$\geq |m| \sum_{i=1}^{\infty} (2x_i)^{m-1} \left\langle Q^J(dx)Q^J(dy), (xy)^{\lambda/2} \mathbf{1}_{(x_{i+1},x_i]}(x) \mathbf{1}_{(x_{i+1},x_i]}(y) \right\rangle$$

$$\geq 2^{m-1}|m| \sum_{i=1}^{\infty} x_i^{m-1} \left\langle Q^J(dx), x^{\lambda/2} \mathbf{1}_{(x_{i+1},x_i]}(x) \right\rangle^2 . \tag{10.2.98}$$

We now infer from (10.2.98) and the Cauchy–Schwarz inequality that

$$\left\langle Q^J(dx), x^m \mathbf{1}_{(0,1]}(x) \right\rangle = \sum_{i=1}^{\infty} \left\langle Q^J(dx), x^{(2m-\lambda)/2} x^{\lambda/2} \mathbf{1}_{(x_{i+1},x_i]}(x) \right\rangle$$

$$\leq \sum_{i=1}^{\infty} \frac{x_i^{(1-m)/2}}{x_{i+1}^{(\lambda-2m)/2}} x_i^{(m-1)/2} \left\langle Q^J(dx), x^{\lambda/2} \mathbf{1}_{(x_{i+1},x_i]}(x) \right\rangle$$

$$\leq \sqrt{C_2} \left(\sum_{i=1}^{\infty} x_i^{m-1} \left\langle Q^J(dx), x^{\lambda/2} \mathbf{1}_{(x_{i+1},x_i]}(x) \right\rangle^2 \right)^{1/2}$$

$$\leq \sqrt{\frac{2^{1-m}C_2}{|m|}} \sqrt{L_J}$$

with

$$C_2 := \sum_{i=1}^{\infty} \frac{x_i^{1-m}}{x_{i+1}^{\lambda-2m}} < \infty ,$$

the finiteness of C_2 being due to the choice of θ and $(x_i)_{i\geq 1}$ which implies

$$\frac{x_i^{1-m}}{x_{i+1}^{\lambda-2m}} = \frac{(i+1)^{\theta(\lambda-2m)}}{i^{\theta(1-m)}} \leq \frac{2^{\theta(\lambda-2m)}}{i^2} .$$

We have thus shown that

$$\left\langle Q^J(dx), x^m \mathbf{1}_{(0,1]}(x) \right\rangle^2 \leq \frac{2^{1-m}C_2}{|m|} L_J ,$$

hence, since $L_J = R_J$ by Lemma 10.2.19 (with $\vartheta(x) = x^m$, $x \in (0, \infty)$),

$$\left\langle Q^J(\mathrm{d}x), x^m \mathbf{1}_{(0,1]}(x) \right\rangle^2 \leq 2^{1-2m} C_2 \left\langle Q^J(\mathrm{d}x), x^m \right\rangle .$$

Recalling (10.2.95) gives

$$\begin{aligned}
\left\langle Q^J(\mathrm{d}x), x^m \right\rangle^2 &\leq 2 \left\langle Q^J(\mathrm{d}x), x^m \mathbf{1}_{(0,1]}(x) \right\rangle^2 + 2 \left\langle Q^J(\mathrm{d}x), x^m \mathbf{1}_{(1,\infty)}(x) \right\rangle^2 \\
&\leq 4^{1-m} C_2 \left\langle Q^J(\mathrm{d}x), x^m \right\rangle + 2 \left\langle Q^J(\mathrm{d}x), 1 \right\rangle^2 \\
&\leq 4^{1-m} C_2 \left\langle Q^J(\mathrm{d}x), x^m \right\rangle + 2 ,
\end{aligned}$$

from which Lemma 10.2.21 follows with the help of Young's inequality. $\qquad\square$

We now rule out the escape of mass for large sizes and handle differently the coagulation kernels according to their boundedness or unboundedness for small sizes.

Lemma 10.2.22. *Assume that $k \in \{k_4, k_5\}$ and let $m > 0$. There exists $\mu_{1,m} > 0$, depending only on k and m, such that*

$$\left\langle Q^J(\mathrm{d}x), x^m \right\rangle \leq \mu_{1,m} , \qquad J \geq 1 .$$

Proof. Let $J \geq 1$ and $m \geq 1$. Since $m \geq 1$, it follows from the convexity of $x \mapsto x^m$ that

$$x^m - (x - 1/J)^m \geq \frac{m}{J}(x - 1/J)^{m-1} , \qquad x \in (0, \infty) , \tag{10.2.99}$$

$$x^m - (x + y)^m \geq -my(x + y)^{m-1} , \qquad (x, y) \in (0, \infty)^2 . \tag{10.2.100}$$

On the one hand, it follows from (10.2.99) that

$$\begin{aligned}
R_J &:= \left\langle Q^J(\mathrm{d}x), (Jx - 1)\left[x^m - (x - 1/J)^m\right] \right\rangle \\
&\geq m \left\langle Q^J(\mathrm{d}x), (x - 1/J)^m \right\rangle .
\end{aligned}$$

Consequently, thanks to (10.2.95) and Lemma 7.4.1,

$$\begin{aligned}
\left\langle Q^J(\mathrm{d}x), x^m \right\rangle &\leq 2^{m-1} \left[\left\langle Q^J(\mathrm{d}x), (x - 1/J)^m \right\rangle + \frac{\left\langle Q^J(\mathrm{d}x), 1 \right\rangle}{J^m} \right] \\
&\leq \frac{2^{m-1}}{m} R_J + \frac{2^{m-1}}{J^m} \\
&\leq 2^m R_J + 2^m . \tag{10.2.101}
\end{aligned}$$

On the other hand, by Lemma 7.4.1 for k_4, and Young's inequality for k_5, there is $C_3 > 0$, depending only on k, such that

$$k(x, y) \leq C_3 \left(x^\lambda + y^\lambda\right) , \qquad (x, y) \in (0, \infty)^2 ,$$

which gives, together with (10.2.100) and Lemma 7.4.1,

$$\begin{aligned}
L_J &:= \left\langle Q^J(\mathrm{d}x) Q^J(\mathrm{d}y), \frac{k^J(x, y)}{y}\left[(x + y)^m - x^m\right] \right\rangle \\
&\leq mC_3 \left\langle Q^J(\mathrm{d}x) Q^J(\mathrm{d}y), (x^\lambda + y^\lambda)(x + y)^{m-1} \right\rangle \\
&\leq 2^{m+1} C_3 \left\langle Q^J(\mathrm{d}x) Q^J(\mathrm{d}y), x^\lambda \left(x^{m-1} + y^{m-1}\right) \right\rangle .
\end{aligned}$$

Since

$$x^\lambda \left(x^{m-1} + y^{m-1} \right) \le \left(1 + \frac{\lambda}{\lambda + m - 1} \right) x^{\lambda+m-1} + \frac{m-1}{\lambda + m - 1} y^{\lambda+m-1}$$

by Young's inequality, we further obtain, using also (10.2.95),

$$L_J \le 2^{m+2} C_3 \left\langle Q^J(\mathrm{d}x), x^{\lambda+m-1} \right\rangle \left\langle Q^J(\mathrm{d}y), 1 \right\rangle \le 2^{m+2} C_3 \left\langle Q^J(\mathrm{d}x), x^{\lambda+m-1} \right\rangle . \qquad (10.2.102)$$

Recalling that $L_J = R_J$ by Lemma 10.2.19 (with $\vartheta(x) = x^m$, $x \in (0, \infty)$), we deduce from (10.2.101) and (10.2.102) that

$$\left\langle Q^J(\mathrm{d}x), x^m \right\rangle \le 4^{m+1} C_3 \left\langle Q^J(\mathrm{d}x), x^{\lambda+m-1} \right\rangle + 2^m .$$

Since $\lambda + m - 1 < m$, a further use of Young's inequality and (10.2.95) lead us to

$$\left\langle Q^J(\mathrm{d}x), x^m \right\rangle \le \frac{\lambda + m - 1}{m} \left\langle Q^J(\mathrm{d}x), x^m \right\rangle + \frac{1-\lambda}{m} \left(4^{m+1} C_3 \right)^{m/(1-\lambda)} \left\langle Q^J(\mathrm{d}x), 1 \right\rangle + 2^m$$

$$\le \frac{\lambda + m - 1}{m} \left\langle Q^J(\mathrm{d}x), x^m \right\rangle + \frac{1-\lambda}{m} \left(4^{m+3} C_3 \right)^{m/(1-\lambda)} + 2^m ,$$

from which Lemma 10.2.22 follows for $m \ge 1$. Recalling that $Q^J(\mathrm{d}x)$ is a probability measure by (10.2.95), $\left\langle Q^J(\mathrm{d}x), 1 \right\rangle = 1$, Lemma 10.2.22 extends to $m \in (0, 1)$ by interpolation. $\qquad \square$

We now establish the analogue of Lemma 10.2.22 for $\{k_1, k_2, k_3\}$. Its proof differs from that of Lemma 10.2.22 in that the moments of negative order come into play due to the possible negativity of λ and have to be controlled with the help of Lemma 10.2.20.

Lemma 10.2.23. *Assume that $k \in \{k_1, k_2, k_3\}$ and let $m > 0$. There exists $\mu_{1,m} > 0$, depending only on k and m, such that*

$$\left\langle Q^J(\mathrm{d}x), x^m \right\rangle \le \mu_{1,m} , \qquad J \ge 1 .$$

Proof. We first recall that, owing to Lemma 10.2.20, there holds

$$\left\langle Q^J(\mathrm{d}x), x^{\lambda-1} \right\rangle \le \mu_{1,\lambda-1} , \qquad J \ge 1 . \qquad (10.2.103)$$

Let $J \ge 1$ and $m \ge 1$. Arguing as in the proof of (10.2.101), we obtain

$$\left\langle Q^J(\mathrm{d}x), x^m \right\rangle \le 2^m R_J + 2^m . \qquad (10.2.104)$$

with

$$R_J := \left\langle Q^J(\mathrm{d}x), (Jx - 1) \left[x^m - (x - 1/J)^m \right] \right\rangle .$$

Also, it follows from (10.2.100) that

$$L_J := \left\langle Q^J(\mathrm{d}x) Q^J(\mathrm{d}y), \frac{k^J(x,y)}{y} \left[(x+y)^m - x^m \right] \right\rangle$$

$$\le m \left\langle Q^J(\mathrm{d}x) Q^J(\mathrm{d}y), k^J(x,y)(x+y)^{m-1} \right\rangle . \qquad (10.2.105)$$

Case 1: $k \in \{k_2, k_3\}$. In this case, the negativity of λ implies that $k(x,y) \le (xy)^{\lambda/2}$ for $(x,y) \in (0, \infty)^2$, which allows us to deduce from (10.2.105) and Lemma 7.4.1 that

$$L_J \le m 2^{m-1} \left\langle Q^J(\mathrm{d}x) Q^J(\mathrm{d}y), (xy)^{\lambda/2} \left(x^{m-1} + y^{m-1} \right) \right\rangle$$

$$\le m 2^m \left\langle Q^J(\mathrm{d}x) Q^J(\mathrm{d}y), x^{(2m-2+\lambda)/2} y^{\lambda/2} \right\rangle$$

$$= m2^m \left\langle Q^J(\mathrm{d}x), x^{(2m-2+\lambda)/2} \right\rangle \left\langle Q^J(\mathrm{d}y), y^{\lambda/2} \right\rangle .$$

Now, since $\lambda/2 \in (\lambda - 1, 0)$ and $(2m - 2 + \lambda)/2 \in (\lambda - 1, m)$, we infer from (10.2.95), (10.2.103) and Hölder's inequality that

$$\left\langle Q^J(\mathrm{d}x), x^{(2m-2+\lambda)/2} \right\rangle \le \left\langle Q^J(\mathrm{d}x), x^m \right\rangle^{(2m-\lambda)/2(m+1-\lambda)} \left\langle Q^J(\mathrm{d}x), x^{\lambda-1} \right\rangle^{(2-\lambda)/2(m+1-\lambda)}$$

$$\le \left\langle Q^J(\mathrm{d}x), x^m \right\rangle^{(2m-\lambda)/2(m+1-\lambda)} \mu_{1,\lambda-1}^{(2-\lambda)/2(m+1-\lambda)} ,$$

and

$$\left\langle Q^J(\mathrm{d}y), y^{\lambda/2} \right\rangle \le \left\langle Q^J(\mathrm{d}y), y^{\lambda-1} \right\rangle^{-\lambda/2(1-\lambda)} \left\langle Q^J(\mathrm{d}y), 1 \right\rangle^{(2-\lambda)/2(1-\lambda)}$$

$$\le \mu_{1,\lambda-1}^{-\lambda/2(1-\lambda)} .$$

Consequently, there is $C_4(m) > 0$, depending only on k and m, such that

$$L_J \le C_4(m) \left\langle Q^J(\mathrm{d}x), x^m \right\rangle^{(2m-\lambda)/2(m+1-\lambda)} .$$

Since $L_J = R_J$ by Lemma 10.2.19, we combine the previous estimate with (10.2.104) to conclude that

$$\left\langle Q^J(\mathrm{d}x), x^m \right\rangle \le 2^m C_4(m) \left\langle Q^J(\mathrm{d}x), x^m \right\rangle^{(2m-\lambda)/2(m+1-\lambda)} + 2^m .$$

Since $0 < 2m - \lambda < 2(m + 1 - \lambda)$, Lemma 10.2.23 for $m \ge 1$ readily follows from the above estimate and Young's inequality. Since $Q^J(\mathrm{d}x)$ is a probability measure by (10.2.95), the validity of Lemma 10.2.23 is finally extended to $m \in (0, 1)$ by interpolation.

Case 2: $k = k_1$. The peculiarity of this kernel is that it mixes positive and negative powers and it thus requires to be handled in a slightly different way. Let $J \ge 1$ and $m \ge 1$. Since $x^\alpha + y^\alpha \le 2^{1-\alpha}(x + y)^\alpha$ for $(x, y) \in (0, \infty)^2$ by Lemma 7.4.1 and $\alpha \in [0, 1)$, we infer from (10.2.95), (10.2.105) and Lemma 7.4.1 that

$$L_J \le m2^{1-\alpha} \left\langle Q^J(\mathrm{d}x)Q^J(\mathrm{d}y), (x + y)^{m+\alpha-1} \left(x^{-\beta} + y^{-\beta}\right) \right\rangle$$

$$\le m2^{2-\alpha} \left\langle Q^J(\mathrm{d}x)Q^J(\mathrm{d}y), (x + y)^{m+\alpha-1} x^{-\beta} \right\rangle$$

$$\le m2^{m+1} \left\langle Q^J(\mathrm{d}x)Q^J(\mathrm{d}y), \left(x^{m+\alpha-1} + y^{m+\alpha-1}\right) x^{-\beta} \right\rangle$$

$$\le m2^{m+1} \left\langle Q^J(\mathrm{d}x), x^{m+\lambda-1} \right\rangle \left\langle Q^J(\mathrm{d}y), 1 \right\rangle$$

$$\quad + m2^{m+1} \left\langle Q^J(\mathrm{d}x), x^{-\beta} \right\rangle \left\langle Q^J(\mathrm{d}y), y^{m+\alpha-1} \right\rangle$$

$$\le m2^{m+1} \left[\left\langle Q^J(\mathrm{d}x), x^{m+\lambda-1} \right\rangle + \left\langle Q^J(\mathrm{d}x), x^{-\beta} \right\rangle \left\langle Q^J(\mathrm{d}y), y^{m+\alpha-1} \right\rangle \right] .$$

Now, since $-\beta \in (\lambda - 1, 0)$, $m + \lambda - 1 \in (\lambda - 1, m)$, and $m + \alpha - 1 \in [0, m)$, it follows from (10.2.95), (10.2.103) and Hölder's inequality that

$$\left\langle Q^J(\mathrm{d}x), x^{-\beta} \right\rangle \le \left\langle Q^J(\mathrm{d}x), x^{\lambda-1} \right\rangle^{\beta/(1-\lambda)} \left\langle Q^J(\mathrm{d}x), 1 \right\rangle^{(1-\alpha)/(1-\lambda)}$$

$$\le \mu_{1,\lambda-1}^{\beta/(1-\lambda)} ,$$

$$\left\langle Q^J(\mathrm{d}x), x^{m+\lambda-1} \right\rangle \le \left\langle Q^J(\mathrm{d}x), x^m \right\rangle^{m/(m+1-\lambda)} \left\langle Q^J(\mathrm{d}x), x^{\lambda-1} \right\rangle^{(1-\lambda)/(m+1-\lambda)}$$

$$\le \left\langle Q^J(\mathrm{d}x), x^m \right\rangle^{m/(m+1-\lambda)} \mu_{1,\lambda-1}^{(1-\lambda)/(m+1-\lambda)} ,$$

and

$$\left\langle Q^J(\mathrm{d}y), y^{m+\alpha-1}\right\rangle \leq \left\langle Q^J(\mathrm{d}y), y^m\right\rangle^{(m+\alpha-1)/m} \left\langle Q^J(\mathrm{d}y), 1\right\rangle^{(1-\alpha)/m}$$
$$\leq \left\langle Q^J(\mathrm{d}y), y^m\right\rangle^{(m+\alpha-1)/m} .$$

Therefore, there is $C_5(m) > 0$ depending only on k and m such that

$$L_J \leq C_5(m) \left[\left\langle Q^J(\mathrm{d}x), x^m\right\rangle^{m/(m+1-\lambda)} + \left\langle Q^J(\mathrm{d}x), x^m\right\rangle^{(m+\alpha-1)/m}\right] .$$

Since $m < m + 1 - \lambda$ and $m + \alpha - 1 < m$, we complete the proof of Lemma 10.2.23 for $k = k_1$, as in the previous case. $\square$

We finally deal with the remaining moments with negative exponents which, in accordance with [263], are expected to be finite for coagulation kernels k involving negative powers in a suitable way, such as k_1 or k_2 (but not k_3). Indeed, according to Definition 10.2.2, k_1 and k_2 belong to class III, while k_3 lies in class II, and the mass-conserving self-similar profiles associated with the former are expected to vanish exponentially fast for small sizes, see (10.2.17a).

Lemma 10.2.24. *Assume that $k \in \{k_1, k_2\}$ and let $m \in (-\infty, \lambda-1)$. There exist $\mu_{1,m} > 0$, depending only on k and m, and $\bar{J}$, depending only on k, such that*

$$\left\langle Q^J(\mathrm{d}x), x^m\right\rangle \leq \mu_{1,m} , \qquad J \geq \bar{J} .$$

Proof. Let $m \leq \min\{-1, \lambda - 1\}$ and $J \geq 1$. Since $m < 0$, we first recall that, by (10.2.97),

$$R_J := \left\langle Q^J(\mathrm{d}x), (Jx - 1)\left[(x - 1/J)^m - x^m\right]\right\rangle \leq |m|2^{-m} \left\langle Q^J(\mathrm{d}x), x^m\right\rangle . \qquad (10.2.106)$$

We next use the symmetry of k to obtain

$$L_J := \left\langle Q^J(\mathrm{d}x)Q^J(\mathrm{d}y), \frac{k^J(x,y)}{y}\left[x^m - (x+y)^m\right]\right\rangle$$
$$= \left\langle Q^J(\mathrm{d}x)Q^J(\mathrm{d}y), \frac{k^J(x,y)}{xy}\left[x^{m+1} - x(x+y)^m\right]\right\rangle$$
$$= \frac{1}{2}\left\langle Q^J(\mathrm{d}x)Q^J(\mathrm{d}y), \frac{k^J(x,y)}{xy}\left[x^{m+1} + y^{m+1} - (x+y)^{m+1}\right]\right\rangle$$
$$= \left\langle Q^J(\mathrm{d}x)Q^J(\mathrm{d}y), \frac{k^J(x,y)}{xy}\left[x^{m+1} - \frac{(x+y)^{m+1}}{2}\right]\right\rangle .$$

Therefore

$$L_J \geq \frac{1}{2}\left\langle Q^J(\mathrm{d}x)Q^J(\mathrm{d}y), k^J(x,y)x^m y^{-1}\right\rangle . \qquad (10.2.107)$$

Case 1: $k = k_2$. It readily follows from (10.2.107) and the definition of k_2 that

$$L_J \geq \frac{1}{2}\left\langle Q^J(\mathrm{d}x)Q^J(\mathrm{d}y), x^{(2m+\lambda)/2}y^{(\lambda-2)/2}\mathbf{1}_{(0,J]}(x+y)\right\rangle$$
$$\geq \frac{1}{2}\left\langle Q^J(\mathrm{d}x), x^{(2m+\lambda)/2}\right\rangle\left\langle Q^J(\mathrm{d}y), y^{(\lambda-2)/2}\right\rangle$$
$$- \frac{1}{2}\left\langle Q^J(\mathrm{d}x)Q^J(\mathrm{d}y), x^{(2m+\lambda)/2}y^{(\lambda-2)/2}\mathbf{1}_{(J,\infty)}(x+y)\right\rangle .$$

On the one hand, since

$$\{(x,y) \in (0,\infty)^2 \;:\; x+y > J\} \subset \{(x,y) \in (0,\infty)^2 \;:\; \max\{x,y\} > J/2\}\,, \qquad (10.2.108)$$

and $(\lambda - 2)/2 \in (\lambda - 1, 0)$, it follows from (10.2.95) and Lemma 10.2.20 that

$$\left\langle Q^J(dx)Q^J(dy), x^{(2m+\lambda)/2}y^{(\lambda-2)/2}\mathbf{1}_{(J,\infty)}(x+y)\right\rangle$$
$$\le \left\langle Q^J(dx)Q^J(dy), x^{(2m+\lambda)/2}y^{(\lambda-2)/2}\mathbf{1}_{(J/2,\infty)}(x)\right\rangle$$
$$+ \left\langle Q^J(dx)Q^J(dy), x^{(2m+\lambda)/2}y^{(\lambda-2)/2}\mathbf{1}_{(J/2,\infty)}(y)\right\rangle$$
$$\le \left(\frac{2}{J}\right)^{(-\lambda-2m)/2}\left\langle Q^J(dx), 1\right\rangle\left\langle Q^J(dy), y^{(\lambda-2)/2}\right\rangle$$
$$+ \left(\frac{2}{J}\right)^{(2-\lambda)/2}\left\langle Q^J(dx), x^{(2m+\lambda)/2}\right\rangle\left\langle Q^J(dy), 1\right\rangle$$
$$\le 2^{(-\lambda-2m)/2}\mu_{1,(\lambda-2)/2} + \left(\frac{2}{J}\right)^{(2-\lambda)/2}\left\langle Q^J(dx), x^{(2m+\lambda)/2}\right\rangle\,.$$

On the other hand, we infer from (10.2.95), Lemma 10.2.23 and Hölder's inequality that

$$1 = \left\langle Q^J(dy), 1\right\rangle^2 \le \left\langle Q^J(dy), y^{(\lambda-2)/2}\right\rangle\left\langle Q^J(dy), y^{(2-\lambda)/2}\right\rangle$$
$$\le \left\langle Q^J(dy), y^{(\lambda-2)/2}\right\rangle\mu_{1,(2-\lambda)/2}\,.$$

Collecting the above estimates, we find

$$L_J \ge \frac{1}{2}\left[\frac{1}{\mu_{1,(2-\lambda)/2}} - \left(\frac{2}{J}\right)^{(2-\lambda)/2}\right]\left\langle Q^J(dx), x^{(2m+\lambda)/2}\right\rangle$$
$$- 2^{(-\lambda-2m-2)/2}\mu_{1,(2-\lambda)/2}$$
$$\ge \frac{1}{4\mu_{1,(2-\lambda)/2}}\left\langle Q^J(dx), x^{(2m+\lambda)/2}\right\rangle - 2^{(-\lambda-2m-2)/2}\mu_{1,(2-\lambda)/2}$$

for $J \ge \bar{J}$ large enough. Recalling that $R_J = L_J$ by Lemma 10.2.19, and using (10.2.106), we end up with

$$\left\langle Q^J(dx), x^{(2m+\lambda)/2}\right\rangle \le C_6(m)\left[1 + \left\langle Q^J(dx), x^m\right\rangle\right]\,, \qquad J \ge \bar{J}\,,$$

for some constant $C_6(m) > 0$ depending only on k and m. Since $m \in ((2m+\lambda)/2, 0)$, Hölder's inequality and (10.2.95) give

$$\left\langle Q^J(dx), x^{(2m+\lambda)/2}\right\rangle$$
$$\le C_6(m)\left[1 + \left\langle Q^J(dx), x^{(2m+\lambda)/2}\right\rangle^{2m/(2m+\lambda)}\left\langle Q^J(dx), 1\right\rangle^{\lambda/(2m+\lambda)}\right]$$
$$\le C_6(m)\left[1 + \left\langle Q^J(dx), x^{(2m+\lambda)/2}\right\rangle^{2m/(2m+\lambda)}\right]\,, \qquad J \ge \bar{J}\,.$$

Using once more Young's inequality provides a bound on $\left\langle Q^J(dx), x^{(2m+\lambda)/2}\right\rangle$ for $J \ge \bar{J}$, which only depends on k and m, and thus proves Lemma 10.2.24 for sufficiently negative values of m. One then closes the gap by interpolation with the help of (10.2.95).

Case 2: $k = k_1$. We only sketch the proof as it follows quite closely that of the previous case. Let $m \leq -1$ and $J \geq 1$. We deduce from (10.2.107) and the definition of k_1 that

$$L_J \geq \frac{1}{2} \left\langle Q^J(\mathrm{d}x) Q^J(\mathrm{d}y), x^{m-\beta} y^{\alpha-1} \mathbf{1}_{(0,J]}(x+y) \right\rangle$$

$$\geq \frac{1}{2} \left\langle Q^J(\mathrm{d}x), x^{m-\beta} \right\rangle \left\langle Q^J(\mathrm{d}y), y^{\alpha-1} \right\rangle$$

$$- \frac{1}{2} \left\langle Q^J(\mathrm{d}x) Q^J(\mathrm{d}y), x^{m-\beta} y^{\alpha-1} \mathbf{1}_{(J,\infty)}(x+y) \right\rangle .$$

Since $\alpha - 1 \in (\lambda - 1, 0)$, it follows from (10.2.95), (10.2.108) and Lemma 10.2.20 that

$$\left\langle Q^J(\mathrm{d}x) Q^J(\mathrm{d}y), x^{m-\beta} y^{\alpha-1} \mathbf{1}_{(J,\infty)}(x+y) \right\rangle$$

$$\leq \left\langle Q^J(\mathrm{d}x) Q^J(\mathrm{d}y), x^{m-\beta} y^{\alpha-1} \mathbf{1}_{(J/2,\infty)}(x) \right\rangle$$

$$+ \left\langle Q^J(\mathrm{d}x) Q^J(\mathrm{d}y), x^{m-\beta} y^{\alpha-1} \mathbf{1}_{(J/2,\infty)}(y) \right\rangle$$

$$\leq 2^{\beta-m} \mu_{1,\alpha-1} + \left(\frac{2}{J} \right)^{1-\alpha} \left\langle Q^J(\mathrm{d}x), x^{m-\beta} \right\rangle ,$$

while

$$1 \leq \left\langle Q^J(\mathrm{d}y), y^{\alpha-1} \right\rangle \mu_{1,\alpha-1}$$

by (10.2.95) and Lemma 10.2.23. Consequently, there is $\bar{J} \geq 2$ large enough such that, for $J \geq \bar{J}$,

$$L_J \geq \frac{1}{4\mu_{1,\alpha-1}} \left\langle Q^J(\mathrm{d}x), x^{m-\beta} \right\rangle - 2^{\beta-m-1} \mu_{1,\alpha-1} .$$

Together with Lemma 10.2.19 and (10.2.106), this ensures the existence of a constant $C_7(m) > 0$, depending only on k and m, such that

$$\left\langle Q^J(\mathrm{d}x), x^{m-\beta} \right\rangle \leq C_7(m) \left[1 + \left\langle Q^J(\mathrm{d}x), x^m \right\rangle \right] , \qquad J \geq \bar{J} .$$

Since $m - \beta < m < 0$, the end of the proof is similar to that of the previous case.   □

Let us summarise the outcome of this section in the following proposition.

Proposition 10.2.25. *Assume that $k \in \{k_1, k_2, k_3, k_4, k_5\}$. Given $m \in \mathbb{R}$, there is $\mu_{1,m} > 0$, depending on k and m, and $\bar{J} \geq 1$, depending on k, such that*

$$\left\langle Q^J(\mathrm{d}x), x^m \right\rangle \leq \mu_{1,m} , \qquad J \geq \bar{J} ,$$

for all

(a) $m \in \mathbb{R}$ if $k \in \{k_1, k_2\}$,

(b) $m \in [\lambda - 1, \infty)$ if $k \in \{k_3, k_4\}$,

(c) $m \in (\lambda - 1, \infty)$ if $k = k_5$.

10.2.4.3 Integrability Estimates

The analysis performed in the previous section guarantees that the sequence of probability measures $(Q^J(\mathrm{d}x))_{J \geq 1}$ is tight in $(0, \infty)$ in the sense of [44, p. 59], and thus that there is no loss of compactness as $x \to 0$ and $x \to \infty$. Since we expect self-similar profiles to be functions (and not measures), we now turn to "integrability" properties of $Q^J(\mathrm{d}x)$. Since $Q^J(\mathrm{d}x)$ is a discrete measure, such properties obviously cannot be satisfied by $Q^J(\mathrm{d}x)$ but rather by the piecewise constant function P^J defined by $P^J(x) = 0$ for $x \in (J, \infty)$ and

$$P^J(x) := (i-1) q_i^J , \qquad x \in ((i-1)/J, i/J] , \qquad 1 \leq i \leq I . \tag{10.2.109}$$

Proposition 10.2.26. *(a) If $k \in \{k_1, k_2, k_3\}$, then P^J belongs to $L_\infty(0, \infty)$ and there is $\mu_{2,\infty} > 0$, and $\mathcal{G}_0 \geq \bar{J}$, depending only on k, such that*

$$\left\| P^J \right\|_\infty \leq \mu_{2,\infty} , \qquad J \geq \mathcal{G}_0 .$$

(b) If $k \in \{k_4, k_5\}$ and $p \in (1, 1/\lambda)$, then P^J belongs to $L_p((0, \infty), x^{\lambda-1}\mathrm{d}x)$ and there is $\mu_{2,p} > 0$, depending only on k and p, such that

$$\int_0^\infty x^{\lambda-1} \left| P^J(x) \right|^p \, \mathrm{d}x \leq \mu_{2,p} , \qquad J \geq 2 .$$

Proof. Consider $\zeta \in C([0, \infty))$ and $m \in \mathbb{R}$. We take

$$\vartheta(x) = \int_0^x y^m \zeta(y) \, \mathrm{d}y , \qquad x \in (0, \infty) ,$$

as a test function in Lemma 10.2.19. Since

$$\langle Q^J(\mathrm{d}x), (Jx - 1)\left[\vartheta(x) - \vartheta(x - 1/J)\right]\rangle = \int_0^\infty x^m \zeta(x) P^J(x) \, \mathrm{d}x$$

and, by Fubini's theorem,

$$\left| \left\langle Q^J(\mathrm{d}x)Q^J(\mathrm{d}y), \frac{k^J(x,y)}{y}\left[\vartheta(x+y) - \vartheta(x)\right] \right\rangle \right|$$

$$\leq \sum_{i,j=1}^I \frac{J}{j} k^J(i/J, j/J) q_i^J q_j^J \int_{i/J}^{(i+j)/J} z^m |\zeta(z)| \, \mathrm{d}z$$

$$\leq \int_0^\infty z^m |\zeta(z)| \sum_{i,j=1}^I \frac{J}{j} k^J(i/J, j/J) q_i^J q_j^J \mathbf{1}_{(i/J,(i+j)/J)}(z) \, \mathrm{d}z$$

$$\leq \int_0^\infty z^m |\zeta(z)| \left\langle Q^J(\mathrm{d}x)Q^J(\mathrm{d}y), \frac{k^J(x,y)}{y} \mathbf{1}_{(x,x+y)}(z) \right\rangle \, \mathrm{d}z ,$$

we deduce from Lemma 10.2.19 and the obvious bound $k^J \leq k$ that

$$\left| \int_0^\infty x^m \zeta(x) P^J(x) \, \mathrm{d}x \right|$$

$$\leq \int_0^\infty z^m |\zeta(z)| \left\langle Q^J(\mathrm{d}x)Q^J(\mathrm{d}y), \frac{k(x,y)}{y} \mathbf{1}_{(x,x+y)}(z) \right\rangle \, \mathrm{d}z . \qquad (10.2.110)$$

Case 1: $k = k_1$. We take $m = 0$ in (10.2.110) and use (10.2.95) together with Proposition 10.2.25 to obtain, for J large enough,

$$\left| \int_0^\infty \zeta(x) P^J(x) \, \mathrm{d}x \right| \leq \|\zeta\|_1 \left\langle Q^J(\mathrm{d}x)Q^J(\mathrm{d}y), \frac{k(x,y)}{y} \right\rangle \leq C_8 \|\zeta\|_1 ,$$

with

$$C_8 := \mu_{1,\lambda}\mu_{1,-1} + \mu_{1,\alpha}\mu_{1,-\beta-1} + \mu_{1,-\beta}\mu_{1,\alpha-1} + \mu_{1,\lambda-1} .$$

Hence, by duality, $\|P^J\|_\infty \leq C_8$ and we have shown Proposition 10.2.26 in this case.

Case 2: $k \in \{k_2, k_3\}$. Since $k(x,y) \leq (xy)^{\lambda/2}$ for $(x,y) \in (0, \infty)^2$, it follows from (10.2.110) with $m = 0$ that

$$\left| \int_0^\infty \zeta(x) P^J(x) \, \mathrm{d}x \right| \leq \|\zeta\|_1 \left\langle Q^J(\mathrm{d}x)Q^J(\mathrm{d}y), \frac{k(x,y)}{y} \right\rangle$$

$$\leq \|\zeta\|_1 \left\langle Q^J(\mathrm{d}x), x^{\lambda/2} \right\rangle \left\langle Q^J(\mathrm{d}y), y^{(\lambda-2)/2} \right\rangle .$$

Since $\lambda - 1 < (\lambda - 2)/2 < \lambda/2 < 0$, we infer from Proposition 10.2.25 that

$$\left| \int_0^\infty \zeta(x) P^J(x) \, \mathrm{d}x \right| \leq \|\zeta\|_1 \mu_{1,\lambda/2} \mu_{1,(\lambda-2)/2} .$$

As above, a duality argument completes the proof in this case.

Case 3: $k \in \{k_4, k_5\}$. Let $p \in (1, 1/\lambda)$ and fix $\varepsilon > 0$ such that

$$0 < \varepsilon < \frac{1}{p} - \lambda = \frac{\lambda}{p} + 1 - 2\lambda - \frac{(1-\lambda)(p-1)}{p} < \frac{\lambda}{p} + 1 - 2\lambda . \tag{10.2.111}$$

Since $k(x,y) \leq C_3(x^\lambda + y^\lambda)$ for $(x,y) \in (0,\infty)^2$, for some constant $C_3 > 0$ depending only on k, we infer from (10.2.110) with $m = \lambda - 1$ that

$$\left| \int_0^\infty x^{\lambda-1} \zeta(x) P^J(x) \, \mathrm{d}x \right|$$

$$\leq C_3 \int_0^\infty z^{\lambda-1} |\zeta(z)| \left\langle Q^J(\mathrm{d}x) Q^J(\mathrm{d}y), \frac{x^\lambda y^{\lambda-1+\varepsilon} + y^{2\lambda-1+\varepsilon}}{y^{\lambda+\varepsilon}} \mathbf{1}_{(x,x+y)}(z) \right\rangle \mathrm{d}z$$

$$\leq C_3 \int_0^\infty z^{\lambda-1} |\zeta(z)| \left\langle Q^J(\mathrm{d}x) Q^J(\mathrm{d}y), \frac{x^\lambda y^{\lambda-1+\varepsilon} + y^{2\lambda-1+\varepsilon}}{(z-x)^{\lambda+\varepsilon}} \mathbf{1}_{(x,\infty)}(z) \right\rangle \mathrm{d}z . \tag{10.2.112}$$

Assume now that

$$\Lambda_\zeta := \left(\int_0^\infty z^{\lambda-1} |\zeta(z)|^{p/(p-1)} \, \mathrm{d}z \right)^{(p-1)/p} < \infty .$$

On the one hand, since $\lambda - 1 + \varepsilon \in (\lambda - 1, 0)$ by (10.2.111), it follows from Proposition 10.2.25, Fubini's theorem and Hölder's inequality that

$$\int_0^\infty z^{\lambda-1} |\zeta(z)| \left\langle Q^J(\mathrm{d}x) Q^J(\mathrm{d}y), \frac{x^\lambda y^{\lambda-1+\varepsilon}}{(z-x)^{\lambda+\varepsilon}} \mathbf{1}_{(x,\infty)}(z) \right\rangle \mathrm{d}z$$

$$\leq \left\langle Q^J(\mathrm{d}y), y^{\lambda-1+\varepsilon} \right\rangle \int_0^\infty z^{\lambda-1} |\zeta(z)| \left\langle Q^J(\mathrm{d}x), \frac{x^\lambda}{(z-x)^{\lambda+\varepsilon}} \mathbf{1}_{(x,\infty)}(z) \right\rangle \mathrm{d}z$$

$$\leq \mu_{1,\lambda-1+\varepsilon} \left\langle Q^J(\mathrm{d}x), x^\lambda \int_x^\infty \frac{z^{\lambda-1}}{(z-x)^{\lambda+\varepsilon}} |\zeta(z)| \, \mathrm{d}z \right\rangle$$

$$\leq \mu_{1,\lambda-1+\varepsilon} \Lambda_\zeta \left\langle Q^J(\mathrm{d}x), x^\lambda \left(\int_x^\infty \frac{z^{\lambda-1}}{(z-x)^{p(\lambda+\varepsilon)}} \, \mathrm{d}z \right)^{1/p} \right\rangle$$

$$\leq C_9(p) \mu_{1,\lambda-1+\varepsilon} \Lambda_\zeta \left\langle Q^J(\mathrm{d}x), x^{-\varepsilon+(\lambda/p)} \right\rangle$$

with

$$C_9(p,\varepsilon) := \left(\int_1^\infty \frac{z^{\lambda-1}}{(z-1)^{p(\lambda+\varepsilon)}} \, \mathrm{d}z \right)^{1/p} < \infty .$$

Note that the choice of p and ε, see (10.2.111), guarantees not only that $C_9(p,\varepsilon) < \infty$ but also that $-\varepsilon + (\lambda/p) > \lambda - 1$. It then follows from the previous estimate and Proposition 10.2.25 that

$$\int_0^\infty z^{\lambda-1} |\zeta(z)| \left\langle Q^J(\mathrm{d}x) Q^J(\mathrm{d}y), \frac{x^\lambda y^{\lambda-1+\varepsilon}}{(z-x)^{\lambda+\varepsilon}} \mathbf{1}_{(x,\infty)}(z) \right\rangle \mathrm{d}z$$

$$\leq C_9(p,\varepsilon) \mu_{1,\lambda-1+\varepsilon} \mu_{1,(\lambda-\varepsilon p)/p} \Lambda_\zeta . \tag{10.2.113}$$

Similarly, by (10.2.111) and Proposition 10.2.25,

$$\int_0^\infty z^{\lambda-1}|\zeta(z)| \left\langle Q^J(\mathrm{d}x) Q^J(\mathrm{d}y), \frac{y^{2\lambda-1+\varepsilon}}{(z-x)^{\lambda+\varepsilon}} \mathbf{1}_{(x,\infty)}(z) \right\rangle \, \mathrm{d}z \tag{10.2.114}$$
$$\leq C_9(p,\varepsilon) \mu_{1,2\lambda-1+\varepsilon} \mu_{1,(\lambda-\lambda p-\varepsilon p)/p} \Lambda_\zeta \ .$$

Gathering (10.2.112), (10.2.113) and (10.2.114), we end up with

$$\left| \int_0^\infty x^{\lambda-1} \zeta(x) P^J(x) \, \mathrm{d}x \right| \leq C_3 C_9(p,\varepsilon) \mu_{1,\lambda-1+\varepsilon} \mu_{1,(\lambda-\varepsilon p)/p} \Lambda_\zeta$$
$$+ C_3 C_9(p,\varepsilon) \mu_{1,2\lambda-1+\varepsilon} \mu_{1,(\lambda-\lambda p-\varepsilon p)/p} \Lambda_\zeta \ .$$

Since Λ_ζ is nothing but the norm of ζ in $L_{p/(p-1)}((0,\infty), x^{\lambda-1}\mathrm{d}x)$, a duality argument completes the proof. □

10.2.4.4 Existence of Mass-Conserving Self-Similar Profiles

Thanks to the estimates derived in the previous sections, we are now in a position to carry out the proof of the existence of mass-conserving self-similar profiles to the coagulation equation (10.2.1).

Proof of Theorem 10.2.17. Three steps are needed to complete the proof: we first deduce from Proposition 10.2.25 that there is a cluster point $Q(\mathrm{d}x)$ as $J \to \infty$ of the sequence $(Q^J(\mathrm{d}x))_{J\geq 1}$, and that $Q(\mathrm{d}x)$ is also a probability measure on $(0,\infty)$. We next deduce from Proposition 10.2.26 that this measure has actually a density Q with respect to the Lebesgue measure on $(0,\infty)$. We finally show that the function φ defined by $\varphi(x) := Q(x)/x$ for $x \in (0,\infty)$ is a sought self-similar profile which possesses the properties listed in Theorem 10.2.17.

Step 1: Convergence. Let $m_0 \in (\lambda - 1, 0)$. Since the sequence of probability measures $(Q^J(\mathrm{d}x))_{J\geq 1}$ has bounded moments of order m_0 and 1 according to Proposition 10.2.25, it is a tight sequence of probability measures on $[0,\infty)$. Consequently, there is a subsequence of $(Q^J(\mathrm{d}x))_{J\geq 1}$ (not relabelled), and a probability measure $Q(\mathrm{d}x)$ on $[0,\infty)$, such that $(Q^J(\mathrm{d}x))_{J\geq 1}$ converges narrowly to $Q(\mathrm{d}x)$ [44, Theorem 5.1]; that is,

$$\lim_{J\to\infty} \langle Q^J(\mathrm{d}x), \vartheta \rangle = \langle Q(\mathrm{d}x), \vartheta \rangle \tag{10.2.115}$$

for all $\vartheta \in C([0,\infty)) \cap L_\infty(0,\infty)$. In particular, $\langle Q(\mathrm{d}x), 1 \rangle = 1$. Furthermore, the boundedness of the moment of order $m_0 < 0$ of $(Q^J(\mathrm{d}x))_{J\geq 1}$ guarantees that $Q(\{0\}) = 0$ and thus that $Q(\mathrm{d}x)$ is a probability measure on $(0,\infty)$. Also, it follows from Proposition 10.2.25 and Fatou's lemma that $\langle Q(\mathrm{d}x), x^m \rangle \leq \mu_{1,m}$ for m ranging in the sets described in Proposition 10.2.25 which depend on k.

Step 2: Absolute continuity. Let us now check that $Q(\mathrm{d}x)$ is absolutely continuous with respect to the Lebesgue measure on $(0,\infty)$.

We first consider $k \in \{k_1, k_2, k_3\}$. Then $(P^J)_{J\geq 1}$ is bounded in $L_\infty(0,\infty)$ by Proposition 10.2.26 and we may thus assume without loss of generality that there is $P \in L_\infty(0,\infty)$ such that $(P^J)_{J\geq 1}$ converges weakly-$\star$ to P in $L_\infty(0,\infty)$. Consider now $\vartheta \in C_0^\infty(0,\infty)$ and $r > 1$ such that $\operatorname{supp} \vartheta \subset (1/r, r)$. Since

$$\int_0^\infty \vartheta(x) P^J(x) \, \mathrm{d}x = \left\langle Q^J(\mathrm{d}x), (Jx-1) \int_{x-1/J}^x \vartheta(y) \, \mathrm{d}y \right\rangle$$

and

$$\left| (Jx - 1) \int_{x-1/J}^{x} \vartheta(y) \; \mathrm{d}y - x\vartheta(x) \right| \le \frac{1}{2J} \left(x \left\| \frac{d\vartheta}{dx} \right\|_{\infty} + 2\|\vartheta\|_{\infty} \right) \mathbf{1}_{(1/r, r+1)}(x) \; ,$$

we may pass to the limit as $J \to \infty$ in the above identity and find that

$$\int_0^{\infty} \vartheta(x) P(x) \; \mathrm{d}x = \langle Q(\mathrm{d}x), x\vartheta(x) \rangle \; .$$

Therefore, the measure $xQ(\mathrm{d}x)$ and the essentially bounded function P coincide in the sense of distributions which, together with the fact that $Q(\mathrm{d}x)$ has no atom at $x = 0$, implies that there is $Q \in L_1(0, \infty)$ such that $Q(\mathrm{d}x) = Q(x) \; \mathrm{d}x$. In addition, $xQ(x) = P(x)$ for a.e. $x \in (0, \infty)$ so that $x \mapsto xQ(x)$ belongs to $L_{\infty}(0, \infty)$.

We next consider $k \in \{k_4, k_5\}$ and fix $p \in (1, 1/\lambda)$. According to Proposition 10.2.26, the sequence $(P^J)_{J \ge 1}$ is bounded in $L_p((0, \infty), x^{\lambda-1}\mathrm{d}x)$ so that we may assume that there is $P \in L_p((0, \infty), x^{\lambda-1}\mathrm{d}x)$ such that $(P^J)_{J \ge 1}$ converges weakly in $L_p((0, \infty), x^{\lambda-1}\mathrm{d}x)$ to P. Consider again $\vartheta \in C_0^{\infty}(0, \infty)$ and $r > 1$ such that $\operatorname{supp} \vartheta \subset (1/r, r)$. Since

$$\int_0^{\infty} x^{\lambda-1}\vartheta(x)P^J(x) \; \mathrm{d}x = \left\langle Q^J(\mathrm{d}x), (Jx - 1) \int_{x-1/J}^{x} y^{\lambda-1}\vartheta(y) \; \mathrm{d}y \right\rangle$$

and

$$\left| (Jx - 1) \int_{x-1/J}^{x} y^{\lambda-1}\vartheta(y) \; \mathrm{d}y - x^{\lambda}\vartheta(x) \right|$$

$$\le \frac{(2r)^{1-\lambda}}{2J} \left[x \left\| \frac{d\vartheta}{dx} \right\|_{\infty} + 2(1 + rx)\|\vartheta\|_{\infty} \right] \mathbf{1}_{(1/r, r+1)}(x)$$

for $J \ge 2r$, we may let $J \to \infty$ in the above identity and deduce that

$$\int_0^{\infty} \vartheta(x) x^{\lambda-1} P(x) \; \mathrm{d}x = \langle Q(\mathrm{d}x), x^{\lambda}\vartheta(x) \rangle \; .$$

We then conclude as in the previous case that $Q(\mathrm{d}x)$ is absolutely continuous with respect to the Lebesgue measure and that its density $Q \in L_1(0, \infty)$ satisfies $xQ(x) = P(x)$ for a.e. $x \in (0, \infty)$, and thus belongs to $L_p((0, \infty), x^{\lambda+p-1}\mathrm{d}x)$.

Step 3: Limit equation. In the final step, we identify the equation solved by Q. Let us first consider $\vartheta \in W_{\infty}^2(0, \infty)$. Since

$$\left| x\frac{d\vartheta}{dx}(x) - (Jx - 1)[\vartheta(x) - \vartheta(x - 1/J)] \right| \le \frac{1}{J} \left(x \left\| \frac{d^2\vartheta}{dx^2} \right\|_{\infty} + \left\| \frac{d\vartheta}{dx} \right\|_{\infty} \right)$$

and $\langle Q^J(\mathrm{d}x), x \rangle \le \mu_{1,1}$ by Proposition 10.2.25, it readily follows from the convergence (10.2.115) and the continuity of $x \mapsto x(d\vartheta/dx)(x)$ on $[0, \infty)$ that

$$\lim_{J \to \infty} \langle Q^J(\mathrm{d}x), (Jx - 1)[\vartheta(x) - \vartheta(x - 1/J)] \rangle = \int_0^{\infty} xQ(x)\frac{d\vartheta}{dx}(x) \; \mathrm{d}x \; . \qquad (10.2.116)$$

Next, owing to the continuity of ϑ and the explicit algebraic growth of $k \in \{k_1, k_2, k_3, k_4, k_5\}$ at large and small sizes, we infer from (10.2.115) and Proposition 10.2.25 that

$$\lim_{J \to \infty} \left\langle Q^J(\mathrm{d}x)Q^J(\mathrm{d}y), \frac{k(x, y)}{y}[\vartheta(x + y) - \vartheta(x)] \right\rangle$$

$$= \int_0^{\infty} \int_0^{\infty} \frac{k(x, y)}{y}[\vartheta(x + y) - \vartheta(x)]Q(x)Q(y) \; \mathrm{d}y\mathrm{d}x \; . \qquad (10.2.117)$$

Finally, recalling (10.2.108) and using the symmetry of k,

$$\left| \left\langle Q^J(\mathrm{d}x)Q^J(\mathrm{d}y), \frac{k(x,y) - k^J(x,y)}{y}[\vartheta(x+y) - \vartheta(x)] \right\rangle \right|$$

$$\leq \left\| \frac{\mathrm{d}\vartheta}{\mathrm{d}x} \right\|_\infty \left\langle Q^J(\mathrm{d}x)Q^J(\mathrm{d}y), k(x,y)\mathbf{1}_{(J,\infty)}(x+y) \right\rangle$$

$$\leq 2 \left\| \frac{\mathrm{d}\vartheta}{\mathrm{d}x} \right\|_\infty \left\langle Q^J(\mathrm{d}x)Q^J(\mathrm{d}y), k(x,y)\mathbf{1}_{(J/2,\infty)}(x) \right\rangle$$

$$\leq \frac{4}{J} \left\| \frac{\mathrm{d}\vartheta}{\mathrm{d}x} \right\|_\infty \left\langle Q^J(\mathrm{d}x)Q^J(\mathrm{d}y), xk(x,y) \right\rangle ,$$

and Proposition 10.2.25 ensures that $\left(\left\langle Q^J(\mathrm{d}x)Q^J(\mathrm{d}y), xk(x,y)\right\rangle\right)_{J\geq 1}$ is bounded. Consequently,

$$\lim_{J\to\infty} \left\langle Q^J(\mathrm{d}x)Q^J(\mathrm{d}y), \frac{k(x,y) - k^J(x,y)}{y}[\vartheta(x+y) - \vartheta(x)] \right\rangle = 0 . \qquad (10.2.118)$$

Combining (10.2.116), (10.2.117), (10.2.118) and Lemma 10.2.19 gives

$$\int_0^\infty xQ(x)\frac{\mathrm{d}\vartheta}{\mathrm{d}x}(x) \, \mathrm{d}x = \int_0^\infty \int_0^\infty \frac{k(x,y)}{y}[\vartheta(x+y) - \vartheta(x)]Q(x)Q(y) \, \mathrm{d}y\mathrm{d}x , \qquad (10.2.119)$$

which is valid for all $\vartheta \in W_\infty^2(0,\infty)$. We next use a density argument to extend the validity of (10.2.119) to arbitrary functions belonging to $W_\infty^1(0,\infty)$, so that the function φ defined by $\varphi(x) := Q(x)/x$ for $x \in (0,\infty)$ satisfies (10.2.83). The properties (10.2.81) as well as the moment and integrability estimates listed in Theorem 10.2.17 have been already established in Steps 1 and 2 and ensure that (10.2.82) holds true. $\qquad \square$

Besides the coagulation kernels (10.2.85) dealt with in Theorem 10.2.17, the existence of at least one mass-conserving self-similar profile is known for

$$k(x,y) = x^\alpha y^\beta + x^\beta y^\alpha , \qquad (x,y) \in (0,\infty)^2 , \qquad (10.2.120)$$

with $\alpha \in [-1,\beta]$, $\beta \in [0,1)$, and $\lambda = \alpha + \beta \in [0,1)$, see [113, Section 5] and [111, Theorem 4.1], and for coagulation kernels k satisfying (10.2.78) and (10.2.79) with $\lambda = 0$ as well as

$$k(x,y) \leq K_0 \left[\left(\frac{x}{y}\right)^\alpha + \left(\frac{y}{x}\right)^\alpha \right] , \qquad (x,y) \in (0,\infty)^2 , \qquad (10.2.121)$$

$$\min_{|x-y|\leq\kappa(x+y)} \{k(x,y)\} \geq K_1 , \qquad (10.2.122)$$

for some $\alpha \in [0,1)$, $\kappa \in (0,1]$, $K_0 > 0$, and $K_1 > 0$ [219, Proposition 1.1]. Note that some particular cases of the kernels $\{k_1, k_4, k_5\}$ given by (10.2.85) are included in (10.2.120), while k_1 with $\alpha = \beta \in (0,1)$ is homogeneous of degree zero and satisfies (10.2.121) and (10.2.122).

Finally, coagulation kernels with a linearly growing term with respect to x or y, such as

$$k(x,y) = x^{\lambda-1}y + xy^{\lambda-1} , \qquad (x,y) \in (0,\infty)^2 , \qquad (10.2.123)$$

for some $\lambda \in (-\infty, 1)$, are not included in the above results. In fact, as mentioned earlier, formal asymptotics performed in [261, Section 5] do not provide a definitive identification of the large size behaviour of mass-conserving self-similar profiles φ for kernels such as (10.2.123). This issue is revisited in [45], the outcome being that there are mass-conserving self-similar profiles decaying algebraically fast for large sizes and a continuum of different algebraic tails is found. Though this is in contrast to the common expectation (10.2.16), it does not exclude the existence of mass-conserving self-similar profiles decaying exponentially fast for large sizes. More precisely, the following existence result, which encompasses the particular case (10.2.123), is shown in [45]. The method of proof is completely different from that of Theorem 10.2.17.

Theorem 10.2.27. *[45] Let $k \in C((0,\infty)^2)$ be a nonnegative function satisfying (10.2.78) and (10.2.79) with homogeneity degree $\lambda \in (-\infty, 1)$. Assume further that there is $K_0 > 0$, and $\delta > 0$ such that*

$$|x^{1-\lambda}k(x,1) - 1| \leq K_0 x^\delta , \qquad x \in (0,2) . \tag{10.2.124}$$

There is $\omega_* > 0$ such that, for each $\omega \in (0, \omega_*)$, there is a mass-conserving self-similar profile φ_ω to the coagulation equation in the sense of Definition 10.2.14 which is positive and continuous in $(0,\infty)$, and possesses the following properties:

$$M_\lambda(\varphi_\omega) = \frac{\omega}{1+\omega} , \tag{10.2.125}$$

there are $C_i > 0$, $i \in \{1,2,3\}$, such that

$$\frac{\omega C_1}{x^2} e^{-2x^{\lambda-1}/(1-\lambda)} \leq \varphi_\omega(x) \leq \frac{\omega C_1}{x^2} e^{-2x^{\lambda-1}/(1-\lambda)} , \qquad x \in (0,1) , \tag{10.2.126}$$

and

$$\varphi_\omega(x) \sim \omega C_3 x^{-2-\omega} \quad as \quad x \to \infty . \tag{10.2.127}$$

According to (10.2.126), φ_ω vanishes very rapidly for small sizes, which complies with the prediction (10.2.17a): indeed, the assumption (10.2.124) implies that k belongs to class III. In contrast, the decay of φ_ω for large sizes may be arbitrarily slow as long as it is compatible with φ_ω having a finite first moment. In particular, each of these profiles is associated with a mass-conserving self-similar solution and thus there is a continuum of such solutions in this case, a situation which is closer to that encountered for the additive coagulation kernel k_+ (see Section 10.2.2) than to that expected for coagulation kernels with homogeneity degree strictly below one. Finally, while formal asymptotics performed in [45] seem to exclude the existence of mass-conserving self-similar profiles decaying faster than algebraically, whether there is one decaying like $x^{-2-\omega}$ for large sizes for all $\omega > 0$ seems to be an open question.

We conclude this section with the uniqueness issue. As already mentioned, this is still a widely open problem and, besides the results available for the constant and additive coagulation kernels, the only uniqueness results we are aware of deal with either small perturbations of the constant kernel [214, 220] or the coagulation kernels $k(x,y) = (xy)^{-\alpha}$, $(x,y) \in (0,\infty)^2$, for some $\alpha > 0$ [163].

Theorem 10.2.28. *[214] Let k be a coagulation kernel satisfying (10.2.78) and (10.2.79) with $\lambda = 0$ and set $W := k - 2$. Assume that there is $\varepsilon > 0$, and $\alpha \in [0, 1/2)$ such that*

$$0 \leq W(x,y) \leq \varepsilon \left[\left(\frac{x}{y}\right)^\alpha + \left(\frac{y}{x}\right)^\alpha \right] , \qquad (x,y) \in (0,\infty)^2 .$$

Assume further that

- $x \mapsto W(x,1)$ *has an analytic extension to $\mathbb{C} \setminus (-\infty, 0]$,*

- $\operatorname{Re} W(z,1) \geq 0$ *for* $z \in \mathbb{C}$ *such that* $\operatorname{Re} z \geq 0$,

- *there is* $C_1 > 0$ *such that*

$$W(z,1) \leq C \left(|z|^{-\alpha} + |z|^{\alpha} \right) , \qquad z \in \mathbb{C} \setminus (-\infty, 0] .$$

- *Furthermore, defining* $W_+(z) = W(z,1)$ *on* $\{z \in \mathbb{C} : \operatorname{Im} z > 0\}$ *and* $W_-(z) = W(z,1)$ *on* $\{z \in \mathbb{C} : \operatorname{Im} z < 0\}$, *the functions* W_+ *and* W_- *have a* $C^{1,\omega}$-*extension to, respectively,* $\{z \in \mathbb{C} \setminus \{0\} : \operatorname{Im} z \geq 0\}$ *and* $\{z \in \mathbb{C} \setminus \{0\} : \operatorname{Im} z \leq 0\}$, *for some* $\omega > 0$,

- *there is* $C_2 > 0$ *such that*

$$|W_\pm(z)| + |zW'_\pm(z)| \leq C_2 \left(|z|^{-\alpha} + |z|^{\alpha} \right) , \qquad z \in \mathbb{C} , \ \pm \operatorname{Im} z > 0 ,$$

- *and there is* $C_W > 0$ *such that*

$$\limsup_{r \to 0} \left\{ |z^\alpha W(z,1) - C_W| \ : \ z \in \mathbb{C} \setminus (-\infty, 0] , \ |z| = r \right\} = 0 .$$

If ε *is small enough, then there is a unique mass-conserving self-similar profile in the sense of Definition 10.2.14 to the coagulation equation (10.2.1).*

The rather strong regularity required on the perturbation W is related to the use of norms involving the Laplace transform. The use of the latter to show Theorem 10.2.28 is motivated by its efficiency in proving the uniqueness of self-similar profiles for the constant kernel [198].

A different approach to uniqueness is developed in [163] for the coagulation kernel $k(x,y) = (xy)^{-\alpha}$, $(x,y) \in (0,\infty)^2$, with $\alpha > 0$. It is actually inspired by [120] and relies on the study of the weighted indefinite integral

$$x \longmapsto \int_x^\infty \frac{\varphi(x_*)}{x_*^\alpha} \, \mathrm{d}x_* , \qquad x \in (0,\infty) ,$$

of the mass-conserving profile φ.

Theorem 10.2.29. *[163] Let* $\alpha > 0$ *and* $k(x,y) = (xy)^{-\alpha}$, $(x,y) \in (0,\infty)^2$. *There is a unique mass-conserving self-similar profile* φ *in the sense of Definition 10.2.14 to the coagulation equation (10.2.1), satisfying*

$$\varphi \in C^1((0,\infty)) \cap \bigcap_{m \in \mathbb{R}} X_m .$$

We finally mention the partial uniqueness results obtained in [63].

Proposition 10.2.30. *[63] Assume that the coagulation kernel* k *is given by (10.2.120) with* $-1 < \alpha \leq \beta < 1$ *and* $\lambda = \alpha + \beta \in (-1,1)$.

(i) *If* $\alpha = 0$ *and* φ_1, φ_2 *are two mass-conserving self-similar profiles, in the sense of Definition 10.2.14, that satisfy* $M_\lambda(\varphi_1) = M_\lambda(\varphi_2)$, *then* $\varphi_1 = \varphi_2$.

(ii) *If* $\alpha < 0$ *and* φ_1, φ_2 *are two mass-conserving self-similar profiles, in the sense of Definition 10.2.14, that satisfy* $M_\alpha(\varphi_1) = M_\alpha(\varphi_2)$, $M_\beta(\varphi_1) = M_\beta(\varphi_2)$, *and also* $\varphi_1(x) \sim \varphi_2(x)$ *as* $x \to 0$, *then* $\varphi_1 = \varphi_2$.

In other words, Proposition 10.2.30 guarantees uniqueness of mass-conserving self-similar profiles once some specific information on the profile has been identified.

10.2.5   Regularity of Mass-Conserving Self-Similar Profiles

Since the behaviour for small and large sizes of mass-conserving self-similar profiles varies greatly with the properties of the coagulation kernel, see Theorem 10.2.17, it is not surprising that additional information is only available for some specific kernels, and the aim of this section is to describe the state of the art in that direction. We begin with some coagulation kernels in class III.

Proposition 10.2.31. *[63, 111] Consider $\alpha \in (-1,0)$, $\beta \in [\alpha, 1)$ with $\lambda := \alpha + \beta \in (-1,1)$, and $k(x,y) = x^\alpha y^\beta + x^\beta y^\alpha$, $(x,y) \in (0,\infty)^2$. If φ is a mass-conserving self-similar profile to the coagulation equation (10.2.1) (in the sense of Definition 10.2.14), then $\varphi \in C^\infty((0,\infty))$, $x \mapsto \varphi(x) e^{\Lambda(x)}$ is decreasing, and there is $A_0 > 0$ such that*

$$\varphi(x) \sim A_0 e^{-\Lambda(x)} \quad as \;\; x \to 0 \;,$$

where

$$\Lambda(x) := 2\log x - (1-\lambda)\left[\frac{M_\beta(\varphi)}{\alpha} x^\alpha + \frac{M_\alpha(\varphi)}{\beta} x^\beta\right] \;, \qquad x \in (0,\infty) \;.$$

Assume further that $\alpha \in [-\beta, 0)$ and $\beta \in [0,1)$. Then there are $\zeta_2 \geq \zeta_1 > 0$ such that

$$e^{-\zeta_2 x} \leq \varphi(x) \leq e^{-\zeta_1 x} \;, \qquad x \in (1,\infty) \;. \tag{10.2.128}$$

According to Proposition 10.2.31, the mass-conserving self-similar profile behaves for small sizes exactly as predicted in (10.2.17a). The derived large size behaviour is less precise than the expected decay (10.2.16) but reveals that it cannot be faster than exponential. More refined information in that direction is available for some kernels with homogeneity zero.

Proposition 10.2.32. *[219] Consider $\alpha \in (0,1)$ and let $k(x,y) = x^\alpha y^{-\alpha} + x^{-\alpha} y^\alpha$ for $(x,y) \in (0,\infty)^2$. If φ is a mass-conserving self-similar profile to the coagulation equation (10.2.1), then*

$$\omega_* := -\lim_{x \to \infty} \frac{\log \varphi(x)}{x} \quad exists.$$

Proposition 10.2.32 complies with (10.2.16) (with $\lambda = 0$) and is in fact valid for a larger class of coagulation kernels with homogeneity zero. We refer to [219] for precise assumptions. Also, some information on the cluster points as $R \to \infty$ of the family of functions $\{x \mapsto e^{\omega_* R x} \varphi(Rx) \; : \; R > 0\}$ is obtained. In particular, the set of cluster points is reduced to a single point if k is a small perturbation of the constant kernel.

We now turn to some kernels of class II for which the local behaviour for small sizes is also well-understood.

Proposition 10.2.33. *[63, 111, 119] Consider $\lambda \in (0,1)$ and let $k(x,y) = x^\lambda + y^\lambda$ for $(x,y) \in (0,\infty)^2$. If φ is a mass-conserving self-similar profile to the coagulation equation (10.2.1), then $\varphi \in C^\infty((0,\infty))$ and there is $A_0 > 0$ such that*

$$\varphi(x) \sim A_0 x^{-2+M_\lambda(\varphi)} \quad as \;\; x \to 0 \;, \qquad 2 - M_\lambda(\varphi) \in (1, \min\{3/2, 1+\lambda\}) \;.$$

In addition,

$$x \longmapsto x^{1-M_\lambda(\varphi)} \int_x^\infty \varphi(x_*) \, \mathrm{d}x_* \quad is \; decreasing,$$

and there are $\zeta_2 \geq \zeta_1 > 0$ such that (10.2.128) holds true.

Thus, for the sum kernel $k(x,y) = x^\lambda + y^\lambda$, the small size behaviour of mass-conserving self-similar profiles is completely determined and in accordance with (10.2.17b), though the exponent of the singularity of φ is not *a priori* given, as expected.

The most intricate case turns out to be that of coagulation kernels in class I, which includes, in particular, the product kernels $(xy)^{\lambda/2}$, $(x,y) \in (0,\infty)^2$, for $\lambda \in (0,1)$, and their variants.

Proposition 10.2.34. *[63, 111, 215] Consider $\beta \in (0,1)$, $\alpha \in (0,\beta]$ with $\lambda := \alpha+\beta \in (0,1)$, and $k(x,y) = x^\alpha y^\beta + x^\beta y^\alpha$, $(x,y) \in (0,\infty)^2$. If φ is a mass-conserving self-similar profile to the coagulation equation (10.2.1), then $\varphi \in C^\infty((0,\infty))$ and has the exponential decay property (10.2.128). Furthermore,*

$$x \mapsto x^m \varphi(x) \in L_\infty(0,\infty) \quad for \ \ m > 1+\lambda \ ,$$
$$x \mapsto x^m \varphi(x) \notin L_\infty(0,\infty) \quad for \ \ m < 1+\lambda \ , \tag{10.2.129}$$

and, if $\beta < 1/2$, then

$$\liminf_{x \to 0} \varphi(x) x^{1+\lambda} > 0 \quad and \quad \sup_{x \in (0,\infty)} \left\{ \varphi(x) x^{1+\lambda} \right\} < \infty \ . \tag{10.2.130}$$

Surprisingly, though the small size behaviour of mass-conserving self-similar profiles is expected to be explicitly determined, see (10.2.17c) and (10.2.17d), the properties (10.2.129) and (10.2.130) only state that $\varphi(x)$ is of order $x^{-1-\lambda}$ as $x \to 0$. The difficulty in proving that $\varphi(x) \sim A_0 x^{-1-\lambda}$ as $x \to 0$ with A_0 given by (10.2.17d) seems to be due to an oscillatory behaviour of φ for small sizes. Indeed, such a behaviour was observed numerically in [115, 178], and formal asymptotics performed in [193] show that when $\alpha = \beta = \lambda/2 \in (0,1/2)$, the small size behaviour of φ is expected to be

$$x^{1+\lambda}\varphi(x) \sim A_* \left[1 + C x^{\omega_1} \cos(\omega_2 \log(x) + \omega_3) \right] \quad as \ \ x \to 0 \ ,$$

for some suitable positive constants ω_1 and ω_2, and real constants C and ω_3.

Summarising, the small and large size behaviours of mass-conserving self-similar profiles to the coagulation equation are still not fully understood and, as their precise determination might be helpful to investigate the uniqueness issue, some more work is clearly required.

10.2.6 Other Self-Similar Profiles

We end our discussion on self-similar solutions with a couple of additional results dealing with coagulation kernels with homogeneity degree equal to one and with self-similar profiles with infinite mass, the latter corresponding to the choice $\tau \neq 2$ in the ansatz (10.0.4), according to the discussion in the preamble of Chapter 10.

10.2.6.1 Self-Similar Profiles: $\lambda = 1$

Besides the complete classification of self-similar solutions available for the additive coagulation kernel k_+, which is described in Section 10.2.2, for other coagulation kernels satisfying (10.2.79) with homogeneity degree $\lambda = 1$, such as

$$k(x,y) = (x^\alpha + y^\alpha)^{1/\alpha} \ , \qquad (x,y) \in (0,\infty)^2 \ , \tag{10.2.131}$$

for some $\alpha \in (0,\infty) \setminus \{1\}$, or

$$k(x,y) = x^\alpha y^{1-\alpha} + x^{1-\alpha} y^\alpha, \qquad (x,y) \in (0,\infty)^2 \ , \tag{10.2.132}$$

for some $\alpha \in (0, 1/2]$, the problem is still far from being completely understood. Indeed, several difficulties arise when $\lambda = 1$. On the one hand, the separation constant w cannot be set to an arbitrary value independently of the total mass as a consequence of (10.0.8), and has thus to be determined jointly with the profile φ. On the other hand, according to [180, 263], for coagulation kernels lying in class I such as (10.2.132), the scaling ansatz (10.0.1) with $\tau = 2$ for mass-conserving self-similar solutions leads to the prediction that $\varphi(x) \sim Ax^{-2}$ as $x \to 0$ for the small size behaviour of the scaling profile φ and is clearly not compatible with $\varphi \in X_1$. Consequently, in this case, self-similar solutions of the form (10.0.1) with $\tau = 2$ (if any) have infinite mass and their role in the description of the large time dynamics of weak solutions in the sense of Definition 8.2.1 is unclear. However, recent breakthroughs [46, 141, 221] shed some new light on this borderline case (recall that gelation is expected to occur for coagulation kernels having a homogeneity degree strictly exceeding one). The outcome, which we summarise below, depends heavily on the class to which k belongs (in the sense of Definition 10.2.2), namely class I as (10.2.132), or class II as (10.2.131). The analysis performed in [46, 141, 221] relies on the following change of the size variable and the unknown function

$$u(t, X) := e^{2X} f(t, e^X) , \qquad (t, X) \in [0, \infty) \times \mathbb{R} . \tag{10.2.133}$$

Then

$$\int_{\mathbb{R}} u(t, X) \, dX = M_1(f(t)) , \qquad t \in [0, \infty) , \tag{10.2.134}$$

and it follows from the coagulation equation (10.2.1) and the homogeneity of k that u solves

$$\partial_t u(t, X) = \frac{1}{2} \int_{-\infty}^{X} \frac{e^{2X}}{(e^X - e^Y)^2} l \left(e^{X-Y} - 1\right) u \left(t, \ln(e^X - e^Y)\right) u(t, Y) \, dY$$
$$- \int_{\mathbb{R}} l \left(e^{X-Y}\right) u(t, X) u(t, Y) \, dY \tag{10.2.135}$$

for $(t, X) \in (0, \infty) \times \mathbb{R}$, recalling that $l(z) = k(z, 1)$, $z \in (0, \infty)$, see (10.2.13a). Also observe that the change of variable $Y \to \ln \left(e^X - e^Y\right)$ in the first integral on the right-hand side of (10.2.135) leads to the alternative formula

$$\int_{-\infty}^{X} \frac{e^{2X}}{(e^X - e^Y)^2} l \left(e^{X-Y} - 1\right) u \left(t, \ln(e^X - e^Y)\right) u(t, Y) \, dY$$
$$= \int_{-\infty}^{X} \frac{e^X}{e^X - e^Y} l \left(e^{X-Y} - 1\right) u \left(t, \ln(e^X - e^Y)\right) u(t, Y) \, dY$$

for $X \in \mathbb{R}$. Taking (10.2.134) into account, the right-hand side of (10.2.135) is actually a derivative and an alternative formulation of (10.2.135) is given by

$$\partial_t u(t, X) = -\partial_X \left[\int_{-\infty}^{X} \int_{\ln(e^X - e^Y)}^{\infty} l \left(e^{Y-Z}\right) u(t, Y) u(t, Z) \, dZ dY \right] \tag{10.2.136}$$

for $(t, X) \in (0, \infty) \times \mathbb{R}$.

Now, let f_s be a mass-conserving self-similar solution to the coagulation equation of the form (10.0.4) with $\tau = 2$; that is, there is $w > 0$ such that

$$f_s(t, x) = e^{-2wt} \varphi(x e^{-wt}) , \qquad (t, x) \in [0, \infty) \times (0, \infty) ,$$

according to (10.0.7b). In the new variables (10.2.133), it is transformed into a travelling wave u_s, with velocity w, which is given by

$$u_s(t, X) = G(X - wt) , \qquad G(X) = e^{2X} \varphi(e^X) , \qquad (t, X) \in (0, \infty) \times \mathbb{R} . \tag{10.2.137}$$

Since the above-mentioned defect of integrability of φ occurs for small sizes, we only consider the situation where φ decays sufficiently rapidly for large sizes, thus assuming that

$$\lim_{X \to \infty} G(X) = 0 . \qquad (10.2.138)$$

We then infer from (10.2.136), (10.2.137) and (10.2.138) that, for $X \in \mathbb{R}$,

$$
\begin{aligned}
wG(X) &= \int_{-\infty}^{X} \int_{\ln(e^X - e^Y)}^{\infty} l\left(e^{Y-Z}\right) G(Y) G(Z) \, \mathrm{d}Z \mathrm{d}Y \\
&= \int_{-\infty}^{0} \int_{\ln(1 - e^Y)}^{\infty} l\left(e^{Y-Z}\right) G(Y+X) G(Z+X) \, \mathrm{d}Z \mathrm{d}Y .
\end{aligned}
\qquad (10.2.139)
$$

Introducing

$$\mathcal{I}(l) := \int_{-\infty}^{0} \int_{\ln(1 - e^Y)}^{\infty} l\left(e^{Y-Z}\right) \, \mathrm{d}Z \mathrm{d}Y \qquad (10.2.140\mathrm{a})$$

and observing that it can also be expressed as

$$\mathcal{I}(l) = \int_{0}^{\infty} \frac{l(z)}{z} \ln\left(\frac{1+z}{z}\right) \, \mathrm{d}z = \int_{0}^{\infty} \frac{l(z)}{z^2} \ln(1+z) \, \mathrm{d}z , \qquad (10.2.140\mathrm{b})$$

due to the scaling properties of k, we realise that this integral is finite when k lies in class I, thanks to (10.2.13b) and (10.2.14). This property, along with (10.2.139), allows the function G to have a finite positive limit as $X \to -\infty$ which in turn implies that G is not integrable. In other words, the corresponding self-similar profile φ, given by (10.2.137), does not belong to X_1. We refer to [141, 221] for the existence of such travelling waves and the consequences of the non-existence of self-similar solutions with a profile in X_1 for the large time dynamics of the coagulation equation.

Next, if k belongs to class II, then $\mathcal{I}(l) = \infty$ and the validity of (10.2.139) requires that $G(X) \to 0$ as $X \to -\infty$, a property which does not exclude *a priori* the integrability of G over $\mathbb{R}$. Formal asymptotics performed in [141] reveal that, in this case, if there is $\omega > 0$, and $g_\infty > 0$ such that $G(X) \sim g_\infty e^{-\omega X}$ as $X \to \infty$, then $w = l_0(1+\omega)/\omega$, with l_0 defined in (10.2.13b), and $G(X) \sim g_0 e^{\omega X/(1+\omega)}$ as $X \to -\infty$ for some $g_0 > 0$. On the one hand, if there exists such a solution G to (10.2.139), then the associated self-similar profile φ, given by (10.2.137), satisfies

$$x\varphi(x) \underset{x \to \infty}{\sim} g_\infty x^{-1-\omega} \quad \text{and} \quad x\varphi(x) \underset{x \to 0}{\sim} g_0 x^{-1/(1+\omega)} .$$

In particular, φ belongs to X_1. On the other hand, it is conjectured that no such solution exists when ω is too large. In this regard, we report the following existence result [46].

Theorem 10.2.35. *[46] Let $k \in C([0,\infty)^2)$ be a nonnegative and symmetric function satisfying (10.2.8) with homogeneity degree $\lambda = 1$. Assume further that there is $\delta > 0$, and $K_0 > 0$ such that*

$$|k(x,1) - 1| \le K_0 x^\delta , \qquad x \in [0,1] . \qquad (10.2.141)$$

Then there exists $\omega_* > 0$ such that, for every $\omega \in (0, \omega_*)$, there is a positive function $\varphi_\omega \in X_1 \cap C^1(0,\infty)$ satisfying $M_1(\varphi_\omega) = 1$ and

$$\frac{1+\omega}{\omega} x^2 \varphi_\omega(x) = \int_{0}^{x} \int_{x-y}^{\infty} y k(y,z) \varphi_\omega(y) \varphi_\omega(z) \, \mathrm{d}z \mathrm{d}y , \qquad x \in (0,\infty) , \qquad (10.2.142)$$

and such that

$$x\varphi_\omega(x) \underset{x \to \infty}{\sim} A_\infty x^{-1-\omega} \quad \text{and} \quad x\varphi_\omega(x) \underset{x \to 0}{\sim} A_0 x^{-1/(1+\omega)} , \qquad (10.2.143)$$

for some $A_0 > 0$ and $A_\infty > 0$.

It follows from (10.0.6) (with $w = (1 + \omega)/\omega$), (10.0.7b), (10.2.142) and Proposition 10.2.35 that, for every $\omega \in (0, \omega_*)$, the function f_ω defined by

$$f_\omega(t, x) := e^{-2(1+\omega)t/\omega} \varphi_\omega \left(xe^{-(1+\omega)t/\omega} \right) , \qquad (t, x) \in [0, \infty) \times (0, \infty) ,$$

is a mass-conserving self-similar solution to the coagulation equation (10.2.1). Thus there is a continuum of mass-conserving self-similar solutions with algebraic decay for large sizes, as for the additive coagulation kernel k_+ [42, 198]. In this connection, we point out that $\omega_* = 1$ for k_+, see Section 10.2.2.

10.2.6.2   Self-Similar Profiles with Infinite Mass

We finally report existence results for self-similar solutions f_s to the coagulation equation (10.2.1) with homogeneous coagulation kernel when the scaling exponent $\tau \neq 2$ in (10.0.4); that is,

$$f_s(t, x) = \frac{1}{\sigma(t)^\tau} \varphi \left(\frac{x}{\sigma(t)} \right) , \qquad (t, x) \in (0, \infty)^2 , \tag{10.2.144}$$

the mean size $\sigma(t)$ at time $t > 0$ being given by (10.0.7a) if $\tau > \lambda + 1$, or by (10.0.7b) if $\tau = \lambda + 1$. According to the discussion in the preamble to Chapter 10, in this case, the profile φ does not belong to X_1 due to its slow decay at infinity. The existence of such self-similar solutions is proved in [198] for the constant coagulation kernel. This result gave impetus to further research and it has turned out that self-similar solutions of the form (10.2.144) with $\tau \neq 2$ exist for a large class of coagulation kernels [213, 216, 217, 218]. Since the analysis is more involved for coagulation kernels with a singularity for small sizes, a more complete result is available for locally bounded coagulation kernels [217, 218].

Theorem 10.2.36. *[217, 218] Consider a nonnegative, symmetric function $k \in C((0, \infty)^2)$ which satisfies (10.2.79) for some $\lambda \in [0, 1)$. Assume further that there is $K_0 > 0$ such that*

$$k(x, y) \leq K_0(x^\lambda + y^\lambda) , \qquad (x, y) \in (0, \infty)^2 . \tag{10.2.145}$$

Given $\tau \in (\lambda + 1, 2)$, there is a nonnegative function $\varphi \in C(0, \infty)$ with the following properties:

$$x\varphi(x) \sim (2 - \tau)(\tau - \lambda - 1)x^{1-\tau} \quad as \quad x \to \infty , \tag{10.2.146}$$

$$x^{\tau-2} \int_0^x y\varphi(y) \, dy \leq \tau - \lambda - 1 , \qquad x \in (0, \infty) , \tag{10.2.147}$$

$$\sup_{x \in (0, \infty)} \left\{ x^{\tau-1} \int_x^\infty \varphi(y) \, dy + x^{\tau-\lambda-1} \int_x^\infty y^\lambda \varphi(y) \, dy \right\} < \infty , \tag{10.2.148}$$

and

$$\int_0^\infty \frac{d\vartheta}{dx}(x) \int_0^x \int_{x-y}^\infty yk(y, z)\varphi(y)\varphi(z) \, dzdydx$$

$$= \int_0^\infty x^2\varphi(x)\frac{d\vartheta}{dx}(x) \, dx - (\tau - 2) \int_0^\infty x\varphi(x)\vartheta(x) \, dx , \tag{10.2.149}$$

for all $\vartheta \in C^1(0, \infty)$ with compact support in $(0, \infty)$.

According to (10.2.145) and (10.2.148),

$$x \mapsto \int_0^x \int_{x-y}^\infty yk(y, z)\varphi(y)\varphi(z) \, dzdy \in L_1(0, R)$$

for any $R > 0$, so that (10.2.149) is meaningful. Also, it readily follows from (10.2.146) that $\varphi \notin X_1$, since $\tau < 2$. Similarly to the proof of Theorem 10.2.17, that of Theorem 10.2.36 relies on a dynamical systems approach and φ is constructed as a steady-state solution of the coagulation equation in self-similar variables

$$\partial_t g(t, x) = x \partial_x g(t, x) + \tau g(t, x) + \mathcal{C} g(t, x) , \qquad (t, x) \in (0, \infty)^2 .$$

Here, however, the dynamical systems approach is considerably more difficult to implement. Indeed, since there is no conserved quantity, one has to ensure that the stationary solution constructed in this way is not identically zero, which requires a positive lower bound to be found which is preserved throughout time evolution. In addition, since the sought profile has a rather slow algebraic decay for large sizes, the control for large sizes requires more care, which is in sharp contrast with the situation in Theorem 10.2.17. We refer to [217, 218] for a complete proof, as well as to the survey article [212].

An extension of Theorem 10.2.36 is established in [213] for the case of coagulation kernels which are unbounded for small sizes, such as the Smoluchowski coagulation kernel $k(x, y) = \left(x^{1/3} + y^{1/3} \right) \left(x^{-1/3} + y^{-1/3} \right)$.

Theorem 10.2.37. *[213] Consider a nonnegative and symmetric function $k \in C^1((0, \infty)^2)$ satisfying (10.2.79) for some $\lambda \in (-\infty, 1)$ and for which there is $\alpha \in (0, 1 - \lambda)$, $C_1 > 0$, and $C_2 > 0$ such that*

$$C_1 \left(x^{-\alpha} y^{\lambda + \alpha} + x^{\lambda + \alpha} y^{-\alpha} \right) \leq k(x, y) \leq C_2 \left(x^{-\alpha} y^{\lambda + \alpha} + x^{\lambda + \alpha} y^{-\alpha} \right) \qquad (10.2.150)$$

for $(x, y) \in (0, \infty)^2$. Assume further that, for each $r > 1$, there is $C_3(r) > 0$ such that

$$|\partial_x k(x, y)| \leq C_3(r) \left(y^{-\alpha} + y^{\lambda + \alpha} \right) , \qquad (x, y) \in (1/r, r) \times (0, \infty) . \qquad (10.2.151)$$

Given $\tau \in (1 + (\lambda + \alpha)_+, 2)$, there is a nonnegative function $\varphi \in C(0, \infty)$ which satisfies (10.2.149) and

$$x \varphi(x) \sim (2 - \tau) x^{1 - \tau} \quad as \quad x \to \infty . \qquad (10.2.152)$$

In contrast to Theorem 10.2.36, the parameter τ is required to be above $1 + (\lambda + \alpha)_+$ in Theorem 10.2.37 and it is yet unclear whether Theorem 10.2.37 extends to the expected range $(\lambda + 1, 2)$. That this might indeed be true, at least for some classes of coagulation kernels, is suggested by the existence results established in [216] for the diagonal coagulation kernel $k(x, y) = x^{\lambda + 1} \delta_{x - y}$ when $\lambda \in (-\infty, 1)$, see Proposition 10.2.12. In addition, when $\lambda = 0$ and k is a small perturbation of the constant kernel, Theorem 10.2.37 is supplemented with a uniqueness result in the spirit of Theorem 10.2.28 [256]. Finally, it is shown in [257] that if φ is a nonnegative solution to (10.2.149) having suitable integrability properties for a fairly general class of coagulation kernels, then its behaviour for large sizes is prescribed by (10.2.146) or (10.2.152).

10.3 Coagulation-Fragmentation

Having investigated the long-term behaviour of solutions to the fragmentation and coagulation equations separately, though with limited success for the latter, we now turn to the situation when both coagulation and fragmentation are taken into account. Since these mechanisms act in opposite directions, we might expect them to reach an equilibrium as time goes by, and thus that stationary solutions exist. This is clearly the case when

$$k(x,y) = k(y,x) = 2 , \quad a(x) = x , \quad \text{and} \quad b(x,y) = \frac{2}{y} , \qquad 0 < x < y , \tag{10.3.1}$$

for which the C-F equation has a one-parameter family of explicit stationary solutions given by $\{e^{x \ln z} : z \in (0, \infty)\} \cup \{0\}$. Indeed, with the choice (10.3.1), the C-F equation can also be written

$$\partial_t f(t,x) = \int_0^x \left(f(t, x-y)f(t,y) - f(t,x) \right) \, dy$$
$$- 2 \int_0^\infty \left(f(t,x)f(t,y) - f(t, x+y) \right) \, dy ,$$

from which it is readily seen that any time-independent function f having the property that $f(x)f(y) = f(x+y)$, for $(x,y) \in (0,\infty)^2$, is a stationary solution. Clearly $x \mapsto e^{x \ln z}$ has this property for any $z \in (0, \infty)$, but has finite total mass $|\ln z|^{-2}$ only for $z \in (0,1)$. Consequently, the C-F equation with coefficients given by (10.3.1) has at least a stationary solution for each mass. Fortunately, the coefficients (10.3.1) are not the only ones for which stationary solutions can be obtained in such a simple way, and a similar construction can be performed when the coagulation and fragmentation coefficients satisfy the so-called detailed balance condition (6.1.20), already discussed in Section 2.3.3 in Volume I: there is a nonnegative symmetric function F defined in $(0,\infty)^2$, and a positive function Q defined in $(0,\infty)$ such that

$$a(x) = \frac{1}{2} \int_0^x F(x_*, x - x_*) \, dx_* , \qquad x > 0 ,$$
$$a(y)b(x,y) = F(x, y-x) , \qquad 0 < x < y , \tag{10.3.2a}$$

and

$$k(x,y)Q(x)Q(y) = F(x,y)Q(x+y) , \qquad (x,y) \in (0,\infty)^2 . \tag{10.3.2b}$$

Note that the choice (10.3.1) satisfies (10.3.2) with $F \equiv 2$ and $Q \equiv 1$. Thanks to (10.3.2), the C-F equation can be written as

$$\partial_t f(t,x) = \frac{1}{2} \int_0^x [k(x-y,y)f(t, x-y)f(t,y) - F(x-y,y)f(t,x)] \, dy$$
$$- \int_0^\infty [k(x,y)f(t,x)f(t,y) - F(x,y)f(t, x+y)] \, dy , \tag{10.3.3}$$

and (10.3.2b) ensures that Q is a stationary solution. Since $Q_z : x \mapsto Q(x)e^{x \ln z}$ also satisfies (10.3.2b) for all $z \in (0, \infty)$, it is also a stationary solution and we have thus found a one-parameter family $\{Q_z : z \in (0, \infty)\}$ of stationary solutions. However, whether they have finite mass or not depends strongly on the behaviour of Q and the additional requirement that $\{Q_z : z \in (0, \infty)\} \cap X_1$ is non-empty is needed for these particular stationary solutions to play a role in the long-term dynamics of (10.3.3). Specifically, let us define a threshold value z_{th} by

$$z_{th} := \sup\{z \in (0, \infty) : Q_z \in X_1\} \in [0, \infty] , \tag{10.3.4}$$

and, when $z_{th} > 0$, we set

$$\varrho_{th} := \sup_{z \in (0, z_{th})} M_1(Q_z) \in (0, \infty] . \tag{10.3.5}$$

Clearly $z_{th} = 1$ and $\varrho_{th} = \infty$ for the constant coefficients (10.3.1). When $z_{th} > 0$, the function $z \mapsto M_1(Q_z)$ is increasing from $(0, z_{th})$ onto $(0, \varrho_{th})$ and, given $\varrho \in (0, \varrho_{th})$, there is a unique $z(\varrho) \in (0, z_{th})$ such that $M_1(Q_{z(\varrho)}) = \varrho$. The long-term behaviour of a mass-conserving weak solution f to (10.3.3) on $[0, \infty)$, satisfying $M_1(f(t)) = \varrho$ for all $t \geq 0$ and some $\varrho > 0$, is then expected to be the following.

Conjecture 10.3.1. *(a) Assume that $z_{th} > 0$. If $\varrho \in (0, \varrho_{th})$, then $f(t)$ converges to $Q_{z(\varrho)}$ as $t \to \infty$ in a suitable topology. If $\varrho \geq \varrho_{th}$ then $f(t)$ converges weakly to $Q_{z_{th}}$ as $t \to \infty$ in a suitable topology.*

(b) Assume that $z_{th} = 0$. Then $f(t)$ converges weakly to zero as $t \to \infty$ in a suitable topology.

A complete proof of (a) and (b) is unfortunately not yet available and we provide partial results in that direction in Section 10.3.1, see [2], and Section 10.3.2, see [105, 169]. This issue is actually best understood for the discrete C-F equations and in particular for the Becker–Döring equations. Concerning the latter, assertions (a) and (b) are proved in [13, 15, 245] and in [69], respectively, for a fairly large class of coefficients. Further results on the long-term behaviour of solutions to these equations are available, including rates of convergence to zero of $f(t) - Q_{z(\varrho)}$ as $t \to \infty$ [60, 62, 143, 207, 208] and metastable behaviour [228]. We refer to Section 11.1 for a more detailed account of these results. As for the discrete C-F equations, we refer to [58, 60, 65, 67, 91, 92] for contributions to (a) and to [58, Theorem 1.2] and [60, Theorem 1.5] for decay rates of $f(t) - Q_{z(\varrho)}$ in X_1 when $z_{th} \in (0, \infty)$ and $M_1(f^{in}) \in (0, \varrho_{th})$.

Roughly speaking, the building block of the analysis of the long-term behaviour of solutions to C-F equations satisfying the detailed balance condition (10.3.2) is that the latter guarantees the availability of a Lyapunov functional which proves useful when investigating this issue, at least when $z_{th} > 0$. Before providing a more precise account in the forthcoming Sections 10.3.1 and 10.3.2, let us sketch how to proceed towards a proof of assertion (a) when $z_{th} > 0$. Since $z_{th} > 0$, we may assume without loss of generality that $Q \in X_1$ and $z_{th} > 1$. We next define a functional L by

$$L\zeta := \int_0^\infty \left[\zeta(x) \left(\ln \left(\frac{\zeta(x)}{Q(x)} \right) - 1 \right) + Q(x) \right] \, \mathrm{d}x \geq 0 \, , \tag{10.3.6}$$

for any function ζ for which it makes sense. It then formally follows from (10.3.3) that if f is a mass-conserving weak solution to (10.3.3) on $[0, \infty)$ with a nonnegative initial condition $f^{in} \in X_1$, then

$$Lf(t) + \int_0^t \mathscr{D}f(s) \, \mathrm{d}s = Lf^{in} \, , \qquad t \geq 0 \, , \tag{10.3.7}$$

where

$$\mathscr{D}f := \int_0^\infty \int_0^\infty J_0 \left(k(x,y)f(x)f(y), F(x,y)f(x+y) \right) \, \mathrm{d}y\mathrm{d}x \, , \tag{10.3.8}$$

and J_0 is the convex function defined by

$$J_0(r,s) := \begin{cases} (r-s)(\ln r - \ln s) \, , & (r,s) \in (0, \infty)^2 \, , \\ 0 \, , & (r,s) = (0,0) \, , \\ \infty \, , & \text{otherwise.} \end{cases} \tag{10.3.9}$$

Two important consequences stem from (10.3.7) and (10.3.8). On the one hand, $t \mapsto Lf(t)$ is a non-increasing function of time and it is nonnegative according to (10.3.6). Consequently, there is $\bar{L} \in [0, Lf^{in}]$ such that

$$\lim_{t \to \infty} Lf(t) = \bar{L} \, . \tag{10.3.10}$$

On the other hand, $t \mapsto \mathcal{D}f(t)$ belongs to $L_1(0, \infty)$. Thus, $\mathcal{D}f(t)$ somehow converges to zero as $t \to \infty$ and one expects any cluster point f_∞ of $\{f(t) \, : \, t \geq 0\}$ as $t \to \infty$ in a suitable topology to satisfy $\mathcal{D}f_\infty = 0$. Taking this for granted, we combine it with the positivity properties of J_0 to conclude, at least formally, that $k(x, y)f_\infty(x)f_\infty(y) = F(x, y)f_\infty(x + y)$ for $(x, y) \in (0, \infty)^2$; that is, there is $\bar{z} \in (0, z_{th})$ such that $f_\infty = Q_{\bar{z}}$. To complete the proof of assertion (a), it remains to identify the possible values taken by $\bar{z}$. Only one value is expected to be achievable, namely $\bar{z} = z(\varrho^{in})$ when $\varrho^{in} := M_1(f^{in}) \in (0, \varrho_{th})$ and $\bar{z} = z_{th}$ otherwise. Since f is a mass-conserving solution, this is obvious when $\{f(t) \, : \, t \geq 0\}$ is weakly compact in X_1. Indeed, in that case, there is a sequence $(t_j)_{j \geq 1}$ such that $t_j \to \infty$ and $f(t_j) \rightharpoonup Q_{\bar{z}}$ in X_1 as $j \to \infty$, hence

$$M_1(Q_{\bar{z}}) = \lim_{j \to \infty} M_1(f(t_j)) = \varrho^{in} \, ,$$

and $\bar{z} = z(\varrho^{in})$. Obviously, weak compactness in X_1 can only hold when $\varrho^{in} \in (0, \varrho_{th})$, and it is actually only shown to occur when either fragmentation is sufficiently strong or $z_{th} = \infty$. When weak compactness in X_1 of $\{f(t) \, : \, t \geq 0\}$ is not available (or not known), a different route, making use of (10.3.10), might be taken. Introducing the functional

$$L_* \zeta := L\zeta - \ln(z_{th})M_1(\zeta) \, ,$$

we observe that the mass conservation and (10.3.7) imply that $t \mapsto L_* f(t)$ is also a non-increasing function of time which is bounded from below, and that $L_* f(t)$ converges to $\bar{L} - \varrho^{in} \ln(z_{th})$ as $t \to \infty$. Since

$$L_* Q_z = \int_0^\infty \left[(\ln z - \ln z_{th})xe^{x \ln z} - e^{x \ln z} + 1 \right] Q(x) \, \mathrm{d}x \, , \qquad z \in (0, z_{th}) \, ,$$

the function $z \mapsto L_* Q_z$ is decreasing and there is a unique $z_\infty \in (0, z_{th})$ which is such that $L_* Q_{z_\infty} = \bar{L} - \varrho^{in} \ln(z_{th})$. One then expects that $\bar{z} = z_\infty$ and thus that $\{f(t) \, : \, t \geq 0\}$ has a unique cluster point Q_{z_∞}. The drawback of this approach is that z_∞ is not identified, contrary to expectation (a), and it is thus a weaker version of assertion (a). Another difficulty to be faced is that, for the previous argument to be valid, one has to prove that the topology in which the convergence to $Q_{\bar{z}}$ takes place is strong enough to guarantee that $LQ_{\bar{z}} = \bar{L}$. The latter is not obvious and, as already observed in [15], the previous property can only be shown for L_* instead of L. More details are provided in Sections 10.3.1 and 10.3.2 below.

We emphasise here that the detailed balance condition (10.3.2) is far from being universal and that most choices of coagulation and fragmentation coefficients do not have such a property. The behaviour of solutions to the C-F equation is then still widely open in general, though stationary solutions can be constructed for some classes of coefficients, see Section 10.3.3.

Finally, following the discussion on self-similarity presented at the beginning of this chapter, it turns out that there are a few cases where the coagulation and fragmentation terms have the same scaling invariance and one may look for self-similar solutions, see Section 10.3.4.

10.3.1 The Aizenman–Bak Result for Constant Coefficients

We focus in this section on the choice (10.3.1) of the coagulation and fragmentation coefficients; that is,

$$k(x, y) = k(y, x) = 2 \, , \qquad a(x) = x \, , \quad \text{and} \quad b(x, y) = \frac{2}{y} \, , \qquad 0 < x < y \, . \qquad (10.3.11)$$

In this case, the C-F equation (8.0.1a) is

$$\partial_t f(t,x) = \int_0^x [f(t,x-y)f(t,y) - f(t,x)] \, dy$$
$$- 2\int_0^\infty [f(t,x)f(t,y) - f(t,x+y)] \, dy \qquad (10.3.12a)$$

for $(t,x) \in (0,\infty)^2$, and it is supplemented with an initial condition

$$f(0,x) = f^{in}(x) , \qquad x \in (0,\infty) , \qquad (10.3.12b)$$

satisfying

$$f^{in} \in X_{0,1,+} . \qquad (10.3.13)$$

According to Corollary 8.2.59, the initial-value problem (10.3.12) is well posed and has a unique mass-conserving weak solution

$$f \in C([0,\infty), X_{0,1}) , \qquad M_1(f(t)) = \varrho := M_1(f^{in}) , \qquad t \geq 0 . \qquad (10.3.14)$$

Owing to the specific choice (10.3.11) of the coefficients, it further follows from (10.3.12a) that $f \in C^1([0,\infty), X_0)$.

The purpose of this section is to investigate the long-term behaviour of f. As already mentioned, there is a one-parameter family $\{Q_z : z \in [0,\infty)\}$ of stationary solutions to (10.3.12a) given by $Q_0 \equiv 0$ and $Q_z(x) = e^{x \ln z}$, $(x,z) \in (0,\infty)^2$. Clearly, $Q_z \in X_1$ with $M_1(Q_z) = |\ln z|^{-2}$ if and only if $z \in [0,1)$. For further use, we change notation and set $E_m := Q_{e^{-1/\sqrt{m}}}$ for $m \in (0,\infty)$, that is,

$$E_m(x) := e^{-x/\sqrt{m}} , \qquad x \in (0,\infty) , \qquad m > 0 . \qquad (10.3.15)$$

Clearly E_m belongs to $X_{0,1}$ for all $m > 0$ with $M_0(E_m) = \sqrt{m}$ and $M_1(E_m) = m$. These stationary solutions are obvious candidates for attractors for the dynamics of (10.3.12) and a simple guess based on the mass conservation (10.3.14) is that, as $t \to \infty$, $(f(t))$ should converge to E_ϱ with $\varrho = M_1(f^{in})$. To confirm this conjecture, two steps are required: as already mentioned, the detailed balance condition (10.3.2b) guarantees the existence of a Lyapunov functional L which is here defined on X_0 by

$$L\zeta := \int_0^\infty \zeta(x)[\ln(\zeta(x)) - 1] \, dx \text{ if } \zeta \geq 0 \text{ a.e. in } (0,\infty) \text{ and } \zeta \ln \zeta \in X_0 , \qquad (10.3.16)$$

and $L\zeta = \infty$ otherwise. The availability of a Lyapunov functional is a valuable tool when investigating the long-term behaviour as it allows LaSalle invariance principle, or a variant thereof, to be applied. This approach, however, only shows convergence. The second step provides more quantitative information on the convergence, and a more sophisticated argument to achieve it has been devised in [2]. While a natural idea is to compare Lf with its expected limit LE_ϱ, one of the building blocks of the convergence result obtained in [2] is rather to compare the evolution of $Lf(t)$ to that of $LE_{M_0(f(t)),\varrho}$, where $(E_{m,\varrho})_{m,\varrho}$ is a two-parameter family of functions defined by

$$E_{m,\varrho}(x) := \frac{m^2}{\varrho} e^{-mx/\varrho} , \qquad x \in (0,\infty) , \qquad (m,\varrho) \in (0,\infty)^2 . \qquad (10.3.17)$$

Note that $E_{\sqrt{\varrho},\varrho} = E_\varrho$ for $\varrho > 0$. Salient properties of this two-parameter family of functions include a close connection to the Lyapunov functional L and an explicit computation of their dynamics relying on the following lemma.

Lemma 10.3.2. *For $t \geq 0$,*

$$M_0(f(t)) = \sqrt{\varrho} \frac{(M_0(f^{in}) + \sqrt{\varrho})e^{t\sqrt{\varrho}} + (M_0(f^{in}) - \sqrt{\varrho})e^{-t\sqrt{\varrho}}}{(M_0(f^{in}) + \sqrt{\varrho})e^{t\sqrt{\varrho}} - (M_0(f^{in}) - \sqrt{\varrho})e^{-t\sqrt{\varrho}}} ,$$

and

$$\lim_{t\to\infty} M_0(f(t)) = \sqrt{\varrho} .$$

Proof. It readily follows from (10.3.12a) and (10.3.14) that

$$\frac{dM_0(f)}{dt} = \varrho - M_0^2(f) ,$$

from which Lemma 10.3.2 readily follows. □

We may now state the convergence result obtained in [2].

Theorem 10.3.3. *[2] Assume further that $f^{in} \ln f^{in} \in X_0$. Then, for $t \geq 0$,*

$$0 \leq Lf(t) - LE_{M_0(f(t)),\varrho} \leq \left[Lf^{in} - LE_{M_0(f^{in}),\varrho}\right] \exp\left(-\int_0^t M_0(f(s)) \, ds\right) \quad (10.3.18)$$

and

$$\lim_{t\to\infty} \|f(t) - E_\varrho\|_1 = 0 . \quad (10.3.19)$$

As already outlined above, the proof of Theorem 10.3.3 exploits specific features of the coagulation and fragmentation coefficients (10.3.11) and is unfortunately restricted to this case. In particular, it does not seem to extend to other coefficients satisfying the detailed balance condition (10.3.2b).

To begin the proof of Theorem 10.3.3, we first study the connection between the functional L and the family $(E_{m,\varrho})_{m,\varrho}$, thereby establishing the first inequality in (10.3.18).

Lemma 10.3.4. *Consider a nonnegative function $\zeta \in X_{0,1}$ such that $M_0(\zeta) = m > 0$, $M_1(\zeta) = \varrho$, and $\zeta \ln \zeta \in X_0$. Then*

$$L\zeta \geq LE_{m,\varrho} = m \ln\left(\frac{m^2}{\varrho}\right) - 2m .$$

Proof. Introducing the convex function $\psi_1(r) := r(\ln(r) - 1)$, $r \in [0, \infty)$, it follows from the explicit formula for $E_{m,\varrho}$ that

$$L\zeta = \int_0^\infty \zeta(x) \left[\ln(\zeta(x)) - \ln(E_{m,\varrho}(x)) - 1\right] \, dx + \int_0^\infty \zeta(x) \left[\ln\left(\frac{m^2}{\varrho}\right) - \frac{mx}{\varrho}\right] \, dx$$

$$= m \int_0^\infty \psi_1\left(\frac{\zeta(x)}{E_{m,\varrho}(x)}\right) E_{m,\varrho}(x)\frac{dx}{m} + m \ln\left(\frac{m^2}{\varrho}\right) - m .$$

Thanks to the convexity of ψ_1, we deduce from Jensen's inequality that

$$L\zeta \geq m\psi_1\left(\int_0^\infty \zeta(x)\frac{dx}{m}\right) + LE_{m,\varrho} + m = LE_{m,\varrho} ,$$

and the proof of Lemma 10.3.4 is complete. □

We now turn to the proof of the second inequality in (10.3.18) at the heart of which lies the following functional inequality [2, Proposition 4.3].

Proposition 10.3.5. *[2] Consider a nonnegative function $\zeta \in X_0$ such that $\zeta \ln \zeta \in X_0$. Then*

$$\int_0^\infty \int_0^\infty \zeta(x)\zeta(y) \ln \zeta(x+y) \, dydx + M_0^2(\zeta) \leq M_0(\zeta)M_0(\zeta \ln \zeta) \ .$$

Proof. Define

$$\zeta_1(x) := \int_x^\infty \zeta(y) \, dy \quad \text{and} \quad \psi_0(x) := x \ln x \ , \qquad x \in (0, \infty) \ .$$

We may assume without loss of generality that

$$\int_0^\infty \int_0^\infty \zeta(x)\zeta(y) \ln \zeta(x+y) \, dydx > -\infty \ ,$$

as Proposition 10.3.5 is obviously true otherwise. Then,

$$J(\zeta) := M_0(\zeta)M_0(\zeta \ln \zeta) - \int_0^\infty \int_0^\infty \zeta(x)\zeta(y) \ln \zeta(x+y) \, dydx$$

$$= \int_0^\infty \int_0^\infty \zeta(x)\zeta(y) \ln \left(\frac{\zeta(x)}{\zeta(x+y)} \right) \, dydx$$

$$= \int_0^\infty \zeta(x)\zeta_1(x) \int_0^\infty \psi_0 \left(\frac{\zeta(y)}{\zeta(x+y)} \right) \zeta(x+y) \frac{dy}{\zeta_1(x)} dx \ .$$

Since

$$\int_0^\infty \zeta(x+y) \, dy = \zeta_1(x) \ , \qquad x \in (0, \infty) \ ,$$

and ψ_0 is convex, it follows from Jensen's inequality that

$$J(\zeta) \geq \int_0^\infty \zeta(x)\zeta_1(x)\psi_0 \left(\int_0^\infty \frac{\zeta(y)}{\zeta_1(x)} \, dy \right) dx$$

$$= \int_0^\infty \zeta(x)M_0(\zeta) \ln \left(\frac{M_0(\zeta)}{\zeta_1(x)} \right) dx$$

$$= M_0^2(\zeta) \ln(M_0(\zeta)) + M_0(\zeta) \int_0^\infty \frac{d\zeta_1}{dx}(x) \ln \zeta_1(x) \, dx$$

$$= M_0^2(\zeta) \ln(M_0(\zeta)) + M_0(\zeta) \left[-\zeta_1(0) \ln(\zeta_1(0)) + \zeta_1(0) \right]$$

$$= M_0^2(\zeta) \ ,$$

since $\zeta_1(0) = M_0(\zeta)$. $\qquad\square$

The inequality in Proposition 10.3.5 is strict unless $\zeta(x) = e^{-mx}$, $x \in (0, \infty)$, for some $m > 0$. This follows from the strict convexity of $r \mapsto r \ln r$, $r \in [0, \infty)$.

We are now ready to prove that L is a Lyapunov functional and the more precise inequality (10.3.18).

Proposition 10.3.6. *Assume further that $f^{in} \ln f^{in} \in X_0$. Then (10.3.18) holds for $t \geq 0$. Moreover,*

$$\lim_{t \to \infty} LE_{M_0(f(t)),\varrho} = -2\sqrt{\varrho} = LE_\varrho \ , \tag{10.3.20}$$

$$\lim_{t \to \infty} Lf(t) = LE_\varrho \ . \tag{10.3.21}$$

Proof. We first show that $t \mapsto Lf(t)$ is a nonincreasing function of time. Since L involves a logarithm, we proceed in two steps and first handle a more restrictive class of initial data.
Step 1. We further assume that there is $K > 1$, and $\varepsilon > 0$ such that

$$\varepsilon e^{-x} \le f^{in}(x) \le K \;, \qquad x \in (0, \infty) \;, \tag{10.3.22}$$

and claim that

$$\varepsilon e^{-\mu_0(t,x)} \le f(t,x) \le K \;, \qquad (t,x) \in (0,\infty)^2 \;, \tag{10.3.23}$$

where

$$\mu_0(t,x) := (1+t)x + 2 \int_0^t M_0(f(s)) \, \mathrm{d}s \;, \qquad (t,x) \in (0,\infty)^2 \;.$$

Indeed, on the one hand, it follows from (10.3.12a) and the nonnegativity of f that

$$\partial_t \left[f(t,x) \exp \left(tx + 2 \int_0^t M_0(f(s)) \, \mathrm{d}s \right) \right] \ge 0 \;, \qquad (t,x) \in (0,\infty)^2 \;.$$

Integrating the previous inequality with respect to time and using the lower bound (10.3.22) give the lower bound in (10.3.23). On the other hand, it follows from (10.3.12a) and Lemma 8.2.17 (with $\ell \equiv 1$ and $\Phi(r) = (r - K)_+$, $r \in (0,\infty)$) that, for $t \in (0,\infty)$,

$$\frac{d}{dt} \int_0^\infty (f(t,x) - K)_+ \, \mathrm{d}x \le -2K \int_0^\infty \int_x^\infty f(t,y) \mathbf{1}_{(0,\infty)}(f(t,x) - K) \, \mathrm{d}y\mathrm{d}x$$

$$- \int_0^\infty x f(t,x) \mathbf{1}_{(0,\infty)}(f(t,x) - K) \, \mathrm{d}x$$

$$+ 2 \int_0^\infty \int_x^\infty f(t,y) \mathbf{1}_{(0,\infty)}(f(t,x) - K) \, \mathrm{d}y\mathrm{d}x$$

$$\le -2(K-1) \int_0^\infty \int_x^\infty f(t,y) \mathbf{1}_{(0,\infty)}(f(t,x) - K) \, \mathrm{d}y\mathrm{d}x$$

$$\le 0 \;.$$

Since $(f^{in} - K)_+ = 0$ by (10.3.22), we readily deduce from the previous differential inequality that the upper bound in (10.3.23) holds true.
A straightforward consequence of (10.3.23) and Lemma 10.3.2 is that

$$\ln f \in L_\infty \left((0,T) \times (0,\infty), \frac{\mathrm{d}x\mathrm{d}t}{1+x} \right)$$

for all $T > 0$. It then follows from (10.3.12a) that

$$\frac{d}{dt} Lf(t) = \int_0^\infty \int_0^\infty \ln \left(\frac{f(t,x+y)}{f(t,x)f(t,y)} \right) f(t,x)f(t,y) \, \mathrm{d}y\mathrm{d}x$$

$$- \int_0^\infty x f(t,x) \ln f(t,x) \, \mathrm{d}x$$

$$+ 2 \int_0^\infty \int_0^\infty f(t,x+y) \ln f(t,x) \, \mathrm{d}y\mathrm{d}x \;.$$

Since

$$\int_0^\infty x f(t,x) \ln f(t,x) \, \mathrm{d}x = \int_0^\infty \int_0^x f(t,x) \ln f(t,x) \, \mathrm{d}y\mathrm{d}x$$

$$= \int_0^\infty \int_y^\infty f(t,x) \ln f(t,x) \, \mathrm{d}x\mathrm{d}y$$

$$= \int_0^\infty \int_0^\infty f(t, x+y) \ln f(t, x+y) \, dxdy$$

and

$$2 \int_0^\infty \int_0^\infty f(t, x+y) \ln f(t, x) \, dydx = \int_0^\infty \int_0^\infty f(t, x+y) \ln(f(t, x)f(t, y)) \, dydx ,$$

we further obtain

$$\frac{d}{dt} Lf(t) = \int_0^\infty \int_0^\infty \ln(f(t, x+y)) f(t, x) f(t, y) \, dydx - 2M_0(f(t)) M_0(f(t) \ln f(t))$$
$$- \int_0^\infty \int_0^\infty \psi_0 \left(\frac{f(t, x+y)}{f(t, x)f(t, y)} \right) f(t, x) f(t, y) \, dydx ,$$

where $\psi_0(r) = r \ln r$, $r \in (0, \infty)$. We infer from Proposition 10.3.5, the convexity of ψ_0 and Jensen's inequality that

$$\frac{d}{dt} Lf(t) \le -M_0^2(f(t)) - M_0(f(t)) M_0(f(t) \ln f(t))$$
$$- M_0^2(f(t)) \psi_0 \left(\int_0^\infty \int_0^\infty \frac{f(t, x+y)}{M_0^2(f(t))} \, dydx \right) .$$

Since

$$\int_0^\infty \int_0^\infty f(t, x+y) \, dydx = M_1(f(t)) = \varrho$$

by (10.3.14), we end up with

$$\frac{d}{dt} Lf(t) \le -2M_0^2(f(t)) - M_0(f(t)) Lf(t) + \varrho \ln \left(\frac{M_0^2(f(t))}{\varrho} \right) , \qquad t \ge 0 . \qquad (10.3.24)$$

Finally, according to Lemma 10.3.2,

$$\frac{d}{dt} LE_{M_0(f(t)),\varrho} = \left(\varrho - M_0^2(f(t)) \right) \ln \left(\frac{M_0^2(f(t))}{\varrho} \right) , \qquad t \ge 0 . \qquad (10.3.25)$$

Combining (10.3.17), (10.3.24) and (10.3.25) gives (10.3.18) after integration with respect to time.

Step 2. We are left with removing the additional assumption (10.3.22). To this end, we use an approximation argument and define $f_j^{in}(x) := \min\{j, f^{in}(x)\} + j^{-1} e^{-x}$ for $x \in (0, \infty)$ and $j \ge 1$. Denoting the corresponding solution to (10.3.12) on $[0, \infty)$ by f_j, an easy application of Theorem 8.2.11 and the well-posedness of (10.3.12), guaranteed by Corollary 8.2.59, ensure that $(f_j)_{j \ge 1}$ converges to f in $C([0, T], X_{0,w})$ for all $T > 0$. In particular,

$$\lim_{j \to \infty} M_0(f_j(t)) = M_0(f(t)) , \qquad t \in [0, \infty) ,$$

while the convexity of $r \mapsto r(\ln r - 1)$, $r \in [0, \infty)$, implies that

$$Lf(t) \le \liminf_{j \to \infty} Lf_j(t) , \qquad t \in [0, \infty) .$$

Owing to the previous step, f_j satisfies (10.3.18) for all $j \ge 1$. Thus we may pass to the limit as $j \to \infty$ and conclude that f also satisfies (10.3.18).

Step 3. We next deduce (10.3.20) from Lemma 10.3.2 and Lemma 10.3.4, while (10.3.21) follows from (10.3.18), (10.3.20), and Lemma 10.3.2. $\qquad \square$

Proposition 10.3.6 provides the first statement in Theorem 10.3.3 and in particular the convergence of $Lf(t)$ to LE_ϱ as $t \to \infty$. To interpret this result in terms of $f(t) - E_\varrho$, another functional inequality, referred to as the Csiszár–Kullback inequality, is needed. We recall it below [84].

Proposition 10.3.7 (Csiszár–Kullback inequality). *Consider two functions* $\zeta_i \in X_{0,+}$, *$i = 1, 2$, such that* $\|\zeta_1\|_1 = \|\zeta_2\|_1$. *Then*

$$\|\zeta_1 - \zeta_2\|_1 \le \sqrt{2\|\zeta_1\|_1} \sqrt{\int_0^\infty \zeta_1(x) \ln\left(\frac{\zeta_1(x)}{\zeta_2(x)}\right) \, dx} \ .$$

Proof of Theorem 10.3.3. Let $t > 0$. Since

$$\int_0^\infty f(t,x) \ln\left(\frac{f(t,x)}{E_{M_0(f(t)),\varrho}(x)}\right) \, dx = Lf(t) - LE_{M_0(f(t)),\varrho}$$

and $\|f(t)\|_1 = \|E_{M_0(f(t)),\varrho}\|_1$, we infer from the Csiszár–Kullback inequality recalled in Proposition 10.3.7 (with $\zeta_1 = f(t)$ and $\zeta_2 = E_{M_0(f(t)),\varrho}$) that

$$\left\|f(t) - E_{M_0(f(t)),\varrho}\right\|_1 \le \sqrt{2M_0(f(t))} \sqrt{Lf(t) - LE_{M_0(f(t)),\varrho}} \ , \tag{10.3.26}$$

hence

$$\lim_{t \to \infty} \left\|f(t) - E_{M_0(f(t)),\varrho}\right\|_1 = 0$$

by (10.3.21) and Lemma 10.3.2. Recalling Lemma 10.3.2 yields (10.3.19) and completes the proof of Theorem 10.3.3. □

Remark 10.3.8. *Combining (10.3.26) and the outcome of Lemma 10.3.2 actually provides an exponential decay rate to zero of* $\|f(t) - E_\varrho\|_1$ *as* $t \to \infty$. *In the simplest case, when* $M_0(f^{in}) = \sqrt{\varrho}$, *Lemma 10.3.2 implies that* $M_0(f(t)) = \sqrt{\varrho}$ *for all* $t \ge 0$ *and we infer from (10.3.18) and (10.3.26) that*

$$\|f(t) - E_\varrho\|_1 \le \sqrt{2\varrho(Lf^{in} - LE_\varrho)} \, e^{-t\sqrt{\varrho}/2} \ , \qquad t \ge 0 \ .$$

Additional information, including the extension of the convergence (10.3.19) to initial conditions satisfying only (10.3.13) and the study of the linearisation of (10.3.12a) around E_ϱ, are supplied in [2, 153].

10.3.2   Detailed Balance

We now turn to coagulation and fragmentation coefficients satisfying the detailed balance condition (10.3.2) but not necessarily given by (10.3.11). More precisely, we assume throughout this section that there is a nonnegative measurable symmetric function F defined on $(0, \infty)^2$, and a nonnegative function $Q \in X_{0,1}$ such that the overall fragmentation rate a and the daughter distribution function b are given by

$$a(x) := \frac{1}{2} \int_0^x F(y, x - y) \, dy \ , \qquad x \in (0, \infty) \ , \tag{10.3.27a}$$

$$a(y)b(x,y) := F(x, y - x) \ , \qquad 0 < x < y \ , \tag{10.3.27b}$$

while k and F are related by the detailed balance condition

$$k(x,y)Q(x)Q(y) = F(x,y)Q(x+y) , \qquad (x,y) \in (0,\infty)^2 . \tag{10.3.27c}$$

As a consequence of (10.3.27a) and (10.3.27b), there is no loss of mass during fragmentation events and the number of fragments produced equals two; that is,

$$\int_0^y xb(x,y) \, \mathrm{d}x = y , \qquad n_0(y) = \int_0^y b(x,y) \, \mathrm{d}x = 2 , \qquad y \in (0,\infty) . \tag{10.3.27d}$$

Obviously, $n_0 \in L_\infty(0,\infty)$. While the detailed balance condition (10.3.27) guarantees that some features used in the previous section are still available (such as the existence of a suitable Lyapunov functional), several steps of the proof of Theorem 10.3.3 clearly rely on the specific choice (10.3.11) of the coagulation and fragmentation coefficients. It is thus not surprising that the analysis of the long-term dynamics of the C-F equation with coefficients satisfying only the detailed balance condition (10.3.27) requires different arguments and is more involved. Also, as we shall see below, the convergence result is less precise. To proceed further, additional assumptions on k, F, and Q are needed: there is $0 \le \alpha \le \beta \le 1$, $\gamma > 0$, and $K_0 > 0$ such that

$$\lambda := \alpha + \beta \in [0,1] \tag{10.3.28}$$

and

$$0 < k(x,y) \le K_0 \left[(1+x)^\alpha(1+y)^\beta + (1+x)^\beta(1+y)^\alpha\right] , \tag{10.3.29a}$$

$$0 < F(x,y) \le K_0(1+x+y)^{\gamma-1} , \tag{10.3.29b}$$

for $(x,y) \in (0,\infty)^2$. A straightforward consequence of (10.3.28), (10.3.29a) and Young's inequality is the upper bound:

$$k(x,y) \le K_0[(1+x)^\lambda + (1+y)^\lambda] \le K_0(2+x+y) , \qquad (x,y) \in (0,\infty)^2 . \tag{10.3.29c}$$

Finally we require that Q is positive and bounded for small sizes and that $\ln Q$ behaves algebraically for large sizes. Specifically, there is $K_1 > 1$, $K_\infty > 1$, and $q_\infty > 0$ such that

$$\frac{1}{K_1} \le Q(x) \le K_1 , \qquad x \in (0,1) , \tag{10.3.30a}$$

$$\frac{x^{q_\infty}}{K_\infty} \le -\ln Q(x) \le K_\infty x^{q_\infty} , \qquad x \in [1,\infty) . \tag{10.3.30b}$$

Obviously, $Q \in L_\infty(0,\infty)$ as a consequence of (10.3.30). We emphasise that the assumptions (10.3.30) on Q are unlikely to be optimal and are made here for the sake of simplicity. It may, in particular, be possible to extend the analysis presented below to other cases, of course at the expense of some modifications.

Without additional assumptions on F, it is unclear whether the assumptions (8.2.75) and (8.2.76), required to apply Theorem 8.2.23, hold true, so that the analysis of Chapter 8 does not ensure the existence of mass-conserving weak solutions to the C-F equation (10.3.3) in this case. Still, the structure provided by the detailed balance condition (10.3.27) allows us to construct at least one mass-conserving solution to the C-F equation (10.3.3) on $[0,\infty)$ for a suitable class of initial conditions, this solution satisfying additional properties stemming from (10.3.27). Indeed, as already mentioned in the preamble of Section 10.3 and exploited in Section 10.3.1, an important role is played by the functional L defined in (10.3.6), namely,

$$L\zeta = \int_0^\infty \left\{\zeta(x) \left[\ln\left(\frac{\zeta(x)}{Q(x)}\right) - 1\right] + Q(x)\right\} \, \mathrm{d}x , \tag{10.3.31}$$

which is well defined for $\zeta \in \mathcal{Y}$ due to (10.3.30), where

$$\mathcal{Y} := \{\zeta \in X_{0,1,+} \cap X_{q_\infty} \; : \; \zeta \ln \zeta \in X_0\} \; . \tag{10.3.32}$$

Indeed, we have the following upper and lower bounds on L in $\mathcal{Y}$.

Lemma 10.3.9. *For $\zeta \in \mathcal{Y}$, there holds*

$$L\zeta \; \leq \; \|\zeta \ln \zeta\|_1 + \ln (K_1) M_0(\zeta) + K_\infty M_{q_\infty}(\zeta) + M_0(Q) \; , \tag{10.3.33}$$
$$L\zeta \; \geq \; \|\zeta \ln \zeta\|_1 - 4 - 2M_1(\zeta) - M_0(\zeta) \; . \tag{10.3.34}$$

Proof. The inequality (10.3.33) readily follows from (10.3.30) and the nonnegativity of ζ. As for (10.3.34), we first recall that

$$\int_0^\infty \zeta(x) \ln \zeta(x) \, \mathrm{d}x = \int_0^\infty \zeta(x) |\ln \zeta(x)| \, \mathrm{d}x + 2 \int_0^\infty \zeta(x) \ln (\zeta(x)) \mathbf{1}_{(0,1)}(\zeta(x)) \, \mathrm{d}x \; .$$

Now, if $x \in (0,\infty)$ is such that $\zeta(x) \in (0, e^{-x}]$, then

$$\zeta(x) \ln \zeta(x) \geq \sqrt{\zeta(x)} \inf_{r \in (0,1)} \{\sqrt{r} \ln r\} \geq -\frac{2}{e} \sqrt{\zeta(x)} \geq -e^{-x/2} \; .$$

If $x \in (0,\infty)$ and $\zeta(x) \in (e^{-x}, 1)$, then $\zeta(x) \ln (\zeta(x)) \geq -x\zeta(x)$. Combining the previous estimates leads us to

$$\int_0^\infty \zeta(x) \ln \zeta(x) \, \mathrm{d}x \geq \int_0^\infty \zeta(x) |\ln \zeta(x)| \, \mathrm{d}x - 2 \int_0^\infty e^{-x/2} \, \mathrm{d}x - 2 \int_0^\infty x\zeta(x) \, \mathrm{d}x \; .$$

Now, owing to (10.3.30) and the above inequality,

$$L\zeta = \int_0^\infty \zeta(x) \ln \zeta(x) \, \mathrm{d}x - \int_0^\infty \zeta(x) \ln Q(x) \, \mathrm{d}x - M_0(\zeta) + M_0(Q)$$
$$\geq \|\zeta \ln \zeta\|_1 - 4 - 2M_1(\zeta) - M_0(\zeta) \; ,$$

as claimed. $\qquad\square$

The purpose of the next section is to show that the C-F equation (10.3.3) has at least one mass-conserving weak solution f on $[0,\infty)$ for each initial condition $f^{in} \in \mathcal{Y} \cap X_{1+\gamma}$, as well as the monotonicity of $t \mapsto Lf(t)$, the latter being an outstanding consequence of the detailed balance condition (10.3.27). This last property is revisited in Section 10.3.2.2, where we show the validity of the entropy identity (10.3.7), adapting the proof of a similar result established in [186] for the Boltzmann equation. We next turn to the main aim of this section, namely, the long-term behaviour, and prove stabilisation and convergence results in Section 10.3.2.3. The analysis carried out there owes much to that dedicated to the long-term behaviour of the Boltzmann equation and was adapted to C-F equations with coefficients satisfying the detailed balance condition (10.3.27) in [65, 67] for the discrete case and in [165, 169] for the continuous case.

Throughout this section, C and C_i are positive constants that may vary from line to line but depend only on k, F, Q, and f^{in}. Dependence upon additional parameters will be indicated explicitly.

10.3.2.1 Existence

The starting point of the analysis of the C-F equation (10.3.3) with coefficients satisfying the detailed balance condition (10.3.27) is the following existence result.

Proposition 10.3.10. *Assume that the coagulation and fragmentation coefficients k, a and b satisfy (10.3.27), (10.3.28), (10.3.29) and (10.3.30), and consider an initial condition $f^{in} \in \mathcal{Y} \cap X_\mu$ for some $\mu \geq 1 + \gamma$, the set $\mathcal{Y}$ being defined in (10.3.32). There is at least one mass-conserving weak solution f to the C-F equation (10.3.3) on $[0, \infty)$ which possesses the following additional properties: for all $T > 0$,*

$$f \in L_\infty((0,T), \mathcal{Y} \cap X_\mu) , \qquad Lf \in L_\infty(0,T) , \qquad \mathcal{D}f \in L_1((0,T) \times (0,\infty)^2) , \quad (10.3.35a)$$

and

$$Lf(T) + \int_0^T \mathcal{D}f(s) \, ds \leq Lf^{in} . \tag{10.3.35b}$$

where $\mathcal{D}f$ is defined in (10.3.8).

Proof. As in Chapter 8, we consider an approximation of the C-F equation (10.3.3) obtained by truncating k, F, and f^{in}. Specifically, let $j \geq 1$ be a positive integer and set

$$k_j(x,y) := k(x,y)\mathbf{1}_{(0,j)}(x+y) , \qquad F_j(x,y) = F(x,y)\mathbf{1}_{(0,j)}(x+y) , \tag{10.3.36a}$$

for $(x,y) \in (0,\infty)^2$ and

$$a_j(y) := \frac{1}{2} \int_0^y F_j(x, y-x) \, dx = a(y)\mathbf{1}_{(0,j)}(y) , \qquad y \in (0,\infty) , \tag{10.3.36b}$$

the daughter distribution b being still given by (10.3.27b). Since $a_j(y) = F_j(x, y-x) = 0$ for $y \in (j,\infty)$ and $x \in (0,y)$, it follows from (10.3.27b) that

$$a_j(y)b(x,y) = F_j(x, y-x) , \qquad 0 < x < y , \tag{10.3.36c}$$

while (10.3.27c) implies that

$$k_j(x,y)Q(x)Q(y) = F_j(x,y)Q(x+y) , \qquad (x,y) \in (0,\infty)^2 . \tag{10.3.36d}$$

We finally define the approximation f_j^{in} of f^{in} by

$$f_j^{in} := \min\left\{ \max\left\{ f^{in}, \frac{Q}{j} \right\}, j \right\} \mathbf{1}_{(0,j)} . \tag{10.3.36e}$$

Observe that this approximation is slightly more complicated than the one used in Chapter 8 but that it is positive on $(0,j)$, a property which proves useful in the computations involving the logarithm function.

Owing to (10.3.27d), (10.3.29) and (10.3.36), we are in a position to apply Proposition 8.2.7 (with $m_0 = 0$) and infer that there is a unique mass-conserving weak solution f_j on $[0,\infty)$ to the C-F equation with coefficients (k_j, a_j, b) and initial condition f_j^{in}. As usual, we extend f_j to $[0,\infty) \times (j,\infty)$ by zero and note that

$$M_1(f_j(t)) = M_1(f_j^{in}) \leq M_1(f^{in}) + M_1(Q) , \qquad t \in [0,\infty) . \tag{10.3.37}$$

In addition, owing to the specific structure (10.3.36b) and (10.3.36c) of a_j and b and the symmetry of F, we see that, for $\vartheta \in L_\infty(0,\infty)$ and $t > 0$,

$$\int_0^\infty a_j(y)N_\vartheta(y)f_j(t,y) \, dy = \frac{1}{2} \int_0^\infty \int_0^y \vartheta(y)F_j(x, y-x)f_j(t,y) \, dxdy$$

$$-\int_0^\infty \int_0^y \vartheta(x) F_j(x, y-x) f_j(t, y) \, dxdy$$

$$= \frac{1}{2} \int_0^\infty \int_0^\infty \vartheta(x+y) F_j(x, y) f_j(t, x+y) \, dxdy$$

$$-\int_0^\infty \int_0^\infty \vartheta(x) F_j(x, y) f_j(t, x+y) \, dxdy$$

$$= \frac{1}{2} \int_0^\infty \int_0^\infty \chi_\vartheta(x, y) F_j(x, y) f_j(t, x+y) \, dxdy \, ,$$

so that, by Proposition 8.2.7,

$$\frac{d}{dt} \int_0^\infty \vartheta(x) f_j(t, x) \, dx$$
$$= \frac{1}{2} \int_0^\infty \int_0^\infty \chi_\vartheta(x, y) \left[k_j(x, y) f_j(t, x) f_j(t, y) - F_j(x, y) f_j(t, x+y) \right] \, dxdy \, . \tag{10.3.38}$$

Step 1: Moment estimates. Set

$$m := \max\{\mu, q_\infty\} \geq 1 + \gamma > 1 \, . \tag{10.3.39}$$

We take $\vartheta \equiv 1$ in (10.3.38) and deduce from (10.3.29b), (10.3.36a), (10.3.37), the elementary inequality $x(1+x)^{\gamma-1} \leq C(x+x^\gamma)$, $x \in (0, \infty)$, and Young's inequality that, for $t \geq 0$,

$$\frac{d}{dt} M_0(f_j(t)) \leq \frac{1}{2} \int_0^\infty \int_0^\infty F_j(x, y) f_j(t, x+y) \, dydx$$

$$\leq \frac{K_0}{2} \int_0^\infty x(1+x)^{\gamma-1} f_j(t, x) \, dx \leq C \left[M_1(f_j(t)) + M_\gamma(f_j(t)) \right]$$

$$\leq C \left[1 + \frac{\gamma}{m} M_m(f_j(t)) + \frac{m-\gamma}{m} M_0(f_j(t)) \right]$$

$$\leq C \left[1 + M_0(f_j(t)) + M_m(f_j(t)) \right] \, . \tag{10.3.40}$$

Next, by (10.3.29c) and (10.3.36a),

$$k_j(x, y) \leq k(x, y) \leq K_0(2 + x + y) \, , \qquad (x, y) \in (0, \infty)^2 \, ,$$

and we infer from Lemma 7.4.2 and Lemma 7.4.4 that

$$\chi_m(x, y) k_j(x, y) \leq 2K_0 \chi_m(x, y) + K_0(x+y) \chi_m(x, y)$$
$$\leq C \left[xy^{m-1} + x^{m-1}y + x^m y + xy^m \right]$$

for $(x, y) \in (0, \infty)^2$. Since $\chi_m F_j \geq 0$, we deduce from (10.3.37), (10.3.38) and Young's inequality that

$$\frac{d}{dt} M_m(f_j(t)) \leq C M_1(f_j(t)) \left[M_{m-1}(f_j(t)) + M_m(f_j(t)) \right]$$

$$\leq C \left[\frac{m-1}{m} M_m(f_j(t)) + \frac{1}{m} M_0(f_j(t)) + M_m(f_j(t)) \right]$$

$$\leq C \left[M_0(f_j(t)) + M_m(f_j(t)) \right] \, . \tag{10.3.41}$$

Combining (10.3.40) and (10.3.41) gives

$$\frac{d}{dt} \left[M_0(f_j(t)) + M_m(f_j(t)) \right] \leq C \left[1 + M_0(f_j(t)) + M_m(f_j(t)) \right]$$

for $t \geq 0$. Since $f^{in} \in \mathcal{Y}$ and Q satisfies (10.3.30), we note that

$$M_0(f_j^{in}) + M_m(f_j^{in}) \leq M_0(f^{in}) + M_0(Q) + M_m(f^{in}) + M_m(Q) < \infty .$$

Integrating the previous differential inequality, we conclude that, for all $T > 0$ and $j \geq 1$, there is $C_1(T) > 0$ such that

$$M_0(f_j(t)) + M_m(f_j(t)) \leq C_1(T) , \qquad t \in [0, T] . \tag{10.3.42}$$

Step 2: Entropy identity. We now aim at deriving an analogue of the entropy inequality (10.3.35b) for f_j. To this end, we first recall that $f_j \in C^1([0, \infty), X_0)$ according to Proposition 8.2.7. Also, since

$$0 < w_j^{in} := \min\left\{\frac{1}{jK_1}, \frac{e^{-K_\infty j^{q_\infty}}}{j}, j\right\} \leq f_j^{in} \leq j \text{ in } (0, j)$$

by (10.3.30) and (10.3.36e) and

$$\int_x^j a(y)b(x, y) \; \mathrm{d}y \leq \frac{K_0(1+j)^\gamma}{\gamma} , \qquad 0 < x < j ,$$

by (10.3.29b), it follows from Proposition 8.2.10 that there are functions $w_j \in C^1([0, \infty))$ and $W_j \in C^1([0, \infty))$ depending on k, a, b, Q, j, and f_j such that

$$0 < w_j(t) \leq f_j(t, x) \leq W_j(t) , \qquad (t, x) \in [0, \infty) \times (0, j) . \tag{10.3.43}$$

In particular, the function $r \mapsto r \ln r$ is Lipschitz continuous in the range of f_j so that $\partial_t[f_j \ln(f_j) - f_j] = \ln(f_j)\partial_t f_j$ a.e. in $(0, \infty) \times (0, j)$. Furthermore, the bounds (10.3.30) on Q guarantee that $f_j \in C^1([0, \infty), L_1((0, j), |\ln Q(x)|\mathrm{d}x))$. Consequently, it follows from (10.3.36d) and (10.3.38) that

$$\frac{d}{dt} Lf_j(t) = \int_0^j (\ln f_j(t, x) - \ln Q(x)) \, \partial_t f_j(t, x) \; \mathrm{d}x$$

$$= \frac{1}{2} \int_0^j \int_0^{j-x} \left[\ln\left(\frac{f_j(t, x+y)}{Q(x+y)}\right) - \ln\left(\frac{f_j(t, x)f_j(t, y)}{Q(x)Q(y)}\right) \right]$$

$$\cdot [k_j(x, y)f_j(t, x)f_j(t, y) - F_j(x, y)f_j(t, x+y)] \; \mathrm{d}y\mathrm{d}x$$

$$= -\frac{1}{2} \int_0^j \int_0^{j-x} J_0(k_j(x, y)f_j(t, x)f_j(t, y), F_j(x, y)f_j(t, x+y)) \; \mathrm{d}y\mathrm{d}x ,$$

the function J_0 being defined in (10.3.9). Integrating with respect to time, we conclude that

$$Lf_j(t) + \frac{1}{2} \int_0^t \int_0^j \int_0^{j-x} J_0(k_j(x, y)f_j(s, x)f_j(s, y), F_j(x, y)f_j(s, x+y)) \; \mathrm{d}y\mathrm{d}x\mathrm{d}s$$

$$= Lf_j^{in} \tag{10.3.44}$$

for $t \geq 0$.

Step 3: Uniform integrability. A useful consequence of (10.3.44) is the uniform integrability of the sequence $(f_j)_{j \geq 1}$ which turns out to be a consequence of the detailed balance condition in this particular case. Indeed, for $t \in (0, \infty)$, we infer from (10.3.44), Lemma 10.3.9 and the nonnegativity of J_0 that

$$\|f_j(t) \ln f_j(t)\|_1 - 4 - 2M_1(f_j(t)) - M_0(f_j(t)) \leq Lf_j(t) \leq Lf_j^{in} , \qquad t \in [0, \infty) ,$$

and
$$Lf_j^{in} \leq \left\|f_j^{in} \ln f_j^{in}\right\|_1 + \ln(K_1)M_0(f_j^{in}) + K_\infty M_{q_\infty}(f_j^{in}) + M_0(Q) \ .$$

Consequently, on using (10.3.30), (10.3.37), (10.3.42), and the properties $f^{in} \in X_{q_\infty}$ and $f_j^{in} \leq f^{in} + Q$, we obtain that, for $t \in [0,\infty)$,

$$\|f_j(t) \ln f_j(t)\|_1 \leq C + \left\|f_j^{in} \ln f_j^{in}\right\|_1 \ .$$

In addition, owing to (10.3.36e) and the monotonicity of $r \mapsto r \ln r$ on (e^{-1}, ∞),

$$
\begin{aligned}
\left\|f_j^{in} \ln f_j^{in}\right\|_1 &\leq \int_0^j \min\{f^{in}(x), j\} \left|\ln(\min\{f^{in}(x), j\})\right| \, dx \\
&\quad + \int_0^j \min\left\{\frac{Q(x)}{j}, j\right\} \left|\ln\left(\min\left\{\frac{Q(x)}{j}, j\right\}\right)\right| \, dx \\
&\leq \|f^{in} \ln f^{in}\|_1 + \int_0^j \frac{Q(x)}{j}\left|\ln\left(\frac{Q(x)}{j}\right)\right| \mathbf{1}_{(0,j^2]}(Q(x)) \, dx \\
&\quad + \int_0^j j\ln(j)\mathbf{1}_{(j^2,\infty)}(Q(x)) \, dx \\
&\leq \|f^{in} \ln f^{in}\|_1 + \|Q \ln Q\|_1 + \frac{\ln j}{j}M_0(Q) \\
&\quad + \int_0^j Q(x)\ln(Q(x))\mathbf{1}_{(j^2,\infty)}(Q(x)) \, dx \\
&\leq C \ .
\end{aligned}
$$

Gathering the previous two estimates we end up with

$$\|f_j(t) \ln f_j(t)\|_1 \leq C_2 \ , \qquad t \in [0,\infty) \ , \ j \geq 1 \ . \tag{10.3.45}$$

Step 4: Time equicontinuity. We next turn to the time equicontinuity of the sequence $(f_j)_{j\geq 1}$, its proof being quite close to that of similar results in Chapter 8 due to (10.3.29), (10.3.37), and (10.3.42). We denote the coagulation and fragmentation operators associated with k_j, a_j and b by $\mathcal{C}_j$ and $\mathcal{F}_j$, respectively. Let $T > 0$. On the one hand, it follows from (10.3.29c), (10.3.36a), (10.3.37), (10.3.42) and Fubini's theorem that, for $t \in [0,T]$,

$$
\begin{aligned}
\int_0^j |\mathcal{C}_j f(t,x)| \, dx &\leq 2\int_0^\infty \int_0^\infty k_j(x,y)f_j(t,x)f_j(t,y) \, dydx \\
&\leq 2K_0 \int_0^\infty \int_0^\infty (2+x+y)f_j(t,x)f_j(t,y) \, dydx \\
&\leq 4K_0 \left[M_0(f_j(t)) + M_1(f_j(t))\right]M_0(f_j(t)) \leq C(T) \ .
\end{aligned}
$$

On the other hand, (10.3.29b), (10.3.36a), (10.3.37), (10.3.39), (10.3.42) and Fubini's theorem imply that, for $t \in [0,T]$,

$$
\begin{aligned}
\int_0^j |\mathcal{F}_j f(t,x)| \, dx &\leq 2\int_0^\infty \int_0^x F_j(y, x-y)f_j(t,x) \, dydx \\
&\leq 2K_0 \int_0^\infty x(1+x)^{\gamma-1}f_j(t,x) \, dx \\
&\leq C\left[M_1(f_j(t)) + M_{1+\gamma}(f_j(t))\right] \leq C(T) \ .
\end{aligned}
$$

Therefore,

$$\int_0^\infty |\partial_t f_j(t,x)| \, dx \le C_3(T) \,, \qquad t \in [0,T] \,, \; j \ge 1 \,, \tag{10.3.46}$$

which, in particular, ensures the time weak equicontinuity of the sequence $(f_j)_{j\ge 1}$.

Step 5: Convergence. Since $m > 1$ by (10.3.39), we infer from (10.3.42), (10.3.45) and the Dunford–Pettis theorem (Theorem 7.1.3) that, for every $T > 0$, there is a weakly compact subset $\mathcal{E}_T$ of $X_{0,1}$ such that

$$f_j(t) \in \mathcal{E}_T \,, \qquad t \in [0,T] \,, \qquad j \ge 1 \,. \tag{10.3.47}$$

Since $(f_j)_{j\ge 1}$ is weakly equicontinuous in X_0 at each $t \in [0,T]$ by (10.3.46), we are in a position to apply the variant of the Arzelà–Ascoli theorem stated in Theorem 7.1.16 to conclude that there is a subsequence of $(f_j)_{j\ge 1}$ (not relabelled), and $f \in C([0,\infty), X_{0,w})$ such that

$$f_j \longrightarrow f \quad \text{in } C([0,T], X_{0,w}) \tag{10.3.48a}$$

for all $T > 0$. Owing to (10.3.42) and $m > 1$, we argue as in the proof of (8.2.109) to improve (10.3.48a) to

$$f_j \longrightarrow f \quad \text{in } C([0,T], X_{1,w}) \tag{10.3.48b}$$

for all $T > 0$. Immediate consequences of (10.3.36e), (10.3.37), (10.3.42), and (10.3.48) are the nonnegativity of f, the mass conservation $M_1(f(t)) = M_1(f^{in})$, and the property $f \in L_\infty((0,t), X_0 \cap X_m)$ for all $t \ge 0$. Moreover, it follows from (10.3.29), (10.3.36), (10.3.42), (10.3.48), and $m > \max\{1,\gamma\}$ (recall (10.3.39)) that the condition (8.2.36) is satisfied (with $m_0 = 0$) and hence we deduce from Theorem 8.2.11 that f is a weak solution to the C-F equation (10.3.3) on $[0,\infty)$ with coefficients (k,a,b). We have thus shown that f is a mass-conserving weak solution to the C-F equation (10.3.3) on $[0,\infty)$ with coefficients (k,a,b) and it belongs to $L_\infty((0,T), X_0 \cap X_m)$ for all $T > 0$.

Step 6: Entropy inequality. The last step of the proof is devoted to the integrability properties (10.3.35) of both Lf and $\mathcal{D}f$, which are inherited from the entropy identity (10.3.44) satisfied by f_j. More precisely, we infer from (10.3.48) and the convexity of the function $r \mapsto r \ln r$ that

$$\int_0^\infty f(t,x) \ln f(t,x) \, dx \le \liminf_{j\to\infty} \int_0^\infty f_j(t,x) \ln f_j(t,x) \, dx \,, \qquad t \ge 0 \,,$$

while (10.3.30), (10.3.42) and (10.3.48) imply that

$$\lim_{j\to\infty} \int_0^\infty f_j(t,x) \ln Q(x) \, dx = \int_0^\infty f(t,x) \ln Q(x) \, dx \,, \qquad t \ge 0 \,.$$

Consequently,

$$Lf(t) \le \liminf_{j\to\infty} Lf_j(t) \,, \qquad t \ge 0 \,. \tag{10.3.49}$$

We next observe that it follows from (10.3.29), (10.3.36a), (10.3.42), (10.3.48) and Proposition 7.1.12 that

$$[(s,x,y) \mapsto k_j(x,y)f_j(s,x)f_j(s,y)] \rightharpoonup [(s,x,y) \mapsto k(x,y)f(s,x)f(s,y)]$$

and

$$[(s,x,y) \mapsto F_j(x,y)f_j(s,x+y)] \rightharpoonup [(s,x,y) \mapsto F(x,y)f(s,x+y)]$$

in $L_1((0,T) \times (0,\infty)^2)$ for all $T > 0$. Since J_0 is a lower semicontinuous convex function in $\mathbb{R}^2$, we conclude that

$$\int_0^t \mathscr{D}f(s) \, \mathrm{d}s$$

$$\le \liminf_{j \to \infty} \int_0^t \int_0^j \int_0^{j-x} J_0(k_j(x,y)f_j(s,x)f_j(s,y), F_j(x,y)f_j(s,x+y)) \, \mathrm{d}y\mathrm{d}x\mathrm{d}s .$$

Owing to the above analysis, we may let $j \to \infty$ in (10.3.44) and thereby obtain (10.3.35b). We finally combine (10.3.35b), (10.3.37), (10.3.42) and Lemma 10.3.9 to deduce (10.3.35a), recalling that we have already established in Step 5 that $f \in L_\infty((0,T), X_0 \cap X_m)$ for all $T > 0$. $\square$

10.3.2.2   Entropy Identity

As observed in [186] for the Boltzmann equation, the entropy identity (10.3.7) can be recovered from the C-F equation and the regularity of the solution constructed in Proposition 10.3.10 by a suitable approximation argument, provided the initial condition decays sufficiently fast for large sizes.

Proposition 10.3.11. *[169] Assume that the coagulation and fragmentation coefficients k, a and b satisfy (10.3.27), (10.3.28), (10.3.29), and (10.3.30), and consider an initial condition $f^{in} \in \mathcal{Y} \cap X_\mu$ for some $\mu \ge 2 + (\gamma - 1)_+ + q_\infty$. Let f be a mass-conserving weak solution to the C-F equation (10.3.3) on $[0,\infty)$ satisfying (10.3.35). Then*

$$Lf(t) + \int_0^t \mathscr{D}f(s) \, \mathrm{d}s = Lf^{in} , \qquad t \ge 0 .$$

Proof. The proof of Proposition 10.3.11 relies on an approximation argument which is of a different nature than the one used in the existence proof performed in Section 10.3.2.1. For $j \ge 1$, $t \ge 0$ and $(x,y) \in (0,\infty)^2$, we set

$$g_j(t,x) := \min\{f(t,x),j\} + \phi_j(x) , \qquad \phi_j(x) := \frac{Q(x)}{j} ,$$

$$\bar{f}(t,x) := f(t,x) + Q(x) ,$$

$$u(t,x,y) := k(x,y)f(t,x)f(t,y) , \qquad v(t,x,y) := F(x,y)f(t,x+y) ,$$

and $G_j := (f + \phi_j)(\ln(g_j/Q) - 1) + Q$.

Let $T > 0$ and $t \in [0,T]$. On the one hand, it follows from the symmetry of k, (10.3.29c) and Proposition 10.3.10 that

$$\int_0^\infty (1+x)^{q_\infty} |\mathcal{C}f(t,x)| \, \mathrm{d}x$$

$$\le \frac{1}{2} \int_0^\infty \int_0^\infty [(1+x+y)^{q_\infty} + (1+x)^{q_\infty} + (1+y)^{q_\infty}] \, u(t,x,y) \, \mathrm{d}y\mathrm{d}x$$

$$\le C \int_0^\infty \int_0^\infty (1 + x^{q_\infty} + y^{q_\infty})(1+x+y)f(t,x)f(t,y) \, \mathrm{d}y\mathrm{d}x$$

$$\le C \left[M_0^2(f(t)) + M_0(f(t))M_1(f(t)) + M_0(f(t))M_{q_\infty}(f(t)) \right]$$

$$\quad + C \left[M_1(f(t))M_{q_\infty}(f(t)) + M_0(f(t))M_{1+q_\infty}(f(t)) \right]$$

$$\le C(T) . \tag{10.3.50}$$

We also infer from (10.3.27), (10.3.29b) and Proposition 10.3.10 that

$$
\begin{aligned}
\int_0^\infty (1+x)^{q_\infty} |\mathscr{F}f(t,x)| \ dx &\leq 3 \int_0^\infty (1+x)^{q_\infty} a(x) f(t,x) \ dx \\
&\leq \frac{3K_0}{2} \int_0^\infty x(1+x)^{q_\infty+\gamma-1} f(t,x) \ dx \\
&\leq C(T) \ .
\end{aligned}
\tag{10.3.51}
$$

Combining the previous two estimates with the C-F equation (10.3.3) ensures that

$$
\partial_t f \in L_\infty((0,T), L_1((0,\infty), (1+x)^{q_\infty} dx)) \ . \tag{10.3.52}
$$

On the other hand, for $j \geq 1$ and $x \in (0,\infty)$, the nonnegativity of f gives

$$
\ln\left(\frac{g_j(t,x)}{Q(x)}\right) \geq -\ln j \ ,
$$

while (10.3.30) ensures that

$$
\begin{aligned}
\ln\left(\frac{g_j(t,x)}{Q(x)}\right) &\leq \ln\left(\frac{j+\phi_j(x)}{Q(x)}\right) \leq \ln\left(\frac{j+Q(x)}{Q(x)}\right) \\
&\leq \ln(j+Q(x)) - \ln Q(x) \\
&\leq \ln(2Q(x)) \mathbf{1}_{(j,\infty)}(Q(x)) + \ln(2j) \mathbf{1}_{(0,j]}(Q(x)) - \ln Q(x) \\
&\leq \ln 2 + [\ln(j) - \ln(Q(x))]\mathbf{1}_{(0,j]}(Q(x)) \\
&\leq \ln 2 + \ln j + \ln K_1 + K_\infty x^{q_\infty} \ .
\end{aligned}
$$

Consequently,

$$
\ln(g_j/Q) \in L_\infty((0,T), L_1((0,\infty), (1+x)^{-q_\infty} dx)) \ . \tag{10.3.53}
$$

It follows from (10.3.52) and (10.3.53) that $\partial_t G_j(t)$ belongs to X_0 for each $t \in [0,T]$ with

$$
\partial_t G_j(t,x) = \left[\ln\left(\frac{g_j(t,x)}{Q(x)}\right) - \mathbf{1}_{(j,\infty)}(f(t,x))\right] \partial_t f(t,x) \quad \text{for a.e. } x \in (0,\infty) \ .
$$

We then infer from the C-F equation (10.3.3) that

$$
\frac{d}{dt} \int_0^\infty G_j(t,x) \ dx = -\frac{1}{2} \int_0^\infty \int_0^\infty J_j(t,x,y) \ dydx - R_j(t) \ ,
$$

where

$$
J_j(t,x,y) := (u-v)(t,x,y) \left[\ln\left(\frac{g_j(t,x)g_j(t,y)}{Q(x)Q(y)}\right) - \ln\left(\frac{g_j(t,x+y)}{Q(x+y)}\right)\right] \ , \tag{10.3.54}
$$

$$
R_j(t) := \int_0^\infty [\mathcal{C}f(t,x) + \mathscr{F}f(t,x)] \mathbf{1}_{(j,\infty)}(f(t,x)) \ dx \ , \tag{10.3.55}
$$

for $(t,x,y) \in [0,T] \times (0,\infty)^2$ and $j \geq 1$. Integrating with respect to time gives, for $t \in [0,T]$,

$$
\begin{aligned}
\int_0^\infty G_j(t,x) \ dx = \int_0^\infty G_j(0,x) \ dx &- \frac{1}{2} \int_0^t \int_0^\infty \int_0^\infty J_j(s,x,y) \ dydxds \\
&- \int_0^t R_j(s) \ ds \ .
\end{aligned}
\tag{10.3.56}
$$

We are left with passing to the limit as $j \to \infty$ in (10.3.56) and identifying the limit of the four terms involved. Let $T > 0$. Introducing

$$\bar{J}(s, x, y) := (1 + \|Q\|_\infty)[k(x, y) + F(x, y)] \left[\bar{f}(s, x + y) \left(1 + \bar{f}(s, x) + \bar{f}(s, y)\right)\right]$$
$$+ (1 + \|Q\|_\infty)[k(x, y) + F(x, y)]\bar{f}(s, x)\bar{f}(s, y)$$

for $(s, x, y) \in (0, T) \times (0, \infty)^2$, and observing that

$$k(x, y) + F(x, y) \le C\left[1 + x + y + (1 + x + y)^{\gamma - 1}\right] \le C(1 + x + y)^{1 + (\gamma - 1)_+}$$

for $(x, y) \in (0, \infty)^2$ by (10.3.29), we infer from Proposition 10.3.10 that, for $s \in (0, T)$,

$$\int_0^\infty \int_0^\infty \bar{J}(s, x, y) \, dy dx \le C \int_0^\infty \int_0^\infty (1 + x + y)^{1 + (\gamma - 1)_+} \bar{f}(s, x + y) \, dy dx$$
$$+ C \int_0^\infty \int_0^\infty (1 + x + y)^{1 + (\gamma - 1)_+} \bar{f}(s, x + y)\bar{f}(s, x) \, dy dx$$
$$+ C \int_0^\infty \int_0^\infty (1 + x + y)^{1 + (\gamma - 1)_+} \bar{f}(s, x)\bar{f}(s, y) \, dy dx$$
$$\le C\left[M_1(\bar{f}(s)) + M_{2 + (\gamma - 1)_+}(\bar{f}(s))\right]$$
$$+ C\left[M_0(\bar{f}(s)) + M_{1 + (\gamma - 1)_+}(\bar{f}(s))\right] M_0(\bar{f}(s))$$
$$\le C(T) ,$$

so that

$$\bar{J} \in L_\infty((0, T), L_1((0, \infty)^2)) . \tag{10.3.57}$$

We next estimate J_j in terms of $\bar{J}$ and $J_0(u, v)$, the latter being in $L_1((0, T) \times (0, \infty)^2)$ according to (10.3.35a). To this end, we consider $(s, x, y) \in (0, T) \times (0, \infty)^2$ and split the analysis into four cases.

Case 1: If $u(s, x, y) \ge v(s, x, y)$ and

$$\frac{g_j(s, x)g_j(s, y)}{Q(x)Q(y)} \ge \frac{g_j(s, x + y)}{Q(x + y)} ,$$

then $J_j(s, x, y) \ge 0$ and it follows from the inequality $\ln r \le r$, $r \in [0, \infty)$, and the detailed balance condition (10.3.27c) that

$$0 \le J_j(s, x, y) = J_0(u(s, x, y), v(s, x, y))$$
$$+ (u - v)(s, x, y) \ln \left(\frac{f(s, x + y)g_j(s, x)g_j(s, y)}{f(s, x)f(s, y)g_j(s, x + y)}\right)$$
$$\le J_0(u(s, x, y), v(s, x, y)) + (u - v)(s, x, y)\frac{f(s, x + y)g_j(s, x)g_j(s, y)}{f(s, x)f(s, y)g_j(s, x + y)}$$
$$\le J_0(u(s, x, y), v(s, x, y)) + u(s, x, y)\frac{f(s, x + y)g_j(s, x)g_j(s, y)}{f(s, x)f(s, y)g_j(s, x + y)}$$
$$\le J_0(u(s, x, y), v(s, x, y)) + k(x, y)\frac{f(s, x + y)g_j(s, x)g_j(s, y)}{g_j(s, x + y)} .$$

Now, either $f(s, x + y) \le j$ and

$$\frac{f(s, x + y)g_j(s, x)g_j(s, y)}{g_j(s, x + y)} = \frac{f(s, x + y)g_j(s, x)g_j(s, y)}{f(s, x + y) + \phi_j(x + y)} \le g_j(s, x)g_j(s, y)$$

$$\leq \bar{f}(s,x)\bar{f}(s,y) ,$$

or $f(s, x + y) > j$ and

$$\frac{f(s,x+y)g_j(s,x)g_j(s,y)}{g_j(s,x+y)} = \frac{f(s,x+y)g_j(s,x)g_j(s,y)}{j+\phi_j(x+y)} \leq \frac{j+\|Q\|_\infty}{j}f(s,x+y)g_j(s,y)$$

$$\leq (1+\|Q\|_\infty)\bar{f}(s,x+y)\bar{f}(s,y) .$$

In both situations, we conclude that

$$0 \leq J_j(s,x,y) \leq J_0(u(s,x,y),v(s,x,y)) + \bar{J}(s,x,y) . \tag{10.3.58}$$

Case 2: If $u(s, x, y) \leq v(s, x, y)$ and

$$\frac{g_j(s,x)g_j(s,y)}{Q(x)Q(y)} \geq \frac{g_j(s,x+y)}{Q(x+y)} ,$$

then $J_j(s, x, y) \leq 0$ and the inequality $\ln r \leq r$, $r \in [0, \infty)$, and the detailed balance condition (10.3.27c) give

$$0 \leq -J_j(s,x,y) = (v-u)(s,x,y)\ln\left(\frac{Q(x+y)g_j(s,x)g_j(s,y)}{Q(x)Q(y)g_j(s,x+y)}\right)$$

$$\leq (v-u)(s,x,y)\frac{Q(x+y)g_j(s,x)g_j(s,y)}{Q(x)Q(y)g_j(s,x+y)}$$

$$\leq v(s,x,y)\frac{Q(x+y)g_j(s,x)g_j(s,y)}{Q(x)Q(y)g_j(s,x+y)}$$

$$= k(x,y)\frac{f(s,x+y)g_j(s,x)g_j(s,y)}{g_j(s,x+y)} .$$

We now argue as in the previous case to show that

$$0 \leq -J_j(s,x,y) \leq \bar{J}(s,x,y) . \tag{10.3.59}$$

Case 3: If $u(s, x, y) \geq v(s, x, y)$ and

$$\frac{g_j(s,x)g_j(s,y)}{Q(x)Q(y)} \leq \frac{g_j(s,x+y)}{Q(x+y)} ,$$

then $J_j(s, x, y) \leq 0$ and similar arguments give

$$0 \leq -J_j(s,x,y) = (u-v)(s,x,y)\ln\left(\frac{Q(x)Q(y)g_j(s,x+y)}{Q(x+y)g_j(s,x)g_j(s,y)}\right)$$

$$\leq (u-v)(s,x,y)\frac{Q(x)Q(y)g_j(s,x+y)}{Q(x+y)g_j(s,x)g_j(s,y)}$$

$$\leq u(s,x,y)\frac{Q(x)Q(y)g_j(s,x+y)}{Q(x+y)g_j(s,x)g_j(s,y)}$$

$$= F(x,y)\frac{f(s,x)f(s,y)g_j(s,x+y)}{g_j(s,x)g_j(s,y)} .$$

Now, if $f(s, x) \in (0, j)$ and $f(s, y) \in (0, j)$, then

$$\frac{f(s,x)f(s,y)g_j(s,x+y)}{g_j(s,x)g_j(s,y)} \leq g_j(s,x+y) \leq \bar{f}(s,x+y) .$$

If $f(s,x) \geq j \geq f(s,y)$, then

$$\frac{f(s,x)f(s,y)g_j(s,x+y)}{g_j(s,x)g_j(s,y)} \leq \frac{f(s,y)}{f(s,y)+\phi_j(y)} \frac{f(s,x)g_j(s,x+y)}{j+\phi_j(x)} \leq \bar{f}(s,x)\bar{f}(s,x+y) \ .$$

Similarly, if $f(s,y) \geq j \geq f(s,x)$, then

$$\frac{f(s,x)f(s,y)g_j(s,x+y)}{g_j(s,x)g_j(s,y)} \leq \bar{f}(s,y)\bar{f}(s,x+y) \ .$$

Finally if $f(s,x) \in [j,\infty)$ and $f(s,y) \in [j,\infty)$ then

$$\frac{f(s,x)f(s,y)g_j(s,x+y)}{g_j(s,x)g_j(s,y)} \leq f(s,x)f(s,y)\frac{j+\phi_j(x+y)}{(j+\phi_j(x))(j+\phi_j(y))}$$
$$\leq (1+\|Q\|_\infty)\bar{f}(s,x)\bar{f}(s,y) \ .$$

Summarising, we have shown that (10.3.59) holds true in this case as well.

Case 4: If $u(s,x,y) \leq v(s,x,y)$ and

$$\frac{g_j(s,x)g_j(s,y)}{Q(x)Q(y)} \leq \frac{g_j(s,x+y)}{Q(x+y)} \ ,$$

then $J_j(s,x,y) \geq 0$ and we proceed as above to establish that (10.3.58) also holds true in this case.

Collecting the analysis just carried out, we deduce from (10.3.58) and (10.3.59) that, for $(s,x,y) \in (0,T) \times (0,\infty)^2$,

$$\begin{aligned} [J_j(s,x,y)]_+ &\leq J_0(u(s,x,y),v(s,x,y)) + \bar{J}(s,x,y) \ , \\ [-J_j(s,x,y)]_+ &\leq \bar{J}(s,x,y) \ . \end{aligned} \tag{10.3.60}$$

Since

$$\lim_{j\to\infty}[J_j(s,x,y)]_+ = J_0(u(s,x,y),v(s,x,y)) \quad \text{and} \quad \lim_{j\to\infty}[-J_j(s,x,y)]_+ = 0$$

for $(s,x,y) \in (0,T) \times (0,\infty)^2$, and both $J_0(u,v)$ and $\bar{J}$ belong to $L_1((0,T) \times (0,\infty)^2)$ by (10.3.35b) and (10.3.57), we infer from (10.3.60) and the Lebesgue dominated convergence theorem that, for all $t \in (0,T)$,

$$\begin{aligned} \lim_{j\to\infty}\int_0^t\int_0^\infty\int_0^\infty [J_j(s,x,y)]_+ \, dydxds &= \int_0^t\int_0^\infty\int_0^\infty J_0(u(s,x,y),v(s,x,y)) \, dydxds \\ &= \int_0^t \mathcal{D}f(s) \, ds \end{aligned}$$

and

$$\lim_{j\to\infty}\int_0^t\int_0^\infty\int_0^\infty [-J_j(s,x,y)]_+ \, dydxds = 0 \ .$$

Recalling that $J_j = [J_j]_+ - [-J_j]_+$, we have thus established that

$$\lim_{j\to\infty}\int_0^t\int_0^\infty\int_0^\infty J_j(s,x,y) \, dydxds = \int_0^t \mathcal{D}f(s) \, ds \ , \qquad t \in (0,T) \ . \tag{10.3.61}$$

Moreover, it readily follows from (10.3.50), (10.3.51) and the Lebesgue dominated convergence theorem that

$$\lim_{j\to\infty} \int_0^t R_j(s)\, \mathrm{d}s = 0 \ . \tag{10.3.62}$$

Finally, we observe that, for $j \geq e\|Q\|_\infty$ and $x \in (0,\infty)$, there holds

$$\frac{(r + \phi_j(x))}{Q(x)} \left[\ln\left(\frac{\min\{r,j\} + \phi_j(x)}{Q(x)} \right) - 1 \right] + 1 \geq 0 \ , \qquad r \in [0,\infty) \ ,$$

as this function attains its minimum at $r = Q(x) - \phi_j(x)$. Consequently, $G_j \geq 0$ and we infer from the convexity of $\psi : r \mapsto r(\ln r - 1) + 1$ that

$$0 \leq G_j \leq (f + \phi_j) \left[\ln\left(\frac{f + \phi_j}{Q} \right) - 1 \right] + Q = \psi\left(\frac{f}{Q} + \frac{1}{j} \right)$$

$$\leq \frac{Q}{2} \left[\psi\left(\frac{2f}{Q} \right) + \psi\left(\frac{2}{j} \right) \right] \leq \ln(2)(f + Q) + Q\psi\left(\frac{f}{Q} \right) \ .$$

Since $(G_j(t))_{j\geq 1}$ converges almost everywhere to $f(t)[\ln(f(t)/Q) - 1] + Q$ in $(0,\infty)$ as $j \to \infty$ for all $t \in [0,T]$ and the right-hand side of the previous inequality belongs to X_0 by Proposition 10.3.10, we use once more the Lebesgue dominated convergence theorem to conclude that

$$\lim_{j\to\infty} \int_0^\infty G_j(t,x)\, \mathrm{d}x = Lf(t) \ , \qquad t \in [0,T] \ . \tag{10.3.63}$$

Owing to (10.3.61), (10.3.62) and (10.3.63), we may pass to the limit as $j \to \infty$ in (10.3.56) to complete the proof of Proposition 10.3.11. □

10.3.2.3 Stabilisation

We are now ready to study the long-term behaviour of the mass-conserving weak solutions f on $[0,\infty)$ to the C-F equation (10.3.3) we constructed in Proposition 10.3.10. To this end, we follow the approach developed in [165, Section 6 and Appendix C] and split the proof into several steps. As already mentioned, introducing

$$Q_z(x) := Q(x)e^{x \ln z} \ , \qquad (z,x) \in (0,\infty) \times (0,\infty) \ , \qquad Q_0 \equiv 0 \ , \tag{10.3.64}$$

it readily follows from (10.3.27c) that Q_z also satisfies (10.3.27c) and is thus a stationary solution to the C-F equation. The main goal being to show that the cluster points of $f(t)$ as $t \to \infty$ belong to the set

$$\mathcal{Q} := \{Q_z \ : \ z \in [0,\infty)\} \tag{10.3.65}$$

with the help of the entropy inequality (10.3.35b), we begin with the characterisation of the functions for which $\mathcal{D}$ vanishes.

Lemma 10.3.12. *Assume that the coagulation and fragmentation coefficients k, a and b satisfy (10.3.27) and assume further that $1/Q \in L_\infty(0,r)$ for all $r > 0$. If ζ is a nonnegative function in X_0 such that*

$$k(x,y)\zeta(x)\zeta(y) = F(x,y)\zeta(x+y) \ \text{ for a.e. } (x,y) \in (0,\infty)^2 \ ,$$

then $\zeta \in \mathcal{Q}$.

Proof. Since Q is positive in $(0,\infty)$, the function $g := \zeta/Q$ is well defined and satisfies

$$g(x)g(y) = g(x+y) \ \text{ for a.e. } (x,y) \in (0,\infty)^2 \ . \tag{10.3.66}$$

In addition, since $1/Q \in L_\infty(0,r)$ for all $r > 0$, we see that $g \in L_1(0,r)$ for all $r > 0$, so that its integral

$$\bar{g}(x) := \int_0^x g(y) \, \mathrm{d}y , \qquad x \in [0, \infty) ,$$

is well defined and continuous. Integrating (10.3.66) with respect to y over $(0,1)$ gives

$$\bar{g}(1)g(x) = \bar{g}(x+1) - \bar{g}(x) \quad \text{for a.e. } x \in (0,\infty) . \tag{10.3.67}$$

Either $\bar{g}(1) = 0$ and we infer from (10.3.67) that $\bar{g}(i) = \bar{g}(1) = 0$ for all integers $i \geq 1$, hence $g = 0$ a.e. in $(0,\infty)$, or $\bar{g}(1) > 0$ and (10.3.67) implies that $g \in C([0,\infty))$. Since g solves (10.3.66), we readily deduce that there is $z \in (0,\infty)$ such that $g(x) = e^{x \ln z}$ for $x \in (0,\infty)$. □

The next step is to look for estimates which do not depend on time. At first glance, the only ones supplied by Proposition 10.3.10 are the mass conservation and (10.3.35b), the latter providing only an upper bound on Lf thanks to the nonnegativity of $\mathcal{D}$. As observed in [165, Lemma 3.1], it nevertheless leads to the following estimates which, in particular, yield the weak compactness of the trajectories in X_0, see (10.3.70) below.

Lemma 10.3.13. *[165] Assume that the coagulation and fragmentation coefficients k, a and b satisfy (10.3.27), (10.3.28), (10.3.29) and (10.3.30), and consider an initial condition $f^{in} \in \mathcal{Y} \cap X_\mu$ for some $\mu \geq 1 + \gamma$. Let f be a mass-conserving weak solution to the C-F equation (10.3.3) on $[0,\infty)$ satisfying (10.3.35). Then, for $t \geq 0$, $R > 1$ and any measurable subset E of $(0,\infty)$,*

$$M_1(f(t)) = \varrho^{in} := M_1(f^{in}) , \tag{10.3.68}$$

$$M_0(f(t)) + \int_0^\infty f(t,x) \left| \ln\left(\frac{f(t,x)}{Q(x)}\right) \right| \, \mathrm{d}x \leq C_4 , \tag{10.3.69}$$

$$\int_E f(t,x) \, \mathrm{d}x \leq R \int_E Q(x) \, \mathrm{d}x + \frac{C_4}{\ln R} , \tag{10.3.70}$$

and

$$\int_0^\infty \mathcal{D}f(s) \, \mathrm{d}s \leq C_4 . \tag{10.3.71}$$

Proof. Let $t \geq 0$. On the one hand, since $r \ln r \geq r|\ln r| - 2/e$ for $r \in [0,\infty)$, it follows from (10.3.35b) that

$$Lf^{in} \geq Lf(t) \geq \int_0^\infty f(t,x) \left| \ln\left(\frac{f(t,x)}{Q(x)}\right) \right| \, \mathrm{d}x - \frac{2}{e}M_0(Q) - M_0(f(t)) + M_0(Q) ,$$

hence

$$\int_0^\infty f(t,x) \left| \ln\left(\frac{f(t,x)}{Q(x)}\right) \right| \, \mathrm{d}x \leq Lf^{in} + \frac{2}{e}M_0(Q) + M_0(f(t)) . \tag{10.3.72}$$

On the other hand, if $R > 1$ and E is a measurable subset of $(0,\infty)$, then

$$\int_E f(t,x) \, \mathrm{d}x = \int_E f(t,x)\mathbf{1}_{[0,\infty)}(RQ(x) - f(t,x)) \, \mathrm{d}x$$

$$+ \int_E f(t,x)\mathbf{1}_{(0,\infty)}(f(t,x) - RQ(x)) \, \mathrm{d}x$$

$$\leq R \int_E Q(x) \, \mathrm{d}x + \frac{1}{\ln R} \int_0^\infty f(t,x) \left| \ln\left(\frac{f(t,x)}{Q(x)}\right) \right| \, \mathrm{d}x . \tag{10.3.73}$$

We now infer from (10.3.72) and (10.3.73) (with $E = (0, \infty)$) that

$$M_0(f(t)) \le R M_0(Q) + \frac{1}{\ln R}\left[L f^{in} + M_0(f(t)) + \frac{2}{e}M_0(Q)\right] .$$

Taking $R = e^2$ in the previous inequality gives

$$M_0(f(t)) \le \left(2e^2 + \frac{2}{e}\right) M_0(Q) + L f^{in} ,$$

which, together with (10.3.72), implies (10.3.69). Next, (10.3.70) readily follows from (10.3.69) and (10.3.73), while (10.3.71) is a straightforward consequence of (10.3.35b), (10.3.69) and (10.3.72). Finally, (10.3.68) is nothing but the mass-conserving property of the solution f. $\qquad\square$

An additional estimate is available in the case $z_{th} = \infty$, where

$$z_{th} = \sup \{z \in (0, \infty) \; : \; Q_z \in X_1\} , \qquad (10.3.74)$$

the proof relying on an argument similar to that used to prove (10.3.73).

Lemma 10.3.14. *Assume that the coagulation and fragmentation coefficients k, a and b satisfy (10.3.27), (10.3.28), (10.3.29) and (10.3.30), and consider an initial condition $f^{in} \in \mathcal{Y} \cap X_\mu$ for some $\mu \ge 1 + \gamma$. Let f be a mass-conserving weak solution to the C-F equation (10.3.3) on $[0, \infty)$ satisfying (10.3.35). If $z_{th} = \infty$ then*

$$\lim_{r \to \infty} \sup_{t \ge 0} \left\{\int_r^\infty x f(t, x) \, \mathrm{d}x\right\} = 0 .$$

Proof. Let $t \ge 0$, $r > 1$ and $z > 1$. We infer from (10.3.69) that

$$\int_r^\infty x f(t, x) \, \mathrm{d}x = \int_r^\infty x f(t, x) \mathbf{1}_{[0, \infty)}(Q_z(x) - f(t, x)) \, \mathrm{d}x$$
$$+ \int_r^\infty x f(t, x) \mathbf{1}_{(0, \infty)}(f(t, x) - Q_z(x)) \, \mathrm{d}x$$
$$\le \int_r^\infty x Q_z(x) \, \mathrm{d}x + \frac{1}{\ln z}\int_0^\infty f(t, x) \left|\ln\left(\frac{f(t, x)}{Q(x)}\right)\right| \, \mathrm{d}x$$
$$\le \int_r^\infty x Q_z(x) \, \mathrm{d}x + \frac{C_4}{\ln z} .$$

Since $Q_z \in X_1$, we may let $r \to \infty$ in the previous inequality and find

$$\limsup_{r \to \infty} \sup_{t \ge 0} \left\{\int_r^\infty x f(t, x) \, \mathrm{d}x\right\} \le \frac{C_4}{\ln z} .$$

We then let $z \to \infty$ to complete the proof of Lemma 10.3.14. $\qquad\square$

Having established these preliminary results, we are now in a position to state and prove the main result of this section.

Theorem 10.3.15. *Assume that the coagulation and fragmentation coefficients k, a and b satisfy (10.3.27), (10.3.28), (10.3.29) and (10.3.30), and consider an initial condition $f^{in} \in \mathcal{Y} \cap X_\mu$ for some $\mu \geq 1 + \gamma$. Let f be a mass-conserving weak solution to the C-F equation (10.3.3) on $[0, \infty)$ satisfying (10.3.35) and define its ω-limit set in $X_{0,w}$ by*

$$\omega(f) := \left\{ \zeta \in X_{0,1} : \begin{array}{c} \text{there is } (t_j)_{j \geq 1} \text{ such that } t_j \to \infty \\ \text{and } f(t_j) \rightharpoonup \zeta \text{ in } X_0. \end{array} \right\} . \tag{10.3.75}$$

(a) *The set $\omega(f)$ is non-empty and, in addition, if $\zeta \in \omega(f)$, then $\zeta \in \mathcal{Q} \cap X_1$ and $M_1(\zeta) \leq \varrho^{in} := M_1(f^{in})$.*

(b) *If $\lambda < 1$, see (10.3.28), then the set $\omega(f)$ is compact in X_0.*

While the first statement of Theorem 10.3.15 is mainly a consequence of the entropy inequality (10.3.35b), Lemma 10.3.12 and Lemma 10.3.13, its proof following the lines of that of LaSalle invariance principle, the proof of the compactness of $\omega(f)$ in X_0 is more involved and is adapted from a similar result for the Boltzmann equation, established in [185]. One of the building blocks of the forthcoming proof is the following inequality due to Arkeryd [10].

Lemma 10.3.16. *[10, Eq. (12)] For $(r, s) \in (0, \infty)^2$ and $R > 1$, there holds*

$$s \leq Rr + \frac{1}{\ln R}(r - s)(\ln r - \ln s) , \tag{10.3.76}$$

$$|s - r| \leq (R - 1)r + \frac{1}{\ln R}(r - s)(\ln r - \ln s) . \tag{10.3.77}$$

Proof. Since $(r - s)(\ln r - \ln s) \geq 0$, the inequality (10.3.76) is obvious when $s \leq Rr$. Next, if $s > Rr$, then $s > r$ and

$$\frac{(r - s)(\ln r - \ln s)}{\ln R} = \frac{s - r}{\ln R} \ln\left(\frac{s}{r}\right) \geq s - r \geq s - Rr ,$$

which completes the proof of (10.3.76). Next, either $s \geq r$ and (10.3.77) readily follows from (10.3.76), or $s < r$ whereupon applying (10.3.76) to (s, r) instead of (r, s) gives

$$r \leq Rs + \frac{1}{\ln R}(s - r)(\ln s - \ln r) \leq s + (R - 1)r + \frac{1}{\ln R}(s - r)(\ln s - \ln r) ,$$

hence (10.3.77) follows. $\qquad \square$

Proof of Theorem 10.3.15 (a). Let us first prove that $\omega(f)$ is non-empty. For that purpose, set

$$\mathcal{E} := \{f(t) : t \in [0, \infty)\} .$$

It follows from (10.3.68) and (10.3.70) that,

$$M_1(f(t)) = \varrho^{in} , \qquad t \geq 0 , \tag{10.3.78}$$

$$\eta\{\mathcal{E}; X_0\} \leq R\eta\{\{Q\}; X_0\} + \frac{C_4}{\ln R} , \qquad R > 1 , \tag{10.3.79}$$

the modulus of integrability $\eta\{\mathcal{E}; X_0\}$ of $\mathcal{E}$ in X_0 being defined in Definition 7.1.2. Since $Q \in X_0$, we have $\eta\{\{Q\}; X_0\} = 0$ and letting $R \to \infty$ in (10.3.79) gives

$$\eta\{\mathcal{E}; X_0\} = 0 . \tag{10.3.80}$$

As a consequence of (10.3.78), (10.3.80) and the Dunford–Pettis theorem (Theorem 7.1.3), we conclude that

$$\mathcal{E} \text{ is relatively sequentially weakly compact in } X_0 . \tag{10.3.81}$$

Thus there are cluster points in $X_{0,w}$ of $f(t)$ as $t \to \infty$, so that $\omega(f)$ is non-empty.

Consider now $\zeta \in \omega(f)$. Then there is a sequence $(t_j)_{j \geq 1}$ such that $t_j \to \infty$ and $f(t_j) \rightharpoonup \zeta$ in X_0. Without loss of generality we may assume that $t_j \geq 1$ for all $j \geq 1$, so that the function $f_j(s) := f(s + t_j)$ is well defined for all $s \in [-1, 1]$ and $j \geq 1$. According to (10.3.81),

$$\{f_j(s) \ : \ s \in [-1, 1] \ , \ j \geq 1\} \text{ is relatively sequentially weakly compact in } X_0 . \tag{10.3.82}$$

Consider next $j \geq 1$ and $-1 \leq s_1 \leq s_2 \leq 1$. We infer from (10.3.29c), (10.3.69) and (10.3.78) that

$$\int_{s_1}^{s_2} \int_0^\infty \int_0^\infty k(x,y) f_j(s,x) f_j(s,y) \, dy dx ds$$

$$\leq 2K_0 \int_{s_1}^{s_2} \left[M_0^2(f_j(s)) + M_0(f_j(s)) M_1(f_j(s)) \right] \, ds$$

$$\leq C(s_2 - s_1) . \tag{10.3.83}$$

It next follows from (10.3.76) that, for $R > 1$, $s \in [s_1, s_2]$ and $(x,y) \in (0,\infty)^2$,

$$F(x,y) f_j(s, x+y) \leq R k(x,y) f_j(s,x) f_j(s,y)$$

$$+ \frac{1}{\ln R} J_0(k(x,y) f_j(s,x) f_j(s,y), F(x,y) f_j(s,x+y)) .$$

Integrating the previous inequality with respect to (s,x,y) over $(s_1, s_2) \times (0,\infty)^2$ gives

$$\int_{s_1}^{s_2} \int_0^\infty \int_0^\infty F(x,y) f_j(s, x+y) \, dy dx ds$$

$$\leq R \int_{s_1}^{s_2} \int_0^\infty \int_0^\infty k(x,y) f_j(s,x) f_j(s,y) \, dy dx ds + \frac{1}{\ln R} \int_{-1+t_j}^{1+t_j} \mathcal{D}f(s) \, ds ,$$

hence, thanks to (10.3.71) and (10.3.83),

$$\int_{s_1}^{s_2} \int_0^\infty \int_0^\infty F(x,y) f_j(s, x+y) \, dy dx ds \leq CR(s_2 - s_1) + \frac{C_4}{\ln R} .$$

Consequently,

$$\int_{s_1}^{s_2} \int_0^\infty \int_0^\infty F(x,y) f_j(s, x+y) \, dy dx ds \leq C \inf_{R \in (1,\infty)} \left\{ R(s_2 - s_1) + \frac{1}{\ln R} \right\} . \tag{10.3.84}$$

Since

$$\|\mathcal{C} f_j(s)\|_1 \leq \frac{3}{2} \int_0^\infty \int_0^\infty k(x,y) f_j(s,x) f_j(s,y) \, dy dx$$

and

$$\|\mathcal{F} f_j(s)\|_1 \leq \frac{3}{2} \int_0^\infty \int_0^\infty F(x,y) f_j(s, x+y) \, dy dx$$

for $s \in (s_1, s_2)$, we infer from (10.3.3), (10.3.83) and (10.3.84) that

$$\|f_j(s_1) - f_j(s_2)\|_1 \leq \int_{s_1}^{s_2} \|\partial_s f_j(s)\|_1 \, ds$$

$$\le C \left[s_2 - s_1 + \inf_{R \in (1,\infty)} \left\{ R(s_2 - s_1) + \frac{1}{\ln R} \right\} \right] . \tag{10.3.85}$$

Since the right-hand side of (10.3.85) converges to zero as $s_2 - s_1 \to 0$, it follows from (10.3.82), (10.3.85) and the variant of the Arzelà–Ascoli theorem stated in Theorem 7.1.16, that $(f_j)_{j\ge 1}$ is relatively compact in $C([-1,1], X_{0,w})$. Therefore, there is a subsequence of $(f_j)_{j\ge 1}$ (not relabelled), and $f_\infty \in C([-1,1], X_{0,w})$ such that

$$f_j \longrightarrow f_\infty \quad \text{in } C([-1,1], X_{0,w}) . \tag{10.3.86}$$

A first consequence of (10.3.86) is that $f_\infty(s) \in \omega(f)$ for all $s \in [-1,1]$ and $f_\infty(0) = \zeta$. Furthermore, by (10.3.71),

$$\lim_{j\to\infty} \int_{-1}^{1} \mathscr{D} f_j(s) \, \mathrm{d}s = \lim_{j\to\infty} \int_{-1+t_j}^{1+t_j} \mathscr{D} f(s) \, \mathrm{d}s = 0 ,$$

while (10.3.86), the convexity and lower semicontinuity of J_0, and an argument similar to the one used in Step 6 of the proof of Proposition 10.3.10 imply that

$$\int_{-1}^{1} \int_{0}^{r} \int_{0}^{r} J_0(k(x,y) f_\infty(s,x) f_\infty(s,y), F(x,y) f_\infty(s, x+y)) \, \mathrm{d}y \mathrm{d}x \mathrm{d}s$$

$$\le \liminf_{j\to\infty} \int_{-1}^{1} \int_{0}^{r} \int_{0}^{r} J_0(k(x,y) f_j(s,x) f_j(s,y), F(x,y) f_j(s, x+y)) \, \mathrm{d}y \mathrm{d}x \mathrm{d}s$$

$$\le \liminf_{j\to\infty} \int_{-1}^{1} \mathscr{D} f_j(s) \, \mathrm{d}s$$

for all $r > 1$. Combining the previous two properties with the nonnegativity of J_0 leads us to

$$J_0(k(x,y) f_\infty(s,x) f_\infty(s,y), F(x,y) f_\infty(s, x+y)) = 0$$

for a.e. $(s,x,y) \in (-1,1) \times (0,\infty)^2$. In other words, owing to the time continuity of f_∞, there holds

$$k(x,y) f_\infty(s,x) f_\infty(s,y) = F(x,y) f_\infty(s, x+y) \quad \text{for a.e. } (x,y) \in (0,\infty)^2$$

and all $s \in [-1,1]$. Hence we deduce from (10.3.30) and Lemma 10.3.12 that

$$f_\infty(s) \in \mathscr{Q} \quad \text{for all } s \in [-1,1] . \tag{10.3.87}$$

In addition, it readily follows from (10.3.78) and (10.3.86) that $f_\infty(s)$ belongs to X_1 with $M_1(f_\infty(s)) \le \varrho^{in}$ for $s \in [-1,1]$. Recalling that $\zeta = f_\infty(0)$ completes the proof. $\qquad \square$

Proof of Theorem 10.3.15 (b). We keep the notation used in the proof of Theorem 10.3.15 (a) and assume in addition that $\lambda < 1$. Our aim is to prove that the sequence $(f(t_j))_{j\ge 1}$ converges to ζ in X_0 as $j \to \infty$. To this end, several steps are needed which we shortly describe prior to providing the proof. We begin with the control of the contributions of coagulation and fragmentation for large sizes, the assumption $\lambda < 1$ and the detailed balance condition (10.3.27c) being at the heart of the proof (Step 1). Once this is done, we combine it with the previously established weak convergence (10.3.86) of $(f_j)_{j\ge 1}$ to prove that some integrals depending linearly on f_j converge strongly in $L_1((-1,1) \times (0,\infty))$, see (10.3.91) below (Step 2). Using again the detailed balance condition as well as an argument from [185] developed for the Boltzmann equation, allows us to extend this strong convergence to terms depending on f_j in a nonlinear way and eventually obtain the convergence of $(f_j)_{j \geq 1}$ in $L_1((-1,1) \times (0,\infty))$ (Step 3). The last step is devoted to the pointwise convergence of $(f_j)_{j \geq 1}$ with respect to time (Step 4). This shows the compactness of $\{f(t) : t \geq 0\}$ in X_0 and thus that of $\omega(f)$ in the same space, as claimed.

Step 1: Tail control. Let $r > 1$. It follows from (10.3.29c) and Lemma 10.3.13 that

$$\int_{-1}^{1} \int_{r}^{\infty} \int_{0}^{\infty} k(x,y)f_j(s,x)f_j(s,y) \, dydxds$$

$$\leq K_0 \int_{-1}^{1} \int_{r}^{\infty} \int_{0}^{\infty} (2 + x^\lambda + y)f_j(s,x)f_j(s,y) \, dydxds$$

$$\leq K_0 \int_{-1}^{1} \left(\frac{2}{r} + \frac{1}{r^{1-\lambda}} \right) M_0(f_j(s))M_1(f_j(s)) \, ds$$

$$+ \frac{K_0}{r} \int_{-1}^{1} M_1^2(f_j(s)) \, ds$$

$$\leq C r^{\lambda - 1} . \tag{10.3.88}$$

We next argue as in the proof of Theorem 10.3.15 (a) and infer from (10.3.27b), (10.3.71), (10.3.76) and (10.3.88) that, for $R > 1$,

$$\int_{-1}^{1} \int_{r}^{\infty} a(x)f_j(s,x) \, dxds = \frac{1}{2} \int_{-1}^{1} \int_{0}^{\infty} \int_{(r-y)_+}^{\infty} F(x,y)f_j(s,x+y) \, dxdyds$$

$$\leq \frac{R}{2} \int_{-1}^{1} \int_{0}^{\infty} \int_{(r-y)_+}^{\infty} k(x,y)f_j(s,x)f_j(s,y) \, dxdyds$$

$$+ \frac{1}{2 \ln R} \int_{-1}^{1} \mathscr{D}f_j(s) \, ds$$

$$\leq \frac{R}{2} \int_{-1}^{1} \int_{0}^{r/2} \int_{r/2}^{\infty} k(x,y)f_j(s,x)f_j(s,y) \, dxdyds$$

$$+ \frac{R}{2} \int_{-1}^{1} \int_{r/2}^{\infty} \int_{0}^{\infty} k(x,y)f_j(s,x)f_j(s,y) \, dxdyds + \frac{C_4}{2 \ln R}$$

$$\leq C R r^{\lambda - 1} + \frac{C_4}{2 \ln R} .$$

Choosing $R = r^{(1-\lambda)/2} > 1$ in the previous inequality gives

$$\int_{-1}^{1} \int_{r}^{\infty} a(x)f_j(s,x) \, dxds \leq C \left(r^{(\lambda-1)/2} + \frac{1}{\ln r} \right) . \tag{10.3.89}$$

Owing to (10.3.86), (10.3.88), and (10.3.89), we deduce from Theorem 8.2.11 that f_∞ is a weak solution to the C-F equation (10.3.3) on $[-1,1]$. Moreover, since $\mathcal{C}f_\infty + \mathscr{F}f_\infty = 0$ in $(-1,1) \times (0,\infty)$ by (10.3.27c) and (10.3.87), we conclude that $\partial_s f_\infty = 0$ in $(-1,1) \times (0,\infty)$, hence there is $z \in [0, z_{th}] \cap [0,\infty)$ such that

$$f_\infty(s) = \zeta = Q_z , \qquad s \in [-1,1] . \tag{10.3.90}$$

Step 2: Strong convergence. Introducing

$$\mathcal{B}f_j(s,x) := \int_{x}^{\infty} a(y)b(x,y)f_j(s,y) \, dy \quad \text{and} \quad \mathcal{K}f_j(s,x) := \int_{0}^{\infty} k(x,y)f_j(s,y) \, dy \tag{10.3.91}$$

for $(s, x) \in [-1, 1] \times (0, \infty)$ and $j \geq 1$, the next step is to establish the strong convergence in $L_1((-1, 1) \times (0, \infty))$ of these two terms to their respective limits

$$\mathcal{B} Q_z(x) = \int_x^\infty a(y) b(x, y) Q_z(y) \ \mathrm{d}y \ \text{ and } \ \mathcal{K} Q_z(x) = \int_0^\infty k(x, y) Q_z(y) \ \mathrm{d}y \ ,$$

the parameter $z \geq 0$ being defined in (10.3.90). Since $M_0(Q_z) \leq C_4$ by (10.3.69) and (10.3.86), and $M_1(Q_z) \leq \varrho^{in}$, we first argue as in the proof of (10.3.88) and (10.3.89) to obtain that, for $r > 1$,

$$\int_r^\infty \int_0^\infty k(x, y) Q_z(x) Q_z(y) \ \mathrm{d}y\mathrm{d}x \leq C(z) r^{\lambda - 1} \tag{10.3.92}$$

and

$$\int_r^\infty a(x) Q_z(x) \ \mathrm{d}x \leq C(z) \left(r^{(\lambda - 1)/2} + \frac{1}{\ln r} \right) \ , \tag{10.3.93}$$

for some constant $C(z) > 0$ depending on z. Now, let $r > 1$ and set

$$I_j(r) := \int_{-1}^1 \int_0^r \left| \int_x^r F(x, y - x)[f_j(s, y) - Q_z(y)] \ \mathrm{d}y \right| \ \mathrm{d}x\mathrm{d}s \ .$$

Thanks to (10.3.27b), (10.3.27d), (10.3.89) and (10.3.93),

$$\int_{-1}^1 \|\mathcal{B} f_j(s) - \mathcal{B} Q_z\|_1 \ \mathrm{d}s$$

$$\leq I_j(r) + \int_{-1}^1 \int_0^r \int_r^\infty F(x, y - x)[f_j(s, y) + Q_z(y)] \ \mathrm{d}y\mathrm{d}x\mathrm{d}s$$

$$+ \int_{-1}^1 \int_r^\infty \int_x^\infty F(x, y - x)[f_j(s, y) + Q_z(y)] \ \mathrm{d}y\mathrm{d}x\mathrm{d}s$$

$$\leq I_j(r) + 4 \int_{-1}^1 \int_r^\infty a(y)[f_j(s, y) + Q_z(y)] \ \mathrm{d}y\mathrm{d}s$$

$$\leq I_j(r) + C(z) \left(r^{(\lambda - 1)/2} + \frac{1}{\ln r} \right) \ .$$

We infer from (10.3.29b), (10.3.86) and (10.3.90) that, for a.e. $x \in (0, r)$,

$$\lim_{j \to \infty} \int_x^r F(x, y - x)[f_j(s, y) - Q_z(y)] \ \mathrm{d}y$$

$$= \lim_{j \to \infty} \int_0^\infty \mathbf{1}_{(x,r)}(y) F(x, y - x)[f_j(s, y) - Q_z(y)] \ \mathrm{d}y = 0 \ .$$

Owing to (10.3.29b) and (10.3.69), we are in a position to apply the Lebesgue dominated convergence theorem and deduce that $I_j(r)$ converges to zero as $j \to \infty$. Consequently,

$$\limsup_{j \to \infty} \int_{-1}^1 \|\mathcal{B} f_j(s) - \mathcal{B} Q_z\|_1 \ \mathrm{d}s \leq C(z) \left(r^{(\lambda - 1)/2} + \frac{1}{\ln r} \right) \ .$$

As $r > 1$ is arbitrary and $\lambda < 1$, we may pass to the limit as $r \to \infty$ in the previous inequality and conclude that

$$\lim_{j \to \infty} \int_{-1}^1 \|\mathcal{B} f_j(s) - \mathcal{B} Q_z\|_1 \ \mathrm{d}s = 0 \ . \tag{10.3.94}$$

Similarly, for $r > 1$, it follows from (10.3.88) and (10.3.92) that

$$\int_{-1}^{1} \|\mathcal{K}f_j(s) - \mathcal{K}Q_z\|_1 \, ds$$

$$\leq \int_{-1}^{1} \int_{0}^{r} \left| \int_{0}^{r} k(x,y)[f_j(s,y) - Q_z(y)] \, dy \right| dx ds$$

$$+ \int_{-1}^{1} \int_{0}^{r} \int_{r}^{\infty} k(x,y)[f_j(s,y) + Q_z(y)] \, dy dx ds$$

$$+ \int_{-1}^{1} \int_{r}^{\infty} \int_{0}^{\infty} k(x,y)[f_j(s,y) + Q_z(y)] \, dy dx ds$$

$$\leq \int_{-1}^{1} \int_{0}^{r} \left| \int_{0}^{r} k(x,y)[f_j(s,y) - Q_z(y)] \, dy \right| dx ds + C(z)r^{\lambda-1} \, .$$

As above, the first term in the last line of the previous inequality converges to zero as $j \to \infty$ due to (10.3.29a), (10.3.86) and (10.3.90), and we deduce that

$$\limsup_{j\to\infty} \int_{-1}^{1} \|\mathcal{K}f_j(s) - \mathcal{K}Q_z\|_1 \, ds \leq C(z)r^{\lambda-1} \, .$$

Letting $r \to \infty$ gives

$$\lim_{j\to\infty} \int_{-1}^{1} \|\mathcal{K}f_j(s) - \mathcal{K}Q_z\|_1 \, ds = 0 \, . \tag{10.3.95}$$

Step 3: Improved strong convergence. We now adapt an argument from [185] to show that (10.3.94) implies that $(f_j \mathcal{K}f_j)_{j\geq 1}$ converges to $Q_z \mathcal{K}Q_z$ in $L_1((-1,1)\times(0,\infty))$. Indeed, for $R > 1$, it follows from (10.3.27b) and (10.3.77) that

$$\int_{-1}^{1} \|f_j(s)\mathcal{K}f_j(s) - \mathcal{B}f_j(s)\|_1 \, ds$$

$$= \int_{-1}^{1} \int_{0}^{\infty} \left| \int_{0}^{\infty} [k(x,y)f_j(s,x)f_j(s,y) - F(x,y)f_j(s,x+y)] \, dy \right| dx ds$$

$$\leq (R-1) \int_{-1}^{1} \int_{0}^{\infty} \int_{0}^{\infty} F(x,y)f_j(s,x+y) \, dy dx ds + \frac{1}{\ln R} \int_{-1}^{1} \mathcal{D}f_j(s) \, ds$$

$$= (R-1) \int_{-1}^{1} \int_{0}^{\infty} \mathcal{B}f_j(s,x) \, dx ds + \frac{1}{\ln R} \int_{-1+t_j}^{1+t_j} \mathcal{D}f(s) \, ds \, .$$

Owing to (10.3.71) and (10.3.94), we may pass to the limit as $j \to \infty$ in the previous inequality to find

$$\limsup_{j\to\infty} \int_{-1}^{1} \|f_j(s)\mathcal{K}f_j(s) - \mathcal{B}f_j(s)\|_1 \, ds \leq 2(R-1) \int_{0}^{\infty} \mathcal{B}Q_z(x) \, dx ds \, .$$

Letting $R \to 1$ in the previous inequality we conclude that

$$\lim_{j\to\infty} \int_{-1}^{1} \|f_j(s)\mathcal{K}f_j(s) - \mathcal{B}f_j(s)\|_1 \, ds = 0 \, . \tag{10.3.96}$$

Recalling the identity $\mathcal{B}Q_z = Q_z \mathcal{K}Q_z$, which stems from the detailed balance condition (10.3.27c), and using once more (10.3.94) allow us to deduce from (10.3.96) that

$$\lim_{j\to\infty} \int_{-1}^{1} \|f_j(s)\mathcal{K}f_j(s) - Q_z \mathcal{K}Q_z\|_1 \, ds = 0 \, . \tag{10.3.97}$$

An immediate consequence of (10.3.95) and (10.3.97) is the existence of a subsequence of $(f_j)_{j\geq1}$ (not relabelled) such that

$$\lim_{j\to\infty} \mathcal{K}f_j(s,x) = \mathcal{K}Q_z(x) \quad \text{and} \quad \lim_{j\to\infty} f_j(s,x)\mathcal{K}f_j(s,x) = Q_z(x)\mathcal{K}Q_z(x) \tag{10.3.98}$$

for a.e. $(s,x) \in (-1,1) \times (0,\infty)$.

At this point, we consider two cases: $z = 0$ and $z > 0$. In the former, the nonnegativity of f_j together with the weak convergence (10.3.86) imply that $(f_j(s))_{j\geq1}$ converges to zero in X_0 as $j \to \infty$ for all $s \in [-1,1]$, which completes the proof of Theorem 10.3.15 (b) in this case. Otherwise $z > 0$ and then $\mathcal{K}Q_z(x) > 0$ for all $x \in (0,\infty)$ due to the positivity of k and Q_z, guaranteed by (10.3.29a) and (10.3.30). Then, given $(s,x) \in (-1,1) \times (0,\infty)$ satisfying (10.3.98), there is j_0 large enough, and $\delta > 0$ small enough (which are likely to depend on (s,x)) such that $\mathcal{K}f_j(s,x) \geq \delta$ for $j \geq j_0$. This property, together with (10.3.98), implies that $(f_j(s,x))_{j\geq1}$ converges to $Q_z(x)$ as $j \to \infty$. This convergence being true for a.e. $(s,x) \in (-1,1) \times (0,\infty)$, we have established that $(f_j)_{j\geq1}$ converges almost everywhere in $(-1,1) \times (0,\infty)$ to Q_z. Combining the latter with (10.3.86) and Vitali's theorem (Theorem 7.1.10) implies that

$$f_j \longrightarrow Q_z \quad \text{in } L_1((-1,1) \times (0,\infty)) \ . \tag{10.3.99}$$

Step 4: Compactness of $\omega(f)$ in X_0. We are left with improving the convergence (10.3.99) to pointwise convergence with respect to time. On the one hand, it easily follows from (10.3.89), (10.3.93), (10.3.99) and the local boundedness of a (which is due to (10.3.27a) and (10.3.29b)) that

$$\lim_{j\to\infty} \int_{-1}^{1} \int_{0}^{\infty} |a(x)f_j(s,x) - a(x)Q_z(x)| \ dx ds = 0 \ . \tag{10.3.100}$$

On the other hand, a similar argument based on (10.3.29a), (10.3.88), (10.3.92) and (10.3.99) gives

$$\lim_{j\to\infty} \int_{-1}^{1} \int_{0}^{\infty} \left| \int_{0}^{x} k(y,x-y) \left[f_j(s,y)f_j(s,x-y) - Q_z(y)Q_z(x-y) \right] \ dy \right| \ dx ds = 0 \ . \tag{10.3.101}$$

Combining (10.3.94), (10.3.97), (10.3.100) and (10.3.101), and recalling that Q_z satisfies $\mathcal{C}Q_z + \mathcal{F}Q_z = 0$, we have shown that

$$\lim_{j\to\infty} \int_{-1}^{1} \|\mathcal{C}f_j(s) + \mathcal{F}f_j(s)\|_1 \ ds = \lim_{j\to\infty} \int_{-1}^{1} \|\mathcal{C}f_j(s) + \mathcal{F}f_j(s) - \mathcal{C}Q_z - \mathcal{F}Q_z\|_1 \ ds = 0 \ ,$$

and hence, by the C-F equation (10.3.3),

$$\lim_{j\to\infty} \int_{-1}^{1} \|\partial_s f_j(s)\|_1 \ ds = 0 \ . \tag{10.3.102}$$

Consequently, for $s \in (-1,0)$,

$$\|f_j(0) - Q_z\|_1 \leq \|f_j(s) - Q_z\|_1 + \int_{s}^{0} \|\partial_s f_j(s_*)\|_1 \ ds_* \ .$$

Integrating the previous inequality with respect to s over $(-1,0)$ leads us to

$$\|f_j(0) - Q_z\|_1 \leq \int_{-1}^{0} \|f_j(s) - Q_z\|_1 \ ds + \int_{-1}^{0} \int_{s}^{0} \|\partial_s f_j(s_*)\|_1 \ ds_* ds$$

$$\le \int_{-1}^{1} \|f_j(s) - Q_z\|_1 \, ds + \int_{-1}^{1} \|\partial_s f_j(s)\|_1 \, ds \ ,$$

and we deduce from (10.3.99) and (10.3.102) that $(f(t_j))_{j\ge 1} = (f_j(0))_{j\ge 1}$ converges to $\zeta = Q_z$ in X_0. In other words, the ω-limit set $\omega(f)$ is compact in X_0. $\qquad\square$

10.3.2.4 Convergence

Up to now, we have proved that $\omega(f)$ is a compact subset of X_0 which contains only elements in $\mathcal{Q} \cap X_1$ with total mass smaller than, or equal to, ϱ^{in}. As already mentioned, the natural conjecture is that $\omega(f)$ is reduced to a single element $Q_{z(\varrho^{in})}$, provided the latter exists; that is, $\varrho^{in} \in [0, \varrho_{th}] \cap [0, \infty)$. A positive answer is available in a few cases which we list now.

Theorem 10.3.17. *Assume that the coagulation and fragmentation coefficients k, a and b satisfy (10.3.27), (10.3.28), (10.3.29) and (10.3.30), and consider an initial condition $f^{in} \in \mathcal{Y} \cap X_\mu$ for some $\mu \ge 1 + \gamma$. Let f be a mass-conserving weak solution to the C-F equation (10.3.3) on $[0, \infty)$ satisfying (10.3.35), and define its ω-limit set $\omega(f)$ by (10.3.75).*

(a) *If $z_{th} = \infty$ in (10.3.74), then $\omega(f) = \{Q_{z(\varrho^{in})}\}$, where $\varrho^{in} := M_1(f^{in})$. In addition, $f(t) \rightharpoonup Q_{z(\varrho^{in})}$ in $X_{0,1}$ as $t \to \infty$.*

(b) *Assume further that $\mu \ge 2 + (\gamma - 1)_+ + q_\infty$, $\lambda < 1$,*

$$z_{th} \in (0, \infty) \ , \qquad \lim_{x\to\infty} Q(x)^{1/x} = \frac{1}{z_{th}} \ , \qquad (10.3.103)$$

and

$$\{f(t) \ln f(t) \ : \ t \ge 0\} \ \text{is relatively sequentially weakly compact in } X_0 \ . \quad (10.3.104)$$

Then there is $z_\infty \in [0, z_{th}] \cap [0, \infty)$ such that $\omega(f) = \{Q_{z_\infty}\}$ and $M_1(Q_{z_\infty}) \le \varrho^{in}$, where, as in (a), $\varrho^{in} := M_1(f^{in})$. In addition, $f(t) \longrightarrow Q_{z_\infty}$ in X_0 as $t \to \infty$.

Owing to the decay assumption (10.3.30b) of Q for large sizes, the value of z_{th} varies with q_∞; indeed, $z_{th} = \infty$ when $q_\infty > 1$, $z_{th} = 1$ when $q_\infty \in (0, 1)$, and $z_{th} \in [e^{1/K_\infty}, e^{K_\infty}]$ when $q_\infty = 1$.

Proof. (a) Let $\zeta \in \omega(f)$. Then there is $(t_j)_{j\ge 1}$ such that $t_j \to \infty$ and $f(t_j) \rightharpoonup \zeta$ in X_0 as $j \to \infty$. On the one hand, by Theorem 10.3.15 (a), there is $z \in [0, \infty)$ such that $\zeta = Q_z$. On the other hand, we infer from the weak convergence of $(f(t_j))_{j\ge 1}$ in X_0 and Lemma 10.3.14 that $f(t_j) \rightharpoonup Q_z$ in X_1, hence

$$M_1(Q_z) = \lim_{j\to\infty} M_1(f(t_j)) = \varrho^{in} \ .$$

Consequently, $z = z(\varrho^{in})$, where $z(\varrho^{in}) \in [0, \infty)$ is uniquely determined from the relation $M_1(Q_{z(\varrho^{in})}) = \varrho^{in}$, and $\zeta = Q_{z(\varrho^{in})}$. The weak convergence of $f(t)$ in $X_{0,1}$ as $t \to \infty$ then readily follows from the weak compactness (10.3.81) of the trajectory, Lemma 10.3.14, and the just established uniqueness of the limit.

(b) Let $\zeta \in \omega(f)$. First, by Theorem 10.3.15 (a), there is $z \in [0, \infty)$ such that $\zeta = Q_z$ and $M_1(Q_z) \le \varrho^{in}$. Next, since $\omega(f)$ is compact in X_0 according to Theorem 10.3.15 (b), there is $(t_j)_{j\ge 1}$ such that $t_j \to \infty$ and $f(t_j) \longrightarrow Q_z$ in X_0 and a.e. in $(0, \infty)$ as $j \to \infty$. Moreover, combining the almost everywhere convergence of $(f(t_j))_{j\ge 1}$ with the relative sequential weak compactness (10.3.104) of $\{f(t)\ln f(t) \; : \; t \geq 0\}$ and Vitali's theorem (Theorem 7.1.10), we may assume, after possibly extracting a subsequence, that

$$\lim_{j\to\infty} \int_0^\infty f(t_j, x)\ln f(t_j, x)\, \mathrm{d}x = \int_0^\infty Q_z(x)\ln Q_z(x)\, \mathrm{d}x \; . \tag{10.3.105}$$

Also, for $r > 1$, it follows from (10.3.30) that

$$\left| \int_0^\infty [f(t_j, x) - Q_z(x)]\ln Q_{z_{th}}(x)\, \mathrm{d}x \right|$$

$$= \left| \int_0^\infty x[f(t_j, x) - Q_z(x)]\ln\left(z_{th}Q(x)^{1/x} \right)\, \mathrm{d}x \right|$$

$$\leq \int_0^r x|f(t_j, x) - Q_z(x)|\left|\ln\left(z_{th}Q(x)^{1/x} \right)\right|\, \mathrm{d}x$$

$$+ \sup_{y\in(r,\infty)}\left\{ \left|\ln\left(z_{th}Q(y)^{1/y} \right)\right| \right\} \int_r^\infty x[f(t_j, x) + Q_z(x)]\, \mathrm{d}x$$

$$\leq (r|\ln z_{th}| + \ln K_1 + K_\infty r^{q_\infty})\|f(t_j) - Q_z\|_1$$

$$+ 2\varrho^{in} \sup_{y\in(r,\infty)}\left\{ \left|\ln\left(z_{th}Q(y)^{1/y} \right)\right| \right\} \; .$$

Thanks to the strong convergence of $(f(t_j))_{j\geq 1}$ to Q_z in X_0, we may pass to the limit as $j \to \infty$ in the previous inequality and obtain

$$\limsup_{j\to\infty}\left| \int_0^\infty [f(t_j, x) - Q_z(x)]\ln Q_{z_{th}}(x)\, \mathrm{d}x \right|$$

$$\leq 2\varrho^{in} \sup_{y\in(r,\infty)}\left\{ \left|\ln\left(z_{th}Q(y)^{1/y} \right)\right| \right\} \; .$$

We next use (10.3.103) to let $r \to \infty$ in the previous inequality and end up with

$$\lim_{j\to\infty} \int_0^\infty f(t_j, x)\ln Q_{z_{th}}(x)\, \mathrm{d}x = \int_0^\infty Q_z(x)\ln Q_{z_{th}}(x)\, \mathrm{d}x \; . \tag{10.3.106}$$

Summing (10.3.105) and (10.3.106) gives

$$\lim_{j\to\infty} L_* f(t_j) = L_* Q_z \; , \tag{10.3.107}$$

where $L_* := L - \varrho^{in}\ln z_{th}$.

Finally, the entropy identity stated in Proposition 10.3.11 and mass conservation ensure that $t \mapsto L_* f(t)$ is a nonincreasing function of time which is bounded from below and

$$L_\infty := \inf_{t\geq 0}\{L_* f(t)\} = \lim_{j\to\infty} L_* f(t_j) \in \left[-\varrho^{in}\ln z_{th}, Lf^{in} - \varrho^{in}\ln z_{th} \right] \; . \tag{10.3.108}$$

We then infer from (10.3.107) and (10.3.108) that $L_* Q_z = L_\infty$. Since there is a unique $z_\infty \in [0, z_{th})$ satisfying $L_* Q_{z_\infty} = L_\infty$, we have established that $z = z_\infty$ and thus that $\omega(f) = \{Q_{z_\infty}\}$. The convergence of $f(t)$ in X_0 as $t \to \infty$ then follows. $\qquad\square$

As already mentioned, the outcome of Theorem 10.3.17 (b) does not provide a complete description of the long-term dynamics as the value of z_∞ is unfortunately left unidentified. Let us emphasise here that the two assumptions needed to apply Theorem 10.3.17 (b) are of a different nature; indeed, (10.3.103) is a structural assumption on Q and thus on the coagulation and fragmentation coefficients and can be made *a priori*. In contrast, the weak compactness assumption (10.3.104) depends on the solution to the C-F equation (10.3.3) and has to be checked beforehand. As reported in [169], it is satisfied when the coagulation and fragmentation coefficients satisfy the monotonicity and growth conditions

$$k(y, x - y) \leq k(x, y) , \qquad F(y, x - y) \leq K_2 k(x, y) , \qquad 0 < y < x ,$$

for some $K_2 > 0$, besides (10.3.27), (10.3.28), (10.3.29) and (10.3.30). These additional assumptions allow one to use Lemma 8.2.18 along with the de la Vallée-Poussin theorem (Theorem 7.1.6) and derive a bound of the form

$$\sup_{t \geq 0} \left\{ \int_0^\infty \Phi(f(t, x)|\ln f(t, x)|) \; dx \right\} < \infty$$

for some function $\Phi \in \mathcal{C}_{VP,\infty}$.

As a final comment, in this section we have only considered coagulation kernels growing at most linearly. Another situation where the convergence to $Q_{z(\varrho^{in})}$ (as in Theorem 10.3.17 (a)) can be shown corresponds to strong fragmentation; that is, when assumption (10.3.28) is replaced by

$$\gamma > \lambda - 1 > 0 \quad \text{and} \quad F(x, y) \geq 2a_0(x + y)^{\gamma - 1} , \qquad (x, y) \in (0, \infty)^2 , \qquad (10.3.109)$$

where $a_0 > 0$. In particular, it follows from (10.3.27a) and (10.3.109) that $a(x) \geq a_0 x^\gamma$ for $x \in (0, \infty)$, a lower bound which, together with the requirement $\gamma > \lambda - 1$, is at the heart of the analysis performed in Section 8.2.2.4. We refer to [65] and [165, 169] where this case is handled in the discrete and continuous settings, respectively.

10.3.3 Stationary Solutions

As already mentioned, the detailed balance condition (10.3.27c) occurs rather seldomly in C-F equations. It is thus of interest to investigate whether stationary solutions also exist in a more general setting, and provide a description of the long-term behaviour in such cases. It turns out that the basic question of the existence of stationary solutions is already a challenging issue and, as far as we know, it has not yet received a definitive answer. This is reflected in the results we present now and which focus on specific classes of coagulation and fragmentation coefficients.

10.3.3.1 Additive Coagulation and Constant Fragmentation

To illustrate the technique used to establish the existence of stationary solutions, let us begin with the particular case of the constant fragmentation kernel

$$a(x) = a_0 x , \quad b(x, y) = \frac{2}{y} , \qquad 0 < x < y , \qquad (10.3.110a)$$

where $a_0 > 0$, and the additive coagulation kernel

$$k(x, y) = k_+(x, y) = x + y , \qquad (x, y) \in (0, \infty)^2 . \qquad (10.3.110b)$$

For any $\varrho > 0$, the existence of a stationary solution f_s with total mass $M_1(f_s) = \varrho$ to the C-F equation (8.0.1) with this choice of coefficients is shown in [105] by a fixed-point argument. Below we provide an alternative proof which relies on a dynamical systems approach. More precisely, we prove the following result.

Theorem 10.3.18. *Assume that the coagulation and fragmentation coefficients are given by* (10.3.110) *and consider* $\varrho > 0$. *There exists a stationary solution* $f_s \in X_{1,+}$ *to the C-F equation* (8.0.1) *satisfying* $M_1(f_s) = \varrho$ *and*

$$f_s \in L_\infty(0,\infty) \cap \bigcap_{m \geq 0} X_m \ .$$

To prove Theorem 10.3.18 we fix $\varrho > 0$ and first recall the well-posedness of the C-F equation (8.0.1) when k, a and b are given by (10.3.110).

Lemma 10.3.19. *Let* $f^{in} \in X_{0,2,+}$ *be such that* $M_1(f^{in}) = \varrho$. *Then there is a unique mass-conserving weak solution* f *to the C-F equation* (8.0.1) *on* $[0,\infty)$ *such that* f *belongs to* $L_\infty((0,T), X_2)$ *for all* $T > 0$ *and*

$$M_1(f(t)) = M_1(f^{in}) = \varrho \ , \qquad t \geq 0 \ . \tag{10.3.111}$$

Proof. According to Theorem 8.2.23 there is at least one mass-conserving weak solution f to the C-F equation (8.0.1) on $[0,\infty)$ with coefficients given by (10.3.110) which also belongs to $L_\infty((0,T), X_2)$ for all $T > 0$. Uniqueness of such a solution follows from Theorem 8.2.55 (with $\ell(x) = 1 + x$ and $\zeta = 1$). $\square$

We next derive additional moment estimates.

Lemma 10.3.20. *Consider* $f^{in} \in X_{0,2,+}$ *such that* $M_1(f^{in}) = \varrho$ *and let* f *be the corresponding solution to the C-F equation* (8.0.1) *on* $[0,\infty)$ *given by Lemma 10.3.19. For each* $m > 1$ *there is* $\mu_{1,m} > 0$ *depending only on* a_0, ϱ *and* m *such that, if* $f^{in} \in X_m$, *then*

$$M_m(f(t)) \leq \max\{M_m(f^{in}), \mu_{1,m}\} \ , \qquad t \geq 0 \ .$$

Moreover,

$$M_0(f(t)) \leq \max\big\{M_0(f^{in}), a_0\big\} \ , \qquad t \geq 0 \ .$$

Proof. We first consider $m > 1$. It follows from (10.3.110b) and Lemma 7.4.4 that

$$\chi_m(x,y)k(x,y) \leq C_m \left(xy^m + x^m y\right) \ , \qquad (x,y) \in (0,\infty)^2 \ ,$$

and we deduce from (10.3.111) that

$$\frac{d}{dt} M_m(f(t)) \leq C_m M_1(f(t)) M_m(f(t)) - a_0 \frac{m-1}{m+1} M_{m+1}(f(t))$$

$$\leq C_m \varrho M_m(f(t)) - a_0 \frac{m-1}{m+1} M_{m+1}(f(t)) \ .$$

Using again (10.3.111) and Hölder's inequality gives

$$M_m(f(t)) \leq M_1(f(t))^{1/m} M_{m+1}(f(t))^{(m-1)/m}$$

$$\leq \varrho^{1/m} M_{m+1}(f(t))^{(m-1)/m} \ ,$$

and we combine the above inequalities to obtain

$$\frac{d}{dt} M_m(f(t)) \leq C_m \varrho M_m(f(t))$$

$$- \frac{a_0(m-1)}{(m+1)\varrho^{1/(m-1)}} M_m(f(t))^{m/(m-1)}$$

for $t \geq 0$. A comparison argument then implies that

$$M_m(f(t)) \leq \max \left\{ M_m(f^{in}), \left(\frac{(m+1)C_m \varrho^{m/(m-1)}}{a_0(m-1)} \right)^{m-1} \right\}$$

for $t \geq 0$ and completes the proof of the first statement of Lemma 10.3.20. Next, by (8.2.4d) (with $\vartheta \equiv 1$) and (10.3.111),

$$\frac{d}{dt} M_0(f(t)) = -M_1(f(t))M_0(f(t)) + a_0 M_1(f(t))$$
$$= \varrho[a_0 - M_0(f(t))]$$

for $t \geq 0$, from which the second statement of Lemma 10.3.20 readily follows. □

We next turn to the uniform integrability and actually establish a bound in $L_\infty(0, \infty)$.

Lemma 10.3.21. *Consider $f^{in} \in X_{0,2,+}$ such that $M_1(f^{in}) = \varrho$, and let f be the corresponding weak solution to the C-F equation (8.0.1) on $[0, \infty)$ given by Lemma 10.3.19. If $f^{in} \in L_\infty(0, \infty)$, then*

$$\|f(t)\|_\infty \leq \max \left\{ \|f^{in}\|_\infty, \frac{4a_0}{\varrho} \sup_{s \geq 0} \{M_0(f(s))\} \right\} , \qquad t \geq 0 .$$

Proof. For $A > 0$ and $r \geq 0$ we define $\Phi_A(r) := (r - A)_+$. Then $\Phi'_A(r) = \mathbf{1}_{(0,\infty)}(r - A)$ and

$$\Phi_A(r) \leq r , \quad r\Phi'_A(r) - \Phi_A(r) = A\Phi'_A(r) \leq r\Phi'_A(r) .$$

Let $t > 0$. On the one hand, we infer from (10.3.111) and the convexity of Φ_A that

$$\int_0^\infty \mathcal{C}f(t,x)\Phi'_A(f(t,x)) \, dx$$

$$= \frac{1}{2} \int_0^\infty f(t,y) \int_y^\infty x\Phi'_A(f(t,x))f(t,x-y) \, dx dy$$

$$\quad - M_0(f(t)) \int_0^\infty xf(t,x)\Phi'_A(f(t,x)) \, dx$$

$$\quad - M_1(f(t)) \int_0^\infty f(t,x)\Phi'_A(f(t,x)) \, dx$$

$$\leq \frac{1}{2} \int_0^\infty f(t,y) \int_y^\infty x \left[A\Phi'_A(f(t,x)) + \Phi_A(f(t,x-y)) \right] \, dx dy$$

$$\quad - M_0(f(t)) \int_0^\infty xf(t,x)\Phi'_A(f(t,x)) \, dx - \varrho \int_0^\infty f(t,x)\Phi'_A(f(t,x)) \, dx$$

$$\leq \frac{A}{2} M_0(f(t)) \int_0^\infty x\Phi'_A(f(t,x)) \, dx$$

$$\quad + \frac{1}{2} \int_0^\infty f(t,y) \int_0^\infty (x+y)\Phi_A(f(t,x)) \, dx dy$$

$$\quad - M_0(f(t)) \int_0^\infty xf(t,x)\Phi'_A(f(t,x)) \, dx - \varrho \int_0^\infty f(t,x)\Phi'_A(f(t,x)) \, dx$$

$$\leq \frac{1}{2} M_0(f(t)) \int_0^\infty x \left[A\Phi'_A(f(t,x)) + \Phi_A(f(t,x)) - 2f(t,x)\Phi'_A(f(t,x)) \right] \, dx$$

$$\quad - \frac{\varrho}{2} \int_0^\infty f(t,x)\Phi'_A(f(t,x)) \, dx$$

$$\leq -\frac{\varrho}{2} \int_0^\infty f(t,x) \Phi_A'(f(t,x)) \, dx \ .$$

On the other hand,

$$\int_0^\infty \mathscr{F}f(t,x) \Phi_A'(f(t,x)) \, dx$$

$$= 2a_0 \int_0^\infty \Phi_A'(f(t,x)) \int_x^\infty f(t,y) \, dy dx - a_0 \int_0^\infty x f(t,x) \Phi_A'(f(t,x)) \, dx$$

$$\leq \frac{2a_0 M_0(f(t))}{A} \int_0^\infty f(t,x) \Phi_A'(f(t,x)) \, dx \ .$$

Combining the above inequalities we obtain

$$\frac{d}{dt} \int_0^\infty \Phi_A(f(t,x)) \, dx \leq \left[\frac{2a_0 M_0(f(t))}{A} - \frac{\varrho}{2} \right] \int_0^\infty f(t,x) \Phi_A'(f(t,x)) \, dx \leq 0 \ ,$$

provided

$$A \geq \frac{4a_0}{\varrho} \sup_{s \geq 0} \{ M_0(f(s)) \} \ .$$

In that case,

$$\int_0^\infty \Phi_A(f(t,x)) \, dx \leq \int_0^\infty \Phi_A(f^{in}(x)) \, dx \ , \qquad t \geq 0 \ .$$

Now, if $A \geq \|f^{in}\|_\infty$, then the right-hand side of the above inequality vanishes and thus $f(t,x) \leq A$ for $(t,x) \in (0,\infty)^2$. $\qquad\square$

Proof of Theorem 10.3.18. Let $\varrho > 0$.

Step 1: Invariant set. We define a subset Z of $X_{1,+}$ as follows: $\zeta \in Z$ if

$$\zeta \in L_\infty(0,\infty) \cap \bigcap_{m \geq 0} X_m \ , \quad \zeta \geq 0 \ \text{ a.e. } \ , \qquad M_1(\zeta) = \varrho \ ,$$

$$M_0(\zeta) \leq \max\{\varrho, a_0\} \ , \quad \|\zeta\|_\infty \leq \max\left\{ \varrho, \frac{4a_0^2}{\varrho} \right\}$$

$$M_m(\zeta) \leq \max\{\varrho\Gamma(m+1), \mu_{1,m}\} \ , \qquad m > 1 \ .$$

First, Z is non-empty as $x \mapsto \varrho e^{-x}$ belongs to Z. It next follows from the Dunford–Pettis theorem (Theorem 7.1.3) that Z is a closed convex subset of $X_{1,+}$ which is sequentially weakly compact in X_1. In addition, from Lemma 10.3.19 and Lemma 10.3.20, we infer that any mass-conserving weak solution f to the C-F equation (8.0.1) on $[0,\infty)$, emanating from $f^{in} \in Z$, satisfies

$$M_1(f(t)) = \varrho, \quad M_0(f(t)) \leq \max\{\varrho, a_0\}, \quad \text{and} \quad M_m(f(t)) \leq \max\{\varrho\Gamma(m+1), \mu_{1,m}\}$$

for $m > 1$ and $t \geq 0$. Furthermore, Lemma 10.3.21 ensures that $\|f(t)\|_\infty \leq \max\{\varrho, 4a_0^2/\varrho\}$. We have thus established that

$$f^{in} \in Z \implies f(t) \in Z \ \text{ for all } \ t \geq 0 \ . \tag{10.3.112}$$

Step 2: Dynamical system in Z equipped with the weak topology of X_1. Let $(f_j^{in})_{j \geq 1}$ be a sequence of initial conditions in Z which converges weakly in X_1 to $f^{in} \in Z$.

For $j \geq 1$, let f_j be the solution to the C-F equation (8.0.1) with initial condition f_j^{in} provided by Lemma 10.3.19. According to (10.3.112), the set

$$\mathcal{E} := \{f_j(t) \ : \ t \geq 0 \ , \ j \geq 1\}$$

is included in Z and thus is sequentially weakly compact in X_1. Furthermore, it readily follows from (10.3.110), (10.3.112) and the C-F equation (8.0.1) that

$$
\begin{aligned}
M_1(|\partial_t f_j(t)|) &\leq \int_0^\infty \int_0^\infty (x+y)k(x,y)f_j(t,x)f_j(t,y) \, \mathrm{d}y\mathrm{d}x + 2\int_0^\infty xa(x)f_j(t,x) \, \mathrm{d}x \\
&\leq 2[M_0(f_j(t))M_2(f_j(t)) + M_1^2(f_j(t)) + a_0 M_2(f_j(t))] \\
&\leq 2\left[2a_0\mu_{1,2} + \varrho^2\right]
\end{aligned}
$$

for $t > 0$ and $j \geq 1$. Therefore, $(f_j(t))_{j \geq 1}$ is equicontinuous in X_1 for each $t > 0$ and we infer from the sequential weak compactness of $\mathcal{E}$ in X_1 and the variant of the Arzelà–Ascoli theorem stated in Theorem 7.1.16 that $(f_j)_{j \geq 1}$ is compact in $C([0,T], X_{1,w})$ for all $T > 0$. There is thus a subsequence $(j_l)_{l \geq 1}$, and $f \in C([0,\infty), X_{1,w})$ such that

$$f_{j_l} \longrightarrow f \quad \text{in } C([0,T], X_{1,w}) \tag{10.3.113}$$

for all $T > 0$. Clearly, $f(t) \in Z$ for all $t \geq 0$ and, in particular, belongs to $L_\infty((0,\infty), X_2)$. Also, the estimates (10.3.112) and the convergence (10.3.113) allow us to apply Theorem 8.2.11 to conclude that f is a mass-conserving weak solution to the C-F equation (8.0.1) on $[0,\infty)$. According to Lemma 10.3.19, it is the only mass-conserving weak solution to the C-F equation (8.0.1) on $[0,\infty)$ and the convergence (10.3.113) actually holds true for the whole sequence $(f_j)_{j \geq 1}$. We have thus shown that $\{f^{in} \mapsto f(t)\}_{t \geq 0}$ is a dynamical system on Z equipped with the weak topology of X_1.

Summarising, the map $\{f^{in} \mapsto f(t)\}_{t \geq 0}$ is a dynamical system on Z equipped with the weak topology of X_1 and it maps Z into itself, so that Z is a positively invariant set for the flow. In addition, Z is a closed convex set which is weakly compact in X_1. The existence of a stationary solution $f_s \in Z$ is then a consequence of Theorem 7.3.7. $\qquad \square$

The analysis carried out in [105] actually goes beyond the existence of at least one stationary solution with total mass ϱ, as proved in Theorem 10.3.18. It also provides the uniqueness of such a solution as well as its local stability in X_0.

10.3.3.2 Singular Coagulation and Strong Fragmentation

We next turn to the case studied in [113, Section 4] which involves a coagulation kernel with a singularity as $x \to 0$ and $y \to 0$ of the same kind as the Smoluchowski coagulation kernel $k(x,y) = \left(x^{1/3} + y^{1/3}\right)\left(x^{-1/3} + y^{-1/3}\right)$. Specifically, we assume that k is given by

$$k(x,y) = x^\alpha y^{\lambda-\alpha} + x^{\lambda-\alpha}y^\alpha \ , \qquad (x,y) \in (0,\infty)^2 \ , \tag{10.3.114}$$

where

$$\alpha \in [-1,0] \ , \qquad \lambda \in [0, 1+\alpha) \ . \tag{10.3.115}$$

Note that (10.3.115) implies that $\lambda < 1$. Concerning the fragmentation coefficients we assume that there is $a_0 > 0$, and $\gamma > 0$ such that

$$a(y) = a_0 y^\gamma \ , \qquad y \in (0,\infty) \ , \tag{10.3.116a}$$

and

$$b(x,y) = \frac{1}{y}h\left(\frac{x}{y}\right) \ , \qquad 0 < x < y \ , \tag{10.3.116b}$$

where h is a nonnegative function in $L_1((0,1), z\mathrm{d}z)$ satisfying

$$\int_0^1 zh(z) \, \mathrm{d}z = 1 \quad \text{and} \quad h \in L_{p_0}(0,1) \cap L_1((0,1), z^{m_0}\mathrm{d}z) \qquad (10.3.116c)$$

for some $p_0 \in (1, 1 + \gamma - \alpha]$ and $m_0 < \alpha$. Recall that, owing to (10.3.116c), the parameter $\mathfrak{h}_m$, defined by

$$\mathfrak{h}_m := \int_0^1 (z - z^m)h(z) \, \mathrm{d}z = 1 - \int_0^1 z^m h(z) \, \mathrm{d}z \qquad (10.3.116d)$$

for $m \geq m_0$, is finite and satisfies $\mathfrak{h}_m(m - 1) \geq 0$. Let us also emphasise here that the parameter m_0 is negative, so that h may only be weakly singular as $z \to 0$.

Theorem 10.3.22. *Let k, a and b be coagulation and fragmentation coefficients satisfying (10.3.114), (10.3.115) and (10.3.116), and consider $\varrho > 0$. There exists a stationary solution $f_s \in X_{1,+}$ to the C-F equation (8.0.1) such that $M_1(f_s) = \varrho$ and*

$$f_s \in L_{p_0}(0,\infty) \cap \bigcap_{m \geq m_0} X_m \ .$$

Let us first mention that the case studied in Section 10.3.3.1 corresponds to the choice $\alpha = 0$ and $\lambda = 1$ in (10.3.114), and $\gamma = 1$, $a_0 = 1/K_1$ and $h \equiv 2$ in (10.3.116), but it is not a particular case of Theorem 10.3.22 as it does not satisfy (10.3.115). However, the general scheme of the proof is similar to that of Theorem 10.3.18, though the possibly negative exponent α in the coagulation kernel k prevents us from estimating separately moments of order higher than one and moments of order smaller than one. We shall instead combine estimates of the moments of order λ and $2 - \lambda + \alpha > 1$ to begin the proof of Theorem 10.3.22, see Lemma 10.3.24 below.

Let us thus fix $\varrho > 0$ and coagulation and fragmentation coefficients (k, a, b) satisfying (10.3.114), (10.3.115) and (10.3.116). We first investigate the well-posedness of the C-F equation (8.0.1) for this particular choice of coefficients.

Lemma 10.3.23. *[113] Consider $f^{in} \in X_{m_0,+} \cap X_{1+\gamma}$ such that $M_1(f^{in}) = \varrho$. There is a unique mass-conserving weak solution f to the C-F equation (8.0.1) on $[0, \infty)$ which belongs to $L_\infty((0,T), X_{m_0} \cap X_{1+\gamma})$ for all $T > 0$. In particular,*

$$M_1(f(t)) = \varrho \ , \qquad t \geq 0 \ . \qquad (10.3.117)$$

Also, if $f^{in} \in X_m$ for some $m > 1$, then $f \in L_\infty((0,\infty), X_m)$.

We refer to [113, Theorem 4.1] for the proof of Lemma 10.3.23, as we cannot apply directly Theorem 8.2.48 to derive the existence statement. Indeed, the assumptions required to apply Theorem 8.2.48 are more restrictive than (10.3.114), (10.3.115) and (10.3.116). This is due to the fact that the singularity of k for small sizes, allowed by Theorem 8.2.48, is stronger than the one in (10.3.114), (10.3.115), see the discussion in Remark 8.2.49. Nevertheless, one may combine the proof of Theorem 8.2.48 with the estimates derived in Lemma 10.3.24, Lemma 10.3.25 and Lemma 10.3.26 below to obtain the existence of at least one mass-conserving weak solution to the C-F equation (8.0.1) on $[0, \infty)$ which belongs to $L_\infty((0,\infty), X_{m_0} \cap X_{1+\gamma})$. The uniqueness statement in Lemma 10.3.23 is proved in [113, Lemma 2.9] and relies on the same arguments as the proofs of Theorems 8.2.55 and 8.2.61 with $\ell(x) = x^\alpha + x^{1+\gamma}$, $x \in (0, \infty)$.

The next step is the derivation of uniform moment and integrability estimates. Since the ultimate goal is to construct a sequentially weakly compact convex subset of X_1, we keep track of the dependence of the estimates on the initial data. Throughout the remainder of this section, C and $(C_i)_{i \geq 1}$ denote positive constants which may vary from line to line and depend only on α, λ, a_0, γ, h, m_0, p_0, and ϱ. Dependence upon additional parameters will be indicated explicitly.

As already mentioned, we start with a combined estimate of the moments of order λ and $2 - \lambda + \alpha > 1$.

Lemma 10.3.24. *Consider $f^{in} \in X_{m_0,+} \cap X_2$ such that $M_1(f^{in}) = \varrho$ and let f be the corresponding mass-conserving weak solution to the C-F equation (8.0.1) on $[0, \infty)$ given by Lemma 10.3.23. There is $\mu_* > 0$, depending only on α, λ, a_0, γ, h and ϱ, such that*

$$M_\lambda(f(t)) + M_{2-\lambda+\alpha}(f(t)) \leq \max\left\{ M_\lambda(f^{in}) + M_{2-\lambda+\alpha}(f^{in}), \mu_* \right\} , \qquad t \geq 0 . \quad (10.3.118)$$

Moreover, for $m > 1$, there is $\mu_{1,m} > 0$, depending only on α, λ, a_0, γ, h, ϱ and m, such that, if $f^{in} \in X_m$, then

$$M_m(f(t)) \leq \max\left\{ M_m(f^{in}), \mu_{1,m} \left(\sup_{s \geq 0} M_\lambda(f(s)) \right)^{\theta_m} \right\} , \qquad t \geq 0 , \quad (10.3.119)$$

where

$$\theta_m := \left(\frac{|\alpha|}{1 - \lambda} + \frac{1 + \alpha - \lambda}{m - \lambda} \right) \frac{(m - \lambda)(m - 1)}{[(m-1)(1 + \alpha - \lambda) + \gamma(m - \lambda)]} > 0 . \quad (10.3.120)$$

Proof. It first follows from (10.3.116) that, for all $m \geq m_0$ and $t \geq 0$,

$$\int_0^\infty x^m \mathscr{F} f(t, x) \, \mathrm{d}x = -a_0 \mathfrak{h}_m M_{m+\gamma}(f(t)) . \quad (10.3.121)$$

Handling the coagulation term requires separate procedures depending on the value of m. Let us begin with $m > 1$ and consider $(x, y) \in (0, \infty)^2$. Introducing $X := \min\{x, y\}$ and $Y := \max\{x, y\}$, we infer from the symmetry of k and χ_m, and the convexity of $r \mapsto r^m$, that

$$k(x, y)\chi_m(x, y) = k(X, Y)\chi_m(X, Y)$$

$$= Y^{\lambda+m} \left[\left(\frac{X}{Y} \right)^\alpha + \left(\frac{X}{Y} \right)^{\lambda-\alpha} \right] \left[\left(1 + \frac{X}{Y} \right)^m - 1 - \left(\frac{X}{Y} \right)^m \right]$$

$$\leq Y^{\lambda+m} \, 2 \left(\frac{X}{Y} \right)^\alpha m \left(\frac{X}{Y} \right) \left(1 + \frac{X}{Y} \right)^{m-1}$$

$$\leq m 2^m X^{1+\alpha} Y^{m+\lambda-\alpha-1}$$

$$\leq m 2^m \left(X^{1+\alpha} Y^{m+\lambda-\alpha-1} + X^{m+\lambda-\alpha-1} Y^{1+\alpha} \right)$$

$$= m 2^m \left(x^{1+\alpha} y^{m+\lambda-\alpha-1} + x^{m+\lambda-\alpha-1} y^{1+\alpha} \right) .$$

Therefore,

$$\int_0^\infty x^m \mathcal{C} f(t, x) \, \mathrm{d}x \leq m 2^m M_{1+\alpha}(f(t)) M_{m-1+\lambda-\alpha}(f(t)) ,$$

which, together with (8.2.4d) and (10.3.121), gives

$$\frac{d}{dt} M_m(f(t)) \leq m 2^m M_{1+\alpha}(f(t)) M_{m-1+\lambda-\alpha}(f(t)) - a_0 \mathfrak{h}_m M_{m+\gamma}(f(t)) . \quad (10.3.122)$$

Observe that the differential inequality (10.3.122) for $M_m(f)$ is not closed as it involves $M_{1+\alpha}(f)$ which is a moment of order smaller than one (when $\alpha < 0$). To estimate the latter we derive a differential inequality for $M_\lambda(f)$. To this end, consider $(x, y) \in (0, \infty)^2$ and define again $X = \min\{x, y\}$ and $Y = \max\{x, y\}$. The symmetry of χ_λ and the concavity of $r \mapsto r^\lambda$ imply that

$$-\chi_\lambda(x, y) = -\chi_\lambda(X, Y) \geq X^\lambda - \lambda XY^{\lambda-1} = X^\lambda \left(1 - \lambda \left(\frac{X}{Y}\right)^{1-\lambda}\right) \geq (1 - \lambda)X^\lambda ,$$

so that, using also the symmetry of k and the non-positivity of α,

$$-\chi_\lambda(x, y)k(x, y) \geq (1 - \lambda)X^\lambda k(X, Y) \geq (1 - \lambda)X^{\lambda+\alpha}Y^{\lambda-\alpha}$$

$$\geq (1 - \lambda)(XY)^\lambda \left(\frac{X}{Y}\right)^\alpha \geq (1 - \lambda)(XY)^\lambda = (1 - \lambda)(xy)^\lambda .$$

Consequently,

$$\int_0^\infty x^\lambda \mathcal{C}f(t, x) \, dx \leq -\frac{1-\lambda}{2} M_\lambda(f(t))^2 ,$$

which, together with (8.2.4d), (10.3.116d) and (10.3.121), gives

$$\frac{d}{dt}M_\lambda(f(t)) \leq -\frac{1-\lambda}{2} M_\lambda(f(t))^2 + a_0|\mathfrak{h}_\lambda|M_{\lambda+\gamma}(f(t)) . \qquad (10.3.123)$$

Adding (10.3.122), with $m = 2 - \lambda + \alpha \in (1, 2]$ (by (10.3.115)), and (10.3.123), and setting

$$P(t) := M_\lambda(f(t)) + M_{2-\lambda+\alpha}(f(t)) , \qquad t \geq 0 ,$$

we obtain

$$\frac{dP}{dt}(t) \leq -\frac{1-\lambda}{2} M_\lambda(f(t))^2 - a_0\mathfrak{h}_{2-\lambda+\alpha}M_{2+\gamma-\lambda+\alpha}(f(t))$$
$$+ a_0|\mathfrak{h}_\lambda|M_{\lambda+\gamma}(f(t)) + 8M_{1+\alpha}(f(t))M_1(f(t)) . \qquad (10.3.124)$$

Now, since $1 + \alpha \in (\lambda, 1]$ and $\gamma + \lambda \in (\lambda, \gamma + 2 - \lambda + \alpha)$ by (10.3.115), it follows from (10.3.117) and Hölder's and Young's inequalities that

$$M_{1+\alpha}(f(t)) \leq M_\lambda(f(t))^{|\alpha|/(1-\lambda)}M_1(f(t))^{(1-\lambda+\alpha)/(1-\lambda)} \leq \varrho + M_\lambda(f(t))$$
$$\leq \frac{1-\lambda}{64} M_\lambda(f(t))^2 + \frac{16}{1-\lambda} + \varrho ,$$

and

$$a_0|\mathfrak{h}_\lambda|M_{\lambda+\gamma}(f(t))$$
$$\leq a_0|\mathfrak{h}_\lambda|M_{2-\lambda+\alpha+\gamma}(f(t))^{\gamma/(\gamma+\alpha+2-2\lambda)}M_\lambda(f(t))^{(\alpha+2-2\lambda)/(\gamma+\alpha+2-2\lambda)}$$
$$\leq \frac{a_0\mathfrak{h}_{2-\lambda+\alpha}}{2} M_{2-\lambda+\alpha+\gamma}(f(t)) + C_1 M_\lambda(f(t))$$
$$\leq \frac{a_0\mathfrak{h}_{2-\lambda+\alpha}}{2} M_{2-\lambda+\alpha+\gamma}(f(t)) + \frac{1-\lambda}{8} M_\lambda(f(t))^2 + C_1 .$$

Inserting these estimates in (10.3.124), we obtain

$$\frac{dP}{dt}(t) \leq C_1 - \frac{1-\lambda}{4} M_\lambda(f(t))^2 - \frac{a_0\mathfrak{h}_{2-\lambda+\alpha}}{2} M_{2-\lambda+\alpha+\gamma}(f(t)) .$$

Now, the positivity of γ, (10.3.117) and Hölder's inequality ensure that

$$M_{2-\lambda+\alpha}(f(t)) \le \varrho^{\gamma/(1-\lambda+\alpha+\gamma)} M_{2-\lambda+\alpha+\gamma}(f(t))^{(1-\lambda+\alpha)/(1-\lambda+\alpha+\gamma)} ,$$

hence

$$\varrho^{-\gamma/(1-\lambda+\alpha)} M_{2-\lambda+\alpha}(f(t))^{(1-\lambda+\alpha+\gamma)/(1-\lambda+\alpha)} \le M_{2-\lambda+\alpha+\gamma}(f(t)) ,$$

and the differential inequality for P becomes

$$\frac{dP}{dt}(t) \le C_1 - \frac{1-\lambda}{4} M_\lambda(f(t))^2 - C_2 M_{2-\lambda+\alpha}(f(t))^{(1-\lambda+\alpha+\gamma)/(1-\lambda+\alpha)} .$$

Finally, introducing

$$q := \min\left\{1, \frac{\gamma}{1-\lambda+\alpha}\right\} > 0 ,$$

and observing that

$$P(t)^{q+1} \le \frac{1}{C_3}\left[1 + \frac{1-\lambda}{4} M_\lambda(f(t))^2 + C_2 M_{2-\lambda+\alpha}(f(t))^{(1-\lambda+\alpha+\gamma)/(1-\lambda+\alpha)}\right]$$

by repeated applications of Young's inequality, we end up with

$$\frac{dP}{dt}(t) \le 1 + C_1 - C_3 P(t)^{1+q} , \qquad t \ge 0 .$$

The comparison principle then ensures that

$$P(t) \le \max\left\{P(0), \left(\frac{1+C_1}{C_3}\right)^{1/(q+1)}\right\} , \qquad t \ge 0 ,$$

which completes the proof of (10.3.118).

Consider next $m > 1 \ge 1 + \alpha$. Since

$$M_{1+\alpha}(f(t)) \le \varrho^{(1-\lambda+\alpha)/(1-\lambda)} M_\lambda(f(t))^{|\alpha|/(1-\lambda)} ,$$

$$M_{m-1+\lambda-\alpha}(f(t)) \le M_\lambda^{(1-\lambda+\alpha)/(m-\lambda)} M_m(f(t))^{(m-1-\alpha)/(m-\lambda)} ,$$

$$M_m(f(t))^{(m+\gamma-1)/(m-1)} \le \varrho^{\gamma/(m-1)} M_{m+\gamma}(f(t))$$

by Hölder's inequality and (10.3.117), and

$$0 < \frac{m-1-\alpha}{m-\lambda} = \frac{m+\gamma-1}{m-1} - \frac{\gamma}{m-1} - \frac{1+\alpha-\lambda}{m-\lambda} < \frac{m+\gamma-1}{m-1} ,$$

we deduce from (10.3.122) and Young's inequality that

$$\frac{d}{dt} M_m(f(t)) \le C_4(m) M_\lambda(f(t))^{|\alpha|/(1-\lambda)+(1+\alpha-\lambda)/(m-\lambda)} M_m(f(t))^{(m-1-\alpha)/(m-\lambda)}$$

$$- a_0 \mathfrak{h}_m \varrho^{-\gamma/(m-1)} M_m(f(t))^{(m+\gamma-1)/(m-1)}$$

$$\le C_4(m) M_\lambda(f(t))^{\theta_m (m+\gamma-1)/(m-1)}$$

$$- \frac{a_0 \mathfrak{h}_m}{2\varrho^{\gamma/(m-1)}} M_m(f(t))^{(m+\gamma-1)/(m-1)}$$

$$\le C_4(m) \left(\sup_{s\ge 0} M_\lambda(f(s))\right)^{\theta_m (m+\gamma-1)/(m-1)}$$

$$- \frac{a_0 \mathfrak{h}_m}{2\varrho^{\gamma/(m-1)}} M_m(f(t))^{(m+\gamma-1)/(m-1)} ,$$

recalling that the exponent θ_m is defined in (10.3.120). The estimate (10.3.119) now follows from the previous differential inequality by the comparison principle. $\square$

A first consequence of Lemma 10.3.24 is that it prevents escape of matter towards large sizes as times goes by. We next turn to moments of negative order, in order to rule out concentration of matter into particles of small sizes in the long term.

Lemma 10.3.25. *Consider* $f^{in} \in X_{m_0,+} \cap X_2$ *such that* $M_1(f^{in}) = \varrho$ *and let* f *be the corresponding mass-conserving weak solution to the C-F equation (8.0.1) on* $[0,\infty)$ *given by Lemma 10.3.23. Given* $m \in [m_0, 0]$, *there exists* $\mu_{1,m}$, *depending only on* α, λ, a_0, γ, h, ϱ *and* m, *such that, if* $f^{in} \in X_{1+\gamma}$, *then*

$$M_m(f(t)) \leq \max\left\{ M_m(f^{in}), \mu_{1,m}\left(\sup_{s \geq 0} M_{1+\gamma}(f(s)) \right)^{\theta'_m} \right\}, \qquad t \geq 0,$$

where

$$\theta'_m := \frac{1-m}{1-m-\alpha}\left(\frac{\gamma-\alpha}{\gamma-\alpha+1-m} + \frac{\alpha+1-\lambda}{\gamma} \right) > 0.$$

Proof. We define

$$Y_{1+\gamma} := \sup_{s \geq 0}\{M_{1+\gamma}(f(s))\},$$

which is finite according to Lemma 10.3.24, and consider $m \in [m_0, 0]$. Owing to (10.3.114),

$$k(x,y)\chi_m(x,y) \leq -x^m k(x,y) \leq -x^{m+\alpha}y^{\lambda-\alpha}, \qquad (x,y) \in (0,\infty)^2,$$

and it follows from (8.2.4d) and (10.3.121) that, for $t \geq 0$,

$$\frac{d}{dt}M_m(f(t)) \leq -\frac{1}{2}M_{m+\alpha}(f(t))M_{\lambda-\alpha}(f(t)) + a_0|\mathfrak{h}_m|M_{m+\gamma}(f(t)).$$

Since $1 \in (\lambda-\alpha, \gamma+1)$ and $m+\gamma \in (m+\alpha, \gamma+1)$, we deduce from (10.3.117) and Hölder's inequality that

$$\varrho = M_1(f(t)) \leq M_{\gamma+1}(f(t))^{(\alpha+1-\lambda)/(\gamma+\alpha+1-\lambda)}M_{\lambda-\alpha}(f(t))^{\gamma/(\gamma+\alpha+1-\lambda)}$$
$$\leq Y_{1+\gamma}^{(\alpha+1-\lambda)/(\gamma+\alpha+1-\lambda)}M_{\lambda-\alpha}(f(t))^{\gamma/(\gamma+\alpha+1-\lambda)}$$

and

$$M_{m+\gamma}(f(t)) \leq M_{\gamma+1}(f(t))^{(\gamma-\alpha)/(\gamma-\alpha+1-m)}M_{m+\alpha}(f(t))^{(1-m)/(\gamma-\alpha+1-m)}$$
$$\leq Y_{1+\gamma}^{(\gamma-\alpha)/(\gamma-\alpha+1-m)}M_{m+\alpha}(f(t))^{(1-m)/(\gamma-\alpha+1-m)}.$$

Consequently,

$$\frac{d}{dt}M_m(f(t)) \leq a_0|\mathfrak{h}_m|Y_{1+\gamma}^{(\gamma-\alpha)/(\gamma-\alpha+1-m)}M_{m+\alpha}(f(t))^{(1-m)/(\gamma-\alpha+1-m)}$$
$$- \frac{\varrho^{(\gamma+\alpha+1-\lambda)/\gamma}}{2}Y_{1+\gamma}^{-(\alpha+1-\lambda)/\gamma}M_{m+\alpha}(f(t))$$
$$\leq Y_{1+\gamma}^{-(\alpha+1-\lambda)/\gamma}\left[C_5(m)Y_{1+\gamma}^{(1-m-\alpha)\theta'_m/(1-m)} - \frac{\varrho^{(\gamma+\alpha+1-\lambda)/\gamma}}{2}M_{m+\alpha}(f(t)) \right].$$

We use once more (10.3.117) and Hölder's inequality to obtain that

$$M_m(f(t))^{1-m-\alpha} \leq \varrho^{|\alpha|}M_{m+\alpha}(f(t))^{1-m},$$

so that

$$\frac{d}{dt}M_m(f(t)) \leq Y_{1+\gamma}^{-(\alpha+1-\lambda)/\gamma} \Big[C_5(m)Y_{1+\gamma}^{(1-m-\alpha)\theta'_m/(1-m)}$$

$$-C_6(m)M_m(f(t))^{(1-m-\alpha)/(1-m)} \Big] \ .$$

Since $1 - m - \alpha \geq 1 - m > 0$, Lemma 10.3.25 follows by the comparison principle. $\qquad\square$

Having ruled out loss of matter at infinity and at zero, we are left with preventing its concentration at some positive size. To this end, we derive an L_{p_0}-estimate, the exponent p_0 being defined in (10.3.116c) and characterizing the integrability of h in $(0,1)$.

Lemma 10.3.26. *Consider $f^{in} \in X_{m_0,+} \cap X_2$ such that $M_1(f^{in}) = \varrho$ and let f be the corresponding mass-conserving weak solution to the C-F equation (8.0.1) on $[0,\infty)$ given by Lemma 10.3.23. There is $\mu_{2,p_0} > 0$, depending only on α, λ, a_0, γ, h, p_0 and ϱ, such that, if $f^{in} \in L_{p_0}(0,\infty)$, then*

$$\|f(t)\|_{p_0}^{p_0} \leq \max \Big\{ \|f^{in}\|_{p_0}^{p_0}, \mu_{2,p_0} Y_{\gamma+1}^{p_0\theta_*} Y_{\gamma}^{p_0-1} Y_{\gamma+1-p_0} \Big\} \ , \qquad t \geq 0 \ ,$$

where

$$Y_m := \sup_{s \geq 0}\{M_m(f(s))\} \ , \qquad m \geq m_0 \ ,$$

are finite according to Lemma 10.3.24 and Lemma 10.3.25, and

$$\theta_* := \frac{1+\alpha-\lambda}{\gamma-\alpha} > 0 \ .$$

Proof. Let $t \geq 0$. On the one hand, by the symmetry of k, the nonnegativity of $\lambda - \alpha$, and Young's inequality,

$$\int_0^\infty \mathcal{C}f(t,x)f^{p_0-1}(t,x) \ dx$$

$$= \int_0^\infty \int_0^\infty x^\alpha y^{\lambda-\alpha} f^{p_0-1}(t,x+y)f(t,x)f(t,y) \ dydx$$

$$\quad - M_\alpha(f(t)) \int_0^\infty y^{\lambda-\alpha} f^{p_0}(t,y) \ dy - M_{\lambda-\alpha}(f(t)) \int_0^\infty x^\alpha f^{p_0}(t,x) \ dx$$

$$\leq \int_0^\infty \int_0^\infty x^\alpha y^{\lambda-\alpha} \Big[\frac{p_0-1}{p_0} f^{p_0}(t,x+y) + \frac{1}{p_0} f^{p_0}(t,y) \Big] f(t,x) \ dydx$$

$$\quad - M_\alpha(f(t)) \int_0^\infty y^{\lambda-\alpha} f^{p_0}(t,y) \ dy - M_{\lambda-\alpha}(f(t)) \int_0^\infty x^\alpha f^{p_0}(t,x) \ dx$$

$$\leq \frac{p_0-1}{p_0} \int_0^\infty \int_x^\infty x^\alpha (y_* - x)^{\lambda-\alpha} f^{p_0}(t,y_*)f(t,x) \ dy_*dx$$

$$\quad - \frac{p_0-1}{p_0} M_\alpha(f(t)) \int_0^\infty y^{\lambda-\alpha} f^{p_0}(t,y) \ dy - M_{\lambda-\alpha}(f(t)) \int_0^\infty x^\alpha f^{p_0}(t,x) \ dx$$

$$\leq -M_{\lambda-\alpha}(f(t)) \int_0^\infty x^\alpha f^{p_0}(t,x) \ dx \ .$$

On the other hand, it follows from (10.3.116) and Young's inequality that, for $\delta > 0$,

$$\int_0^\infty \mathscr{F} f(t,x) f^{p_0-1}(t,x) \, \mathrm{d}x$$

$$= a_0 \int_0^\infty y^\gamma f(t,y) \int_0^y b(x,y) f^{p_0-1}(t,x) \, \mathrm{d}x \mathrm{d}y - a_0 \int_0^\infty y^\gamma f^{p_0}(t,y) \, \mathrm{d}y$$

$$\leq a_0 \int_0^\infty y^\gamma f(t,y) \int_0^y \left(\frac{\delta(p_0-1)}{p_0} f^{p_0}(t,x) + \frac{\delta^{1-p_0}}{p_0} b^{p_0}(x,y) \right) \, \mathrm{d}x \mathrm{d}y$$

$$- a_0 \int_0^\infty y^\gamma f^{p_0}(t,y) \, \mathrm{d}y$$

$$\leq \frac{a_0 \delta (p_0-1)}{p_0} M_\gamma(f(t)) \|f(t)\|_{p_0}^{p_0} + \frac{a_0 \delta^{1-p_0}}{p_0} M_{\gamma+1-p_0}(f(t)) \|h\|_{L_{p_0}(0,1)}^{p_0}$$

$$- a_0 \int_0^\infty y^\gamma f^{p_0}(t,y) \, \mathrm{d}y \ .$$

Since $\gamma + 1 - p_0 \geq \alpha > m_0$ by (10.3.116), we deduce from the C-F equation (8.0.1) and the previous estimates that

$$\frac{1}{p_0} \frac{d}{dt} \|f(t)\|_{p_0}^{p_0} \leq a_0 \delta Y_\gamma \|f(t)\|_{p_0}^{p_0} + a_0 \delta^{1-p_0} Y_{\gamma+1-p_0} \|h\|_{L_{p_0}(0,1)}^{p_0}$$

$$- a_0 \int_0^\infty x^\gamma f^{p_0}(t,x) \, \mathrm{d}x - M_{\lambda-\alpha}(f(t)) \int_0^\infty x^\alpha f^{p_0}(t,x) \, \mathrm{d}x \ . \qquad (10.3.125)$$

We first exploit the positivity of γ and the non-positivity of α to show that the negative terms on the right-hand side of (10.3.125) provide some control on $\|f(t)\|_{p_0}$. It indeed follows from (10.3.117) and Hölder's inequality that

$$\varrho^{\gamma+\alpha+1-\lambda} = M_1(f(t))^{\gamma+\alpha+1-\lambda} \leq M_{1+\gamma}(f(t))^{\alpha+1-\lambda} M_{\lambda-\alpha}(f(t))^\gamma$$

$$\leq Y_{\gamma+1}^{\alpha+1-\lambda} M_{\lambda-\alpha}(f(t))^\gamma \ .$$

Consequently, thanks to Young's inequality,

$$a_0 \int_0^\infty x^\gamma f^{p_0}(t,x) \, \mathrm{d}x + M_{\lambda-\alpha}(f(t)) \int_0^\infty x^\alpha f^{p_0}(t,x) \, \mathrm{d}x$$

$$\geq a_0 \int_0^\infty x^\gamma f^{p_0}(t,x) \, \mathrm{d}x + \varrho^{(\gamma+\alpha+1-\lambda)/\gamma} Y_{\gamma+1}^{-(\alpha+1-\lambda)/\gamma} \int_0^\infty x^\alpha f^{p_0}(t,x) \, \mathrm{d}x$$

$$\geq a_0^{|\alpha|/(\gamma-\alpha)} \varrho^{(\gamma+\alpha+1-\lambda)/(\gamma-\alpha)} Y_{\gamma+1}^{-(\alpha+1-\lambda)/(\gamma-\alpha)} \int_0^\infty f^{p_0}(t,x) \, \mathrm{d}x$$

$$= 2C_7 Y_{\gamma+1}^{-\theta_*} \|f(t)\|_{p_0}^{p_0} \ , \qquad (10.3.126)$$

the exponent θ_* being defined in Lemma 10.3.26. Combining (10.3.125) and (10.3.126) gives

$$\frac{1}{p_0} \frac{d}{dt} \|f(t)\|_{p_0}^{p_0} \leq \left(a_0 \delta Y_\gamma - 2C_7 Y_{\gamma+1}^{-\theta_*} \right) \|f(t)\|_{p_0}^{p_0} + a_0 \delta^{1-p_0} Y_{\gamma+1-p_0} \|h\|_{L_{p_0}(0,1)}^{p_0} \ .$$

We finally choose $\delta = C_7 Y_{\gamma+1}^{-\theta_*} / (a_0 Y_\gamma)$ and thus obtain

$$\frac{1}{p_0} \frac{d}{dt} \|f(t)\|_{p_0}^{p_0} \leq -C_7 Y_{\gamma+1}^{-\theta_*} \|f(t)\|_{p_0}^{p_0} + C_8 Y_{\gamma+1}^{(p_0-1)\theta_*} Y_\gamma^{p_0-1} Y_{\gamma+1-p_0}$$

$$\leq Y_{\gamma+1}^{-\theta_*} \left[C_8 Y_{\gamma+1}^{p_0 \theta_*} Y_\gamma^{p_0-1} Y_{\gamma+1-p_0} - C_7 \|f(t)\|_{p_0}^{p_0} \right] \ .$$

We now infer from the comparison principle that

$$\|f(t)\|_{p_0}^{p_0} \leq \max\left\{\|f^{in}\|_{p_0}^{p_0}, \frac{C_8}{C_7}Y_{\gamma+1}^{p_0\theta_*}Y_{\gamma}^{p_0-1}Y_{\gamma+1-p_0}\right\}$$

for $t \geq 0$, thereby completing the proof of Lemma 10.3.26. $\qquad\square$

Proof of Theorem 10.3.22. Fix $\varrho > 0$ and define $E(x) := \varrho x^{-m_0}e^{-x}/\Gamma(2-m_0)$ for $x \in (0,\infty)$, and

$$\bar{\mu}(m) := M_m(E) = \varrho\frac{\Gamma(m+1-m_0)}{\Gamma(2-m_0)}, \qquad m \geq m_0.$$

Step 1: Invariant set. Let Z be the subset of X_1 defined as follows: $\zeta \in Z$ if

$$\zeta \in X_{m_0} \cap \bigcap_{m\geq 1} X_m, \qquad \zeta \geq 0 \text{ a.e.}, \qquad M_1(\zeta) = \varrho,$$

$$M_\lambda(\zeta) + M_{2-\lambda+\alpha}(\zeta) \leq \mu_{**} := \max\{\mu_*, \bar{\mu}(\lambda) + \bar{\mu}(2-\lambda+\alpha)\},$$

$$M_m(\zeta) \leq \mu_m := \max\{\mu_{1,m}\mu_{**}^{\theta_m}, \bar{\mu}(m)\}, \qquad m \in (1,\infty),$$

$$M_m(\zeta) \leq \mu_m := \max\{\mu_{1,m}\mu_{1,1+\gamma}^{\theta_m}, \bar{\mu}(m)\}, \qquad m \in [m_0, 0],$$

$$\|\zeta\|_{p_0} \leq \Lambda,$$

where

$$\Lambda := \max\left\{\mu_{2,p_0}^{1/p_0}\mu_{1+\gamma}^{\omega}\mu_0^{(p_0-1)/p_0(1+\gamma)}\mu_{m_0}^{1/(1+\gamma-m_0)}, \|E\|_{p_0}\right\},$$

$$\omega := \theta_* + \frac{\gamma(p_0-1)}{p_0(1+\gamma)} + \frac{1+\gamma-m_0-p_0}{p_0(1+\gamma-m_0)},$$

and μ_*, $\mu_{1,m}$, θ_m for $m \in (1,\infty)$ and $m \in [m_0,0]$, and μ_{2,p_0}, θ_* are defined in Lemma 10.3.24, Lemma 10.3.25, and Lemma 10.3.26, respectively. Note that $E \in Z$ so that Z is non-empty.

Let us first check that Z is left invariant by the C-F equation (8.0.1). For that purpose, pick an initial condition $f^{in} \in Z$ and denote the corresponding mass-conserving weak solution to the C-F equation (8.0.1) on $[0,\infty)$, given by Lemma 10.3.23, by f. First, mass conservation guarantees that $M_1(f(t)) = \varrho$ for all $t \geq 0$. In addition, the constraint $M_\lambda(f^{in}) + M_{2-\lambda+\alpha}(f^{in}) \leq \mu_{**}$ along with (10.3.118) ensures that

$$M_\lambda(f(t)) + M_{2-\lambda+\alpha}(f(t)) \leq \mu_{**}, \qquad t \geq 0. \tag{10.3.127}$$

Consider next $m \in (1,\infty)$. Owing to (10.3.127) and the constraint $M_m(f^{in}) \leq \mu_m$, we infer from (10.3.119) that

$$M_m(f(t)) \leq \max\left\{\mu_m, \mu_{1,m}\mu_{**}^{\theta_m}\right\} = \mu_m, \qquad t \geq 0. \tag{10.3.128}$$

Similarly, for $m \in [m_0,0]$, the constraint $M_m(f^{in}) \leq \mu_m$, Lemma 10.3.25, and (10.3.128) (for $M_{\gamma+1}(f)$) imply that

$$M_m(f(t)) \leq \max\left\{\mu_m, \mu_{1,m}\mu_{1+\gamma}^{\theta_m}\right\} = \mu_m, \qquad t \geq 0. \tag{10.3.129}$$

It now follows from (10.3.128), (10.3.129) and Hölder's inequality that, for $t \geq 0$,

$$M_\gamma(f(t)) \leq M_{\gamma+1}(f(t))^{\gamma/(1+\gamma)}M_0(f(t))^{1/(1+\gamma)} \leq \mu_{1+\gamma}^{\gamma/(1+\gamma)}\mu_0^{1/(1+\gamma)}$$

and

$$M_{1+\gamma-p_0}(f(t)) \le M_{\gamma+1}(f(t))^{(1+\gamma-m_0-p_0)/(1+\gamma-m_0)} M_{m_0}(f(t))^{p_0/(1+\gamma-m_0)}$$
$$\le \mu_{1+\gamma}^{(1+\gamma-m_0-p_0)/(1+\gamma-m_0)} \mu_{m_0}^{p_0/(1+\gamma-m_0)} .$$

Combining these estimates with (10.3.128) (for $M_{\gamma+1}(f)$) and Lemma 10.3.26 gives

$$\|f(t)\|_{p_0}^{p_0} \le \max\left\{ \Lambda^{p_0}, \mu_{2,p_0} \mu_{1+\gamma}^{p_0\omega} \mu_0^{(p_0-1)/(1+\gamma)} \mu_{m_0}^{p_0/(1+\gamma-m_0)} \right\} = \Lambda^{p_0} ,$$

for $t \ge 0$. Collecting the previous estimates, we have shown that

$$f^{in} \in Z \implies f(t) \in Z \quad \text{for all } t \ge 0 . \tag{10.3.130}$$

In addition, Z is a closed and convex subset of X_1 and is sequentially weakly compact in X_1 by the Dunford–Pettis theorem (Theorem 7.1.3).

Step 2: Dynamical system in Z equipped with the weak topology of X_1. Consider a nonnegative initial condition $f^{in} \in Z$ and let f be the corresponding mass-conserving weak solution to the C-F equation (8.0.1) on $[0, \infty)$, given by Lemma 10.3.23. We infer from (10.3.114), (10.3.116a), (10.3.130), Hölder's inequality and the C-F equation (8.0.1) that, for $t \ge 0$,

$$M_1(|\partial_t f(t)|) \le \int_0^\infty \int_0^\infty (x+y)k(x,y)f(t,x)f(t,y) \, dydx + 2\int_0^\infty xa(x)f(t,x) \, dx$$
$$\le 2M_{1+\alpha}(f(t))M_{\lambda-\alpha}(f(t)) + 2M_\alpha(f(t))M_{1+\lambda-\alpha}(f(t)) + 2a_0 M_{1+\gamma}(f(t))$$
$$\le 2M_1(f(t))^{1+\alpha}M_0(f(t))^{|\alpha|}M_1(f(t))^{\lambda-\alpha}M_0(f(t))^{1+\alpha-\lambda}$$
$$\quad + 2\mu_\alpha \mu_{1+\lambda-\alpha} + 2a_0 \mu_{1+\gamma}$$
$$\le 2\varrho^{1+\lambda}\mu_0^{1-\lambda} + 2\mu_\alpha \mu_{1+\lambda-\alpha} + 2a_0 \mu_{1+\gamma} .$$

The previous estimate provides the time equicontinuity in X_1 required to argue as in the proof of Theorem 10.3.18 and conclude that $\{f^{in} \mapsto f(t)\}_{t\ge0}$ is a dynamical system on Z endowed with the weak topology of X_1. Since it maps Z into itself and the set Z possesses the needed properties, we are in a position to apply Theorem 7.3.7 and derive the existence of a stationary solution $f_s \in Z$. □

10.3.3.3   Other Stationary Solutions

There are other choices of coagulation and fragmentation coefficients for which the existence of stationary solutions is available and we provide a couple of examples now. The first case to be considered deals with coagulation and fragmentation coefficients similar to those discussed in Section 10.3.3.2, the main difference being that the exponent α in the coagulation kernel is now taken to be positive.

Theorem 10.3.27. *[156] Assume that there is $\gamma > 0$, $a_0 > 0$, $p_0 > 1$, and a nonnegative function $h \in L_1((0,1), zdz)$ such that the fragmentation coefficients satisfy*

$$a(y) = a_0 y^\gamma , \qquad b(x,y) = \frac{1}{y}h\left(\frac{x}{y}\right) , \qquad 0 < x < y , \tag{10.3.131}$$

$$\int_0^1 zh(z) \, dz = 1 , \qquad \int_0^1 h(z)^{p_0} \, dz < \infty . \tag{10.3.132}$$

Assume further that there is $\lambda \in [0,1)$, and $\alpha \in [0, \lambda/2]$ such that

$$k(x,y) = x^\alpha y^{\lambda-\alpha} + x^{\lambda-\alpha}y^\alpha , \qquad (x,y) \in (0,\infty)^2 . \tag{10.3.133}$$

Given $\varrho > 0$ there exists a stationary solution $f_\varrho \in X_{1,+}$ to (8.0.1) satisfying $M_1(f_\varrho) = \varrho$ and there is $p > 1$, and $\mu \in (\lambda, 1)$ such that

$$f_\varrho \in L_p((0, \infty), x^{\mu+\gamma} dx) \cap \bigcap_{m > \lambda} X_m \ .$$

Similarly to Theorems 10.3.22 and 10.3.18, the proof of Theorem 10.3.27 also relies on a dynamical systems approach, but actually requires two steps. Indeed, it does not seem possible to construct directly an invariant set which is weakly compact in X_1, the difficulty being the lack of uniform integrability. To remedy this drawback, we first prove Theorem 10.3.27 with (k, a) in (10.3.131) replaced by $(k + \varepsilon, a + \varepsilon^2)$ for $\varepsilon > 0$ sufficiently small. Throughout the proof, we pay particular attention to the dependence of the various constants upon ε and realise that the ε-dependent family of stationary solutions lies in a weakly compact subset of X_1. We then show that any cluster point in X_1 as $\varepsilon \to 0$ of this family provides the sought stationary solution and refer the reader to [156] for the proof.

Another situation arises in a model of animal grouping considered in [95], where the existence of stationary solutions is shown with the help of the Bernstein function associated with $f \in X_{1,+}$, which is defined by

$$\Lambda f(\xi) = \int_0^\infty \left(1 - e^{-x\xi}\right) f(x) \, dx \ , \qquad \xi \in [0, \infty) \ ,$$

and is closely connected to the Laplace transform. Recall that Bernstein functions have already been used in Sections 10.2.1 and 10.2.2 to study the coagulation equation with constant and additive kernels.

Theorem 10.3.28. *[95] Assume that there is $\lambda \in [0, 2]$, and $a_0 > 0$ such that*

$$k(x, y) = (xy)^{\lambda/2} \ , \qquad a(x) = a_0 x^{\lambda/2} \ , \qquad b(x, y) = \frac{2}{y} \ , \tag{10.3.134}$$

for $(x, y) \in (0, \infty)^2$.

(a) *For all $\lambda \in [0, 2)$ and $\varrho > 0$, there is a unique stationary solution $f_\varrho \in X_{1,+}$ to the C-F equation (8.0.1) satisfying $M_1(f_\varrho) = \varrho$.*

(b) *If $\lambda = 2$, then there is a one-parameter family of stationary solutions $(f_L)_{L>0}$ in $X_{1,+}$ to the C-F equation (8.0.1) satisfying $M_1(f_L) = 2a_0$, and no stationary solution $f_s \in X_1$ to (8.0.1) satisfying $M_1(f_s) = \varrho$ with $\varrho \neq 2a_0$.*

A first remark is that if f_s is a stationary solution to the C-F equation (8.0.1) with coefficients given by (10.3.134) for some $\lambda \in (0, 2]$, then $x \mapsto x^{\lambda/2} f_s(x)$ is a stationary solution to the C-F equation (8.0.1) with coefficients given by (10.3.134) for $\lambda = 0$. The next remark is that when $\lambda = 0$ in (10.3.134), it is possible to convert the C-F equation (8.0.1) to a nonlocal ordinary differential equation for the Bernstein function and it turns out to be possible to perform a complete study of the dynamics of the latter [95]. There is actually a closed-form representation formula for the stationary solutions. Indeed, introducing

$$F(x) := \frac{1}{3x^{2/3}} \sum_{j=0}^\infty \left[\frac{(-1)^j}{\Gamma((4-6j)/3)} \frac{x^j}{(3j)!} + \frac{(-1)^{j+1}}{\Gamma((2-6j)/3)} \frac{x^{(3j+1)/3}}{(3j+1)!} \right] \tag{10.3.135}$$

for $x \in (0, \infty)$, it follows from [95, Section 5] that

$$M_0(F) = M_1(F) = \frac{M_2(F)}{6} = 1 \ . \tag{10.3.136}$$

When $\lambda \in [0, 2)$, we define

$$f_\varrho(x) = \frac{2a_0}{L_\varrho x^{\lambda/2}} F\left(\frac{x}{L_\varrho}\right) , \qquad L_\varrho := \left(\frac{\varrho}{2a_0 M_{(2-\lambda)/2}(F)}\right)^{2/(2-\lambda)} , \qquad (10.3.137)$$

for $x \in (0, \infty)$ and $\varrho > 0$. Then $(f_\varrho)_{\varrho > 0}$ is a one-parameter family of stationary solutions to the C-F equation (8.0.1) with coefficients given by (10.3.134), and $M_1(f_\varrho) = \varrho$. Since $(2 - \lambda)/2 \in (0, 1]$, the parameter $M_{(2-\lambda)/2}(F)$ entering the definition (10.3.137) is finite by (10.3.136), and thus (10.3.137) is meaningful. When $\lambda = 2$, we set

$$f_L(x) = \frac{2a_0}{Lx} F\left(\frac{x}{L}\right) , \qquad x \in (0, \infty) , \qquad L > 0 , \qquad (10.3.138)$$

and $(f_L)_{L>0}$ is a one-parameter family of stationary solutions to the C-F equation (8.0.1) with coefficients given by (10.3.134) and $M_1(f_L) = 2a_0$.

An interesting outcome of Theorem 10.3.28 is the existence of a single value of the total mass for which stationary states exist when $\lambda = 2$ in (10.3.134). This is perfectly consistent with Proposition 9.1.2 which implies that there is no mass-conserving solution with initial total mass bigger than $2a_0$. Additional information about this particular choice of the coefficients is supplied in the next section.

Finally, when $\lambda = 0$ in (10.3.134), the already mentioned use of the Bernstein transform leads to a complete description of the dynamics of the corresponding C-F equation (8.0.1), including the stability of stationary solutions, and is summarised in the next result [95, Theorems 7.1 and 8.1].

Proposition 10.3.29. *[95] Assume that the coagulation and fragmentation coefficients are given by (10.3.134) with $\lambda = 0$. Let $f^{in} \in X_{0,+}$. Then there is a unique weak solution $f \in C([0, \infty), X_0)$ to the C-F equation (8.0.1) on $[0, \infty)$. If f^{in} also belongs to X_1 with $\varrho := M_1(f^{in})$, then $f(t) \in X_1$ and $M_1(f(t)) = \varrho$. Moreover:*

(a) *If $f^{in} \in X_{0,1,+}$ with $\varrho := M_1(f^{in})$, then $f(t) \rightharpoonup f_\varrho$ in X_1, where f_ϱ is the unique stationary solution satisfying $M_1(f_\varrho) = \varrho$ given by Theorem 10.3.28 (a).*

(b) *If $f^{in} \in X_{0,+} \setminus X_1$, then $f(t)$ converges weakly to zero in the sense of measures in $(0, \infty)$ while $M_0(f(t)) \to 1$ as $t \to \infty$.*

The possible existence of steady-state solutions to the C-F equation (8.0.1) is also discussed in the physics literature [194, 232, 249, 266]. Recalling that the additive coagulation kernel k_+ and the multiplicative coagulation kernel $k_\times$ are given for $(x, y) \in (0, \infty)^2$ by $k_+(x, y) = x + y$ and $k_\times(x, y) = xy$, respectively, the specific cases $k \in \{2, k_+, k_\times\}$ and

$$a(x) = a_0 x^\gamma , \qquad b(x, y) = (\nu + 2)\frac{x^\nu}{y^{\nu+1}} , \qquad 0 < x < y ,$$

with $a_0 > 0$, $\gamma \in \{-1, 0, 1\}$, and $\nu \in (-2, 0]$ are considered in [266]. With these (nine) choices of coefficients, the evolution of the moments of order zero and two can be explicitly computed, and thereby provide hints on the dynamical behaviour of the corresponding C-F equation (8.0.1) and the possible occurrence of a stationary regime. Based on these computations, the latter is expected when $k \equiv 2$ and $\gamma \in \{0, 1\}$ whatever the value of $\nu \in (-2, 0]$, and when $k = k_+$, $\gamma = 1$, and $\nu \in (-1, 0]$. This is perfectly consistent with the previous results as the detailed balance condition (10.3.2) is satisfied for the choice $k \equiv 2$, $\gamma = 1$, and $\nu = 0$ (see Section 10.3.1 for a thorough analysis), while Theorem 10.3.18 guarantees the existence of stationary solutions for the choice $k = k_+$, $\gamma = 1$, and $\nu = 0$.

10.3.4 Mass-Conserving Self-Similar Solutions

It is tempting to investigate whether there exist other special solutions to the C-F equation (8.0.1), besides stationary solutions. Looking again at the ansatz for mass-conserving self-similar solutions

$$\frac{1}{\sigma(t)^2}\varphi\left(\frac{x}{\sigma(t)}\right) , \qquad (t,x) \in (0,\infty)^2 , \qquad (10.3.139)$$

see (10.0.1) with $\tau = 2$ for the coagulation equation and (10.0.12) for the fragmentation equation, we realise that the mean sizes σ of the particle distribution given by (10.0.7a) and (10.0.15a) coincide when $\gamma = \lambda - 1$, a fact already observed in [232, 266]. Recall that λ and γ are the homogeneity degrees of the coagulation kernel k and the overall fragmentation rate a, respectively. In that case, one may thus look for solutions to the C-F equation (8.0.1) of the form (10.3.139), assuming further that $\gamma > 0$ to prevent the occurrence of shattering. More specifically, assume that the coagulation and fragmentation coefficients are given by

$$k(x,y) = K_0\left(x^\alpha y^{\lambda-\alpha} + x^{\lambda-\alpha}y^\alpha\right) , \qquad (x,y) \in (0,\infty)^2 , \qquad (10.3.140)$$

and

$$a(x) = a_0 x^\gamma , \qquad b(x,y) = (\nu+2)\frac{x^\nu}{y^{\nu+1}} , \qquad 0 < x < y , \qquad (10.3.141)$$

with $K_0 > 0$, $a_0 > 0$, and

$$\gamma \in (0,1] , \qquad \lambda = \gamma + 1 , \qquad \alpha \in [\gamma, \lambda/2] , \qquad \nu > -2 . \qquad (10.3.142)$$

When

$$\gamma = 1 , \qquad \lambda = 2 , \qquad \alpha = 1 , \qquad \nu = 0 , \qquad (10.3.143)$$

in (10.3.140) and (10.3.141), numerical simulations performed in [232] indicate that self-similar behaviour may be observed if the ratio $\theta_\varrho := a_0/(\varrho K_0)$ is large enough, the parameter ϱ being equal to the total mass of the initial condition f^{in}, $\varrho = M_1(f^{in})$. This restriction is clearly needed as gelation occurs for $\theta_\varrho < 1/2$, see Proposition 9.1.2, and there is a stationary solution with mass a_0/K_0, see Theorem 10.3.28 (b) (with $K_0 = 1/2$).

A first step towards a mathematical justification of these numerical observations is the existence of mass-conserving self-similar solutions with a sufficiently small total mass [164].

Theorem 10.3.30. *[164] Assume that the coagulation and fragmentation coefficients are given by (10.3.140) and (10.3.141) with parameters γ, λ, α and ν satisfying (10.3.142). Assume further that*

$$\alpha \geq \frac{1}{2} , \qquad \nu > -\alpha - 1 , \qquad (10.3.144)$$

and let $\varrho > 0$. If

$$\theta_\varrho = \frac{a_0}{\varrho K_0} > 2(\nu+2)\ln 2 , \qquad (10.3.145)$$

then there exists a mass-conserving self-similar solution to the C-F equation (8.0.1)

$$(t,x) \longmapsto (1+\gamma t)^{2/\gamma}\varphi_\varrho\left(x(1+\gamma t)^{1/\gamma}\right) , \qquad (t,x) \in (0,\infty)^2 ,$$

satisfying the following properties: $M_1(\varphi_\varrho) = \varrho$ and there is $m_0 \in (-\nu-1,\alpha) \cap [0,1)$, and $p_1 > 1$ such that

$$\varphi_\varrho \in L_{p_1}((0,\infty), x^{m_1}dx) \cap \bigcap_{m>m_0} X_m , \qquad m_1 := \max\{m_0, 2-\lambda\} .$$

Observe that when the parameters in (10.3.140) and (10.3.141) are given by (10.3.143), Theorem 10.3.30 applies and provides the existence of a mass-conserving self-similar solution with total mass ϱ as soon as $\theta_\varrho > 4\ln 2 > 1$, which is perfectly consistent with the numerical simulations in [232].

A preliminary step towards the proof of Theorem 10.3.30 is the construction of a mass-conserving weak solution f to the C-F equation (8.0.1) on $[0, \infty)$ for a suitable class of initial data, the above choice of coagulation and fragmentation coefficients being not covered by the analysis performed in Section 8.2.2. A rough sketch of the proof is the following: consider a nonnegative function $f^{in} \in X_{m_1} \cap X_2$ and set $\varrho := M_1(f^{in})$. Denoting the solution to the C-F equation (8.0.1) with truncated coefficients

$$k_j(x,y) := k(x,y)\mathbf{1}_{(0,j)}(x)\mathbf{1}_{(0,j)}(y) \,, \quad a_j(x) := a(x)\mathbf{1}_{(0,j)}(x) \,, \quad (x,y) \in (0,\infty)^2 \,,$$

and initial condition $f_j^{in} := f^{in}\mathbf{1}_{(0,j)}$ by f_j, $j \geq 1$, we begin with a bound on $(M_m(f_j))_{j\geq 1}$ for all $m \in [m_1, 1]$. The next step is the cornerstone of the proof; we show that if ϱ satisfies (10.3.145), then $(M_{L\ln L}(f_j))_{j\geq 1}$ is bounded, where

$$M_{L\ln L}(\zeta) := \int_0^\infty \zeta(x)x\ln x \, \mathrm{d}x \,.$$

Here an important role is played by the inequality stated in Corollary 7.4.6. This bound in turn allows us to derive a bound on $(f_j)_{j\geq 1}$ in X_2 (and, in fact, in X_m for any $m \geq 2$, provided $f^{in} \in X_m$). The final step is the derivation of an estimate in $L_{p_0}((0,\infty), x^{m_0}\mathrm{d}x)$. Thanks to this analysis, we are in a position to apply Theorem 8.2.11 to complete the proof of the existence of at least one mass-conserving solution to (8.0.1), provided ϱ satisfies (10.3.145) [155].

Next, the proof of Theorem 10.3.30 relies on a dynamical systems approach, as that of Theorem 10.1.10. We consider the C-F equation in self-similar variables

$$\partial_s g(s,y) + y\partial_y g(s,y) + 2g(s,y) = \mathcal{C}g(s,y) + \mathcal{F}g(s,y) \,, \quad (s,y) \in (0,\infty)^2 \,, \quad (10.3.146)$$

$$g(0,y) = f^{in}(y) \,, \quad y \in (0,\infty) \,, \quad (10.3.147)$$

which is solved by the rescaled size distribution function

$$g(s,y) := e^{-2s} f\left(\frac{e^{\gamma s} - 1}{\gamma}, ye^{-s}\right) \,, \quad (s,y) \in (0,\infty)^2 \,,$$

with the self-similar variables given by

$$s = \frac{1}{\gamma}\ln(1 + \gamma t) \,, \quad y = x(1 + \gamma t)^{1/\gamma} \,.$$

Equivalently,

$$f(t,x) = (1 + \gamma t)^{2/\gamma}g\left(\frac{1}{\gamma}\ln(1 + \gamma t), x(1 + \gamma t)^{1/\gamma}\right) \,, \quad (t,x) \in (0,\infty)^2 \,.$$

The same steps as for the existence of mass-conserving solutions outlined above are needed to construct an invariant set in X_1 for (10.3.146), (10.3.147) which is weakly compact in X_1. The additional drift stemming from the transformation to self-similar variables is useful in getting rid of some time dependence in the estimates. In fact, an additional approximation argument is needed when $\nu \in (-2, -1]$: we first construct a stationary solution $\varphi_{\varrho,\varepsilon}$ to (10.3.146)–(10.3.147) with a suitable ε-dependent approximation of the daughter distribution b given by (10.3.141). The proof is next completed with the help of a compactness argument [164].

A different set of the parameters γ, λ, α and ν in (10.3.140) and (10.3.141) is dealt with in [173], namely,

$$\gamma = 1 \,, \qquad \lambda = 2 \,, \qquad \alpha = 1 \,, \qquad \nu = -1 \,, \qquad (10.3.148)$$

but with a completely different approach based on the Laplace transform. Before describing precisely the results obtained and providing a short account of the proof, let us mention that Theorem 10.3.30 does not apply to this particular case, which is actually a borderline case (as $\nu = \max\{\gamma - 2, -\alpha - 1\} = -1$).

Proposition 10.3.31. *[173] Assume that the coagulation and fragmentation coefficients are given by (10.3.140) and (10.3.141), with parameters γ, λ, α and ν satisfying (10.3.142). For any $\varrho > 0$ there exists a mass-conserving self-similar solution to the C-F equation (8.0.1)*

$$(t, x) \longmapsto (1 + t)^2 \varphi_\varrho \left(x(1 + t) \right) \,, \qquad (t, x) \in (0, \infty)^2 \,,$$

possessing the following properties: $M_1(\varphi_\varrho) = \varrho$. The Laplace transform $\mathcal{L}\varphi_\varrho$ of $x \mapsto x\varphi_\varrho(x)$ is given explicitly by

$$\mathcal{L}\varphi_\varrho(\xi) := \int_0^\infty x e^{-x\xi} \varphi_\varrho(x) \,\mathrm{d}x = \frac{a_0}{2K_0} \left[\frac{2\varrho K_0 + \xi}{a_0} - W\left(\frac{\xi}{a_0} e^{(2\varrho K_0 + \xi)/a_0} \right) \right]$$

for $\xi \in [0, \infty)$, where W is the Lambert W-function; that is, the inverse function of $z \mapsto ze^z$ on $(0, \infty)$.

As for the previous result, an intermediate step in the proof of Proposition 10.3.31 is the existence of a mass-conserving weak solution f to the C-F equation (8.0.1) for any nonnegative initial condition $f^{in} \in X_1$. The proof is based on the following observation: if f is a mass-conserving weak solution to the C-F equation (8.0.1) for this particular choice of coefficients with initial condition $f^{in} \in X_1$ satisfying $\varrho := M_1(f^{in})$, then the Laplace transform $\mathcal{L}f$ of $(t, x) \mapsto xf(t, x)$, defined by

$$\mathcal{L}f(t, \xi) := \int_0^\infty x e^{-x\xi} f(t, x) \,\mathrm{d}x \,, \qquad (t, \xi) \in [0, \infty) \times (0, \infty) \,,$$

solves

$$\partial_t \mathcal{L}f(t, \xi) = (a_0 + 2K_0\varrho - 2K_0\mathcal{L}f(t, \xi)) \,\partial_\xi \mathcal{L}f(t, \xi) + a_0 \frac{\varrho - \mathcal{L}f(t, \xi)}{\xi} \qquad (10.3.149)$$

for $(t, \xi) \in (0, \infty)^2$ with initial condition

$$\mathcal{L}f(0, \xi) = \mathcal{L}f^{in}(\xi) := \int_0^\infty x e^{-x\xi} f^{in}(x) \,\mathrm{d}x \,, \qquad \xi \in (0, \infty) \,. \qquad (10.3.150)$$

Consequently, $\mathcal{L}f$ is a solution to a scalar conservation law with a linear, but singular, source term. Owing to this property, we first use the method of characteristics to establish that (10.3.149)–(10.3.150) has a solution $\bar{L}$ defined for all times and derive a representation formula for this solution. In particular, it has two important properties: $\bar{L}(t, 0) = \varrho$ and $\xi \mapsto \bar{L}(t, \xi)$ is completely monotone for all $t > 0$. Consequently, for each $t > 0$, the function $\xi \mapsto \bar{L}(t, \xi)$ is the Laplace transform of a bounded measure $\mu(t, \mathrm{d}x)$ and $\mu(t, (0, \infty)) = \varrho$. In addition, μ solves the C-F equation (8.0.1) in an appropriate weak sense, see [173] for precise statements. Furthermore, a careful analysis of the representation formula for $\bar{L}$ shows that

$$\lim_{t \to \infty} \bar{L}(t, t\xi) = \mathcal{L}\varphi_\varrho(\xi) \,, \qquad \xi \in (0, \infty) \,. \qquad (10.3.151)$$

In other words, introducing

$$P(t,x) := \mu(t,(0,x)) , \qquad (t,x) \in (0,\infty)^2 ,$$

there holds

$$\lim_{t\to\infty} P\left(t, \frac{x}{t}\right) = \int_0^x x_* \varphi_\varrho(x_*)\, \mathrm{d}x_* , \qquad x \in (0,\infty) ,$$

this last convergence being a consequence of (10.3.151) by [114, XIII.1, Theorem 2]. Hence, the large time behaviour is self-similar whatever the value of the initial total mass ϱ. As a final remark, we emphasise here that, in this particular case, there is a mass-conserving self-similar solution for all values of the total mass ϱ, which contrasts markedly with the situation depicted in Theorem 10.3.30.

Chapter 11

Miscellanea

This final chapter is devoted to a survey of results that have been obtained for two C-F models, namely the Becker–Döring cluster equations and the C-F equation with diffusion. Both topics have received considerable attention in the past years, and it is only the fact that the book is already of a substantial length that prevents us from providing the more exhaustive treatments that they undoubtedly deserve.

11.1 The Becker–Döring Equations

When comparing the outcome of Theorem 10.3.17 to Conjecture 10.3.1, where the expected long-term behaviour of solutions to C-F equations satisfying the detailed balance condition (10.3.2) is described, it is clear that only the tip of the iceberg has been uncovered and that much remains to be done before a satisfactory answer is achieved.

A noticeable exception is the Becker–Döring (B-D) cluster equations (2.2.8)–(2.2.9) which are a particular case of the discrete C-F equations where clusters only gain or shed a single monomer. More precisely, denoting the number density of particles of size $n \geq 1$ by f_n, the B-D equations as introduced by Burton [55] read

$$\frac{df_n}{dt}(t) = J_{n-1}(f(t)) - J_n(f(t)) , \qquad t > 0 , \ n \geq 2 , \qquad (11.1.1a)$$

$$\frac{df_1}{dt}(t) = -2J_1(f(t)) - \sum_{i=2}^{\infty} J_i(f(t)) , \qquad t > 0 , \qquad (11.1.1b)$$

and are supplemented with (nonnegative) initial conditions

$$f_n(0) = f_n^{in} , \qquad n \geq 1 . \qquad (11.1.2)$$

In (11.1.1), $f = (f_n)_{n \geq 1}$,

$$J_n(f) = k_{n,1} f_n f_1 - F_{n,1} f_{n+1} , \qquad n \geq 1 , \qquad (11.1.3)$$

and $(k_{n,1})_{n \geq 1}$ and $(F_{n,1})_{n \geq 1}$ are the coagulation and fragmentation coefficients, respectively. When both are assumed to be positive,

$$k_{n,1} > 0 , \qquad F_{n,1} > 0 , \qquad n \geq 1 , \qquad (11.1.4)$$

we set

$$Q_1 = 1 , \qquad Q_n = \prod_{i=1}^{n-1} \frac{k_{i,1}}{F_{i,1}} , \qquad n \geq 2 , \qquad (11.1.5)$$

and readily observe that

$$k_{n,1} Q_1 Q_n = F_{n,1} Q_{n+1} , \qquad n \geq 1 , \qquad (11.1.6)$$

so that the detailed balance condition (10.3.2b) is satisfied. It turns out that, due to the specific structure of the B-D equations, a complete description of the long-term behaviour of its solutions is available and we provide below a short account of the known results, referring as well to the recent survey [142].

11.1.1   Well-Posedness

To begin with, we recall the definition of the space X_m, $m \in \mathbb{R}$, in the discrete setting

$$X_m := \{f = (f_n)_{n \geq 1} \ : \ \sum_{n=1}^{\infty} n^m |f_n| < \infty\} \ ,$$

$$X_{m,+} := \{f \in X_m \ : \ f_n \geq 0 \ \text{for all} \ n \in \mathbb{N}\} \ ,$$

with

$$\|f\|_{[m]} = \sum_{n=1}^{\infty} n^m |f_n| \ , \quad M_m(f) = \sum_{n=1}^{\infty} n^m f_n \ , \qquad f \in X_m \ ,$$

and provide a definition of a solution to (11.1.1)–(11.1.2).

Definition 11.1.1. *Let $T \in (0, \infty]$ and $f^{in} = (f_n^{in})_{n \geq 1} \in X_{1,+}$. A solution to (11.1.1)–(11.1.2) on $[0, T)$ is a function $f : [0, T) \to X_{1,+}$ such that*

(i) $f_n \in C([0, T))$ for all $n \in \mathbb{N}$ and $\sup_{t \in [0,T)} \|f(t)\|_{[1]} < \infty$;

(ii) for all $t \in [0, T)$,

$$\sum_{n=1}^{\infty} k_{n,1} f_n \in L_1(0, t) \ , \qquad \sum_{n=1}^{\infty} F_{n,1} f_{n+1} \in L_1(0, t) \ ;$$

(iii) for all $t \in [0, T)$,

$$f_n(t) = f_n^{in} + \int_0^t \left[J_{n-1}(f(s)) - J_n(f(s)) \right] \, ds \ , \qquad n \geq 2 \ ,$$

$$f_1(t) = f_1^{in} - \int_0^t \left[J_1(f(s)) + \sum_{i=1}^{\infty} J_i(f(s)) \right] \, ds \ .$$

For coagulation coefficients growing at most linearly, the existence of mass-conserving solutions to (11.1.1)–(11.1.2) is established in [15, Corollary 2.3 and Corollary 2.6].

Proposition 11.1.2. *Assume that there is $\kappa_0 > 0$ such that*

$$0 \leq k_{n,1} \leq \kappa_0 n \ , \qquad 0 \leq F_{n,1} \ , \qquad n \in \mathbb{N} \ , \tag{11.1.7}$$

and consider $f^{in} \in X_{1,+}$. Then there exists at least one solution f to (11.1.1)–(11.1.2) on $[0, \infty)$. Furthermore, $f \in C([0, \infty), X_1)$ and

$$M_1(f(t)) = \sum_{n=1}^{\infty} n f_n(t) = M_1(f^{in}) \ , \qquad t \geq 0 \ . \tag{11.1.8}$$

The growth condition (11.1.7) on $(k_{n,1})_{n \geq 1}$ is nearly optimal as there are non-existence results when it is not satisfied, see [15, Theorem 2.7] and [157]. In contrast, the fragmentation coefficients may grow arbitrarily fast with size. We next turn to the uniqueness issue for which the following results are available.

Proposition 11.1.3. *Let* $f^{in} \in X_{1,+}$.

(a) *Assume that there is* $\kappa_1 > 0$ *such that either*

$$k_{n,1} + F_{n,1} \leq \kappa_1 \sqrt{n} \, , \qquad n \in \mathbb{N} \, ,$$

or

$$k_{n+1,1} - k_{n,1} \leq \kappa_1 \quad \text{and} \quad F_{n,1} - F_{n+1,1} \leq \kappa_1 \, , \qquad n \in \mathbb{N} \, .$$

Then (11.1.1)–(11.1.2) *has a unique solution on* $[0, \infty)$.

(b) *Assume that* $(k_{n,1})_{n \geq 1}$ *satisfies* (11.1.7) *and there is* $\delta > 0$, $\kappa_2 > 0$, *and a positive sequence* $(g_n)_{n \geq 1}$ *such that*

$$g_{n+1} - g_n \geq \delta \, , \quad (g_{n+1} - g_n)k_{n,1} + nk_{n,1} \leq \kappa_2 g_n \, , \qquad n \in \mathbb{N} \, .$$

Assume further that $\sum_{n=1}^{\infty} g_n f_n^{in} < \infty$. *Then* (11.1.1)–(11.1.2) *has a unique solution* f *on* $[0, \infty)$ *and*

$$\sum_{n=1}^{\infty} g_n f_n \in L_{\infty}(0, T) \quad \text{for all } T > 0 \, .$$

While the existence of a solution is shown in [15, Theorem 2.2], the uniqueness statements in Proposition 11.1.3 (a) are established in [15, Theorem 3.7] (with $k = 0$) and in [166, Theorem 2.1], respectively, while that in Proposition 11.1.3 (b) is to be found in [15, Theorem 3.6]. Proposition 11.1.3 (b) with $g_n = n^2$, $n \in \mathbb{N}$, implies in particular that, if $f^{in} \in X_{2,+}$, then the B-D equations (11.1.1)–(11.1.2) have a unique solution on $[0, \infty)$ which belongs to $L_{\infty}((0, T), X_2)$ for all $T > 0$.

11.1.2 Long-Term Asymptotics

Having settled the issue of well-posedness, we now focus on the long-term behaviour of solutions and assume from now on that the coagulation and fragmentation coefficients are positive, see (11.1.4). As already observed above, due to (11.1.6) the coagulation and fragmentation coefficients in the B-D equations satisfy the detailed balance condition (10.3.2) with $Q = (Q_n)_{n \geq 1}$ defined in (11.1.5). It follows that the sequence $Q_z = (Q_n z^n)_{n \geq 1}$ also satisfies (11.1.6), and so is a stationary solution to (11.1.1) for all $z > 0$. We then define

$$z_{th} := \sup\{z \in (0, \infty) \, : \, Q_z \in X_1\} \in [0, \infty] \, , \tag{11.1.9}$$

and, when $z_{th} > 0$, we set

$$\varrho_{th} := \sup_{z \in (0, z_{th})} M_1(Q_z) \in (0, \infty] \, . \tag{11.1.10}$$

Note that, for $z > 0$,

$$Q_z \in X_1 \iff \sum_{n=1}^{\infty} nQ_n z^n < \infty \, ,$$

so that the Cauchy–Hadamard criterion for power series ensures that

$$\frac{1}{z_{th}} = \limsup_{n \to \infty} (nQ_n)^{1/n} = \limsup_{n \to \infty} Q_n^{1/n} \in [0, \infty] \, . \tag{11.1.11}$$

It is also obvious that $\varrho_{th} = \infty$ when $z_{th} = \infty$. Since $z \mapsto M_1(Q_z)$ is increasing from $(0, z_{th})$ onto $(0, \varrho_{th})$ when $z_{th} > 0$, for each $\varrho \in (0, \varrho_{th})$, there is a unique $z(\varrho) \in (0, \varrho_{th})$ such that $M_1(Q_{z(\varrho)}) = \varrho$.

We next define the Lyapunov functional

$$L(f) := \sum_{n=1}^{\infty} f_n \left(\ln \left(\frac{f_n}{Q_n} \right) - 1 \right) , \qquad (11.1.12)$$

and the corresponding dissipation functional

$$D(f) = \sum_{n=1}^{\infty} (k_{n,1} f_1 f_n - F_{n,1} f_{n+1}) \ln \left(\frac{k_{n,1} f_1 f_n}{F_{n,1} f_{n+1}} \right) \geq 0 , \qquad (11.1.13)$$

whenever they make sense. With these notations we are in a position to describe the long-term behaviour of solutions to (11.1.1)–(11.1.2) and the starting point is the counterpart of Theorem 10.3.17 for the B-D equations which is valid for a large class of coefficients and is proved in [15, Theorems 5.4 and 5.5] and [245, Theorem 5.10].

Theorem 11.1.4. *Assume that the coagulation and fragmentation coefficients satisfy* (11.1.4) *and* (11.1.7) *and consider*

$$f^{in} \in X_{1,+} , \quad f^{in} \not\equiv 0 , \quad \text{with } \varrho := M_1(f^{in}) \text{ and such that } L(f^{in}) < \infty . \qquad (11.1.14)$$

Let f be a solution to (11.1.1)–(11.1.2) *on* $[0, \infty)$ *which satisfies the energy inequality*

$$L(f(t)) + \int_0^t D(f(s)) \, ds \leq L(f^{in}) , \qquad t \geq 0 . \qquad (11.1.15)$$

(a) If

$$\lim_{n \to \infty} Q_n^{1/n} = 0 , \qquad (11.1.16)$$

then

$$\lim_{t \to \infty} \| f(t) - Q_{z(\varrho)} \|_{[1]} = 0 . \qquad (11.1.17)$$

(b) If

$$\lim_{n \to \infty} Q_n^{1/n} = \frac{1}{z_{th}} \in (0, \infty) , \qquad (11.1.18)$$

and

$$F_{n,1} \leq \kappa_0 n , \qquad n \in \mathbb{N} , \qquad (11.1.19)$$

then there is $\varrho_\infty \in [0, \min\{\varrho, \varrho_{th}\}]$ such that

$$\lim_{t \to \infty} \left| f_n(t) - Q_{z(\varrho_\infty),n} \right| = 0 , \qquad n \in \mathbb{N} . \qquad (11.1.20)$$

A few comments on the statement of Theorem 11.1.4 are in order. First, according to [15, Theorem 4.8] when $z_{th} > 0$, the B-D equations have at least one solution on $[0, \infty)$ satisfying the energy inequality (11.1.15) under the assumptions (11.1.4), (11.1.7), and (11.1.14) on the coagulation and fragmentation coefficients and the initial condition, so that Theorem 11.1.4 applies to that solution. In fact, according to [15, Theorem 4.7], when $z_{th} \in (0, \infty)$, the energy identity

$$L(f(t)) + \int_0^t D(f(s)) \, ds = L(f^{in}) , \qquad t \geq 0 , \qquad (11.1.21)$$

is valid for any solution to (11.1.1)–(11.1.2) as soon as $(k_{n,1})_{n \geq 1}$ and $(F_{n,1})_{n \geq 1}$ satisfy (11.1.4) and there is $\kappa_4 > 0$ such that

$$k_{n,1} + F_{n,1} \leq \kappa_4 \frac{n}{\ln n} , \qquad n \geq 2 .$$

The proof of Theorem 11.1.4 relies on dynamical systems arguments and exploits fully the energy inequality (11.1.15) and the continuity properties of the Lyapunov functional L. Let us recall here that the proof of Theorem 10.3.17 is directly inspired by that of Theorem 11.1.4.

An interesting feature of the B-D equations (11.1.1) is that their structure is much simpler than that of the discrete C-F equation, in the sense that only the dynamics of monomers is coupled to that of the other particles. This observation is at the heart of the following improved version of Theorem 11.1.4 which provides a complete proof of Conjecture 10.3.1 for B-D equations in the case when fragmentation dominates coagulation.

Theorem 11.1.5. *[13, 245] Consider coagulation and fragmentation coefficients satisfying* (11.1.4), (11.1.7), (11.1.18), *and* (11.1.19), *and an initial condition f^{in} satisfying* (11.1.14). *Assume that for each $z \in (0, z_{th})$*

$$k_{n,1} z \leq F_{n,1} \quad \text{for } n \text{ large enough,} \tag{11.1.22}$$

and that there is a unique solution f to the B-D equations (11.1.1)–(11.1.2) *on* $[0, \infty)$.

(a) If $\varrho \in (0, \varrho_{th}]$, then

$$\lim_{t \to \infty} \left\| f(t) - Q_{z(\varrho)} \right\|_{[1]} = 0 \quad \text{and} \quad \lim_{t \to \infty} L(f(t)) = L\left(Q_{z(\varrho)}\right) .$$

(b) If $\varrho > \varrho_{th}$, then

$$\lim_{t \to \infty} |f_n(t) - Q_{z_{th}, n}| = 0 , \qquad n \in \mathbb{N} ,$$

and

$$\lim_{t \to \infty} L(f(t)) = L\left(Q_{z_{th}}\right) + (\varrho - \varrho_{th}) \ln z_{th} .$$

The proof relies on a control of the tail of $f(t)$ which is uniform with respect to $t \geq 0$ and it is derived by a comparison argument, see [13] and [245, Theorem 5.11]. Tail control may also be derived from moment estimates and a study of the propagation of both algebraic and exponential moments is performed in [61], comparison arguments playing also an important role there.

The situation where there is no nonzero equilibrium with finite mass, corresponding to $z_{th} = 0$, is excluded from the previous results and is dealt with in the final result of this section [69].

Theorem 11.1.6. *[69, Theorem 3.3] Assume that the coagulation and fragmentation coefficients satisfy* (11.1.4), (11.1.7), (11.1.19), *and*

$$\lim_{n \to \infty} \frac{F_{n,1}}{k_{n,1}} = \lim_{n \to \infty} \frac{F_{n+1,1}}{k_{n,1}} = 0 . \tag{11.1.23}$$

Consider an initial condition f^{in} satisfying (11.1.14) *and let f be a solution to* (11.1.1)–(11.1.2) *on* $[0, \infty)$. *Then $z_{th} = 0$ and*

$$\lim_{t \to \infty} f_n(t) = 0 , \qquad n \in \mathbb{N} .$$

The proof of Theorem 11.1.6 requires a different approach since the property $z_{th} = 0$ renders the Lyapunov functional L useless in that case. In fact, as shown in [69, Theorem 4.4], in that case, under additional assumptions on the coagulation and fragmentation coefficients, there holds

$$\lim_{t \to \infty} L(f(t)) = -\infty .$$

11.1.3　Decay Rates

In this section we return to the case $z_{th} > 0$ and provide a refined analysis of the convergence of the solutions to the B-D equations to their expected limit as described in Theorem 11.1.5. In fact, once the convergence is established, it is worth investigating the rate of convergence and look for temporal estimates of $f(t) - Q_{z(\varrho)}$.

When $\varrho = M_1(f^{in}) > \varrho_{th}$, this issue is likely to be delicate as only componentwise convergence is available, see Theorem 11.1.5 (b). In addition, metastable behaviour may take place in that case as shown in [228], see also [68] for numerical simulations. Roughly speaking, for a suitable class of coagulation and fragmentation coefficients and $z > z_{th}$, there are bounded steady states $(s_n(z))_{n \geq 1}$ to the B-D equations satisfying $s_1(z) = z$,

$$k_{n-1,1} z s_{n-1}(z) - (k_{n,1} z + F_{n,1}) s_n(z) + F_{n+1,1} s_{n+1}(z) = 0 , \qquad n \geq 2 .$$

It is shown in [228] that, for $z > z_{th}$ close to z_{th}, there are solutions to the B-D equations which stay close to $(s_n(z))_{n \geq 1}$ for an exponentially large time.

We thus focus on the case $\varrho = M_1(f^{in}) < \varrho_{th}$ and recall that

$$\lim_{t \to \infty} \left\| f(t) - Q_{z(\varrho)} \right\|_{[1]} = 0$$

by Theorem 11.1.5, provided that $(k_{n,1})_{n \geq 1}$ and $(F_{n,1})_{n \geq 1}$ satisfy (11.1.4), (11.1.7), (11.1.18), (11.1.19), and (11.1.22). The rate of this convergence is investigated in [143], where the following temporal decay rate is derived.

Theorem 11.1.7. *[143, Theorem 2.1 and Corollary 2.2] Assume that the coagulation and fragmentation coefficients satisfy (11.1.4), (11.1.7), (11.1.18), (11.1.19), as well as*

$$k_{n,1} \geq 1 , \ F_{n,1} \geq 1 , \ k_{n,1} z_{th} \leq \min\{F_{n,1}, F_{n+1,1}\} , \qquad n \in \mathbb{N} .$$

Consider an initial condition f^{in} satisfying (11.1.14) with $\varrho \in (0, \varrho_{th})$ and

$$P_\mu := \sum_{n=1}^{\infty} \mu^n f_n^{in} < \infty \ \text{for some} \ \mu > 1 , \tag{11.1.24}$$

and let f be a solution to the B-D equations (11.1.1)–(11.1.2). There are constants $\theta_0 > 0$, depending on ϱ, μ, and P_μ, and $C_0 > 0$, depending on ϱ^{in}, such that

$$L\left[f(t) | Q_{z(\varrho)}\right] \leq L\left[f^{in} | Q_{z(\varrho)}\right] e^{-\theta_0 t^{1/3}} , \qquad t \geq 0 , \tag{11.1.25}$$

and

$$\left\| f(t) - Q_{z(\varrho)} \right\|_{[1]} \leq C_0 e^{-\theta_0 t^{1/3}} , \qquad t \geq 0 , \tag{11.1.26}$$

where the so-called relative energy is defined by

$$L\left[f(t) | Q_{z(\varrho)}\right] := L(f(t)) - L\left(Q_{z(\varrho)}\right) \geq 0 , \qquad t \geq 0 . \tag{11.1.27}$$

We first recall that the nonnegativity of the relative energy follows from the convexity of $r \mapsto r \ln r$, Jensen's inequality, and the property $M_1(f(t)) = \varrho = M_1(Q_{z(\varrho)})$. The proof of the temporal decay rate (11.1.25) relies on a careful analysis of the time evolution of f_1 and different arguments are needed for small and large values of f_1. Steps of the proof include the following: on the one hand, the fast decay (11.1.24) for large n of f^{in} is almost preserved by the dynamics, in the sense that there is $\bar{\mu} \in (1, \mu]$ such that $\sum_{n=1}^{\infty} \bar{\mu}^n f_n$ is bounded uniformly with respect to time. On the other hand, an inequality of the form

$$CD(f) \geq \frac{L\left[f | z_{Q(\varrho)}\right]}{\left(\ln L\left[f | Q_{z(\varrho)}\right]\right)^2} \tag{11.1.28}$$

is derived for suitable sequences $f = (f_n)_{n\geq 1}$ (with $C > 0$) and is combined with the energy identity (11.1.21) to obtain an upper bound on the relative energy $L\left[f|Q_{z(\varrho)}\right]$. The decay rate (11.1.26) in X_1 then follows from (11.1.25) by an appropriate variant of the Csiszár–Kullback inequality [84]

Theorem 11.1.7 provided impetus to subsequent investigations which either improved, or made more precise, the decay rates (11.1.25) and (11.1.26), according to the assumptions on $(k_{n,1})_{n\geq 1}$, $(F_{n,1})_{n\geq 1}$, and f^{in} [60, 62, 207, 208]. In particular, for initial conditions satisfying (11.1.24), exponential convergence to equilibrium can be obtained in a stronger norm [62].

Theorem 11.1.8. *[62, Theorem 1.1] Assume that the coagulation and fragmentation coefficients satisfy (11.1.4), (11.1.7), (11.1.19), and*

$$\inf_{n\geq 1} k_{n,1} > 0 , \qquad \lim_{n\to\infty} \frac{k_{n+1,1}}{k_{n,1}} = 1 , \tag{11.1.29}$$

$$\lim_{n\to\infty} \frac{Q_{n+1}}{Q_n} = \frac{1}{z_{th}} \in (0,\infty) . \tag{11.1.30}$$

Consider an initial condition f^{in} satisfying (11.1.14) with $\varrho \in (0,\varrho_{th})$ and (11.1.24) for some $\mu > 1$ and let f be a solution to the B-D equations (11.1.1)–(11.1.2). Then there are $\mu_\star \in (1,\mu)$ and $\Lambda_\star > 0$ such that for any $\nu \in (1,\mu_\star)$ there is a constant $C > 0$, depending only on ϱ, P_μ in (11.1.24), and ν, such that

$$\sum_{n=1}^{\infty} \nu^n \left|f_n(t) - Q_{z(\varrho),n}\right| \leq Ce^{-\Lambda_\star t} , \qquad t \geq 0 . \tag{11.1.31}$$

Furthermore, if $k_{n,1} \to \infty$ as $n \to \infty$, then

$$\frac{1}{\Lambda_\star} \leq \sup_{n\geq 1} \left\{ \left(\sum_{m=n+1}^{\infty} Q_m z^m(\varrho) \right) \left(\sum_{m=1}^{n} \frac{1}{k_{m,1} Q_m z^m(\varrho)} \right) \right\} < \infty .$$

Since $\nu^n \geq n\ln\nu$ for $n \in \mathbb{N}$ and $\nu > 1$, it readily follows from (11.1.31) that $\|f - Q_{z(\varrho)}\|_{[1]}$ also decays as $e^{-\Lambda_\star t}$ for large times and hence improves the decay estimate (11.1.26). While the proof of Theorem 11.1.8 still makes use of the inequality (11.1.28) to control the behaviour of $f(t)$ when it is far from the equilibrium $Q_{z(\varrho)}$, a different strategy is required to estimate $f(t)$ in the vicinity of $Q_{z(\varrho)}$. Specifically, a careful study of the linearisation of (11.1.1) near $Q_{z(\varrho)}$ is performed and a spectral gap in the Hilbert space

$$\left\{ h = (h_n)_{n\geq 1} : \sum_{n=1}^{\infty} Q_n z^n(\varrho)h_n^2 < \infty , \sum_{n=1}^{\infty} nQ_n z^n(\varrho)h_n = 0 \right\}$$

is identified. More recently, the approach from [143] was revisited in [60] and led to the following refined version of Theorem 11.1.7.

Theorem 11.1.9. *[60, Theorems 1.3 and 1.4] Assume that the coagulation and fragmentation coefficients satisfy (11.1.4), (11.1.30), and that there are $\lambda \in \mathbb{R}$, $\kappa_5 > 0$, and $\kappa_6 > 0$ such that*

$$\kappa_5 n^\lambda \leq k_{n,1} \leq \kappa_6 n^\lambda , \qquad n \in \mathbb{N} . \tag{11.1.32}$$

Assume also that the sequence $(Q_n z_{th}^n)_{n\geq 1}$ is nonincreasing. Consider next an initial condition f^{in} satisfying (11.1.14) with $\varrho \in (0,\varrho_{th})$ and let f be a solution to the B-D equations (11.1.1)–(11.1.2). Recalling the definition (11.1.27) of the relative energy, there holds:

(a) *if* $\lambda = 1$, *then there are constants* $C > 0$ *and* $\theta > 0$ *such that*

$$L\left[f(t)|Q_{z(\varrho)}\right] \leq Ce^{-\theta t}\,, \qquad t \geq 0\,; \tag{11.1.33}$$

(b) *if* $\lambda < 1$ *and* $f^{in} \in X_m$ *for* $m = \max\{2 - \lambda, 1 + \lambda\} > 1$, *then there are constants* $C > 0$ *and* $\theta > 0$ *such that*

$$L\left[f(t)|Q_{z(\varrho)}\right] \leq (C + \theta t)^{-(m-1)/(1-\lambda)}\,, \qquad t \geq 0\,. \tag{11.1.34}$$

Besides providing a more complete and precise description of the dynamics of the B-D equations for initial conditions with a subcritical mass $\varrho = M_1(f^{in}) \in (0, \varrho_{th})$, the outcome of Theorem 11.1.9 reveals also a sensitive dependence of the decay rates upon the growth of the coagulation coefficients $(k_{n,1})_{n\geq 1}$. In particular, for linearly growing coagulation coefficients corresponding to $\lambda = 1$ in (11.1.32), exponential convergence to the equilibrium $Q_{z(\varrho)}$ is a common feature of all solutions starting from an initial condition f^{in} satisfying only (11.1.14) and thus does not require f^{in} to have a finite exponential moment as in Theorem 11.1.8. The situation is different for sublinear coagulation coefficients corresponding to $\lambda < 1$ in (11.1.32), for which the decay is only algebraic for a large class of initial conditions and seems to require a finite exponential moment on f^{in} as (11.1.24) for a temporal exponential decay to equilibrium.

The core of the proof of Theorem 11.1.9 is an improved version of the inequality (11.1.28) relating the relative energy $L\left[f|Q_{z(\varrho)}\right]$ and the dissipation $D(f)$, which takes into account the growth condition (11.1.32) on the coagulation coefficients, see [60, Theorem 1.1]. This improvement implies in particular that, under the assumptions of Theorem 11.1.9, if $\lambda < 1$ and f^{in} satisfies additionally (11.1.24) for some $\mu > 1$, then there are constants $C' > 0$ and $\theta' > 0$ such that

$$L\left[f(t)|Q_{z(\varrho)}\right] + \left\|f(t) - Q_{z(\varrho)}\right\|_{[1]} \leq C'e^{-\theta' t^{1/(2-\lambda)}}\,, \qquad t \geq 0\,,$$

an estimate in the spirit of (11.1.25)–(11.1.26) which is, however, weaker than (11.1.31) for coagulation and fragmentation coefficients to which both Theorem 11.1.8 and Theorem 11.1.9 apply.

An algebraic decay as in (11.1.34), but without assuming the growth control (11.1.32), is also obtained by a detailed study of the linearised operator near an equilibrium in [207]. The relevant result is:

Theorem 11.1.10. *[207, Theorem 1.1] Assume that the coagulation and fragmentation coefficients satisfy* (11.1.4), (11.1.7), (11.1.19), (11.1.29), *and* (11.1.30). *Consider real numbers* $m > 0$ *and* $m' > m + 2$ *and assume that the initial condition* $f^{in} \in X_{1+m'}$ *satisfies* (11.1.14) *with* $\varrho \in (0, \varrho_{th})$. *Then there are* $\delta_{m,m'} > 0$ *and* $C_{m,m'} > 0$ *such that, if* $\|f^{in} - Q_{z(\varrho)}\|_{[1+m']} \leq \delta_{m,m'}$, *then*

$$\|f(t) - Q_{z(\varrho)}\|_{[1+m]} \leq C_{m,m'}\|f^{in} - Q_{z(\varrho)}\|_{[1+m']}(1 + t)^{-(m'-m-1)}\,, \qquad t \geq 0\,.$$

Observe that, since $z(\varrho) < z_{th}$, $f^{in} - Q_{z(\varrho)}$ belongs to X_l if and only if $f^{in} \in X_l$ for $l \in \mathbb{R}$. Further properties of the linearised operator near an equilibrium are derived in [208].

We conclude this section by mentioning that there are strong and deep connections between the B-D equations (11.1.1) and another celebrated model for phase transitions, the Lifshitz–Slyozov–Wagner equation [184, 271]. As far as we are aware, this connection was first uncovered by Penrose [229] via formal asymptotics, and its mathematical justifications may be found in [166, 209, 210, 211, 239]. Diffusive corrections to the Lifshitz–Slyozov–Wagner equation, as proposed in [195, 237], are also derived with the help of the B-D equations in [80, 83, 97, 265].

11.2 Coagulation-Fragmentation with Diffusion

Size might not be the only state variable characterizing the particles, and additional variables, such as spatial position, velocity, charge, or internal structure, may play a non-negligible role. Indeed, in the original derivation of Smoluchowski's coagulation equation for colloidal suspensions performed in [247, 248], besides undergoing pairwise merging, the colloidal particles move in space according to Brownian motion and the modelling leads to the so-called Smoluchowski coagulation equation with diffusion, see Section 2.2.1 in Volume I. Taking into account fragmentation as well, and assuming that the sizes of the particles range in $(0, \infty)$, the C-F equation with diffusion describes the evolution of the size distribution function $f = f(t, x, z)$ of particles of size x at spatial position $z \in \Omega$ and time $t > 0$, and reads

$$\partial_t f(t, x, z) = D(x)\Delta_z f(t, x, z) + \mathcal{C}f(t, x, z) + \mathcal{F}f(t, x, z) \qquad (11.2.1a)$$

for $(t, x, z) \in (0, \infty)^2 \times \Omega$. Here Ω is an open subset of $\mathbb{R}^d$, $d \geq 1$, and Δ_z denotes the standard Laplace operator with respect to the space variable z, while $D(x)$ denotes the diffusion coefficient of particles of size $x \in (0, \infty)$. If Ω is bounded, then equation (11.2.1a) is usually supplemented with homogeneous Neumann boundary conditions

$$D(x)\nabla_z f(t, x, z) \cdot \mathbf{n}(z) = 0 , \qquad (t, x, z) \in (0, \infty)^2 \times \partial\Omega , \qquad (11.2.1b)$$

where $\mathbf{n}(z)$ denotes the outward unit normal vector field to the boundary $\partial\Omega$ of Ω at $z \in \partial\Omega$, so that there is no escape of matter through the boundary of Ω. As in the spatially homogeneous case, conservation of matter is expected and therefore

$$\int_\Omega \int_0^\infty xf(t, x, z) \, \mathrm{d}x\mathrm{d}z = \int_\Omega \int_0^\infty xf(0, x, z) \, \mathrm{d}x\mathrm{d}z , \qquad t \geq 0 . \qquad (11.2.2)$$

Introducing the local total mass

$$\varrho(t, z) := \int_0^\infty xf(t, x, z) \, \mathrm{d}x \qquad (11.2.3)$$

at time $t \geq 0$ and position $z \in \Omega$, an alternative formulation of (11.2.2) reads

$$\|\varrho(t)\|_{L_1(\Omega)} = \|\varrho^{in}\|_{L_1(\Omega)} , \qquad t \geq 0 , \qquad (11.2.4)$$

where $\varrho^{in}(z) := \varrho(0, z)$, $z \in \Omega$. Owing to (11.2.2), the natural functional setting for the C-F equation with diffusion (11.2.1) is

$$\mathcal{X}_1 := L_1((0, \infty) \times \Omega, x\mathrm{d}x\mathrm{d}z)$$

in the continuous setting and, in the discrete size case,

$$\mathcal{X}_1 := \left\{ (f_i)_{i \geq 1} \ : \ f_i \in L_1(\Omega) , \ i \geq 1 , \ \sum_{i=1}^\infty i\|f_i\|_{L_1(\Omega)} < \infty \right\} .$$

The main difference between (11.2.1) and its spatially homogeneous counterpart (6.1.1) is that the coalescence of two particles is local with respect to the space variable, in the sense that both interacting particles are located at the same position $z \in \Omega$. This contrasts markedly with the interactions with respect to the size variable, which are mainly nonlocal, and this difference has far-reaching consequences for the mathematical analysis. Indeed, the coagulation operator behaves somewhat as a convolution with respect to the size variable and, due to the properties of the latter, the mere integrability of the size distribution function f with respect to the size variable (provided by the conservation of mass) is in general sufficient for the quadratic terms involved in $\mathcal{C}f$ to be meaningful. This is no longer true when spatial variations are taken into account and the mere integrability of f with respect to (x, z), provided by mass conservation, is no longer sufficient to give a functional meaning in the state space to the products $f(y, z)f(x - y, z)$ and $f(x, z)f(y, z)$, which are both found in $\mathcal{C}f(x, z)$. This turns out to be a new difficulty to be overcome for handling spatially dependent models. A possible remedy is to derive additional integrability properties of f, such as square integrability with respect to (x, z), which obviously guarantees that the product $(x, y, z) \mapsto f(x, z)f(y, z)$ is integrable on $(0, \infty)^2 \times \Omega$. Clearly, the regularizing properties of the Laplace operator with respect to the space variable $z \in \Omega$ could help but they are known to depend strongly on the space dimension d.

Several techniques have been developed in recent years to show the existence of weak solutions to the C-F equation with diffusion (11.2.1) but usually they require stronger assumptions on the coagulation and fragmentation coefficients than their space-homogeneous counterparts, as well as suitable properties of the diffusion coefficients D.

A straightforward cure is to derive L_∞-estimates, which turns out to be rather simple for the discrete coagulation equation with diffusion. Indeed, in that case, the size distribution function f_1 of the monomers solves

$$\partial_t f_1 - D_1 \Delta_z f_1 = -\sum_{j=1}^\infty k_{1,j} f_1 f_j \le 0 \,, \qquad (t, z) \in (0, \infty) \times \Omega \,,$$

supplemented with homogeneous Neumann boundary conditions and it readily follows from the comparison principle that $\|f_1(t)\|_{L_\infty(\Omega)} \le \|f_1^{in}\|_{L_\infty(\Omega)}$ for all $t \ge 0$. Thanks to this bound, we can deduce that the size distribution function f_2 of the dimers satisfies

$$\partial_t f_2 - D_2 \Delta_z f_2 = \frac{k_{1,1}}{2} f_1^2 - \sum_{j=1}^\infty k_{2,j} f_2 f_j \le k_{1,1} \|f_1^{in}\|_{L_\infty(\Omega)}^2$$

for $(t, z) \in (0, \infty) \times \Omega$, supplemented with homogeneous Neumann boundary conditions and a further application of the comparison principle implies that

$$\|f_2(t)\|_{L_\infty(\Omega)} \le \|f_2^{in}\|_{L_\infty(\Omega)} + k_{1,1} \|f_1^{in}\|_{L_\infty(\Omega)}^2 t$$

for all $t \ge 0$. Arguing by induction, one ends up with a sequence of time-dependent functions $(C_i)_{i \ge 1}$, depending only on the coagulation coefficients such that

$$\|f_i(t)\|_{L_\infty(\Omega)} \le C_i(t) \,, \qquad t \ge 0 \,, \ i \ge 1 \,. \tag{11.2.5}$$

This approach is developed in [41] and further extended to include fragmentation in [277]. It is worth mentioning that this extension is far from being obvious as the presence of fragmentation induces a stronger coupling in (11.2.1a). In particular, the equation for f_1 features a linear source term depending on $(f_i)_{i \ge 2}$ and then rather strong assumptions on the fragmentation coefficients are needed for this approach to work. Later it was realised that, still for the discrete C-F equation with diffusion, improved integrability with respect to the space variable is actually not needed for the existence of weak solutions for a large class of coagulation and fragmentation coefficients [168]. More precisely, assuming that

$$\lim_{j \to \infty} \frac{k_{i,j}}{j} = 0 \,, \qquad \lim_{j \to \infty} \frac{a_{i+j} b_{i,i+j}}{i + j} = 0 \,, \qquad D_i > 0 \,,$$

for all $i \geq 1$, the existence of at least one weak solution to the discrete C-F equation with diffusion is shown in [168, 279] for any initial condition $(f_i^{in})_{i\geq 1}$ with $f_i^{in} \geq 0$, $i \geq 1$. The proof makes use of the properties of the Laplace operator with respect to the weak and strong topologies of $L_1(\Omega)$.

A troublesome feature of (11.2.5) is that the L_∞-estimate on f_i depends on the size i, usually in an unbounded way, which has the following consequences. On the one hand, it is unlikely to provide a suitable control on the local total mass ϱ defined in (11.2.3) or on the infinite series in the right-hand side of (11.2.1a), which would be a first step towards the study of either the conservation of matter, or the uniqueness issue. On the other hand, the extension of (11.2.5) to the continuous setting $x \in (0, \infty)$ does not seem to be straightforward. Still, global existence of weak solutions to (11.2.1) is shown in [165, 203] under more stringent assumptions on the coagulation and fragmentation coefficients: either the detailed balance condition (10.3.2b) is satisfied, or the coagulation coefficients satisfy a monotonicity condition [54] and dominate, in a suitable way, the fragmentation coefficient. In both cases, these assumptions provide either an $L \log L$-estimate or an L_p-estimate, $p > 1$, and pave the way towards a global existence result. When the coagulation kernel k is bounded, a different approach is used in [6, 7] to obtain the global existence and the uniqueness of strong solutions in the following cases: there is no coagulation $k \equiv 0$, or the space dimension d is equal to one, or there is no fragmentation $a \equiv 0$ and the initial condition is sufficiently small, or the diffusion coefficient $D(x) = D_0 > 0$ is constant. The proof relies on a careful study of the properties of the Laplace operator in either $L_p(\Omega, X_{0,1})$ or $L_1((0, \infty), L_p(\Omega))$, $p > 1$, showing that it generates an analytic semigroup and handling the coagulation and fragmentation terms with the help of Duhamel's formula.

Coming back to the questions of conservation of matter and uniqueness, another situation, where an L_∞-estimate on the local total mass ϱ defined in (11.2.3) can be easily obtained, is the case of identical diffusion coefficients $D(x) = D_0 > 0$ for all $x \in (0, \infty)$. Indeed, in that case it is straightforward to deduce from (11.2.1) that ϱ is a subsolution to the linear heat equation with homogeneous Neumann boundary condition, so that the comparison principle ensures that $\|\varrho(t)\|_{L_\infty(\Omega)} \leq \|\varrho^{in}\|_{L_\infty(\Omega)}$ for $t \geq 0$. In fact, assuming that there are $x_0 \geq 0$ and $D_0 > 0$ such that

$$D(x) = D_0 , \qquad x \in (x_0, \infty) , \tag{11.2.6}$$

uniqueness and mass conservation for solutions to the discrete C-F equation with diffusion are shown in [81, 278]. It is actually possible to relax the assumption (11.2.6) and to replace it by

$$0 < D_{\min} := \inf_{x \in (0, \infty)} \{D(x)\} \leq D_{\max} := \sup_{x \in (0, \infty)} \{D(x)\} < \infty ,$$

$$\delta_D := \frac{D_{\max} - D_{\min}}{D_{\max} + D_{\min}} \text{ is suitably small.} \tag{11.2.7}$$

Thanks to (11.2.7), duality techniques [231] can be used to establish bounds in $L_p(\Omega)$ for some $p > 1$ on the local total mass ϱ or even on some other moments [48, 49, 59, 99]. Roughly speaking, this approach relies on the observation that (11.2.1) and the cancelling properties of the coagulation and fragmentation terms guarantee that ϱ solves

$$\partial_t \varrho - \Delta_z (V \varrho) = 0 , \qquad (t, z) \in (0, \infty) \times \Omega , \tag{11.2.8}$$

supplemented with homogeneous Neumann boundary conditions, where

$$V(t, z) := \frac{1}{\varrho(t, x)} \int_0^\infty x D(x) f(t, x, z) \, \mathrm{d}x \in [D_{\min}, D_{\max}] , \qquad (t, z) \in (0, \infty) \times \Omega .$$

The previous positive upper and lower bounds on V and the particular structure of the partial differential equation (11.2.8), along with parabolic regularity results, provide estimates on ϱ in $L_p((0,T) \times \Omega)$ for suitable values of $p \in (1,\infty)$, the admissible values of p being related to the smallness of the ratio δ_D in (11.2.7). Let us point out that $p = 2$ is always admissible since $\delta_D < 1$, see [48, 49, 59, 98, 99] for a more detailed account, and the use of such estimates to obtain uniqueness and mass conservation.

We mention that, on physical grounds, it is expected that $D(x) \to 0$ as $x \to \infty$ (in other words, the larger the particles, the slower they diffuse in space), hence the assumption (11.2.7) on the diffusion coefficients is not realistic from this point of view. However, it shows the importance of "anisotropic estimates" where the integrability requirements are not the same with respect to the size and space variables. Some estimates in this spirit are derived in [138, 225, 235, 236] when $\Omega = \mathbb{R}^d$, $d \geq 1$, and require assumptions connecting the growth of the coagulation kernel k and the decay of the diffusion coefficient D. As already mentioned, such estimates are of utmost importance when investigating uniqueness and mass conservation.

Other contributions to the C-F equation with diffusion include the analysis of the long-term behaviour in some particular cases: detailed balance condition (10.3.2) [70, 82, 98, 165], Becker–Döring model [174, 175], and fragmentation [176]. Furthermore, the shattering transition is studied in [21] for fragmentation processes with diffusion. We also refer to [71, 72, 110] for the fast reaction limit of the C-F equation with diffusion, when the coagulation and fragmentation coefficients satisfy the detailed balance condition (10.3.2).

Finally, besides the issues discussed previously, we recall that the starting point of the original derivation of Smoluchowski's coagulation equation is a system of particles moving in space according to Brownian motions and merging upon collisions [247, 248]. Since the pioneering work [154], the connection between such a system of particles and Smoluchowski's coagulation equation with diffusion has been studied in [136, 137, 226, 280]. Another stochastic approximation to the discrete C-F equation with diffusion is designed in [131].

Bibliography

[1] M. Abramowitz and I. A. Stegun. *Handbook of mathematical functions with formulas, graphs, and mathematical tables*, volume 55 of *National Bureau of Standards Applied Mathematics Series*. For sale by the Superintendent of Documents, U.S. Government Printing Office, Washington, D.C., 1964.

[2] M. Aizenman and T. A. Bak. Convergence to equilibrium in a system of reacting polymers. *Comm. Math. Phys.*, 65(3):203–230, 1979.

[3] D. J. Aldous. Deterministic and stochastic models for coalescence (aggregation and coagulation): A review of the mean-field theory for probabilists. *Bernoulli*, 5(1):3–48, 1999.

[4] H. Amann. *Ordinary differential equations*, volume 13 of *de Gruyter Studies in Mathematics*. Walter de Gruyter & Co., Berlin, 1990. An introduction to nonlinear analysis, Translated from the German by Gerhard Metzen.

[5] H. Amann. *Linear and quasilinear parabolic problems. Vol. I*, volume 89 of *Monographs in Mathematics*. Birkhäuser Boston, Inc., Boston, MA, 1995. Abstract linear theory.

[6] H. Amann. Coagulation-fragmentation processes. *Arch. Ration. Mech. Anal.*, 151(4):339–366, 2000.

[7] H. Amann and C. Walker. Local and global strong solutions to continuous coagulation-fragmentation equations with diffusion. *J. Differential Equations*, 218(1):159–186, 2005.

[8] W. Arendt and A. Rhandi. Perturbation of positive semigroups. *Arch. Math. (Basel)*, 56(2):107–119, 1991.

[9] O. Arino. Some spectral properties for the asymptotic behaviour of semigroups connected to population dynamics. *SIAM Rev.*, 34(3):445–476, 1992.

[10] L. Arkeryd. Loeb solutions of the Boltzmann equation. *Arch. Rational Mech. Anal.*, 86(1):85–97, 1984.

[11] L. Arlotti and J. Banasiak. Strictly substochastic semigroups with application to conservative and shattering solutions to fragmentation equations with mass loss. *J. Math. Anal. Appl.*, 293(2):693–720, 2004.

[12] D. Balagué, J. A. Cañizo, and P. Gabriel. Fine asymptotics of profiles and relaxation to equilibrium for growth-fragmentation equations with variable drift rates. *Kinet. Relat. Models*, 6(2):219–243, 2013.

[13] J. M. Ball and J. Carr. Asymptotic behaviour of solutions to the Becker-Döring equations for arbitrary initial data. *Proc. Roy. Soc. Edinburgh Sect. A*, 108(1-2):109–116, 1988.

[14] J. M. Ball and J. Carr. The discrete coagulation-fragmentation equations: Existence, uniqueness, and density conservation. *J. Statist. Phys.*, 61(1-2):203–234, 1990.

[15] J. M. Ball, J. Carr, and O. Penrose. The Becker-Döring cluster equations: Basic properties and asymptotic behaviour of solutions. *Comm. Math. Phys.*, 104(4):657–692, 1986.

[16] J. Banasiak. On an extension of the Kato-Voigt perturbation theorem for substochastic semigroups and its application. *Taiwanese J. Math.*, 5(1):169–191, 2001. International Conference on Mathematical Analysis and its Applications (Kaohsiung, 2000).

[17] J. Banasiak. On a non-uniqueness in fragmentation models. *Math. Methods Appl. Sci.*, 25(7):541–556, 2002.

[18] J. Banasiak. Conservative and shattering solutions for some classes of fragmentation models. *Math. Models Methods Appl. Sci.*, 14(4):483–501, 2004.

[19] J. Banasiak. On conservativity and shattering for an equation of phytoplankton dynamics. *Comptes Rendus Biologies*, 327(11):1025–1036, 2004.

[20] J. Banasiak. Shattering and non-uniqueness in fragmentation models—an analytic approach. *Phys. D*, 222(1-2):63–72, 2006.

[21] J. Banasiak. Kinetic-type models with diffusion: Conservative and nonconservative solutions. *Transport Theory Statist. Phys.*, 36(1-3):43–65, 2007.

[22] J. Banasiak. Blow-up of solutions to some coagulation and fragmentation equations with growth. *Discrete Contin. Dyn. Syst.*, Suppl. Vol. I:126–134, 2011. Dynamical systems, differential equations and applications. 8th AIMS Conference.

[23] J. Banasiak. On an irregular dynamics of certain fragmentation semigroups. *Rev. R. Acad. Cienc. Exactas Fís. Nat. Ser. A Math. RACSAM*, 105(2):361–377, 2011.

[24] J. Banasiak. Global classical solutions of coagulation-fragmentation equations with unbounded coagulation rates. *Nonlinear Anal. Real World Appl.*, 13(1):91–105, 2012.

[25] J. Banasiak. Transport processes with coagulation and strong fragmentation. *Discrete Contin. Dyn. Syst. Ser. B*, 17(2):445–472, 2012.

[26] J. Banasiak and L. Arlotti. *Perturbations of positive semigroups with applications.* Springer Monographs in Mathematics. Springer-Verlag London, Ltd., London, 2006.

[27] J. Banasiak, L. O. Joel, and S. Shindin. Analysis and simulations of the discrete fragmentation equation with decay. *Math. Methods Appl. Sci.*, 41(16):6530–6545, 2018.

[28] J. Banasiak, L. O. Joel, and S. Shindin. The discrete unbounded coagulation-fragmentation equation with growth, decay and sedimentation. *Kinet. Relat. Models*, 12(5), 2019. doi:10.3934/krm.2019040, arXiv:1809.00046.

[29] J. Banasiak, L. O. Joel, and S. Shindin. Long term dynamics of the discrete growth-decay-fragmentation equation. *J. Evol. Equ.*, 2019. accepted, doi.org/10.1007/s00028-019-00499-4, arXiv:1801.06486.

[30] J. Banasiak and M. Lachowicz. Around the Kato generation theorem for semigroups. *Studia Math.*, 179(3):217–238, 2007.

[31] J. Banasiak and W. Lamb. On the application of substochastic semigroup theory to fragmentation models with mass loss. *J. Math. Anal. Appl.*, 284(1):9–30, 2003.

[32] J. Banasiak and W. Lamb. On a coagulation and fragmentation equation with mass loss. *Proc. Roy. Soc. Edinburgh Sect. A*, 136(6):1157–1173, 2006.

[33] J. Banasiak and W. Lamb. Coagulation, fragmentation and growth processes in a size structured population. *Discrete Contin. Dyn. Syst. Ser. B*, 11(3):563–585, 2009.

[34] J. Banasiak and W. Lamb. Global strict solutions to continuous coagulation-fragmentation equations with strong fragmentation. *Proc. Roy. Soc. Edinburgh Sect. A*, 141(3):465–480, 2011.

[35] J. Banasiak and W. Lamb. Analytic fragmentation semigroups and continuous coagulation-fragmentation equations with unbounded rates. *J. Math. Anal. Appl.*, 391(1):312–322, 2012.

[36] J. Banasiak and W. Lamb. The discrete fragmentation equation: Semigroups, compactness and asynchronous exponential growth. *Kinet. Relat. Models*, 5(2):223–236, 2012.

[37] J. Banasiak, W. Lamb, and M. Langer. Strong fragmentation and coagulation with power-law rates. *J. Engrg. Math.*, 82:199–215, 2013.

[38] J. Banasiak and S. C. O. Noutchie. Controlling number of particles in fragmentation equations. *Phys. D*, 239(15):1422–1435, 2010.

[39] P. K. Barik, A. K. Giri, and Ph. Laurençot. Mass-conserving solutions to the Smoluchowski coagulation equation with singular kernel. Proc. Roy. Soc. Edinburgh Sect. A, to appear (arXiv: 1804.00853).

[40] A. Bátkai, M. K. Fijavž, and A. Rhandi. *Positive operator semigroups: From finite to infinite dimensions*, volume 257. Birkhäuser, 2017.

[41] Ph. Bénilan and D. Wrzosek. On an infinite system of reaction-diffusion equations. *Adv. Math. Sci. Appl.*, 7(1):351–366, 1997.

[42] J. Bertoin. Eternal solutions to Smoluchowski's coagulation equation with additive kernel and their probabilistic interpretations. *Ann. Appl. Probab.*, 12(2):547–564, 2002.

[43] W. Biedrzycka and M. Tyran-Kamińska. Self-similar solutions of fragmentation equations revisited. *Discrete Contin. Dyn. Syst. Ser. B*, 23(1):13–27, 2018.

[44] P. Billingsley. *Convergence of probability measures*. Wiley Series in Probability and Statistics: Probability and Statistics. John Wiley & Sons, Inc., New York, second edition, 1999. A Wiley-Interscience Publication.

[45] M. Bonacini, B. Niethammer, and J. J. L. Velázquez. Self-similar solutions to coagulation equations with time-dependent tails: The case of homogeneity smaller than one. *Comm. Partial Differential Equations*, 43:82–117, 2018.

[46] M. Bonacini, B. Niethammer, and J. J. L. Velázquez. Self-similar solutions to coagulation equations with time-dependent tails: The case of homogeneity one. *Arch. Ration. Mech. Anal.*, 233:1–43, 2019.

[47] F. Bouguet. A probabilistic look at conservative growth-fragmentation equations. In *Séminaire de Probabilités XLIX*, volume 2215 of *Lecture Notes in Math.*, pages 57–74. Springer, Cham, 2018.

[48] M. Breden. Applications of improved duality lemmas to the discrete coagulation-fragmentation equations with diffusion. *Kinet. Relat. Models*, 11(2):279–301, 2018.

[49] M. Breden, L. Desvillettes, and K. Fellner. Smoothness of moments of the solutions of discrete coagulation equations with diffusion. *Monatsh. Math.*, 183(3):437–463, 2017.

[50] G. Breschi and M. A. Fontelos. Self-similar solutions of the second kind representing gelation in finite time for the Smoluchowski equation. *Nonlinearity*, 27(7):1709–1745, 2014.

[51] H. Brezis. *Functional analysis, Sobolev spaces and partial differential equations.* Universitext. Springer, New York, 2011.

[52] E. Buffet and J. V. Pulé. Gelation: The diagonal case revisited. *Nonlinearity*, 2(2):373–381, 1989.

[53] E. Buffet and R. F. Werner. A counterexample in coagulation theory. *J. Math. Phys.*, 32(8):2276–2278, 1991.

[54] A. V. Burobin. Existence and uniqueness of the solution of the Cauchy problem for a spatially nonhomogeneous coagulation equation. *Differentsial'nye Uravneniya*, 19(9):1568–1579, 1983.

[55] J. Burton. Nucleation theory. In *Statistical Mechanics*, pages 195–234. Springer, 1977.

[56] M. J. Cáceres, J. A. Cañizo, and S. Mischler. Rate of convergence to self-similarity for the fragmentation equation in L^1 spaces. *Commun. Appl. Ind. Math.*, 1(2):299–308, 2010.

[57] M. J. Cáceres, J. A. Cañizo, and S. Mischler. Rate of convergence to an asymptotic profile for the self-similar fragmentation and growth-fragmentation equations. *J. Math. Pures Appl. (9)*, 96(4):334–362, 2011.

[58] J. A. Cañizo. Convergence to equilibrium for the discrete coagulation-fragmentation equations with detailed balance. *J. Statist. Phys.*, 129(1):1–26, 2007.

[59] J. A. Cañizo, L. Desvillettes, and K. Fellner. Regularity and mass conservation for discrete coagulation-fragmentation equations with diffusion. *Ann. Inst. H. Poincaré Anal. Non Linéaire*, 27(2):639–654, 2010.

[60] J. A. Cañizo, A. Einav, and B. Lods. Trend to equilibrium for the Becker-Döring equations: An analogue of Cercignani's conjecture. *Anal. PDE*, 10(7):1663–1708, 2017.

[61] J. A. Cañizo, A. Einav, and B. Lods. Uniform moment propagation for the Becker-Döring equation. Proc. Roy. Soc. Edinburgh Sect. A, to appear (arXiv:1706.03524), 2017.

[62] J. A. Cañizo and B. Lods. Exponential convergence to equilibrium for subcritical solutions of the Becker-Döring equations. *J. Differential Equations*, 255(5):905–950, 2013.

[63] J. A. Cañizo and S. Mischler. Regularity, local behavior and partial uniqueness for self-similar profiles of Smoluchowski's coagulation equation. *Rev. Mat. Iberoam.*, 27(3):803–839, 2011.

[64] J. A. Cañizo, S. Mischler, and C. Mouhot. Rate of convergence to self-similarity for Smoluchowski's coagulation equation with constant coefficients. *SIAM J. Math. Anal.*, 41(6):2283–2314, 2009/10.

[65] J. Carr. Asymptotic behaviour of solutions to the coagulation-fragmentation equations. I. The strong fragmentation case. *Proc. Roy. Soc. Edinburgh Sect. A*, 121(3-4):231–244, 1992.

[66] J. Carr and F. P. da Costa. Instantaneous gelation in coagulation dynamics. *Z. Angew. Math. Phys.*, 43(6):974–983, 1992.

[67] J. Carr and F. P. da Costa. Asymptotic behavior of solutions to the coagulation-fragmentation equations. II. Weak fragmentation. *J. Statist. Phys.*, 77(1-2):89–123, 1994.

[68] J. Carr, D. B. Duncan, and C. H. Walshaw. Numerical approximation of a metastable system. *IMA J. Numer. Anal.*, 15(4):505–521, 1995.

[69] J. Carr and R. M. Dunwell. Asymptotic behaviour of solutions to the Becker-Döring equations. *Proc. Edinburgh Math. Soc. (2)*, 42(2):415–424, 1999.

[70] J. A. Carrillo, L. Desvillettes, and K. Fellner. Exponential decay towards equilibrium for the inhomogeneous Aizenman-Bak model. *Comm. Math. Phys.*, 278(2):433–451, 2008.

[71] J. A. Carrillo, L. Desvillettes, and K. Fellner. Fast-reaction limit for the inhomogeneous Aizenman-Bak model. *Kinet. Relat. Models*, 1(1):127–137, 2008.

[72] J. A. Carrillo, L. Desvillettes, and K. Fellner. Rigorous derivation of a nonlinear diffusion equation as fast-reaction limit of a continuous coagulation-fragmentation model with diffusion. *Comm. Partial Differential Equations*, 34(10-12):1338–1351, 2009.

[73] T. Cazenave and A. Haraux. *An introduction to semilinear evolution equations*, volume 13 of *Oxford Lecture Series in Mathematics and its Applications*. The Clarendon Press, Oxford University Press, New York, 1998. Translated from the 1990 French original by Yvan Martel and revised by the authors.

[74] E. Cepeda. Well-posedness for a coagulation multiple-fragmentation equation. *Differential Integral Equations*, 27(1-2):105–136, 2014.

[75] C. Cercignani. *The Boltzmann equation and its applications*, volume 67 of *Applied Mathematical Sciences*. Springer-Verlag, New York, 1988.

[76] C. Cercignani, R. Illner, and M. Pulvirenti. *The mathematical theory of dilute gases*, volume 106 of *Applied Mathematical Sciences*. Springer-Verlag, New York, 1994.

[77] Z. Cheng and S. Redner. Kinetics of fragmentation. *J. Phys. A*, 23(7):1233–1258, 1990.

[78] J. M. C. Clark and V. Katsouros. Stably coalescent stochastic froths. *Adv. in Appl. Probab.*, 31(1):199–219, 1999.

[79] P. Clément, H. J. A. M. Heijmans, S. Angenent, C. J. van Duijn, and B. de Pagter. *One-parameter semigroups*, volume 5 of *CWI Monographs*. North-Holland Publishing Co., Amsterdam, 1987.

[80] J.-F. Collet, T. Goudon, F. Poupaud, and A. Vasseur. The Beker–Döring system and its Lifshitz-Slyozov limit. *SIAM J. Appl. Math.*, 62(5):1488–1500, 2002.

[81] J. F. Collet and F. Poupaud. Existence of solutions to coagulation-fragmentation systems with diffusion. *Transport Theory Statist. Phys.*, 25(3-5):503–513, 1996.

[82] J. F. Collet and F. Poupaud. Asymptotic behaviour of solutions to the diffusive fragmentation-coagulation system. *Phys. D*, 114(1-2):123–146, 1998.

[83] J. G. Conlon. On a diffusive version of the Lifschitz-Slyozov-Wagner equation. *J. Nonlinear Sci.*, 20(4):463–521, 2010.

[84] I. Csiszár. Information-type measures of difference of probability distributions and indirect observations. *Studia Sci. Math. Hungar.*, 2:299–318, 1967.

[85] S. Cueille and C. Sire. Nontrivial polydispersity exponents in aggregation models. *Phys. Rev. E*, 55(5):5465–5478, 1997.

[86] C. Cueto Camejo, R. Gröpler, and G. Warnecke. Regular solutions to the coagulation equations with singular kernels. *Math. Methods Appl. Sci.*, 38(11):2171–2184, 2015.

[87] C. Cueto Camejo and G. Warnecke. The singular kernel coagulation equation with multifragmentation. *Math. Methods Appl. Sci.*, 38(14):2953–2973, 2015.

[88] F. P. da Costa. Existence and uniqueness of density conserving solutions to the coagulation-fragmentation equations with strong fragmentation. *J. Math. Anal. Appl.*, 192(3):892–914, 1995.

[89] F. P. da Costa. On the positivity of solutions to the Smoluchowski equations. *Mathematika*, 42(2):406–412, 1995.

[90] F. P. Da Costa. On the dynamic scaling behaviour of solutions to the discrete Smoluchowski equations. *Proc. Edinburgh Math. Soc. (2)*, 39(3):547–559, 1996.

[91] F. P. da Costa. Asymptotic behaviour of low density solutions to the generalized Becker-Döring equations. *NoDEA Nonlinear Differential Equations Appl.*, 5(1):23–37, 1998.

[92] F. P. da Costa. Convergence to equilibria of solutions to the coagulation-fragmentation equations. In *Nonlinear evolution equations and their applications (Macau, 1998)*, pages 45–56. World Sci. Publ., River Edge, NJ, 1999.

[93] C. De La Vallée Poussin. Sur l'intégrale de Lebesgue. *Trans. Amer. Math. Soc.*, 16(4):435–501, 1915.

[94] M. Deaconu and E. Tanré. Smoluchowski's coagulation equation: Probabilistic interpretation of solutions for constant, additive and multiplicative kernels. *Ann. Scuola Norm. Sup. Pisa Cl. Sci. (4)*, 29(3):549–579, 2000.

[95] P. Degond, J.-G. Liu, and R. L. Pego. Coagulation–fragmentation model for animal group-size statistics. *J. Nonlinear Sci.*, 27(2):379–424, 2017.

[96] C. Dellacherie and P.-A. Meyer. *Probabilités et potentiel.* Hermann, Paris, 1975. Chapitres I à IV, Édition entièrement refondue, Publications de l'Institut de Mathématique de l'Université de Strasbourg, No. XV, Actualités Scientifiques et Industrielles, No. 1372.

[97] J. Deschamps, E. Hingant, and R. Yvinec. Quasi steady state approximation of the small clusters in Becker-Döring equations leads to boundary conditions in the Lifshitz-Slyozov limit. *Comm. Math. Sci.*, 15(5):1353–1384, 2017.

[98] L. Desvillettes and K. Fellner. Large time asymptotics for a continuous coagulation-fragmentation model with degenerate size-dependent diffusion. *SIAM J. Math. Anal.*, 41(6):2315–2334, 2009/10.

[99] L. Desvillettes and K. Fellner. Duality and entropy methods in coagulation-fragmentation models. *Riv. Math. Univ. Parma (N.S.)*, 4(2):215–263, 2013.

[100] R. J. DiPerna and P.-L. Lions. On the Cauchy problem for Boltzmann equations: Global existence and weak stability. *Ann. of Math. (2)*, 130(2):321–366, 1989.

[101] M. Doumic and M. Escobedo. Time asymptotics for a critical case in fragmentation and growth-fragmentation equations. *Kinet. Relat. Models*, 9(2):251–297, 2016.

[102] R. L. Drake. A general mathematical survey of the coagulation equation. In G. Hidy and J. Brock, editors, *Topics in Current Aerosol Research*, International Reviews in Aerosol Physics and Chemistry, pages 201–376. Pergamon, 1972.

[103] P. B. Dubovskii. *Mathematical theory of coagulation*, volume 23 of *Lecture Notes Series*. Seoul National University, Research Institute of Mathematics, Global Analysis Research Center, Seoul, 1994.

[104] P. B. Dubovskiĭ and I. W. Stewart. Existence, uniqueness and mass conservation for the coagulation-fragmentation equation. *Math. Methods Appl. Sci.*, 19(7):571–591, 1996.

[105] P. B. Dubovskiĭ and I. W. Stewart. Trend to equilibrium for the coagulation-fragmentation equation. *Math. Methods Appl. Sci.*, 19(10):761–772, 1996.

[106] K.-J. Engel and R. Nagel. *One-parameter semigroups for linear evolution equations*, volume 194 of *Graduate Texts in Mathematics*. Springer-Verlag, New York, 2000.

[107] K.-J. Engel and R. Nagel. *A short course on operator semigroups.* Universitext. Springer, New York, 2006.

[108] M. H. Ernst, R. M. Ziff, and E. M. Hendriks. Coagulation processes with a phase transition. *J. Colloid Interface Sci.*, 97(1):266–277, 1984.

[109] M. Escobedo, P. Laurençot, S. Mischler, and B. Perthame. Gelation and mass conservation in coagulation-fragmentation models. *J. Differential Equations*, 195(1):143–174, 2003.

[110] M. Escobedo, Ph. Laurençot, and S. Mischler. Fast reaction limit of the discrete diffusive coagulation-fragmentation equation. *Comm. Partial Differential Equations*, 28(5-6):1113–1133, 2003.

[111] M. Escobedo and S. Mischler. Dust and self-similarity for the Smoluchowski coagulation equation. *Ann. Inst. H. Poincaré Anal. Non Linéaire*, 23(3):331–362, 2006.

[112] M. Escobedo, S. Mischler, and B. Perthame. Gelation in coagulation and fragmentation models. *Comm. Math. Phys.*, 231(1):157–188, 2002.

[113] M. Escobedo, S. Mischler, and M. Rodriguez Ricard. On self-similarity and stationary problem for fragmentation and coagulation models. *Ann. Inst. H. Poincaré Anal. Non Linéaire*, 22(1):99–125, 2005.

[114] W. Feller. *An introduction to probability theory and its applications. Vol. II*. Second edition. John Wiley & Sons, Inc., New York-London-Sydney, 1971.

[115] F. Filbet and Ph. Laurençot. Numerical simulation of the Smoluchowski coagulation equation. *SIAM J. Sci. Comput.*, 25(6):2004–2028 (electronic), 2004.

[116] A. F. Filippov. On the distribution of the sizes of particles which undergo splitting. *Theory Probab. Appl.*, 6:275–294, 1961.

[117] I. Fonseca and G. Leoni. *Modern methods in the calculus of variations: L^p spaces*. Springer Monographs in Mathematics. Springer, New York, 2007.

[118] N. Fournier and Ph. Laurençot. Existence of self-similar solutions to Smoluchowski's coagulation equation. *Comm. Math. Phys.*, 256(3):589–609, 2005.

[119] N. Fournier and Ph. Laurençot. Local properties of self-similar solutions to Smoluchowski's coagulation equation with sum kernels. *Proc. Roy. Soc. Edinburgh Sect. A*, 136(3):485–508, 2006.

[120] N. Fournier and Ph. Laurençot. Well-posedness of Smoluchowski's coagulation equation for a class of homogeneous kernels. *J. Funct. Anal.*, 233(2):351–379, 2006.

[121] S. Friedlander and C. Wang. The self-preserving particle size distribution for coagulation by Brownian motion. *J. Colloid Interface Sci.*, 22:126–132, 1966.

[122] P. Gabriel and F. Salvarani. Exponential relaxation to self-similarity for the superquadratic fragmentation equation. *Appl. Math. Lett.*, 27:74–78, 2014.

[123] V. A. Galkin. The existence and uniqueness of the solution of the coagulation equation. *Differencial'nye Uravnenija*, 13(8):1460–1470, 1977.

[124] V. A. Galkin and P. B. Dubovskiĭ. Solutions of a coagulation equation with unbounded kernels. *Differentsial'nye Uravneniya*, 22(3):504–509, 551, 1986.

[125] I. M. Gamba, V. Panferov, and C. Villani. On the Boltzmann equation for diffusively excited granular media. *Comm. Math. Phys.*, 246(3):503–541, 2004.

[126] A. K. Giri. On the uniqueness for coagulation and multiple fragmentation equation. *Kinet. Relat. Models*, 6(3):589–599, 2013.

[127] A. K. Giri, J. Kumar, and G. Warnecke. The continuous coagulation equation with multiple fragmentation. *J. Math. Anal. Appl.*, 374(1):71–87, 2011.

[128] A. K. Giri, Ph. Laurençot, and G. Warnecke. Weak solutions to the continuous coagulation equation with multiple fragmentation. *Nonlinear Anal.*, 75(4):2199–2208, 2012.

[129] A. K. Giri and G. Warnecke. Uniqueness for the coagulation-fragmentation equation with strong fragmentation. *Z. Angew. Math. Phys.*, 62(6):1047–1063, 2011.

[130] A. Golovin. The solution of the coagulation equation for cloud droplets in a rising air current. *Izv. Geophys. Ser.*, 5:482–487, 1963.

[131] F. Guiaş. Convergence properties of a stochastic model for coagulation-fragmentation processes with diffusion. *Stochastic Anal. Appl.*, 19(2):245–278, 2001.

[132] B. Haas. Loss of mass in deterministic and random fragmentations. *Stochastic Process. Appl.*, 106(2):245–277, 2003.

[133] B. Haas. Appearance of dust in fragmentations. *Commun. Math. Sci.*, 2(suppl. 1):65–73, 2004.

[134] B. Haas. Regularity of formation of dust in self-similar fragmentations. *Ann. Inst. H. Poincaré Probab. Statist.*, 40(4):411–438, 2004.

[135] B. Haas. Asymptotic behavior of solutions of the fragmentation equation with shattering: An approach via self-similar Markov processes. *Ann. Appl. Probab.*, 20(2):382–429, 2010.

[136] A. Hammond. Coagulation and diffusion: A probabilistic perspective on the Smoluchowski PDE. *Probab. Surv.*, 14:205–288, 2017.

[137] A. Hammond and F. Rezakhanlou. The kinetic limit of a system of coagulating Brownian particles. *Arch. Ration. Mech. Anal.*, 185(1):1–67, 2007.

[138] A. Hammond and F. Rezakhanlou. Moment bounds for the Smoluchowski equation and their consequences. *Comm. Math. Phys.*, 276(3):645–670, 2007.

[139] O. J. Heilmann. Analytical solutions of Smoluchowski's coagulation equation. *J. Phys. A*, 25(13):3763–3771, 1992.

[140] D. Henry. *Geometric theory of semilinear parabolic equations*, volume 840. Springer, 2006.

[141] M. Herrmann, B. Niethammer, and J. J. L. Velázquez. Instabilities and oscillations in coagulation equations with kernels of homogeneity one. *Quart. Appl. Math.*, 75(1):105–130, 2017.

[142] E. Hingant and R. Yvinec. Deterministic and stochastic Becker-Döring equations: Past and recent mathematical developments. In *Stochastic processes, multiscale modeling, and numerical methods for computational cellular biology*, pages 175–204. Springer, Cham, 2017.

[143] P.-E. Jabin and B. Niethammer. On the rate of convergence to equilibrium in the Becker-Döring equations. *J. Differential Equations*, 191(2):518–543, 2003.

[144] I. Jeon. Existence of gelling solutions for coagulation-fragmentation equations. *Comm. Math. Phys.*, 194(3):541–567, 1998.

[145] I. Jeon. Stochastic fragmentation and some sufficient conditions for shattering transition. *J. Korean Math. Soc.*, 39(4):543–558, 2002.

[146] P. Kapur. Self-preserving size spectra of comminuted particles. *Chem. Eng. Sci.*, 27(2):425–431, 1972.

[147] J. Koch, W. Hackbusch, and K. Sundmacher. H-matrix methods for linear and quasilinear integral operators appearing in population balances. *Comput. Chem. Engng.*, 31(7):745–759, 2007.

[148] N. J. Kokholm. On Smoluchowski's coagulation equation. *J. Phys. A*, 21(3):839–842, 1988.

[149] V. Komornik. *Lectures on functional analysis and the Lebesgue integral.* Universitext. Springer-Verlag, London, 2016. Translated from the 2002 French original.

[150] M. A. Krasnosel'skiĭ and J. B. Rutickiĭ. *Convex functions and Orlicz spaces.* Translated from the first Russian edition by Leo F. Boron. P. Noordhoff Ltd., Groningen, 1961.

[151] M. Kreer and O. Penrose. Proof of dynamical scaling in Smoluchowski's coagulation equation with constant kernel. *J. Statist. Phys.*, 75(3-4):389–407, 1994.

[152] D. S. Krivitsky. Numerical solution of the Smoluchowski kinetic equation and asymptotics of the distribution function. *J. Phys. A*, 28(7):2025–2039, 1995.

[153] W. Lamb and I. W. Stewart. Stability for a class of equilibrium solutions to the coagulation-fragmentation equation. In *Numerical Analysis and Applied Mathematics, International Conference 2008*, pages 942–945. American Institute of Physics, 2008.

[154] R. Lang and X.-X. Nguyen. Smoluchowski's theory of coagulation in colloids holds rigorously in the Boltzmann-Grad-limit. *Z. Wahrsch. Verw. Gebiete*, 54(3):227–280, 1980.

[155] Ph. Laurençot. Mass-conserving solutions to coagulation-fragmentation equations with balanced growth. Colloq. Math., to appear (arXiv: 1901.08313).

[156] Ph. Laurençot. Stationary solutions to coagulation-fragmentation equations. Ann. Inst. H. Poincaré Anal. Non Linéaire, to appear (arXiv 1904.01868).

[157] Ph. Laurençot. Singular behaviour of finite approximations to the addition model. *Nonlinearity*, 12(2):229–239, 1999.

[158] Ph. Laurençot. On a class of continuous coagulation-fragmentation equations. *J. Differential Equations*, 167(2):245–274, 2000.

[159] Ph. Laurençot. The Lifshitz-Slyozov equation with encounters. *Math. Models Methods Appl. Sci.*, 11(4):731–748, 2001.

[160] Ph. Laurençot. The discrete coagulation equations with multiple fragmentation. *Proc. Edinb. Math. Soc. (2)*, 45(1):67–82, 2002.

[161] Ph. Laurençot. Weak compactness techniques and coagulation equations. In *Evolutionary equations with applications in natural sciences*, volume 2126 of *Lecture Notes in Math.*, pages 199–253. Springer, Cham, 2015.

[162] Ph. Laurençot. Mass-conserving solutions to coagulation-fragmentation equations with non-integrable fragment distribution function. *Quart. Appl. Math.*, 76(4):767–785, 2018.

[163] Ph. Laurençot. Uniqueness of mass-conserving self-similar solutions to Smoluchowski's coagulation equation with inverse power law kernels. *J. Statist. Phys.*, 171(3):484–492, 2018.

[164] Ph. Laurençot. Mass-conserving self-similar solutions to coagulation-fragmentation equations. *Comm. Partial Differential Equations*, 44(9):773–800, 2019.

[165] Ph. Laurençot and S. Mischler. The continuous coagulation-fragmentation equations with diffusion. *Arch. Ration. Mech. Anal.*, 162(1):45–99, 2002.

[166] Ph. Laurençot and S. Mischler. From the Becker-Döring to the Lifshitz-Slyozov-Wagner equations. *J. Statist. Phys.*, 106(5-6):957–991, 2002.

[167] Ph. Laurençot and S. Mischler. From the discrete to the continuous coagulation-fragmentation equations. *Proc. Roy. Soc. Edinburgh Sect. A*, 132(5):1219–1248, 2002.

[168] Ph. Laurençot and S. Mischler. Global existence for the discrete diffusive coagulation-fragmentation equations in L^1. *Rev. Mat. Iberoamericana*, 18(3):731–745, 2002.

[169] Ph. Laurençot and S. Mischler. Convergence to equilibrium for the continuous coagulation-fragmentation equation. *Bull. Sci. Math.*, 127(3):179–190, 2003.

[170] Ph. Laurençot and S. Mischler. On coalescence equations and related models. In *Modeling and computational methods for kinetic equations*, Model. Simul. Sci. Eng. Technol., pages 321–356. Birkhäuser Boston, Boston, MA, 2004.

[171] Ph. Laurençot and S. Mischler. Liapunov functionals for Smoluchowski's coagulation equation and convergence to self-similarity. *Monatsh. Math.*, 146(2):127–142, 2005.

[172] Ph. Laurençot, B. Niethammer, and J. J. Velázquez. Oscillatory dynamics in Smoluchowski's coagulation equation with diagonal kernel. *Kinet. Relat. Models*, 11(4):933–952, 2018.

[173] Ph. Laurençot and H. van Roessel. Absence of gelation and self-similar behavior for a coagulation-fragmentation equation. *SIAM J. Math. Anal.*, 47(3):2355–2374, 2015.

[174] Ph. Laurençot and D. Wrzosek. The Becker–Döring model with diffusion. I. Basic properties of solutions. *Colloq. Math.*, 75(2):245–269, 1998.

[175] Ph. Laurençot and D. Wrzosek. The Becker–Döring model with diffusion. II. Long time behavior. *J. Differential Equations*, 148(2):268–291, 1998.

[176] Ph. Laurençot and D. Wrzosek. Fragmentation-diffusion model. Existence of solutions and their asymptotic behaviour. *Proc. Roy. Soc. Edinburgh Sect. A*, 128(4):759–774, 1998.

[177] C. H. Lê. *Etude de la classe des opérateur m-accrétifs de $L^1(\Omega)$ et accrétifs dans $L^\infty(\Omega)$*. PhD thesis, Université de Paris VI, 1977. Thèse de 3ème cycle.

[178] M. H. Lee. A survey of numerical solutions to the coagulation equation. *J. Phys. A*, 34:10219–10241, 2001.

[179] F. Leyvraz. Existence and properties of post-gel solutions for the kinetic equations of coagulation. *J. Phys. A*, 16(12):2861–2873, 1983.

[180] F. Leyvraz. Scaling theory and exactly solved models in the kinetics of irreversible aggregation. *Phys. Rep.*, 383:95–212, 2003.

[181] F. Leyvraz. Rigorous results in the scaling theory of irreversible aggregation kinetics. *J. Nonlinear Math. Phys.*, 12(suppl. 1):449–465, 2005.

[182] F. Leyvraz. Scaling theory for gelling systems: Work in progress. *Phys. D*, 222(1-2):21–28, 2006.

[183] F. Leyvraz and H. R. Tschudi. Singularities in the kinetics of coagulation processes. *J. Phys. A*, 14(12):3389–3405, 1981.

[184] I. Lifshitz and V. Slyozov. The kinetics of precipitation from supersaturated solid solutions. *J. Phys. Chem. Solids*, 19(1):35–50, 1961.

[185] P.-L. Lions. Compactness in Boltzmann's equation via Fourier integral operators and applications. I. *J. Math. Kyoto Univ.*, 34(2):391–427, 1994.

[186] X. Lu. Conservation of energy, entropy identity, and local stability for the spatially homogeneous Boltzmann equation. *J. Statist. Phys.*, 96(3-4):765–796, 1999.

[187] A. Lunardi. *Analytic semigroups and optimal regularity in parabolic problems*, volume 16 of *Progress in Nonlinear Differential Equations and Their Applications*. Birkhäuser Verlag, Basel, 1995.

[188] E. D. McGrady and R. M. Ziff. "Shattering" transition in fragmentation. *Phys. Rev. Lett.*, 58(9):892–895, 1987.

[189] J. B. McLeod. On a recurrence formula in differential equations. *Quart. J. Math. Oxford Ser. (2)*, 13:283–284, 1962.

[190] J. B. McLeod. On an infinite set of non-linear differential equations. *Quart. J. Math. Oxford Ser. (2)*, 13:119–128, 1962.

[191] J. B. McLeod. On an infinite set of non-linear differential equations. II. *Quart. J. Math. Oxford Ser. (2)*, 13:193–205, 1962.

[192] J. B. McLeod. On the scalar transport equation. *Proc. London Math. Soc. (3)*, 14:445–458, 1964.

[193] J. B. McLeod, B. Niethammer, and J. J. L. Velázquez. Asymptotics of self-similar solutions to coagulation equations with product kernel. *J. Statist. Phys.*, 144(1):76–100, 2011.

[194] P. Meakin and M. H. Ernst. Scaling in aggregation with breakup simulations and mean-field theory. *Phys. Rev. Lett.*, 60(24):2503–2506, 1988.

[195] B. Meerson. Fluctuations provide strong selection in Ostwald ripening. *Phys. Rev. E*, 60:3072–3075, Sep 1999.

[196] Z. Melzak. The effect of coalescence in certain collision processes. *Quart. Appl. Math.*, XI(2):231–234, 1953.

[197] Z. A. Melzak. The positivity sets of the solutions of a transport equation. *Michigan Math. J*, 6:331–334, 1959.

[198] G. Menon and R. L. Pego. Approach to self-similarity in Smoluchowski's coagulation equations. *Comm. Pure Appl. Math.*, 57(9):1197–1232, 2004.

[199] G. Menon and R. L. Pego. Dynamical scaling in Smoluchowski's coagulation equations: Uniform convergence. *SIAM J. Math. Anal.*, 36(5):1629–1651 (electronic), 2005.

[200] G. Menon and R. L. Pego. The scaling attractor and ultimate dynamics for Smoluchowski's coagulation equations. *J. Nonlinear Sci.*, 18(2):143–190, 2008.

[201] P. Meyer-Nieberg. *Banach lattices*. Universitext. Springer-Verlag, Berlin, 1991.

[202] Ph. Michel, S. Mischler, and B. Perthame. General relative entropy inequality: An illustration on growth models. *J. Math. Pures Appl. (9)*, 84(9):1235–1260, 2005.

[203] S. Mischler and M. Rodriguez Ricard. Existence globale pour l'équation de Smoluchowski continue non homogène et comportement asymptotique des solutions. *C. R. Math. Acad. Sci. Paris*, 336(5):407–412, 2003.

[204] S. Mischler and J. Scher. Spectral analysis of semigroups and growth-fragmentation equations. *Ann. Inst. H. Poincaré Anal. Non Linéaire*, 33(3):849–898, 2016.

[205] S. Mischler and B. Wennberg. On the spatially homogeneous Boltzmann equation. *Ann. Inst. H. Poincaré Anal. Non Linéaire*, 16(4):467–501, 1999.

[206] M. Mokhtar-Kharroubi and J. Voigt. On honesty of perturbed substochastic C_0-semigroups in L^1-spaces. *J. Operator Theory*, 64(1):131–147, 2010.

[207] R. W. Murray and R. L. Pego. Algebraic decay to equilibrium for the Becker-Döring equations. *SIAM J. Math. Anal.*, 48(4):2819–2842, 2016.

[208] R. W. Murray and R. L. Pego. Cutoff estimates for the linearized Becker-Döring equations. *Commun. Math. Sci.*, 15(6):1685–1702, 2017.

[209] B. Niethammer. On the evolution of large clusters in the Becker-Döring model. *J. Nonlinear Sci.*, 13(1):115–155, 2003.

[210] B. Niethammer. Macroscopic limits of the Becker-Döring equations. *Commun. Math. Sci.*, 2(suppl. 1):85–92, 2004.

[211] B. Niethammer. A scaling limit of the Becker-Döring equations in the regime of small excess density. *J. Nonlinear Sci.*, 14(5):453–468, 2004.

[212] B. Niethammer. Self-similarity in Smoluchowski's coagulation equation. *Jahresber. Dtsch. Math.-Ver.*, 116(1):43–65, 2014.

[213] B. Niethammer, S. Throm, and J. J. L. Velázquez. Self-similar solutions with fat tails for Smoluchowski's coagulation equation with singular kernels. *Ann. Inst. H. Poincaré Anal. Non Linéaire*, 33(5):1223–1257, 2016.

[214] B. Niethammer, S. Throm, and J. J. L. Velázquez. A uniqueness result for self-similar profiles to Smoluchowski's coagulation equation revisited. *J. Statist. Phys.*, 164(2):399–409, 2016.

[215] B. Niethammer and J. J. L. Velázquez. Optimal bounds for self-similar solutions to coagulation equations with product kernel. *Comm. Partial Differential Equations*, 36(12):2049–2061, 2011.

[216] B. Niethammer and J. J. L. Velázquez. Self-similar solutions with fat tails for a coagulation equation with diagonal kernel. *C. R. Math. Acad. Sci. Paris*, 349(9-10):559–562, 2011.

[217] B. Niethammer and J. J. L. Velázquez. Erratum to: Self-similar solutions with fat tails for Smoluchowski's coagulation equation with locally bounded kernels. *Comm. Math. Phys.*, 318(2):533–534, 2013.

[218] B. Niethammer and J. J. L. Velázquez. Self-similar solutions with fat tails for Smoluchowski's coagulation equation with locally bounded kernels. *Comm. Math. Phys.*, 318(2):505–532, 2013.

[219] B. Niethammer and J. J. L. Velázquez. Exponential tail behavior of self-similar solutions to Smoluchowski's coagulation equation. *Comm. Partial Differential Equations*, 39(12):2314–2350, 2014.

[220] B. Niethammer and J. J. L. Velázquez. Uniqueness of self-similar solutions to Smoluchowski's coagulation equations for kernels that are close to constant. *J. Statist. Phys.*, 157(1):158–181, 2014.

[221] B. Niethammer and J. J. L. Velázquez. Oscillatory traveling wave solutions for coagulation equations. *Quart. Appl. Math.*, 76(1):153–188, 2018.

[222] R. Normand and L. Zambotti. Uniqueness of post-gelation solutions of a class of coagulation equations. *Ann. Inst. H. Poincaré Anal. Non Linéaire*, 28(2):189–215, 2011.

[223] J. R. Norris. Smoluchowski's coagulation equation: Uniqueness, nonuniqueness and a hydrodynamic limit for the stochastic coalescent. *Ann. Appl. Probab.*, 9(1):78–109, 1999.

[224] J. R. Norris. Cluster coagulation. *Comm. Math. Phys.*, 209(2):407–435, 2000.

[225] J. R. Norris. Brownian coagulation. *Commun. Math. Sci.*, 2(suppl. 1):93–101, 2004.

[226] J. R. Norris. Notes on Brownian coagulation. *Markov Process. Related Fields*, 12(2):407–412, 2006.

[227] A. Pazy. *Semigroups of linear operators and applications to partial differential equations*, volume 44 of *Applied Mathematical Sciences*. Springer-Verlag, New York, 1983.

[228] O. Penrose. Metastable states for the Becker-Döring cluster equations. *Comm. Math. Phys.*, 124(4):515–541, 1989.

[229] O. Penrose. The Becker-Döring equations at large times and their connection with the LSW theory of coarsening. *J. Statist. Phys.*, 89(1-2):305–320, 1997. Dedicated to Bernard Jancovici.

[230] T. W. Peterson. Similarity solutions for the population balance equation describing particle fragmentation. *Aerosol Sci. Tech.*, 5(1):93–101, 1986.

[231] M. Pierre and D. Schmitt. Blowup in reaction-diffusion systems with dissipation of mass. *SIAM J. Math. Anal.*, 28(2):259–269, 1997.

[232] V. N. Piskunov. The asymptotic behavior and self-similar solutions for disperse systems with coagulation and fragmentation. *J. Phys. A*, 45(23):235003, 17, 2012.

[233] Y. Qin. *Analytic inequalities and their applications in PDEs*, volume 241 of *Operator Theory: Advances and Applications*. Birkhäuser/Springer, Cham, 2017.

[234] M. M. Rao and Z. D. Ren. *Theory of Orlicz spaces*, volume 146 of *Monographs and Textbooks in Pure and Applied Mathematics*. Marcel Dekker, Inc., New York, 1991.

[235] F. Rezakhanlou. Moment bounds for the solutions of the Smoluchowski equation with coagulation and fragmentation. *Proc. Roy. Soc. Edinburgh Sect. A*, 140(5):1041–1059, 2010.

[236] F. Rezakhanlou. Pointwise bounds for the solutions of the Smoluchowski equation with diffusion. *Arch. Ration. Mech. Anal.*, 212(3):1011–1035, 2014.

[237] I. Rubinstein and B. Zaltzman. Diffusional mechanism of strong selection in Ostwald ripening. *Phys. Rev. E (3)*, 61(1):709–717, 2000.

[238] W. Rudin. *Functional analysis*. International Series in Pure and Applied Mathematics. McGraw-Hill, Inc., New York, second edition, 1991.

[239] A. Schlichting. Macroscopic limit of the Becker-Döring equation via gradient flows. arXiv:1607.08735, 2016.

[240] T. Schumann. Theoretical aspects of the size distribution of fog particles. *Q. J. Roy. Meteorol. Soc.*, 66:195–207, 1940.

[241] W. T. Scott. Analytic studies of cloud droplet coalescence I. *J. Atmos. Sci.*, 25:54–65, 1968.

[242] G. R. Sell and Y. You. *Dynamics of evolutionary equations*, volume 143. Springer Science & Business Media, 2013.

[243] M. Shirvani and H. Van Roessel. The mass-conserving solutions of Smoluchowski's coagulation equation: The general bilinear kernel. *Z. Angew. Math. Phys.*, 43(3):526–535, 1992.

[244] M. Shirvani and H. J. van Roessel. Some results on the coagulation equation. *Nonlinear Anal.*, 43(5, Ser. A: Theory Methods):563–573, 2001.

[245] M. Slemrod. Trend to equilibrium in the Becker-Döring cluster equations. *Nonlinearity*, 2(3):429–443, 1989.

[246] D. J. Smit, M. Hounslow, and W. Paterson. Aggregation and gelation. I. Analytical solutions for CST and batch operation. *Chem. Eng. Sci.*, 49(7):1025–1035, 1994.

[247] M. v. Smoluchowski. Drei Vorträge über Diffusion, Brownsche Bewegung und Koagulation von Kolloidteilchen. *Physik. Zeitschr.*, 17:557–571, 585–599, 1916.

[248] M. v. Smoluchowski. Versuch einer mathematischen Theorie der Koagulationskinetik kolloider Lösungen. *Zeitschrift f. phys. Chemie*, 92:129–168, 1917.

[249] C. Sorensen, H. Zhang, and T. T.W. Cluster-size evolution in a coagulation-fragmentation system. *Phys. Rev. Lett.*, 59(3):363–366, 1987.

[250] J. L. Spouge. An existence theorem for the discrete coagulation-fragmentation equations. *Math. Proc. Cambridge Philos. Soc.*, 96(2):351–357, 1984.

[251] R. Srinivasan. Rates of convergence for Smoluchowski's coagulation equations. *SIAM J. Math. Anal.*, 43(4):1835–1854, 2011.

[252] I. W. Stewart. A global existence theorem for the general coagulation-fragmentation equation with unbounded kernels. *Math. Methods Appl. Sci.*, 11(5):627–648, 1989.

[253] I. W. Stewart. A uniqueness theorem for the coagulation-fragmentation equation. *Math. Proc. Cambridge Philos. Soc.*, 107(3):573–578, 1990.

[254] I. W. Stewart. Density conservation for a coagulation equation. *Z. Angew. Math. Phys.*, 42(5):746–756, 1991.

[255] H. R. Thieme and J. Voigt. Stochastic semigroups: Their construction by perturbation and approximation. In *Positivity IV—theory and applications*, pages 135–146. Tech. Univ. Dresden, Dresden, 2006.

[256] S. Throm. Uniqueness of fat-tailed self-similar profiles to Smoluchowski's coagulation equation for a perturbation of the constant kernel. arXiv 1704.01949, 2017.

[257] S. Throm. Tail behaviour of self-similar profiles with infinite mass for Smoluchowski's coagulation equation. *J. Stat. Phys.*, 170(6):1215–1241, 2018.

[258] R. P. Treat. On the similarity solution of the fragmentation equation. *J. Phys. A*, 30(7):2519–2543, 1997.

[259] M. Tyran-Kamińska. Ergodic theorems and perturbations of contraction semigroups. *Studia Math.*, 195(2):147–155, 2009.

[260] P. G. J. van Dongen. On the possible occurrence of instantaneous gelation in Smoluchowski's coagulation equation. *J. Phys. A*, 20(5-6):1889–1904, 1987.

[261] P. G. J. van Dongen. Solutions of Smoluchowski's coagulation equation at large cluster sizes. *Phys. A*, 145(1-2):15–66, 1987.

[262] P. G. J. van Dongen and M. H. Ernst. On the occurrence of a gelation transition in Smoluchowski's coagulation equation. *J. Statist. Phys.*, 44(5-6):785–792, 1986.

[263] P. G. J. van Dongen and M. H. Ernst. Scaling solutions of Smoluchowski's coagulation equation. *J. Statist. Phys.*, 50(1-2):295–329, 1988.

[264] H. J. van Roessel and M. Shirvani. A formula for the post-gelation mass of a coagulation equation with a separable bilinear kernel. *Phys. D*, 222(1-2):29–36, 2006.

[265] J. J. L. Velázquez. The Becker-Döring equations and the Lifshitz-Slyozov theory of coarsening. *J. Statist. Phys.*, 92(1-2):195–236, 1998.

[266] R. D. Vigil and R. M. Ziff. On the stability of coagulation-fragmentation population balances. *J. Colloid Interface Sci.*, 133(1):257–264, 1989.

[267] C. Villani. A review of mathematical topics in collisional kinetic theory. In *Handbook of mathematical fluid dynamics, Vol. I*, pages 71–305. North-Holland, Amsterdam, 2002.

[268] J. Voigt. On the perturbation theory for strongly continuous semigroups. *Math. Ann.*, 229(2):163–171, 1977.

[269] J. Voigt. On resolvent positive operators and positive C_0-semigroups on AL-spaces. *Semigroup Forum*, 38(2):263–266, 1989. Semigroups and differential operators (Oberwolfach, 1988).

[270] I. I. Vrabie. C_0-*semigroups and applications*, volume 191 of *North-Holland Mathematics Studies*. North-Holland Publishing Co., Amsterdam, 2003.

[271] C. Wagner. Theorie der Alterung von Niederschlägen durch Umlösen (Ostwald-Reifung). *Z. Elektrochem.*, 65(7-8):581–591, 1961.

[272] C. Walker. Coalescence and breakage processes. *Math. Methods Appl. Sci.*, 25(9):729–748, 2002.

[273] W. H. White. A global existence theorem for Smoluchowski's coagulation equations. *Proc. Amer. Math. Soc.*, 80(2):273–276, 1980.

[274] P. Wojtaszczyk. *Banach spaces for analysts*, volume 25 of *Cambridge Studies in Advanced Mathematics*. Cambridge University Press, Cambridge, 1991.

[275] C. P. Wong. *Kato's perturbation theorem and honesty theory.* PhD thesis, University of Oxford, 2015.

[276] C. P. Wong. Stochastic completeness and honesty. *J. Evol. Equ.*, 15(4):961–978, 2015.

[277] D. Wrzosek. Existence of solutions for the discrete coagulation-fragmentation model with diffusion. *Topol. Methods Nonlinear Anal.*, 9(2):279–296, 1997.

[278] D. Wrzosek. Mass-conserving solutions to the discrete coagulation-fragmentation model with diffusion. *Nonlinear Anal.*, 49(3, Ser. A: Theory Methods):297–314, 2002.

[279] D. Wrzosek. Weak solutions to the Cauchy problem for the diffusive discrete coagulation-fragmentation system. *J. Math. Anal. Appl.*, 289(2):405–418, 2004.

[280] M. R. Yaghouti, F. Rezakhanlou, and A. Hammond. Coagulation, diffusion and the continuous Smoluchowski equation. *Stochastic Process. Appl.*, 119(9):3042–3080, 2009.

[281] A. C. Zaanen. *Introduction to operator theory in Riesz spaces.* Springer-Verlag, Berlin, 1997.

[282] R. M. Ziff, M. H. Ernst, and E. M. Hendriks. Kinetics of gelation and universality. *J. Phys. A*, 16(10):2293–2320, 1983.

[283] R. M. Ziff, M. H. Ernst, and E. M. Hendriks. A transformation linking two models of coagulation. *J. Colloid Interface Sci.*, 100(1):220–223, 1984.

[284] R. M. Ziff and E. McGrady. Kinetics of polymer degradation. *Macromolecules*, 19(10):2513–2519, 1986.

[285] R. M. Ziff and E. D. McGrady. The kinetics of cluster fragmentation and depolymerisation. *J. Phys. A*, 18(15):3027–3037, 1985.

[286] R. M. Ziff and G. Stell. Kinetics of polymer gelation. *J. Chem. Phys.*, 73(7):3492–3499, 1980.

Index